ENGLISH SEIGNIORIAL AGRICULTURE, 1250–1450

Bruce Campbell's book is the first single-authored treatment of medieval English agriculture at a national scale. Methodologically innovative, it deals comprehensively with the cultivation carried out by or for lords on their demesne farms, the documentation for which is more detailed and abundant than for any other agricultural group either during the medieval period or later. A context is thereby assured for all future work on the medieval and early modern agrarian economies. The book also makes a substantive contribution to on-going historical debates about the dimensions, chronology, and causes of medieval expansion, crisis, and contraction. Topics dealt with include the scale and composition of seigniorial estates, the geography of land-use, pastoral husbandry, arable husbandry, land productivity, levels of commercialisation, and the size of the population in relation to the consumption of food at any given time.

BRUCE M. S. CAMPBELL is Professor of Medieval Economic History at The Queen's University of Belfast.

Cambridge Studies in Historical Geography 31

Series editors:
ALAN R. H. BAKER, RICHARD DENNIS, DERYCK HOLDSWORTH

Cambridge Studies in Historical Geography encourages exploration of the philosophies, methodologies and techniques of historical geography and publishes the results of new research within all branches of the subject. It endeavours to secure the marriage of traditional scholarship with innovative approaches to problems and to sources, aiming in this way to provide a focus for the discipline and to contribute towards its development. The series is an international forum for publication in historical geography which also promotes contact with workers in cognate disciplines.

For a full list of titles in the series, please see end of book.

ENGLISH SEIGNIORIAL AGRICULTURE, 1250–1450

BRUCE M. S. CAMPBELL

Professor of Medieval Economic History
The Queen's University of Belfast

CAMBRIDGE
UNIVERSITY PRESS

CAMBRIDGE UNIVERSITY PRESS
Cambridge, New York, Melbourne, Madrid, Cape Town, Singapore, São Paulo

Cambridge University Press
The Edinburgh Building, Cambridge CB2 2RU, UK

Published in the United States of America by Cambridge University Press, New York

www.cambridge.org
Information on this title: www.cambridge.org/9780521304122

First published 2000
This digitally printed first paperback version 2006

A catalogue record for this publication is available from the British Library

Library of Congress Cataloguing in Publication data

Campbell, B. M. S.
    English seigniorial agriculture, 1250–1450 / Bruce M. S. Campbell.
        p. cm.
    Includes bibliographical references (p.).
    ISBN 0 521 30412 1
    1. Agriculture – England – History. 2. Agriculture – Economic
aspects – England – History. 3. Middle Ages. I. Title.
S455.C26   2000
630′.942 – dc21   99-39384   CIP

ISBN-13 978-0-521-30412-2 hardback
ISBN-10 0-521-30412-1 hardback

ISBN-13 978-0-521-02642-0 paperback
ISBN-10 0-521-02642-3 paperback

For
Mary Campbell
and in remembrance of
Reginald Arthur Mortimer Campbell
who instilled in me their love of the English countryside

Contents

Figures

Tables

Preface and acknowledgements

This is not the book that I originally set out to write; had it been so it would have been completed far sooner. Instead, it is the book to which I have been led by a fortuitous succession of research projects; for research, once begun, has a habit of assuming a momentum of its own. Rather than charting a straight and direct course to a predetermined destination it has been a case of seizing opportunities and following where they lead. My original agenda and techniques have also been overtaken by a fast-changing historiography and the advent of increasingly powerful personal and lap-top computers and menu-driven software which have transformed the potential for data collection and analysis. Evolving an appropriate methodology, including robust methods of classification, has also been a matter of trial and error. With hindsight I can see how more data could have been collected more systematically and analysed and classified more rigorously. Nevertheless, I have resisted the temptation to act like Penelope at her loom. Instead, I offer what I have done, uneven though it is, in the hope that others will improve upon and extend it: there are many unexplored and unresolved issues and the wealth of under-utilised and unexamined archives is great.

My original aim was to write a book about seigniorial agriculture in medieval Norfolk but set in a broader regional and national perspective. The Norfolk accounts database (Appendix 2) was therefore the first to be constructed of the core databases upon which this book is based. Work on it was ongoing throughout the 1970s and early 1980s, aided by periods of study leave and successive grants from the Research and Scholarships Fund of The Queen's University of Belfast. In 1983–4 the tenure of a Personal Research Fellowship awarded by the then Social Science Research Council enabled the Norfolk accounts database to be completed and also made possible a preliminary investigation of the *inquisitiones post mortem* (*IPMs*) at national level. At that time these two databases were intended to form the substance of this book. Then, in 1987, John Langdon, now Professor of History at the University of Alberta, generously put at my disposal the information on crops and livestock which he had transcribed from a national sample of manorial accounts and which constitutes the core of the national accounts database (Appendix 1) which features so prominently in this book. It was from this point that the possibility of writing a book on seigniorial agriculture within England as a whole became a realistic proposition. Before this could be acted upon, however, an invitation from Dr Derek Keene to collaborate in an investigation of the provisioning of London *c.* 1300 proved too good to turn down, and the 'Feeding the city (FTC) 1' project – 'London's food supplies 1270–1339' – was conceived.

'Feeding the city 1' built upon existing experience and knowledge, greatly broadened the range of analysis, and employed lap-top computers for the first time to input data in the archives. The project was funded by the Leverhulme Trust from September 1988 to August 1991, co-directed by Derek Keene, and based at the newly founded Centre for Metropolitan History at the Institute of Historical Research, London, where he was director. Its aim was to investigate the impact of London's demand for food and other supplies on the agriculture and on the distribution systems of the metropolitan hinterland *c.* 1300, when the capital reached an early peak in its population and was one of the largest of European cities. To this end, the 'Feeding the city 1' accounts database was created, covering the years 1288–1315, together with a corresponding *IPM* database covering the years 1270–1339. This project, in turn, begot two others, with further repercussions for work on this book. A second grant from the Leverhulme Trust financed the creation of the national *IPM* database during the period August 1991 to December 1994, in conjunction with the project 'The geography of seigniorial land-ownership and use, 1270–1349', co-directed by Mr John Power, then Lecturer in Geography at The Queen's University of Belfast, and based at QUB. Additional funding from Queen's permitted analysis of the national *IPM* database to be completed over the period 1995–7. Meanwhile, a research grant from the Economic and Social Research Council (grant number R000233157) for the period October 1991 to July 1994 enabled the 'Feeding the city 2' accounts database to be created (Appendix 3) in conjunction with the 'Feeding the city 2' project ('London and its hinterland *c.* 1300–1400'). Like 'Feeding the city 1', this project was based at the Centre for Metropolitan History, where it was co-directed by Dr Derek Keene, Dr James Galloway, and Dr Margaret Murphy. Its aim was to replicate and refine the approach and method of the earlier project with reference to the final years of the fourteenth century, by which time London and the region that supported it contained a much reduced population with a very different pattern of consumption.

The national *IPM* database and the two FTC accounts databases are of an exceptionally high quality and hence have added materially to the depth and scope of this book. I am grateful to the Leverhulme Trust, the ESRC, and The Queen's University of Belfast for providing the funding that made creation and analysis of these databases possible. My thanks are also due to the Rockefeller Foundation for hosting the research workshop 'Agricultural productivity and economic change in the European past' which Mark Overton, now Professor of Economic History at the University of Exeter, and I co-convened at their Bellagio Study Centre, Lake Como, Italy in March 1989. This proved formative for the work on seigniorial land productivity which constitutes Chapter 7 of this book. Publication, in colour, of Figure 3.14 was made possible by a grant from The Scouloudi Foundation in association with the Institute of Historical Research, University of London.

Constructing and analysing these substantial databases would not have been possible without the assistance and expertise of others. Jenitha Orr, then a research officer in the Department of Geography, QUB, helped analyse the Norfolk accounts database and thereby establish the approach followed in much subsequent work. John Power, at that time also a research officer in the Department of Geography, advised on the inputting of the national accounts data and analysed the completed database (to which Dr David Postles of the Department of English Local History in the University of Leicester and Martin Ecclestone, an external MA student at the University of Bath, both contributed data). To John must go the credit for developing and refining the method of classifying land-use and farming systems using cluster analysis which is employed in Chapters 3, 4, and 6. He also developed the inputting systems used in the creation of the FTC1 and FTC2 accounts databases and the national *IPM* database, all of which were mechanised from the outset. James Galloway and Margaret Murphy were responsible for creating both FTC accounts databases, tasks which they undertook with dedication and skill. Technical and administrative support at the Centre for Metropolitan History was provided by Olwen Myhil. Richard Britnell, now Professor of Medieval History at the University of Durham, and Dr Harold Fox of the Department of English Local History at the University of Leicester, as advisers to both FTC projects, offered many useful insights. With Dr Robin Glasscock of the Department of Geography, University of Cambridge, Harold Fox discharged a similar function on the national *IPM* project. The formidable job of extracting data from 9,000 individual *IPM* extents was carried out with commendable care and cheerfulness by Dr Roger Dickinson and Marilyn Livingstone. Development and analysis of the *IPM* database, including its incorporation into a Geographical Information System (GIS) was undertaken by Ken Bartley, then research fellow in the Department of Economic and Social History, QUB. Such was the scale and complexity of this task (and some further work on the national accounts database) that it took four years to complete. This book would have been the poorer without that work, especially the many computer-generated national-scale maps that Ken was instrumental in creating. All the other maps were produced using a variety of technologies by Gill Alexander of the School of Geosciences, QUB, whose work was partially funded by a grant from the QUB Publications Fund. They were photographed by the Queen's Photographic Unit and Moira Concannon of the Ulster Museum. Dr Paul Ell reworked the data upon which Figure 1.01 is based. Emma Touffler checked the format of the footnotes and created the consolidated bibliography.

For permission to consult manuscripts, and for help and advice from the staff concerned, I would like to thank the following public and private institutions and private owners: Bedfordshire Record Office; Berkshire Record Office; Buckinghamshire Record Office; Essex Record Office; Hampshire Record Office; Hertfordshire Record Office; Centre for Kentish Studies, Maidstone; Lancashire Record Office; Corporation of London Records Office; Greater London Record Office; Norfolk Record Office; North Yorkshire Record Office; Northamptonshire Record Office; Nottinghamshire Record Office; Oxfordshire Record Office; Public Record Office; Surrey Record Office; West Suffolk Record Office; Birmingham Reference Library; Bodleian Library, Oxford; British Library; Cambridge University Library; Canterbury Cathedral Archives; Guildhall Library, London; John Rylands Library, Manchester; Joseph Regenstein Library, University of Chicago; Lambeth Palace Library; National Register of Archives; Nottingham University Library; Westminster Abbey Muniments; Christ's College, Cambridge; Eton College; King's College, Cambridge; Magdalen College, Oxford; Merton College, Oxford; New College, Oxford; St George's Chapel, Windsor; Winchester College; Elveden Hall, Suffolk (the Earl of Iveagh); Holkham Hall, Norfolk (the Earl of Leicester); Raynham Hall, Norfolk (the Marquess Townshend); and Pomeroy & Sons, Wymondham. Thanks are due to those friends who have given hospitality on my many and various visits to archives and libraries. In particular, Christine Beavon was an ever-willing landlady in London, while Lyn, Tim, Joanna and Rebecca Atkinson treated me as a member of their family during my long stints in the Norfolk Record Office. While working in the latter office I was alerted to much that I might otherwise have missed by the then Deputy County Archivist, Paul Rutledge.

For permission to draw upon previously published materials, I am grateful to the British Agricultural History Society; the Economic History Association; the Economic History Society; *Histoire et Mesure*; the Historical Geography Research Group; the Institute of British Geographers; the *Journal of Historical Geography*; Manchester University Press; University of Pennsylvania Press; Medieval Institute, Western Michigan University; *Past and Present*; *Transactions in GIS*.

It was Jack Langton who, in 1972, by pressing a question at a Cambridge Occasional Discussion in Historical Geography, prompted me to switch my attention from court rolls to account rolls and thereby start on the quest that has eventually led to this book. Along the way many friends and scholars have helped shape and hone my thinking, including Professor Robert Allen,

Dr Mark Bailey, Professor Mike Baillie, Professor Kathleen Biddick, Professor Ian Blanchard, Professor Richard Britnell, Dr Gregory Clark, Professor Chris Dyer, the late Professor David Farmer, Professor George Grantham, Professor Maryanne Kowaleski, Professor John Langdon, Paul Laxton, Nicholas Poynder, Dr Richard Smith, Dr Christopher Thornton, and Professor, Sir Tony Wrigley. Over many years, commencing with shared post-graduate days in Cambridge and long train journeys to Norwich, I have learnt much from arguing and collaborating with Mark Overton, who will recognise but not necessarily agree with much that is in this book. His work has been material to those sections which endeavour to place medieval seigniorial agriculture in a more securely documented chronological context. Derek Keene similarly helped focus and sharpen my thoughts about the scale and significance of medieval urban demand – especially that of the metropolis – for producers in the countryside. More generally, my thinking about the medieval English agrarian economy has been strongly influenced by the privilege of teaching post-medieval Irish economic history – where many of the same historical issues recur in a different guise – to generations of undergraduates at Queen's.

My immediate family will be as relieved to see the publication of this book as I: their forbearance and encouragement have meant a lot. So, too, have the patience and faith of Dr Alan Baker who, as general editor of this series, has had to wait longer for this volume than any other but never doubted, or at least expressed doubts to me about, its eventual delivery. Vicky Cuthill, formerly history editor of CUP, gave much constructive advice and remained supportive even when it became clear that the finished text would be well over the originally contracted length. Marigold Acland, her successor at the Press, has been similarly positive. Mark Bailey and Richard Britnell read earlier versions of the text and made constructive suggestions on how it could be improved. Virginia Catmur copy-edited the text with care and tact. Above all, this book could never have been written without the secure and well-resourced base provided to me by The Queen's University of Belfast. It is here that I have pursued and brought to fruition the research that I began almost thirty years ago when a postgraduate under Alan Baker's supervision at the University of Cambridge.

BRUCE M. S. CAMPBELL
Michaelmas 1998

Abbreviations

AHEW	*The agrarian history of England and Wales*
AHR	*Agricultural History Review*
BL	British Library, London
BLO	Bodleian Library Oxford
CCA	Canterbury Cathedral Archives
CUL	Cambridge University Library
EcHR	*Economic History Review*, 2nd series
EHR	*English Historical Review*
FTC	'Feeding the city'
GDP	Gross Domestic Product
GIS	Geographical Information System
IPMs	*Inquisitiones post mortem*
JEH	*Journal of Economic History*
JHG	*Journal of Historical Geography*
JLL	J. L. Langdon
NA	*Norfolk Archaeology*
NRO	Norfolk Record Office, Norwich
PP	*Past and Present*
PRO	Public Record Office, London
RO	Record Office
TIBG	*Transactions of the Institute of British Geographers*
TRHS	*Transactions of the Royal Historical Society*
UL	University Library
WACY	Weighted aggregate crop yield
WAGY	Weighted aggregate grain yield
WAM	Westminster Abbey Muniments
WFCP	Wood, forest, chase, park

Weights, measures, values, and boundaries

The units used in this book are those that contemporaries used:

Weight:

English weights are based on the pound avoirdupois:

16 ounces (oz.)	=	1 pound (lb.)	=	0.4536 kilogram (kg)
2,240 lbs.	=	1 ton	=	1.016 tonne
14 lbs.	=	1 stone	=	6.3504 kg
16 stones	=	1 wey	=	101.6064 kg
10 weys	=	1 ton	=	1.016 tonnes

Volume (dry):

8 gallons	=	1 bushel (bus.)	=	35.238 litres (l)
8 bus.	=	1 quarter (qtr.)	=	2.819 hectolitres (hl)

Area:

40 perches (per.)	=	1 rod	=	0.1012 hectares (ha)
4 rods	=	1 acre (ac.)	=	0.4047 ha
640 acres	=	1 square mile	=	259 ha

Volume by area (a measure of yield):

1 bus. per ac.	=	0.8707 hl per ha
1.1485 bus. per ac.	=	1 hl per ha

Length:

3 feet	=	1 yard	=	0.9144 metres (m)
1,760 yards	=	1 mile	=	1.6093 kilometres (km)

Value:

12 pence (d.)   =   1 shilling (s.)
20 s.   =   1 pound (£)

Boundaries:

All counties and their boundaries are as they existed before 1974.

Statute versus non-statute (customary) measures:

The medieval acre was of a variable size. Statute acres were measured with a perch of 16½ feet. Non-statute acres were measured with perches that could be 16, 18, 20, 22 or even 24 feet, yielding customary acres equivalent to 0.94, 1.19, 1.47, 1.78, and 2.12 statute acres. Perches in excess of 30 feet are also sometimes recorded. Large customary acres in excess of 1½ statute acres were most characteristic of Cornwall and parts of northern and north-western England.

The size of the bushel also varied, especially according to whether it was heaped or struck. With a modest amount of heaping, 8 heaped gallons would actually have amounted to 9 struck gallons and 8 heaped bushels to 9 struck bushels; a difference of 12.5 per cent.

Multiples:

Millions and billions are abbreviated to 'm.' and 'b.' A billion is 10^{12}.

1

Introduction: agriculture and the late-medieval English economy

1.1 The seigniorial and non-seigniorial sectors

Between *c.* 1250 and *c.* 1450, for the first time in recorded English history, it becomes possible to reconstruct the development and performance of agriculture in some detail. Several different categories of producer – demesne lords, owners of rectorial glebe, franklins and proto-yeomen, substantial customary tenants, lesser customary tenants, and small freeholders – were involved in shaping the course of agricultural development over this eventful period, but it is the activities of the demesne lords that are the most copiously documented. For demesne farms alone detailed input and output data are available at the level of the individual production unit. It is the analysis of these data that forms the subject of this book. Not only do the insights thereby obtained have implications for agriculture in general but in an overwhelmingly agrarian age any verdict on the agricultural sector – limited and qualified though it may be – has important implications for how the medieval economy is viewed as a whole.

Notwithstanding their many obvious differences, seigniorial and non-seigniorial producers shared much in common. Most conspicuously, they shared a common technology. Indeed, much of the labour-force and know-how used in the management of demesnes came from the non-seigniorial sector with the result that the husbandry documented on demesnes was strongly influenced by local practice.[1] Shaping that practice were common environmental and commercial opportunities. It follows that where peasants led, lords are likely to have followed and *vice versa*. Nor were the stock and crops of lords any more immune to flood and drought or pests and pathogens than those of peasants. Analysis of production patterns and trends within the seigniorial sector can therefore reveal much about those within husbandry in general.[2] At the very least they provide a comparative basis against which the

---

[1] M. Mate, 'Medieval agrarian practices: the determining factors?', *AHR* 33 (1985).
[2] E.g. J. L. Langdon, *Horses, oxen and technological innovation* (Cambridge, 1986).

more fragmentary and indirect evidence available for the larger but far less well-documented peasant sector can be evaluated.

For a customary tenant serving as reeve, managing a demesne under the close supervision of a bailiff, steward or even the lord of the manor, with the annual requirement to render a detailed written account for scrutiny by auditors, was wholly different from running a family farm.[3] This was especially the case on demesnes belonging to perpetual institutions, with their exceptional continuity of management and administration and immunity to death, division, wardship and vacancy. Irrespective of ownership, the average demesne was several times larger than even the most substantial customary holding and usually formed one component within a greater federated estate system of production.[4] The superior scale of their activities and range of resources at their command meant that lords could afford to be far less risk averse than peasants. Thus, whereas small-scale subsistence producers may have lived in dread of dearth, large-scale seigniorial producers were far more likely to 'hang themselves on the expectation of plenty'.[5] The bad harvests that impoverished small-scale producers by suddenly transforming modest grain surpluses into large deficits simultaneously enriched large-scale producers who still had surpluses to sell and could profit greatly from the inflated prices.[6] Famine prices in 1316 and 1317, for instance, delivered bumper profits to many a demesne producer.[7]

Contrasting labour processes went hand in hand with the contrasting scales of seigniorial and peasant producers. Where peasants relied upon family augmented by hired labour, lords were wholly dependent upon a combination of customary and hired workers. For all the historical attention that it has attracted, customary labour was probably more irksome than it was important. The supply of labour services was never equal to the labour requirements of the seigniorial sector, especially on small lay manors and in areas of weak manorialisation.[8] On J. Hatcher's estimation, less than a third of villein house-

---

[3] This, no doubt, was why for thirty-eight years, until his death in 1349, the fellows of Merton College Oxford entrusted Robert Oldman with the office of reeve and, thus, responsibility for supervising the management of their demesne at Cuxham: P. D. A. Harvey, *A medieval Oxfordshire village* (London, 1965), pp. 64, 71–2.

[4] 'The large estates of the great secular or episcopal landowners ... in the late thirteenth century were capitalist concerns: federated grain factories producing largely for cash': M. M. Postan, 'Revisions in economic history: IX. The fifteenth century', *Economic History Review* 9 (1939), 162. [5] *Macbeth*, Act II, Scene III.

[6] M. Overton, *Agricultural revolution in England* (Cambridge, 1996), pp. 20–1.

[7] J. Z. Titow, 'Land and population on the bishop of Winchester's estates 1209–1350', PhD thesis, University of Cambridge (1962), pp. 9–10. The record prices also brought windfall profits to London cornmongers: B. M. S. Campbell, J. A. Galloway, D. J. Keene, and M. Murphy, *A medieval capital and its grain supply*, Historical Geography Research Series 30 (n.p., 1993), pp. 82–4.

[8] H. L. Gray, 'The commutation of villein services in England before the Black Death', *EHR* 29, 116 (1914), 625–56; E. A. Kosminsky, *Studies in the agrarian history of England in the thirteenth*

holds were still regularly performing week-works – the most burdensome of customary services – at the close of the thirteenth century, and performing them, presumably, indifferently.[9] The proportion of seigniorial production actually accounted for by labour services may consequently have been as little as 8 per cent.[10] It was as employers rather than coercers of labour that lords were, therefore, most significant. In fact, lords increasingly substituted hired for customary labour since it was better motivated and incurred lower supervision costs.[11] Harvest works, alone among customary services, tended to be retained to the bitter end since they helped guarantee an adequate workforce in the season of peak labour demand.[12]

For all its intrinsic interest and superior documentation, the seigniorial sector was always of lesser significance than the non-seigniorial sector.[13] Indeed, that minority status became more rather than less pronounced with the passage of time, since the majority 'peasant' sector possessed the more powerful dynamic. Until *c.* 1325 peasants were almost certainly far more active than lords in adding to the agricultural area. Indeed, from the mid-thirteenth century opportunities for extending the agricultural area in lowland demesne-farming contexts were fast running out.[14] Then, from the second quarter of the fourteenth century, tenants gained from the progressive transfer of land and capital from the seigniorial to the non-seigniorial sector via the piecemeal and wholesale leasing of demesne land, stock and buildings, until the point was eventually reached in the mid-fifteenth century when the bulk of all landlords were rentiers rather than direct producers.

1.2 The changing economic context of agricultural production

Over the long span of time from the resumption of direct demesne management in the early thirteenth century until its final demise two and a half centuries later the economic and institutional contexts of seigniorial production

century, trans. R. Kisch, ed. R. H. Hilton (Oxford, 1956), pp. 152–96; R. H. Britnell, 'Commerce and capitalism in late medieval England: problems of description and theory', *Journal of Historical Sociology* 6 (1993), 364.

[9] J. Hatcher, 'English serfdom and villeinage: towards a reassessment', *PP* 90 (1981), 12. On the manors of Ramsey Abbey J. A. Raftis, *Peasant economic development within the English manorial system* (Stroud, 1997), p. 62, reckons 'that only some 10 per cent of the labour resources of the tenant were owed to demesne services'.

[10] Estimated from Britnell, 'Commerce and capitalism', p. 374, n. 35.

[11] D. Stone, 'The productivity of hired and customary labour: evidence from Wisbech Barton in the fourteenth century', *EcHR* 50 (1997), 640–56; R. H. Hilton, 'Peasant movements in England before 1381', *EcHR* 2 (1949), 117–36.

[12] B. M. S. Campbell, 'Agricultural progress in medieval England: some evidence from eastern Norfolk', *EcHR* 36 (1983), 38–9.   [13] Chapter 3, pp. 55–60.

[14] After 1150 the total cultivated area on the estates of the bishops of Worcester grew by only 5–7 per cent: C. C. Dyer, *Lords and peasants in a changing society* (Cambridge, 1980), p. 96. The sown acreage of the Winchester estate grew by 8 per cent between 1209 and 1240, but registered little overall increase thereafter: Titow, 'Land and population', pp. 21–2.

changed significantly.[15] Trends in prices, wages and real wages encapsulate many of these changes and provide a securely documented and precisely calibrated chronology. Figure 1.01 is based on the composite price and wage indices constructed by D. L. Farmer, re-indexed on their respective means for the entire period 1208–1466.[16] It identifies four price-wage phases or eras: the first and longest stretching from the beginning of the thirteenth century until the Great European Famine of 1315–22; the second spanning the twenty-five-year interval that separated the Famine from the Black Death of 1348–9; the third occupying the three decades following the Black Death; and the fourth dating in effect from the Peasants' Revolt of 1381 and lasting until at least the middle of the fifteenth century.

During the first half of the thirteenth century prices and cash wages both registered an inflationary rise, which peaked in the famine year 1257–8. Up to that point real wages do not appear to have suffered any lasting erosion; thereafter they sustained a pronounced and lasting fall as prices moved decisively ahead of wages (Figure 1.01). By the mid-1270s real wages were 50 per cent lower than their level at the start of the century and they were to fluctuate around that same low level for the next fifty years. Years of abundant harvest, such as the late 1280s and early 1300s, brought some temporary recovery in real wages but these gains were invariably wiped out whenever harvests reverted to or sank below normal, as in the mid-1290s and most dramatically during the Great European Famine of 1315–22.[17] The climax of this price-driven divergence between prices and real wages came in 1316 when prices peaked at 150 per cent above and real wages plummeted to 75 per cent below their respective long-term averages.[18]

While a significant increase in the money supply undoubtedly stoked the sustained inflation which took place in both prices and cash wages over the course of the 'long' thirteenth century, it is generally believed that population growth was primarily responsible for the fact that prices rose faster and further than wages from mid-century.[19] These circumstances naturally favoured those able to maintain their economic strength and disadvantaged those who could not. Lords, in particular, stood to gain significantly from the rising prices

---

[15] P. D. A. Harvey, 'The Pipe Rolls and the adoption of demesne farming in England', *EcHR* 27 (1974).

[16] The prices are those of a basket of consumables comprising 4 quarters of barley, 2 quarters of peas, a tenth of an ox, half a wether, half a pig, a quarter of a wey of cheese, a tenth of a quarter of salt, and one stone of wool. The wages are those of both agricultural and building workers: namely, the piece rates for threshing, winnowing, reaping and binding (plus mowing and spreading post-1349), and the day rates for a carpenter, a thatcher and his mate, and a slater/tiler and his mate. Real wages have been calculated by dividing wages by prices.

[17] For a case study of the 1290s famine see P. R. Schofield, 'Dearth, debt and the local land market in a late thirteenth-century village community', *AHR* 45 (1997).

[18] D. L. Farmer, 'Prices and wages', in Hallam (ed.), *AHEW*, vol. II, p. 776.

[19] N. J. Mayhew, 'Modelling medieval monetisation', in R. H. Britnell and B. M. S. Campbell (eds.), *A commercialising economy* (Manchester, 1995), p. 76; R. H. Britnell, *The commercialisation of English society 1000–1500* (Cambridge, 1993), pp. 102–5.

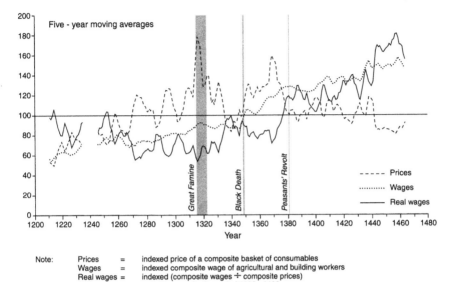

Fig. 1.01. Prices, wages, and real wages in England, 1208–1466 (five-year moving averages) (source: Farmer, 'Prices and wages', pp. 776–7; Farmer, 'Prices and wages, 1350–1500', pp. 520–3).

owing to their inherent tendency to produce in excess of their own consumption requirements. Since they were heavily dependent upon hired workers, they also reaped economic benefit from the rising supply and falling unit cost of labour. Other things being equal, therefore, there were strong incentives for them to expand and intensify their production to the extent that the turn of the thirteenth and fourteenth centuries has sometimes been represented as a time of seigniorial 'high farming', with high inputs being used to achieve high outputs.[20] In contrast, those who sold their labour and bought their food were increasingly squeezed by their weakening purchasing power.[21] Demand, although expanding, became increasingly polarised, especially at times of dearth and famine, as a growing proportion of the population traded down to the cheapest affordable foodstuffs.[22] In effect, that meant a dietary shift from

[20] E.g. E. Miller and J. Hatcher, *Medieval England: rural society and economic change 1086–1348* (London, 1978), pp. 213–24.

[21] A. N. May, 'An index of thirteenth-century peasant impoverishment? Manor court fines', *EcHR* 26 (1973), 389–402. At Halesowen in Worcestershire the poorest socio-economic groups experienced a failure of biological reproduction: Z. Razi, *Life, marriage and death in a medieval parish* (Cambridge, 1980), pp. 94–7.

[22] William Langland describes the foodstuffs that the poor had to make shift with at times of famine: *Visions from Piers Plowman taken from the poem of William Langland*, trans. N. Coghill (London, 1949), pp. 53–4. See also R. W. Frank Jr, 'The "hungry gap", crop failure, and famine: the fourteenth-century agricultural crisis and *Piers Plowman*', in D. Sweeney (ed.), *Agriculture in the Middle Ages* (Philadelphia, 1995), pp. 227–43.

pastoral to arable products and a reduced *per capita* consumption of refined bread and ale. These changes in the relative composition of demand favoured those producers with a strong comparative advantage in arable production.

The Great European Famine inflicted a major demographic shock and in its aftermath prices and wages began to converge.[23] That convergence was most marked in the 1330s, when, for a time, real wages recovered almost to the level of the mid-thirteenth century (Figure 1.01). Several factors were at work here, notably the reduction in population, the tardiness of any demographic recovery, a fortuitous run of good harvests, and a pronounced decrease in the amount of currency in circulation.[24] A slump in the real value of rental property in London's Cheapside demonstrates that the urban and commercial sectors were also in difficulties.[25] Cheapside was the commercial heart of London, as London was of the kingdom: if its pulse was beating slower the nation's commercial prosperity was probably on the wane. In the late 1330s and early 1340s acute bullion famine precipitated the most pronounced and prolonged fall in prices since the onset of inflation well over a century earlier. It was this that delivered windfall gains to wage earners, since wage rates held more or less steady at their customary levels. For smallholders, too, the abundant harvests of the 1330s provided an opportunity to recoup the worst of the losses of land and stock which they had incurred during the famine years.[26] These circumstances were, however, far less auspicious for landlords. Faced by a general recession in trade and with their profit margins squeezed by massively deflated prices, many began to question the wisdom of a policy of direct demesne management.[27] Poor harvests in the mid-1340s nevertheless brought the return of higher prices and for a brief while this must have allayed the worst of their fears, until the outbreak of the Black Death in July 1348 quashed the prospect of any return to the previous economic *status quo*.

Whereas the Great European Famine culled the population by perhaps 10–15 per cent, the mortality precipitated by the Black Death between July

---

[23] I. Kershaw, 'The Great Famine and agrarian crisis in England 1315–22', *PP* 59 (1973); J. D. Chambers, *Population, economy, and society in pre-industrial England* (London, 1972), p. 25; D. B. Grigg, 'Western Europe in the thirteenth and fourteenth centuries: a case of overpopulation?', in Grigg, *Population growth and agrarian change* (Cambridge, 1980), pp. 64–82; W. C. Jordan, *The Great Famine* (Princeton, 1996).

[24] Razi, *Life, marriage and death*, pp. 27–98; B. M. S. Campbell, 'Population pressure, inheritance and the land market in a fourteenth-century peasant community', in R. M. Smith (ed.), *Land, kinship and life-cycle* (Cambridge, 1984), pp. 95–9; L. R. Poos, 'The rural population of Essex in the later Middle Ages', *EcHR* 38 (1985), 521–4; D. G. Watts, 'A model for the early fourteenth century', *EcHR* 20 (1967), 546–7; N. J. Mayhew, 'Numismatic evidence and falling prices in the fourteenth century', *EcHR* 27 (1974).

[25] D. J. Keene, *Cheapside before the Great Fire* (London, 1985), pp. 19–20.

[26] Campbell, 'Population pressure', pp. 115–18.

[27] D. H. Fischer, *The great wave: price revolutions and the rhythm of history* (Oxford, 1996), pp. 30–45; Farmer, 'Prices and wages', pp. 720, 728; Chapter 5, pp. 233–4.

1348 and December 1349 was two to three times greater.[28] Depending upon the size of the total population this amounted to a death toll of at least 1¼ million and as such constitutes the worst crisis of public health in recorded English history.[29] Nor, once the immediate epidemic had passed, was there much prospect of any sustained demographic recovery. Recurrent outbreaks of plague in 1361, 1369, and 1375 continued to drive the population down while changing age structures and nuptiality patterns militated against any compensatory rise in fertility.[30] The massive demographic haemorrhage was all the more profound in its economic impact because it was a Europe-wide phenomenon. With far fewer people to feed, clothe and fuel this amounted to a demand shock of unequalled scale and immediacy and as such presented the agricultural sector with massive and wholly unanticipated problems of read-justment which it took the next thirty years to work out.[31]

The magnitude of the initial demand shock in 1348–9 is manifest in an unprecedented 45 per cent collapse in prices and 24 per cent rise in cash wages, which, together, temporarily sent real wages soaring.[32] Workers were neverthe-less unable to maintain this windfall advantage. Adverse weather and bad har-vests in the 1350s coupled with the massive *per capita* increase in currency brought about by the great reduction in population re-stoked inflation, return-ing prices to their level at the opening of the fourteenth century.[33] Worse was to follow; in 1369–70 the combination of dearth and plague pushed prices up to levels exceeded only in the grimmest years of the Great European Famine (Figure 1.01). Meanwhile, government curbs upon increases in cash wages – hurriedly imposed in 1349 and confirmed by statute as soon as the immediate

[28] R. M. Smith, 'Demographic developments in rural England, 1300–48: a survey', in B. M. S. Campbell (ed.), *Before the Black Death* (Manchester, 1991), pp. 25–78; Jordan, *The Great Famine*, pp. 118–20.

[29] On the impact of plague see P. Ziegler, *The Black Death* (London, 1969), pp. 224–31; J. Hatcher, *Plague, population and the English economy 1348–1530* (London, 1977), pp. 21–5. G. Twigg, *The Black Death* (London, 1984), pp. 70–1, plumps for a lower mortality rate of 20 per cent, while J. F. D. Shrewsbury, *A history of bubonic plague in the British Isles* (Cambridge, 1970), p. 123, argues that 'In all probability the national death-toll from "The Great Pestilence" did not exceed one-twentieth of that population [of 4 million]'. Neither of these lower estimates has found much favour with historians.   [30] Razi, *Life, marriage and death*, pp. 114–51.

[31] The much-discussed late-seventeenth-century demand shift (E. A. Wrigley and R. S. Schofield, *The population history of England, 1541–1871* (Cambridge, 1989), pp. 207–15) was much smaller: see E. L. Jones, 'Agriculture and economic growth in England, 1660–1750: agricultu-ral change', *JEH* 25 (1965), 1–18; A. Kussmaul, 'Agrarian change in seventeenth-century England: the economic historian as paleontologist', *JEH* 45 (1985), 1–30; H. J. Habakkuk, 'The agrarian history of England and Wales: regional farming systems and agrarian change, 1640–1750', *EcHR* 40 (1987), 281–96. Not until the late-nineteenth-century revolution in world food markets was a comparable shock experienced, although on that occasion its origin lay with supply rather than demand.

[32] Farmer, 'Prices and wages', p. 777. Prices and wages were the subjects of much contemporary comment: *The Black Death*, trans. and ed. R. Horrox (Manchester, 1994), pp. 78–9.

[33] D. L. Farmer, 'Prices and wages, 1350–1500', in Miller (ed.), *AHEW*, vol. III, p. 441.

crisis had passed in 1351 – proved remarkably effective.[34] Paradoxically, therefore, prices again pulled ahead of cash wages with the result that for two
decades after the Black Death real wages remained at or below their immediate pre-plague level. Only improved food liveries provided any compensation.[35] For large-scale producers hiring labour in order to produce surpluses
for sale and consumption this was to be an Indian summer conveying a false
illusion that nothing fundamentally had changed.

Such an artificial *status quo* could not last indefinitely. With each successive
plague outbreak the random culling of the population further undermined the
established socio-economic fabric of rural life.[36] As the population dwindled
demand contracted and it was only a matter of time before this translated
itself into falling commodity prices. Heavy export duties and continuing
government interference ensured that the wool export trade declined and
ceased to be the great foreign-exchange earner it had once been; nor did cloth
exports yet provide adequate compensation.[37] Bullion steadily seeped out of
the economy, aided by high crown military expenditure overseas as the
Hundred Years War with France drew expensively but inconclusively on. It
therefore wanted only the bumper harvest of 1376 to send grain prices tum-
bling down, inaugurating an era of price deflation which was to persist for the
next 150 years and return prices in the 1440s, 1450s and 1460s to a level not
much above that of the first quarter of the thirteenth century (Figure 1.01).[38]

Nor could the Statute of Labourers restrain forever the mounting upward
pressure on wages. The plague mortality of 1375 intensified that pressure and
the price collapse of 1376 raised it further, on account of the higher material
expectations that it engendered as cheap foodstuffs brought windfall gains in
living standards.[39] Real wages improved more dramatically during the 1370s
than during any other decade on record and for the first time rose significantly
above their long-term average (Figure 1.01). The more that the lot of wage
earners improved the more that popular discontent with the government's
policy of wage restraint and the justices of labourers who enforced it
mounted.[40] Dissatisfaction surfaced in the Peasants' Revolt of 1381.[41]

---

[34] Farmer, 'Prices and wages, 1350–1500', pp. 483–90.

[35] E.g. C. C. Dyer, 'Changes in diet in the late Middle Ages: the case of harvest workers', *AHR*
36 (1988).

[36] For a case study of a widespread phenomenon see J. A. Raftis, 'Changes in an English village
after the Black Death', *Mediaeval Studies* 29 (1967), 158–77.

[37] E. M. Carus-Wilson and O. Coleman, *England's export trade, 1275–1547* (Oxford, 1963), pp.
122–3, 138; J. L. Bolton, *The medieval English economy, 1150–1500* (London, 1980), pp. 292–3.

[38] Farmer, 'Prices and wages, 1350–1500', pp. 439–43.

[39] Farmer, 'Prices and wages, 1350–1500', pp. 437, 520–1.

[40] E. B. Fryde and N. Fryde, 'Peasant rebellion and peasant discontents', in Miller (ed.), *AHEW*,
vol. III, pp. 753–60.

[41] *The Peasants' Revolt of 1381*, ed. R. B. Dobson, 2nd edition (London, 1983); C. C. Dyer, 'The
social and economic background to the rural revolt of 1381', in R. H. Hilton and T. H. Aston
(eds.), *The English rising of 1381* (Cambridge, 1984), pp. 9–24.

Thenceforth the Statute of Labourers ceased to be enforceable and the door was opened for wages to find their natural market level.[42]

By the final quarter of the fourteenth century the transition to the new post-plague *status quo* had been more or less completed. In the process England had been transformed from a populous to an essentially under-populated country, albeit with the legacy of an extensive agricultural area bequeathed by pre-plague colonisation and reclamation.[43] The relative scarcity of people and surfeit of land confronted the agricultural sector with an entirely new set of challenges, as wages drifted steadily upwards and prices downwards. Slack demand and an inherent tendency towards over-production kept the prices of agricultural commodities low, depressed land values, and discouraged investment, even though interest rates had fallen significantly during the half-century which followed the plague.[44] Vested institutional and proprietorial interests meant that the wholesale withdrawal of land from cultivation took time to gather momentum, delaying the establishment of a more rational economic use of the land and raising the social costs of doing so.[45] Eventually, entire arable-farming villages would be replaced with cattle and sheep ranches, with the structure and strength of lordship often determining whether or not a settlement survived this difficult period.[46]

As labour became ever scarcer and real wages rose the differential steadily narrowed between the wages of unskilled workers and those of skilled craftsmen and officials. As a result the more labour-intensive forms of husbandry rapidly became uneconomic.[47] The disproportionate rise in the remuneration of manual workers may also have been partly in recompense for the improved work performance that resulted from higher standards of nutrition.[48] The army of malnourished and impoverished landless and semi-landless folk which had been so omnipresent before the plague was now no more, so much so that finding and recruiting labour was increasingly difficult and expensive.[49] For lords this problem was exacerbated by the decay of serfdom: insofar as

[42] Farmer, 'Prices and wages, 1350–1500', pp. 488–94.   [43] Hatcher, *Plague, population*.

[44] G. Clark, 'The cost of capital and medieval agricultural technique', *Explorations in Economic History* 25 (1988), 265–94.

[45] C. C. Dyer, *Warwickshire farming 1349–c. 1520* (Oxford, 1981), pp. 9–12; E. King, 'The occupation of the land: the east midlands', in Miller (ed.), *AHEW*, vol. III, pp. 73–6; H. S. A. Fox, 'The occupation of the land: Devon and Cornwall', in Miller (ed.), *AHEW*, vol. III, 152–71.

[46] M. W. Beresford, *The lost villages of England* (Lutterworth, 1963), pp. 177–216.

[47] Farmer, 'Prices and wages, 1350–1500', pp. 478–9; B. M. S. Campbell, 'A fair field once full of folk: agrarian change in an era of population decline, 1348–1500', *AHR* 41 (1993), 63; Campbell, 'Agricultural progress', pp. 38–9; C. Thornton, 'The determinants of land productivity on the bishop of Winchester's demesne of Rimpton, 1208 to 1403', in B. M. S. Campbell and M. Overton (eds.), *Land, labour and livestock* (Manchester, 1991), pp. 204–7; M. Mate, 'Farming practice and techniques: Kent and Sussex', in Miller (ed.), *AHEW*, vol. III, pp. 268–71; B. M. S. Campbell, K. C. Bartley, and J. P. Power, 'The demesne-farming systems of post Black Death England: a classification', *AHR* 44 (1996), 177–8.

[48] Dyer, 'Changes in diet'.   [49] *Black Death*, pp. 318–20.

customary labour had hitherto subsidised production on certain demesnes, that subsidy was effectively withdrawn.[50] The less enlightened lords were often simply deserted by their tenants as migration increased; others sensibly granted or sold their tenants their freedom.[51] How this worked itself out on the ground depended very much upon the structure and strength of lordship with the result that the decay of serfdom could proceed at very different rates and have entirely different outcomes on neighbouring manors. In serfdom's place lords became even more dependent upon the hire of farm servants by the year and casual labourers by the task.

Squeezed by falling prices and rising labour costs, demesne lords found themselves at an increasing disadvantage relative to middling-sized farms worked largely by family labour. Generally, it was farms such as these, with smaller overheads and lower unit costs, which fared best. The temptation for lords to convert their demesnes into leasehold farms therefore grew. In contrast to the heady high-farming days before the plague, getting the land to pay now depended more upon minimising costs than maximising revenues. The land's physical productivity was of less moment than whether it could be got to yield a profit. Here, grassland farming offered cost advantages over arable husbandry because it required only a fifth of the labour force per unit area at a time when labour was becoming the scarcest and most expensive of the factors of production. This swing from corn to horn was further encouraged by higher *per capita* consumption of pastoral products as average purchasing power rose.

There was naturally a strong spatial dimension to all these trends since no seigniorial producer enjoyed an equal comparative advantage in all branches of production.[52] Producers with a strong comparative advantage in arable production tended to fare better than average before the Black Death and worse than average thereafter. Whether or not soils were light or heavy could make all the difference at times when the key to success lay in keeping costs down. Not all soils converted well to grass and in areas of closely regulated commonfields there were often institutional obstacles to the withdrawal of land from tillage. Proximity to concentrated urban demand was always an advantage. Change therefore was always environmentally, institutionally, and locationally specific, with the relative importance of environmental, institutional, and commercial influences itself varying over time.

1.3   Strategies for raising (and reducing) agricultural output

The repertoire of ways in which medieval producers could respond to the expansion and contraction of demand was wider than has often been cred-

---

[50] R. H. Hilton, *The decline of serfdom in medieval England*, 2nd edition (London and Basingstoke, 1983).

[51] J. A. Raftis, *Tenure and mobility* (Toronto, 1964), pp. 153–66; L. R. Poos, *A rural society after the Black Death* (Cambridge, 1991), pp. 159–79; *Black Death*, pp. 326–31; Raftis, *Peasant economic development*, pp. 99–117.   [52] Campbell and others, 'Demesne-farming'.

ited.[53] First, and most obviously, land could be brought into or taken out of agricultural use. Thus, the demographic and economic expansion of the twelfth and thirteenth centuries found physical expression in a widespread and well-documented process of reclamation and colonisation.[54] This was more a peasant than a seigniorial movement, although among lords the contribution of the new religious houses of the age – the Augustinians, Praemonstratensians, and especially the Cistercians – was outstanding.[55] The superior organisational and capital resources of lords also meant that they took the lead in much wetland reclamation, such as the drainage and embanking of the Romney and Walland marshes in Kent, the Essex marshes, the Somerset Levels, and, most spectacularly of all, the silt fens of south Lincolnshire and west Norfolk.[56] Typically, such late reclamations of hill land and marshland added more to the grassland area than they did to the arable.[57] Nevertheless, reclamation was rarely of wholly virgin land. Instead it almost invariably involved upgrading land from a usage of lower productivity and intensity to one of a higher, such as from rough pasture to improved pasture, from woodland to grassland, or from marsh to meadow. The supply of land naturally set limits to the capacity of reclamation to keep delivering output gains but this was compounded by the constraints of available technology and the institutional obstacles that royal and private hunting grounds and common pastures presented to the advance of the plough.[58]

An alternative route to higher output lay in raising the carrying capacity of

[53] M. Overton and B. M. S. Campbell, 'Productivity change in European agricultural development', in Campbell and Overton (eds.), *Land, labour and livestock*, pp. 17–28; Overton, *Agricultural revolution*, pp. 88–121.

[54] R. A. Donkin, 'Changes in the early Middle Ages', in H. C. Darby (ed.), *A new historical geography of England* (Cambridge, 1973), pp. 98–106; B. M. S. Campbell, 'People and land in the Middle Ages, 1066–1500', in R. A. Dodgshon and R. A. Butlin (eds.), *An historical geography of England and Wales*, 2nd edition (London, 1990), pp. 73–7.

[55] T. A. M. Bishop, 'Monastic granges in Yorkshire', *EHR* 51 (1936), 193–214; R. A. Donkin, *The Cistercians* (Toronto, 1978), pp. 104–34; R. R. Davies, *Lordship and society in the March of Wales 1282–1400* (Oxford, 1978), pp. 116–17.

[56] H. C. Darby, *The medieval Fenland* (Cambridge, 1940); L. F. Salzman, 'The inning of Pevensey Levels', *Sussex Archaeological Collections* 52 (1910), 32–60; S. G. E. Lythe, 'The organization of drainage and embankment in medieval Holderness', *Yorkshire Archaeological Journal* 34 (1939), 282–95; R. A. L. Smith, 'Marsh embankment and sea defence in medieval Kent', *Economic History Review* 10 (1940), 29–37; N. Harvey, 'The inning and winning of the Romney marshes', *Agriculture* 62 (1955), 334–8; M. Gardiner, 'Medieval farming and flooding in the Brede valley', in J. Eddison (ed.), *Romney Marsh* (Oxford, 1995), pp. 127–37; B. E. Cracknell, *Canvey Island* (Leicester, 1959); M. Williams, *The draining of the Somerset Levels* (Cambridge, 1970).

[57] E.g. J. McDonnell, 'Medieval assarting hamlets in Bilsdale, north-east Yorkshire', *Northern History* 22 (1986), 269–79; R. I. Hodgson, 'Medieval colonization in northern Ryedale, Yorkshire', *Geographical Journal* 135 (1969), 44–54; A. J. L. Winchester, *Landscape and society in medieval Cumbria* (Edinburgh, 1987), pp. 37–44.

[58] L. Cantor, 'Forests, chases, parks and warrens', in L. Cantor (ed.), *The English medieval landscape* (London, 1982), pp. 56–85. At its maximum extent in the thirteenth century as much as a quarter of the land of England may have been subject to forest law: C. R. Young, *The royal forests of medieval England* (Leicester, 1979), p. vii.

the land by substituting outputs with a higher energy/food yield for those with a lower. Replacing grassland with arable was the most conspicuous means of doing this, since the output of edible calories and protein from crops is far higher than that from animals and their products.[59] This was contingent upon a corresponding adjustment in diets and the adoption of measures that would maintain the fertility of the expanded area of arable. Even without changing the ratio of grassland to arable much could be achieved merely by altering the balance of the animals stocked and crops grown. Horses were more energy productive than oxen, dairying was more food productive than fattening, and grains and legumes grown for bread and pottage yielded higher food-extraction rates than those grown for brewing and fodder. Such simple production shifts delivered significant gains in the rate of food output per unit area and thereby maximised food-extraction rates.

Typically, these shifts in the composition of flocks and herds and the cropped acreage were driven by relative prices as more and more consumers were obliged to trade down to cheaper foodstuffs of a lower dietary preference. Those who traded down to the cheapest foodstuffs of all became the most vulnerable at times of harvest failure and extreme price inflation for they had nowhere left to trade. Within a commercialised economy it could also make sense to substitute crops and animals that yielded industrial raw materials in the form of flax, hemp, dye plants, and wool for those that yielded food since the former were capable of yielding higher cash returns per unit area. The revenues thereby generated could then be used to purchase foodstuffs. Many of these specific changes in the composition of agricultural output were associated with the adoption of agricultural food-chains which delivered higher yields of food and energy per unit area.[60] Merely by feeding animals on a combination of grazing and produced fodder rather than grazing alone raised the proportion of non-working animals that could be carried. That in turn usually meant a greater emphasis upon longer and more flexible rotations coupled with the enhanced importance of fodder cropping and managed hay meadows, all of which required higher factor inputs per unit area.

Raising output was invariably contingent upon raising inputs. Typically that meant putting more people to work on the land. The reclamation process itself required significant quantities of labour and invariably led to the permanent establishment of more labour-intensive forms of land-use. Irrespective of such land-use changes, however, there were gains in productivity to be obtained by managing existing resources more intensively. For instance, the fuel yield of woodland could be maximised by instituting a system of coppicing whereby felling conformed to a regular cycle. At the close of the thirteenth

---

[59] D. B. Grigg, *The dynamics of agricultural change: the historical experience* (London, 1982), pp. 70–1.

[60] B. M. S. Campbell, 'The livestock of Chaucer's reeve: fact or fiction?', in E. B. DeWindt (ed.), *The salt of common life* (Kalamazoo, 1995), p. 290; I. G. Simmons, *The ecology of natural resources* (London, 1974).

century intensively managed and highly valued coppiced woodlands engaged in supplying the lucrative London market employed felling cycles of three to eleven years, with a mean of seven years.[61] The labour requirements of such a regime were naturally far greater than those of woodlands either cropped less frequently or more haphazardly. Grassland, too, could be managed so as to maximise the yield of hay and the number of stock that could be supported. Most meadows were artificial creations – drained, embanked and mown – and among the most intensively managed and vital of medieval resources.[62] Dyking and ditching could similarly enhance the grazing potential of marshes and pastures, while fencing and hedging, or at the very least stinted and supervised grazing, could improve the efficiency of their utilisation. The most conspicuous output gains were nevertheless those that derived from increasing the labour and capital inputs to arable husbandry.

Substitution of legumes for bare fallows, more thorough application of fertilisers, better preparation of the seed-bed, improved weed control, and more scrupulous harvesting were all possible with increased labour inputs and were greedy in their consumption of labour.[63] Where on-the-farm sources of fertiliser were inadequate these could be augmented by supplies from outside – nightsoil from towns and seaweed and sea sand from the coast – although these were expensive and troublesome to obtain and invariably only used when cheap to transport or available close at hand.[64] Such measures were especially important whenever the decision was taken to increase the frequency of cropping and decrease the frequency of fallowing, since without them nitrogen and other essential soil nutrients would become progressively depleted. Provided that fertility could be maintained, cultivating the land more frequently was one of the most effective methods of raising output per unit area.[65] Any improvement in unit output in turn required more labour per unit area to reap, bind, cart, and thresh the harvest. G. Clark has calculated that at the end of the eighteenth century approximately 40 per cent of labour costs were directly dependent, and a further 25 per cent of labour costs partially dependent, upon yields.[66]

---

[61] National *IPM* database; J. A. Galloway, D. J. Keene, and M. Murphy, 'Fuelling the city: production and distribution of firewood and fuel in London's region, 1290–1400', *EcHR* 49 (1996), 454.

[62] E.g. H. S. A. Fox, 'The alleged transformation from two-field to three-field systems in medieval England', *EcHR* 39 (1986), 544–5.

[63] W. Harwood Long, 'The low yields of corn in medieval England', *EcHR* 32 (1979), 464–9; Campbell, 'Agricultural progress', pp. 32–6, 38; D. Postles, 'Cleaning the medieval arable', *AHR* 37 (1989); R. A. L. Smith, *Canterbury Cathedral Priory* (Cambridge, 1943), pp. 135–8.

[64] Chapter 7, pp. 360–2; Campbell, 'Agricultural progress', p. 34; J. Hatcher, 'Farming techniques: south-western England', in Hallam (ed.), *AHEW*, vol. II, p. 388; R. I. Jack, 'Farming techniques: Wales and the Marches', in Hallam (ed.), *AHEW*, vol. II, p. 442.

[65] B. M. S. Campbell, 'Arable productivity in medieval England: some evidence from Norfolk', *JEH* 43 (1983), 390–4.

[66] G. Clark, 'Labour productivity in English agriculture, 1300–1860', in Campbell and Overton (eds.), *Land, labour and livestock*, pp. 222–6.

Higher capital inputs could also raise output. An increase in the iron component of ploughs, spades, scythes, and other implements could greatly facilitate the breaking up and cultivation of more land.[67] Under the right circumstances heavier seeding rates were capable of delivering higher yields per unit area.[68] Fixed capital investment in barns, granaries, byres, stables, sties, and cotes could also improve the storage of crops and housing of livestock.[69] From the thirteenth century investment in such buildings was able to take advantage of contemporary innovations in methods of construction, notably pegged mortises, the abandonment of earthfast construction, and the use of ceramic roof tiles.[70] Although there was an element of prestige and display in this, such buildings were more durable and helped minimise loss and damage. Housing livestock, for instance, helped maximise fertility and minimise mortality rates and sustain more intensive forms of management.

The danger with increasing inputs is that beyond a certain point they become prone to diminishing returns. One means of raising the efficiency as well as the output of agriculture was therefore through greater specialisation. When individual producers made the most of their comparative advantage and concentrated upon what they produced best they maximised not only their own output but also that of the agricultural system as a whole.[71] Because proportionately more of a commodity would have been produced by those with a genuine advantage for doing so, and less by those who did not, *mean* yields of individual crop and livestock products would have risen, although actual yields on individual farms may have remained the same. Additional efficiency gains accrued from a fuller spatial division of labour.

Pastoral husbandry was particularly open to specialisation owing to the ease with which animals and certain of their products – wool, hides, cheese, and butter – could be transported over long distances.[72] Thus, wool, sometimes produced in the remotest locations, was the one agricultural commodity regularly to be traded in local, regional, national and international markets.[73] For pastoral producers the way forward lay in greater specialisation by type of

---

[67] J. Myrdal, *Medieval arable farming in Sweden. Technical change A.D. 1000–1520*, Nordiska museets Handlingar 105 (Stockholm, 1986). On iron production see Bolton, *English economy*, pp. 163–4; G. Astill, 'An archaeological approach to the development of agricultural technologies in medieval England', in G. Astill and J. L. Langdon (eds.), *Medieval farming and technology* (Leiden, 1997), pp. 207–11.   [68] Campbell, 'Arable productivity', pp. 385–92.

[69] National *IPM* database; J. G. Hurst, 'Rural building in England and Wales: England', in Hallam (ed.), *AHEW*, vol. II, pp. 888–98; C. C. Dyer, 'Sheepcotes: evidence for medieval sheepfarming', *Medieval Archaeology* 39 (1995).

[70] Astill, 'Archaeological approach', pp. 212–13.

[71] This describes the principle of absolute advantage. The more complicated case is comparative advantage: R. G. Lipsey, *An introduction to positive economics* (London, 1972), pp. 592–6.

[72] M. Overton and B. M. S. Campbell, 'Norfolk livestock farming 1250–1740: a comparative study of manorial accounts and probate inventories', *JHG* 18 (1992), 393.

[73] E. Power, *The wool trade in English medieval history* (Oxford, 1941); T. H. Lloyd, *The English wool trade in the Middle Ages* (Cambridge, 1977).

livestock, stage of livestock production, and type of livestock product.[74] As more animals changed hands and changed hands more frequently so more livestock fairs were required to service the concomitant growth in trade. Within the arable sector the opportunities for specialising were equally real but geographically more circumscribed, owing to the higher unit transport costs of grain, especially overland. Access to cheap bulk transport by river and sea thus shaped patterns of arable specialisation, especially within the hinterlands of major urban centres.[75] Concentrated urban demand was, in fact, the single greatest spur to all forms of agricultural specialisation. In the relatively highly urbanised continental economies of thirteenth-century Tuscany and the Low Countries K. G. Persson has argued that greater specialisation for market exchange was the principal means by which agricultural output was raised during the thirteenth century and the growing towns and cities fed.[76]

The final process by which output could have been expanded was through technological change. Here, progress was typically more evolutionary than revolutionary since the range of new techniques available to medieval cultivators was limited and their pace of diffusion often painfully slow.[77] Agricultural progress, when it came, invariably entailed advance across a broad front involving a host of minor technological adjustments the individual significance of which is easily overlooked within the overall technological complex.[78] For example, the substitution of horses for oxen was in turn contingent upon many associated improvements in harnessing technique, the shoeing of horses, and design and construction of ploughs and carts.[79] In an essentially organic and animate age the most significant technological breakthrough of all lay in the development of integrated mixed-farming systems in which the arable and pastoral sectors were complementary rather than competitive. Characteristically, these required the development of new types of rotation in conjunction with increased production of leguminous and fodder crops and the reorientation of animal husbandry.[80] Developing and operating

---

[74] Overton and Campbell, 'Norfolk livestock farming'; Campbell, 'Chaucer's reeve'.

[75] B. M. S. Campbell and J. P. Power, 'Mapping the agricultural geography of medieval England', *JHG* 15 (1989); B. M. S. Campbell, 'Ecology versus economics in late thirteenth- and early fourteenth-century English agriculture', in Sweeney (ed.), *Agriculture*, pp. 81–91.

[76] K. G. Persson, *Pre-industrial economic growth, social organization and technological progress in Europe* (Oxford, 1988), pp. 30–1.

[77] Technological progress in this period is discussed in Persson, *Pre-industrial growth*, pp. 24–31; B. M. S. Campbell, 'Constraint or constrained? Changing perspectives on medieval English agriculture', *Neha-Jaarboek voor economische, bedrijfs- en techniekgeschiedenis* 61 (Amsterdam, 1998), 24–7.

[78] J. L. Langdon, G. Astill, and J. Myrdal, 'Introduction', in Astill and Langdon (eds.), *Medieval farming*, p. 6.

[79] G. Raepsaet, 'The development of farming implements between the Seine and the Rhine from the second to the twelfth centuries', in Astill and Langdon (eds.), *Medieval farming*, pp. 41–68.

[80] B. M. S. Campbell and M. Overton, 'A new perspective on medieval and early modern agriculture: six centuries of Norfolk farming c. 1250–c. 1850', *PP* 141 (1993), 88–95.

systems in which the arable sector supported the pastoral by supplying fodder and temporary grazing and the pastoral sector supported the arable by supplying manure and traction required skill and experience and potentially made heavy demands upon labour. Consequently, the commercial success of these systems often depended upon keeping labour costs within reasonable limits. Essentially that meant adopting better forms of organisation which either enabled increased output to be obtained from a given level of inputs, or the same output to be obtained from fewer inputs, thus raising total factor productivity.[81]

None of these five basic strategies for raising (or, when reversed, for lowering) agricultural output were mutually exclusive. Bringing more land into cultivation invariably entailed the outlay of more labour and capital per unit area. Changing the balance and composition of outputs was rarely possible without concomitant changes in inputs. Developing more specialised systems of production was usually contingent upon some alteration of technique. Sustaining productive, manure-intensive, mixed-farming systems required a combination of intensive and innovative methods. Output growth was therefore a multi-faceted phenomenon and invariably assumed different forms in different environmental, institutional, and locational contexts. Producers acted rationally when they optimised output according to their given factor endowment and prevailing levels of economic rent; only rarely did this justify taking available technology to the limit and maximising output.

1.4   Risks, dilemmas, and debates

When an economic system succumbs to crises of subsistence as profound as those of the first half of the fourteenth century it is tempting to regard this as an indictment of the majority agrarian sector. Demographic and economic expansion, to be securely based, needed to be underpinned by levels of agricultural production that were sustainable. Hence Adam Smith's dictum in *The Wealth of Nations* that 'of all the ways in which capital could be employed investment in agriculture was by far the most advantageous to society'.[82] Without adequate investment all pre-industrial agrarian regimes were vulnerable to decline. In the first place there was an agronomic dilemma of how to expand output without jeopardising the fragile ecological equilibrium conditioned by the cycle of nitrogen and availability of potassium, phosphorus, and other essential plant nutrients in the soil. This dilemma was as old as agriculture itself. The problem applied to pastoral as much as arable husbandry for, just as arable soils could be depleted of their fertility if they were over-

---

[81] C. Ritson, *Agricultural economics* (London, 1980), p. 95; A. N. Link, *Technological change and productivity growth* (London, 1987), p. 4.

[82] Adam Smith, *An inquiry into the nature and causes of the wealth of nations*, ed. R. Campbell and A. Skinner (Oxford, 1976), vol. I, p. 364.

cultivated, so, too, over-stocking could degrade pastures and debilitate livestock.[83] Medieval historians have long debated whether soil exhaustion may have depressed grain yields, but as yet there is little unambiguous evidence to support this thesis.[84] In contrast, there is good archaeological evidence that a significant degeneration of livestock had taken place by the fourteenth century. Thus, faunal remains indicate that carcass weights of all the major domesticated animals had greatly diminished since Roman and early Anglo-Saxon times when the availability of pasturage had presumably been far more abundant.[85] Such deficiencies within the pastoral sector undoubtedly exacerbated any problems being experienced by the arable. J. Z. Titow and Farmer both blame the low yields obtained on the demesnes of the bishopric of Winchester and abbey of Westminster upon the 'chronic state of under-manuring' which arose from shortages of livestock.[86]

Compounding this agronomic dilemma was the equally enduring economic dilemma identified by David Ricardo of how – as populations rose – to avoid diminishing returns to land and labour in conditions of a fixed supply of land and in the absence of significant technological progress.[87] So long as rising food requirements could only be met by extending cultivation to progressively inferior soils there was an inevitable tendency, in the absence of much specialisation, for mean arable productivity to fall. If at the same time population growth resulted in the application of increased labour to existing methods of production – a process termed 'involution' by C. Geertz – the marginal

---

[83] R. S. Shiel, 'Improving soil fertility in the pre-fertiliser era', in Campbell and Overton (eds.), *Land, labour and livestock*, pp. 51–77; W. S. Cooter, 'Ecological dimensions of medieval agrarian systems', *Agricultural History* 52 (1978), 458–77; R. S. Loomis, 'Ecological dimensions of medieval agrarian systems: an ecologist responds', *Agricultural History* 52 (1978), 478–83; J. N. Pretty, 'Sustainable agriculture in the Middle Ages: the English manor', *AHR* 38 (1990), 1–19.

[84] See Thornton, ' Determinants of productivity', pp. 183–210, for a critical reassessment of the Winchester evidence. For a scientific analysis of the Cuxham evidence suggesting a progressive depletion of phosphorus see E. I. Newman and P. D. A. Harvey, 'Did soil fertility decline in medieval English farms? Evidence from Cuxham, Oxfordshire, 1320–1340', *AHR* 45 (1997), 119–36.

[85] A. Grant, 'Animal resources', in G. Astill and A. Grant (eds.), *The countryside of medieval England* (Oxford, 1988), pp. 176–7; S. Bökönyi, 'The development of stockbreeding and herding in medieval Europe', in Sweeney (ed.), *Agriculture*, pp. 42–55.

[86] J. Z. Titow, *Winchester yields* (Cambridge, 1972), p. 30; D. L. Farmer, 'Grain yields on Westminster Abbey manors, 1271–1410', *Canadian Journal of History* 18 (1983), 331–47. In fact, there is little correlation between stocking densities and crop yields. A correlation of the weighted aggregate net yield per acre against the number of livestock units per 100 grain acres for Norfolk for the period 1250–1449 gives a correlation coefficient of +0.037. At Felbrigg between 1401 and 1420 a correlation of the weighted aggregate yield per seed and per acre against the number of livestock units per 100 grain acres, using five-year means, gives negative correlation coefficients of −0.71 and −0.57 respectively: NRO, WKC 2/130–31/398×6.

[87] E. A. Wrigley, 'The classical economists and the Industrial Revolution', in Wrigley, *People, cities and wealth* (Oxford, 1987), pp. 21–45.

productivity of labour in agriculture would likewise eventually decline.[88] A complex series of checks upon the continued growth of both the economy and its dependent population would thereby be set in train.[89]

M. M. Postan, in one of the most influential interpretations of the late medieval economy, argued that agriculture resolved neither of these dilemmas.[90] For him rapid population growth was one of the most salient economic facts of the twelfth and thirteenth centuries. Following the ideas of the German historian W. Abel, he argued that pressure of numbers ultimately led to the colonisation of land that was physically marginal for cultivation.[91] Apart from depressing both mean crop yields and mean output per worker in agriculture, as Ricardo predicted, this promoted a growing dependence upon land that was ecologically vulnerable and soils that were easily exhausted. The problem was further compounded by the conversion of pasture to arable, which depressed stocking densities and thereby starved the arable of essential traction and manure. Consequently, supplies of soil nitrogen – probably the single greatest constraint upon yields at that time – became progressively exhausted, with the result that there was a severe cut-back of agricultural output through soil deterioration and falling yields.[92]

It was Postan's belief that arable productivity failed even on the older-settled and intrinsically more fertile lowland soils. He cites as evidence the low yield ratios obtained by many seigniorial demesnes and places particular stress upon the declining yield ratios of spring-sown crops on the estates of the bishops of Winchester during the second half of the thirteenth century.[93] Scarcities of livestock and thus of manure were for him the chief culprits. If the situation was bad on demesnes it was, he believes, far worse on peasant holdings since they were even more prone to under-stocking.[94] Thus, it was not

---

[88] C. Geertz, *Agricultural involution: the process of ecological change in Indonesia* (Berkeley, 1963); Overton and Campbell, 'Productivity change', p. 19.

[89] For a discussion of some of these linkages see Wrigley and Schofield, *Population history*, pp. 454–84.

[90] This thesis was first fully elaborated in M. M. Postan, 'Medieval agrarian society in its prime: England', in M. M. Postan (ed.), *The Cambridge economic history of Europe*, vol. I, *The agrarian life of the Middle Ages*, 2nd edition (Cambridge, 1966), pp. 549–632. It is restated at greater length in M. M. Postan, *The medieval economy and society* (London, 1972). See also H. E. Hallam, 'The Postan thesis', *Historical Studies* 15 (1972), 203–22; C. C. Dyer, 'The past, the present and the future in medieval rural history', *Rural History* 1 (1990), 42–7.

[91] W. Abel, *Agrarkrisen und Agrarkonjunktur in Mitteleuropa vom 13. bis zum 19. Jahrhundert* (Berlin, 1935), trans. O. Ordish, *Agricultural fluctuations in Europe from the thirteenth to the twentieth centuries* (London, 1980); W. Abel, *Die Wüstungen des Ausgehenden Mittelalters*, 2nd edition (Stuttgart, 1955); Postan, 'Agrarian society', p. 559. For a critique see M. Bailey, 'The concept of the margin in the medieval English economy', *EcHR* 42 (1989), 1–17.

[92] Postan, 'Agrarian society', pp. 553–9; Shiel, 'Improving soil fertility'.

[93] Titow, *Winchester yields*.

[94] M. M. Postan, 'Village livestock in the thirteenth century', *EcHR* 15 (1962), 219–49; Postan, 'Agrarian society', p. 557. But see J. Masschaele, *Peasants, merchants, and markets* (New York, 1997), pp. 42–7.

just, as in the conventional Malthusian equation, that food production failed to keep pace with population growth. It was worse. The techniques that had sufficed to enable the population to reach the existing limit were no longer adequate and production of essential bread grains began to decline in absolute as well as relative terms.[95]

Whereas Postan, following in the tradition of the classical economists, stressed the technological inability of medieval agriculture to sustain population growth on a finite supply of land, Marxist historians attach more importance to the disincentives to investment inherent within feudal socio-property relations.[96] Rather than invest, lords preferred to spend 'up to the hilt on personal display, on extravagant living, on the maintenance of a numerous retinue, and on war'.[97] Interest in their estates went little further than the exaction of maximum profit and the notion of reinvesting profits to raise productivity occurred to very few.[98] The net outcome was that 'even highly organised and superficially efficient estates were failing in one quite basic requirement of good husbandry: the keeping of the land in good heart'.[99] Nor could peasants make good this deficiency, for they were deprived of capital by a combination of excessive feudal exactions, ecclesiastical tithes, arbitrary royal purveyancing, and punitive taxation.[100] The upshot in both cases was technological inertia. Agriculture – the producer of vital food and raw materials – thus remained within a low productivity trap, with increments in output dependent upon a process of *extensification* (to which the supply of land set finite limits) rather than one of *intensification* and productivity growth (to which investment and innovation provided the keys). According to this line of argument, it required the replacement of feudal with capitalist socio-property relations before rates of investment could improve and attempts to raise the total output of English agriculture would cease to be bought at the price of diminishing returns to either land or labour – something which did not happen until after 1650.[101]

These negative verdicts upon the performance of thirteenth- and fourteenth-century English agriculture have an impressive historiographic pedigree. It was W. Denton who in 1888 first advanced the thesis that soils became progressively impoverished during the Middle Ages, eventually leading to

[95] Chambers, *Population, economy, and society*, pp. 24–5.
[96] R. Brenner, 'Agrarian class structure and economic development in pre-industrial Europe', *PP* 70 (1976), reprinted in T. H. Aston and C. H. E. Philpin (eds.), *The Brenner debate* (Cambridge, 1985), pp. 30–4. See also G. Bois, *The crisis of feudalism* (Cambridge, 1984); M. Dunford and D. Perrons, *The arena of capital* (Basingstoke and London, 1983), pp. 90–123.
[97] R. H. Hilton, *The English peasantry in the later Middle Ages* (Oxford, 1975), p. 177.
[98] R. H. Hilton, 'Rent and capital formation in feudal society', in *English peasantry*, pp. 177–96.
[99] Miller and Hatcher, *Rural society*, p. 217.
[100] Brenner, 'Agrarian class structure', pp. 31–4.
[101] R. C. Allen, *Enclosure and the yeoman* (Oxford, 1992); Campbell and Overton, 'New perspective', pp. 95–9; Overton, *Agricultural revolution*, pp. 63–132. For a recent reassessment of the decline of feudalism see Britnell, 'Commerce and capitalism'.

falling yields, demographic decline, and the abandonment of land.[102] Among those who subsequently espoused it, R. Prothero (later Lord Ernle) has been particularly influential. For him, 'large improvements in the mediaeval methods of arable farming were impossible until farmers commanded the increased resources of more modern times'.[103] Yet it suited his argument to emphasise the inertia of medieval agriculture since his prime concern was to highlight the post-medieval march of progress. Given the key role of enclosure in his account of the 'agricultural revolution', it was to him axiomatic that little technological progress could have been possible during the Middle Ages owing to the predominance of communal agriculture in subdivided fields. Nor were the nitrogenous and root crops yet available which subsequently would enable husbandmen to diversify rotations, raise vital soil-nitrogen levels, and increase fodder output. The problem was not just that farmers lacked clover, sainfoin, and turnips, it was also that mounting demand for bread and pottage grains led to over-expansion of arable at the expense of pasture thereby driving down stocking densities and starving the land of manure.[104] On this diagnosis, key structural and institutional changes combined with improvements in agricultural know-how had to take place before English farmers could escape from the low productivity trap which had been their lot during the greater part of the Middle Ages.

These pessimistic verdicts upon medieval agriculture all dwell upon the supply-side obstacles to agricultural progress. Whether implicitly Ricardian, Malthusian, Marxist, or Whiggish, they attach prime importance to socio-property, institutional, and technological considerations rather than market forces and commercial opportunities. Their arguments are also lent force by the knowledge that the medieval economy eventually succumbed to war, famine, and pestilence on a spectacular scale.[105] Yet such verdicts are unduly harsh and ripe for reassessment.

The achievements of English medieval agriculture are far from unimpressive. By *c.* 1300 domestic agriculture was feeding at least twice as many people as in 1086 (Table 8.06). It was also provisioning a greatly enlarged urban population, whose share of the total may have doubled from a tenth to a fifth.[106]

---

[102] W. Denton, *England in the fifteenth century* (London, 1888). This basic thesis has since been adapted and employed by a number of historians: for a discussion of the relevant historiography see N. Hybel, *Crisis or change*, trans. J. Manley (Aarhus, 1989).

[103] Lord Ernle (formerly R. Prothero), *English farming past and present* (London, 1912; 3rd edition, 1922), p. 33.

[104] V. G. Simkhovitch, 'Hay and history', *Political Science Quarterly* 28 (1913), 385–404; Lord Ernle, 'The enclosures of open-field farms', *Journal of the Ministry of Agriculture* 27 (1920), 831–41. See also G. Clark, 'The economics of exhaustion, the Postan thesis, and the agricultural revolution', *JEH* 52 (1992), 61–84.

[105] Campbell, 'Constraint or constrained', 15–17.

[106] R. H. Britnell, 'Commercialisation and economic development in England, 1000–1300', in Britnell and Campbell (eds.), *Commercialising economy*, pp. 10–11, reckons that 600,000 people lived in towns of 2,000 inhabitants or more *c.* 1300, equivalent to 15 per cent of a pop-

The leading urban centres were all individually much larger. In 1086 London alone contained over 10,000 inhabitants; by *c.* 1300 it had grown to a city of perhaps 60–80,000 inhabitants and had been joined by at least thirteen other towns with populations of at least 10,000.[107] Each of these urban centres drew upon a greatly extended rural hinterland for food and fuel.[108] All these cities, as well as many lesser towns and some rural areas, contained significant numbers of craftsmen and artisans. The latter had grown proportionately as a socio-economic group as the population had increased and the economy had become more differentiated and complex.[109] In almost every case they processed or utilised agriculturally produced raw materials: flax, hemp, wool, dye plants, hides, skins, tallow, grain, straw, timber, and wood. Some of the goods they manufactured were subsequently exported along with impressive quantities of many of these same raw materials and primary products; English wool, for instance, kept many an Italian, Cahorsian, and Flemish spinner and weaver busy at the end of the thirteenth century.[110]

Over the course of the thirteenth century alone the value of England's export trade approximately trebled, outpacing the concurrent increase in population.[111] Inland trade grew commensurately.[112] Products of direct and indirect agricultural provenance dominated both branches of trade.[113] By the opening of the fourteenth century wool, cloth, hides, grain, and small amounts of firewood comprised over 90 per cent of an export trade possibly worth £302,000.[114] On currently available estimates of national income at the same date (Table 8.07) English exports may already have been worth 6–8 per cent of GDP. Agricultural exports constituted an even greater proportion – probably in excess of 10 per cent – of the nation's gross agricultural production.[115] Never again would so many people be so exclusively dependent upon

ulation of 4m. or 10 per cent of a population of 6m. C. C. Dyer, 'How urbanized was medieval England?', in J.-M. Duvosquel and E. Thoen (eds.), *Peasants and townsmen in medieval Europe* (Ghent, 1995), pp. 169–83, reckons that a fifth of the population lived in towns of one sort or another.

[107] Campbell and others, *Medieval capital*, pp. 9–11; P. Nightingale, 'The growth of London in the medieval English economy', in R. H. Britnell and J. Hatcher (eds.), *Progress and problems in medieval England* (Cambridge, 1996), pp. 95–8.

[108] Campbell and others, *Medieval capital*, pp. 172–3; Galloway and others, 'Fuelling the city', pp. 447–72.

[109] E. Miller and J. Hatcher, *Medieval England: towns, commerce and crafts 1086–1348* (London, 1995), pp. 51–2, 128–34; Britnell, *Commercialisation*, p. 104.

[110] Miller and Hatcher, *Towns, commerce and crafts*, p. 213; Lloyd, *English wool trade*, pp. 43–9.

[111] Miller and Hatcher, *Towns, commerce and crafts*, p. 214.

[112] Miller and Hatcher, *Towns, commerce and crafts*, pp. 135–80; Masschaele, *Peasants, merchants, and markets*.

[113] Miller and Hatcher, *Towns, commerce and crafts*, pp. 210–14; Masschaele, *Peasants, merchants, and markets*, pp. 13–54.   [114] Miller and Hatcher, *Towns, commerce and crafts*, p. 213.

[115] At the end of the seventeenth century exports were still worth only 5–6 per cent of GDP and during most of the first half of the nineteenth century they accounted for 9–11 per cent of national income: P. Deane and W. A. Cole, *British economic growth 1699–1959*, 2nd edition (Cambridge, 1969), pp. 28–9.

domestic agriculture for the bulk of all foodstuffs, fuel, draught power, building materials, raw materials, and export earnings; nor would agriculture be of such overwhelming importance as an employer of labour and source of wealth, power, and status; nor would trade in agricultural produce bulk so large in the commercial life of the nation.[116]

Judged by these criteria medieval agriculture had achieved much, although whether it was capable of sustaining further increments of population and economic activity is a moot point. Certainly, greater allowance needs to be made for the role of exogenous environmental factors in precipitating the succession of natural disasters to which the population and economy both succumbed during the course of the fourteenth century. Dendrochronology identifies the years from 1318 to 1353 as the longest episode of depressed oak growth during the last two millennia. Growth was most depressed during the 1340s, which stands out as the only occasion this millennium when tree growth was simultaneously depressed in Europe, North America, and Australasia.[117] Reconstructed Fenno-Scandian temperatures identify the 1340s as the colder of two early-fourteenth-century spells of unusually low temperatures and the same decade also stands out as a pronounced discontinuity in all currently available Greenland ice-cores (which preserve an annual record of precipitation and associated air quality over the northern ice cap).[118] On this evidence it would appear that farmers in the first half of the fourteenth century were contending with considerably more than an unlucky run of 'bad weather'; they were, in fact, in the grip of a short-term climatic deterioration of global proportions. Even more intriguingly, an apparent discontinuity in Carbon 14 decay rates can also be dated to the self-same period. This anomaly is most explicable in terms of the discharge of significant quantities of dead carbon into the atmosphere, possibly as a result of some kind of tectonic activity or out-gassing event.[119] Significantly, there are many contemporary accounts of earthquakes, corrupted air, and abnormal atmospheric effects in the decades prior to the Black Death, although historians have rarely attached much importance to them.[120]

These thirty to forty years of climatic and tectonic disturbance proved

---

[116] G. Clark, 'A revolution too many: the agricultural revolution, 1700–1850', Agricultural History Center, University of California at Davis, Working Paper Series 91 (1997), pp. 25–9; B. M. S. Campbell, 'The sources of tradable surpluses: English agricultural exports 1250–1350', in L. Berggren, N. Hybel, and A. Landen (eds.), *Trade and transport in northern Europe 1150–1400* (Toronto, forthcoming).

[117] M. G. L. Baillie, 'Dendrochronology provides an independent background for studies of the human past', in D. Cavaciocchi (ed.), *L'uomo e la foresta secc. XIII–XVIII* (Prato, 1995), pp. 99–119.

[118] K. R. Briffa, P. D. Jones, T. S. Bartholin, D. Eckstein, F. H. Schweingruber, W. Karlen, P. Zetterberg, and M. Eronen, 'Fennoscandian summers from AD 500: temperature changes on short and long timescales', *Climate Dynamics* 7 (1992), 111–19.

[119] M. G. L. Baillie, 'Gas hydrate hazards: have human populations been affected?', unpublished manuscript.   [120] *Black Death*, pp. 158–82.

exceptionally hazardous and unhealthy for both humans and domesticated animals. Apart from the Great European Famine itself, which began with three consecutive years of the most agriculturally disastrous weather in the second millennium, there were further serious harvest failures in 1331, 1346, and 1351.[121] In 1319–20 cattle herds were ravaged by disease – probably rinderpest – and over the next thirty years recurrent outbreaks of murrain and scab ensured a high background mortality of sheep.[122] These essentially biological catastrophes can hardly have been unconnected with the disturbed environmental conditions prevailing at the time. The same is probably true of plague, which began its terrible spread from Asia across Europe at precisely the point of greatest environmental stress, presumably because of some ecologically triggered epizootic crisis in the plague bacterium *Pasteurella pestis* and the rat flea *Xenopsylla cheopis* which carried it.[123] In this context, it is worth noting that the previous great plague pandemic to spread from Asia into Europe – the Justinian Plague of AD 541–4 – likewise began its spread in the immediate aftermath of a similarly acute episode of environmental dislocation.[124] Several other lesser pandemics which spread from Asia as far as the Near East and eastern Mediterranean also correlate with abnormal atmospheric and climatic conditions.

It seems likely, therefore, that the extreme weather conditions which caused harvest failure and famine and the various pestilences of animals and humans were separate manifestations of the same prolonged episode of environmental disturbance which commenced in 1314, reached its climax in the 1340s, and was not over until the mid-1350s. Viewed in this light, these exogenous events assume a far greater magnitude than that which historians have been inclined to ascribe to them. No socio-economic system exposed for so long to such a variety of severe shocks could have withstood them unscathed, let alone one at early-fourteenth-century Europe's stage of technological and economic development. Agricultural producers, in particular, had to contend with a series of environmental hazards not of their making, outside their control, and far beyond their comprehension. The first half of the fourteenth century stands out as perhaps the most difficult and hazardous episode in the annals of English agriculture.

For B. Harvey, 'a century divided by a demographic disaster of the order of

---

121 Farmer, 'Prices and wages', pp. 790–1; Campbell, 'Population pressure', pp. 110–19; Schofield, 'Dearth, debt'.
122 Kershaw, 'The Great Famine', pp. 102–11; Farmer, 'Prices and wages', p. 727; Jordan, *The Great Famine*, pp. 35–9; T. H. Lloyd, *The movement of wool prices in medieval England* (Cambridge, 1973), p. 13. Complaints of reduced cropped acreages owing to shortages of plough oxen are recorded in the *Nonae* Rolls of 1340–1: A. R. H. Baker, 'Evidence in the *Nonarum inquisitiones* of contracting arable lands in England during the early fourteenth century', *EcHR* 19 (1966), 523, 530.   123 Shrewsbury, *Bubonic plague*, pp. 1–2.
124 M. G. L. Baillie, 'Dendrochronology raises questions about the nature of the AD 536 dust-veil event', *The Holocene* 4 (1994), 212–17.

magnitude of the Black Death is not easily seen as a single period'.[125] In a very real sense, that *deus ex machina* marks both a historiographic and historical divide. Thereafter, the issues of investment, technological innovation, productivity, and output which loom so large in debate about pre-Black Death English agriculture cease to be of much historical concern. The normal historical assumption seems to be that after 1380 output per unit area fell but output per worker rose, but this has yet to be put to any systematic test.[126] Total agricultural output certainly contracted, although if labour productivity rose the reduction in output would have been less than the reduction in population. The composition of output also undoubtedly changed as the capacity of consumers to indulge their dietary preferences increased.[127] Since the cost of capital fell relative to that of labour there was a strong economic incentive to substitute capital for labour. Nevertheless, apart from the introduction of buckwheat in the last quarter of the fifteenth century, the period is remarkably devoid of agricultural innovations.[128] Instead, the preferred forms of investment appear to have been the expansion of flocks and herds, engrossing of holdings, and functional modification of field systems leading in certain instances to piecemeal or wholesale enclosure.[129] The key agricultural developments of the period were therefore primarily structural, functional, and tenurial. Although market forces were not without influence upon these developments, not least by promoting a fundamental redistribution of population and economic activity, the general slackness of demand throughout the period meant that institutional structures and socio-property arrangements were often of more decisive importance at a local level.[130] Paradoxically, of a period regarded by Whigs and Marxists as marking a decisive stage in the replacement of feudal with capitalist socio-property relations in the countryside, the type of manor or estate, character of lordship, nature of property rights, and form of field system were often of more profound importance in shaping developments after 1380 than they had been in the era of expanding and strengthening market demand before 1315.

---

[125] B. F. Harvey, 'Introduction: the "crisis" of the early fourteenth century', in Campbell (ed.), *Before the Black Death*, p. 3.

[126] Britnell, 'Commercialisation and economic development', p. 24. Nevertheless, G. D. Snooks, *Economics without time* (London, 1992), pp. 256–64, believes that initially at least economic growth – and by implication productivity – was driven into reverse.

[127] C. C. Dyer, 'English diet in the later Middle Ages', in T. H. Aston, P. R. Coss, C. C. Dyer, and J. Thirsk (eds.), *Social relations and ideas* (Cambridge, 1983), pp. 209–14; B. M. S. Campbell, 'Matching supply to demand: crop production and disposal by English demesnes in the century of the Black Death', *JEH* 57 (1997), 832–9.

[128] Campbell and Overton, 'New perspective', p. 60.

[129] B. M. S. Campbell, 'The extent and layout of commonfields in eastern Norfolk', *NA* 38 (1981), 5–32; Campbell, 'Fair field', pp. 64–5.

[130] R. S. Schofield, 'The geographical distribution of wealth in England, 1334–1649', *EcHR* 18 (1965), 483–510; A. R. H. Baker, 'Changes in the later Middle Ages', in Darby (ed.), *New historical geography*; H. C. Darby, R. E. Glasscock, J. Sheail, and G. R. Versey, 'The changing geographical distribution of wealth in England 1086–1334–1525', *JHG* 5 (1979), 256–62.

Medieval English agriculture is a subject of great scope and complexity. A host of demand-side and supply-side influences interacted to shape the course of agricultural development, which possessed important structural, institutional, tenurial, and functional dimensions. Analysis, interpretation and explanation are further compounded by the incompleteness and unrepresentativeness of the available evidence and the methodological difficulties involved in making best sense of it. Historical enquiry to be effective must perforce be selective. Accordingly, this book offers an analysis of the agriculture undertaken by lords on their estates over the period *c.* 1250–*c.* 1450, focusing upon issues of land use, production, productivity and commercialisation.

Chapter 2 reviews the sources available for a study of the seigniorial sector and the ways in which they can be approached. It also describes the four principal databases from which the bulk of the results presented in this volume have been derived. Chapter 3 then considers the relative scales of the seigniorial and non-seigniorial sectors; the scale, value, and land-use composition of individual seigniorial estates and their component demesnes; and the geography of seigniorial land-use within England as a whole. Temporal and spatial trends in the function and composition of seigniorial pastoral husbandry are the subject of Chapter 4, which also examines the varying balance struck between the arable and pastoral sectors. The next four chapters then consider different aspects of the majority arable sector: Chapter 5 focuses upon the attributes of the principal crops and the main temporal trends in their production and use; Chapter 6 analyses spatial trends in arable cropping; and Chapter 7 discusses alternative definitions of arable productivity and their measurement. Building upon the principal results presented in Chapters 5, 6 and 7, Chapter 8 then estimates net national grain output in 1086, *c.* 1300 and *c.* 1375 and considers the total populations which could thereby have been fed. Finally, Chapter 9 returns to the themes of this opening chapter and considers the role of demand in stimulating changes in agricultural methods and output, both during the period *c.* 1250–*c.* 1450 and subsequently.

2

Sources, databases, and typologies

2.1 Sources

Seigniorial producers invite separate historical study because they alone kept
detailed records of their agricultural activities. The size, composition, and
value of demesnes held by lay tenants-in-chief are also the subject of separate
records kept by the crown. The combined documentary legacy which this has
bequeathed is without peer or parallel in the annals of European agrarian
history. Such is the quality and quantity of these sources – notably manorial
accounts, but also extents and a variety of other estate and manorial records
– that the seigniorial sector is the obvious starting point for any systematic
analysis of medieval agriculture as a whole. Perhaps 25–30 per cent of all agricultural land was held by lords in demesne.[1] How that land was worked and
what it produced are consequently issues of considerable significance in their
own right, but spatial and temporal trends within the seigniorial sector also
imply much about developments within the wider agrarian economy, including the technological proficiency and productivity potential of agriculture.
Moreover, the fuller picture that can be reconstructed of the seigniorial sector
helps to make better sense of the more miscellaneous and fragmentary
material pertaining to other classes of producer – glebe owners, large freeholders, virgators and other substantial customary tenants, and the host of lesser
free and unfree peasants.[2]

2.11 Manorial accounts

No single source provides fuller, more systematic, and more precise information on the practice, performance, and profits of husbandry than the annual
accounts rendered at the end of each farming year – sometimes at Lammas

---

[1] Within the highly manorialised area spanned by the Hundred Rolls demesne arable comprised
about 30 per cent of the total: Kosminsky, *Studies*, p. 93.
[2] As exemplified by Langdon, *Horses*.

(1st August) but usually Michaelmas (29th September) – by the reeves and bailiffs of individual manors.[3] For each twelve-month period they invariably provide a financial account of all cash receipts and expenses, usually with a stock or grange account containing corresponding information of all receipts and losses of grain, livestock, and livestock products. Sometimes, too, there is a works account which itemises the labour services due and the uses made of them. Each annual account is therefore a veritable mine of information. Long runs of consecutive accounts for individual manors, usually with supporting manorial documentation, have hitherto attracted most historical attention since they lend themselves to detailed temporal analysis.[4] Three, four or five accounts are nevertheless quite sufficient to reconstruct a robust profile of the husbandry and management of any one manor.[5] Much may be learnt even when only one account survives in isolation; apart from anything else, even a single account will provide two years of livestock data.[6]

Although enrolled accounts are included in the annual Pipe Rolls of the bishops of Winchester from 1208, the earliest individual manorial account dates from 1233–4 and relates to Froyle, Hampshire, a possession of St Mary's Abbey, Winchester.[7] To judge from surviving accounts, estates based in and around the ancient administrative capital of Winchester stood in the van of manorial accounting. They were closely followed by those located in East Anglia. It is to East Anglia that the earliest extant lay accounts relate, namely those of the honours of Clare and Gloucester dating from 1234–7, but with their exception (and the accounts of crown manors and manors temporarily in the king's hands enrolled in the Pipe Rolls) all other known pre-1250 accounts relate exclusively to ecclesiastical estates.[8] This is very much the case in Norfolk, probably the county with more extant accounts than any other, whose earliest accounts date from the late 1230s and mid-1240s and relate to

---

[3] For a discussion of manorial accounts and their development see F. B. Stitt, 'The medieval minister's account', *Society of Local Archivists Bulletin* 11 (1953), 2–8; P. D. A. Harvey, 'Agricultural treatises and manorial accounting in medieval England', *AHR* 20 (1972), 170–82; P. D. A. Harvey, 'Introduction, part ii, accounts and other manorial records', in *Manorial records of Cuxham, Oxfordshire circa 1200–1359*, ed. Harvey (London, 1976), pp. 12–71; P. D. A. Harvey, *Manorial records* (London, 1984).

[4] For case studies of individual well-documented manors, see Harvey, *Oxfordshire village*; D. V. Stern, 'A Hertfordshire manor of Westminster Abbey: an examination of demesne profits, corn yields, and weather evidence', PhD thesis, University of London (1978; to be published); C. Thornton, 'The demesne of Rimpton, 938 to 1412: a study in economic development', PhD thesis, University of Leicester (1989).

[5] Both 'Feeding the city' projects (see below, n. 81) used a maximum of three accounts whenever possible.

[6] The normal convention was to state the total number of livestock at both the start and the end of the accounting year. The cropped acreages, in contrast, relate solely to the accounting year in question.   [7] *Manorial records of Cuxham*, p. 15.

[8] For an analysis of early accounts relating to royal manors and enrolled in the annual Pipe Rolls, see R. C. Stacey, 'Agricultural investment and the management of the royal demesne manors, 1236–1240', *JEH* 46 (1986), 919–34.

the estates of the abbey of St Benet at Holm, an old-established Benedictine house of considerable local significance.[9] These are closely followed in date by the accounts of several manors belonging to Ramsey Abbey, another Benedictine house of similar antiquity.[10]

These early accounts are all rather rudimentary in form and less detailed than came to be usual half a century later. Most were produced centrally and drawn up and enrolled following the audit so as to incorporate any changes made at the audit. They were based on information supplied orally, with the assistance of tallies and other aids to memory.[11] From the 1250s, however, as record keeping spread among the land-owning classes, accounts began to be produced on the manor and handed over at the audit (after which they might be enrolled or a fair copy made), with the result that full manorial accounts become increasingly common (Figure 2.01). The earliest Ramsey accounts are of this type as are those of the manors of Norwich Cathedral Priory, which date from the mid-1250s. The form and contents of the accounts also become increasingly standardised. By the 1260s the accounts of the East Anglian manors of Roger le Bigod, earl of Norfolk, had already assumed the form which accounts were to retain with little modification for the next 200 years.[12] Thereafter, during the final quarter of the thirteenth century, the creation of written manorial accounts rapidly spread to most classes of estate throughout the centre, south and east of the country. Already by the 1280s, as minor and humble a Norfolk lord as Henry le Cat, the holder of a composite sub-manor at Hevingham, was keeping accounts.[13] That in Norfolk the practice of keeping accounts penetrated so far down the social spectrum so soon is consistent with that county's place in the van of this particular administrative innovation.[14]

From the 1290s on accounts become very numerous indeed. In Norfolk they are particularly abundant during the troubled first half of the fourteenth century (although there are curiously few accounts for the years of acute harvest failure from 1314 to 1318), with the 1330s emerging as the single best represented decade of all (Figure 2.01). Thereafter, those with both cash and

---

[9] The St Benet's accounts comprise the greater part of two grouped accounts, the earlier dating from 1238–9, the later from 1245–6, and between them cover some fifteen different manors. In both cases the entries for each manor are enrolled chancery-style upon one side of a continuous roll: NRO, Diocesan Est/1, Est/2/1.

[10] BL, Add. Charters 39669, 39934. NRO, Hare 212 × 1/4207.

[11] M. T. Clanchy, *From memory to written record: England 1066 to 1307* (London, 1979).

[12] These accounts are described in N. Denholm-Young, *Seignorial administration in England* (Oxford, 1937), pp. 123–51. See also F. J. Davenport, *The economic development of a Norfolk manor 1086–1565* (Cambridge, 1906) and M. Lyons, 'The manor of Ballysax 1280–1288', *Retrospect*, new series, 1 (1981), 40–50, for the Bigods and their Irish manors.

[13] B. M. S. Campbell, 'The complexity of manorial structure in medieval Norfolk: a case study', *NA* 39 (1986), 239–42.

[14] Norfolk demesnes also pioneered the introduction of the work-horse: Langdon, *Horses*, pp. 43–5. Other innovations are discussed in Campbell, 'Agricultural progress'. On the other hand, vetches were adopted relatively late on Norfolk demesnes: B. M. S. Campbell, 'The diffusion of vetches in medieval England', *EcHR* 41 (1988), 193–208.

stock accounts decline in number as the direct management of demesnes fitfully gave way to leasing. The decline is gradual at first, but becomes more marked from the 1380s, which was a bad decade for grain producers and inaugurated a spate of leasings. The accounts drawn up during the second half of the fourteenth century nevertheless tend to be the most detailed and informative of all. Most are annotated with the auditor's calculations of yield and many regularly specify the actual fields, furlongs, and parcels being sown along with other previously un-noted aspects of demesne land-use such as the amounts of land bare-fallowed and leased out.[15]

With the opening of the fifteenth century accounts of demesnes which were still in hand become very sparse. By this date most of the big lay estates had been farmed out and stock accounts are increasingly confined to those estates which still found it prudent to retain a few home farms in hand. Norwich Cathedral Priory kept a few of its demesnes in hand as the principal means of provisioning its household until the early 1420s, the nuns of Marham Abbey still had their home farm in hand in 1427, and the demesne at Kempstone in Norfolk was still serving as the home farm of Castle Acre Priory in 1449.[16] On other ecclesiastical estates direct management lasted even longer. It is documented on several of the manors of the bishopric of Winchester until well into the second half of the fifteenth century, and in the remote south-west, on the estates of Tavistock Abbey, direct management and account keeping lasted until the very end of that century.[17] Some minor lay landlords also chose to keep a home farm in hand. Again, it was a way of provisioning the household and the families concerned were often in a position to adopt a direct 'hands-on' management approach. In Norfolk the Cleres did this at Ormesby until 1458, as did the Yelvertons at Rougham until even later, although for the Pastons this was very much a last-ditch measure to be resorted to only when no satisfactory lessee was forthcoming.[18] A few grange accounts can

---

[15] All of these elements appear much earlier – named fields on the demesnes of the abbey of Bury St Edmunds by the close of the thirteenth century, a statement of the total amount of demesne arable and of the area left fallow on the earl of Norfolk's demesnes at South Walsham in 1268–9 and Acle in 1270–1, and marginal notations of yields at East Carleton in 1277–8 – but it is only after 1350 that they become a standard feature of significant numbers of accounts.

[16] NRO, NNAS 5917–18 20 D3; DCN 60/29/46, 60/35/52; L'Estrange IB 3/4; Hare 194×5/2204; WIS 163×2/37. BLO, MS rolls Norfolk 44–5.

[17] J. N. Hare, 'Change and continuity in Wiltshire agriculture: the later Middle Ages', in W. Minchinton (ed.), *Agricultural improvement: medieval and modern* (Exeter, 1981), pp. 1–18; H. P. R. Finberg, *Tavistock Abbey: a study in the social and economic history of Devon* (Cambridge, 1951).

[18] PRO, SC 6/940–41; NRO, MS 21483/1. In 1470 Margaret Paston complained that '. . . I am fayn to takyn Mautby in myn owyn hand and to set up husbondry ther, and how it shall profite me God knowyth. The fermour owyth me lxxx li. and more; whan I shall haue it I wote neuer' (*Paston letters and papers of the fifteenth century*, ed. N. Davis (Oxford, 1971), vol. I, no. 208, lines 32–4, p. 351); R. H. Britnell, 'The Pastons and their Norfolk', *AHR* 36 (1988), 132–44. During the course of the fifteenth century the Paston manor of Guton Hall in Brandiston was sporadically in hand owing to similar difficulties of finding a suitable lessee: Magdalen College, Oxford, Fastolf Papers.

consequently still be found during the second half of the fifteenth century, lingering on like wasps in autumn. Nonetheless, these are very much the exception and by 1450 the era of the manorial account is effectively over.

Manorial accounts span a period in excess of 200 years and are to be numbered in thousands rather than hundreds. Because they are the products of individual estate administrations they are dispersed among a host of public and private archives where many still languish unrecognised and uncalendared. The sole central finding aid is the Register of Manorial Documents maintained by the National Register of Archives, which is only as good as the information supplied to it by the individual record repositories.[19] Just how full and varied the pattern of account survival can sometimes be is illustrated by Norfolk, a county for which all known accounts have been traced (bar those relating to manors in royal hands enrolled in the central pipe rolls of the exchequer). Between 1238 (the date of the earliest account) and 1450 there are almost 2,000 extant accounts for Norfolk alone, which record the details of direct management on some 220 different manors.[20] This amounts to a roughly 15 per cent sample of all manors in the county and provides a degree of geographical coverage with which few other counties can vie.[21]

Notwithstanding this apparently ample documentation, there are very few Norfolk manors and even fewer estates with long, Winchester-style runs of accounts. Sedgeford, a property of the prior of Norwich, is by far the best documented manor in the county with three out of every five accounts (many in regrettably bad condition) surviving from a 175-year period.[22] Martham, another Cathedral Priory property and the next best recorded Norfolk manor, retains just over two out of every five accounts from a 167-year period.[23] These fragmentary series are the longest and best available and, typically, relate to the estate of a major ecclesiastical landlord (the fifth greatest landlord in the county after the bishop of Norwich, the abbot of St Benet, the earl of Pembroke, and the barons Bardolf). Other manors of the prior of Norwich are also among the best recorded.[24] Frustratingly, though, there is no single year for which accounts are simultaneously extant for all the prior's manors, thus precluding the kind of in-depth cross-sectional analysis undertaken by

[19] National Register of Archives, Quality House, Quality Court, Chancery Lane, London. For a detailed listing of all known pre-1350 Kentish manorial accounts, see J. A. Galloway, M. Murphy, and O. Myhill, *Kentish demesne accounts up to 1350* (London, 1993).

[20] The many accounts of manors where the demesne was at farm are excluded from these figures.

[21] On the evidence of the *Nomina villarum* of 1316 there were approximately 1,450 separate headlordships in Norfolk: W. J. Blake, 'Norfolk manorial lords in 1316', *NA* 30 (1952), 261.

[22] NRO, DCN 60/33; L'Estrange IB 1/4, 3/4: it is the rolls in the L'Estrange collection, mostly dating from after 1350, that are in particularly bad condition.

[23] NRO, DCN 60/23; L'Estrange IB 4/4; NNAS 20 D1–3.

[24] For further information on the estates of Norwich Cathedral Priory and associated archives, see H. W. Saunders, *An introduction to the obedientiary and manor rolls of Norwich Cathedral Priory* (Norwich, 1930); E. Stone, 'The estates of Norwich Cathedral Priory 1100–1300', DPhil thesis, University of Oxford (1956).

K. Biddick for the estate of Peterborough Abbey.[25] For the properties of other Norfolk religious houses there is just the occasional reasonably well-recorded manor: Calthorpe and Costessey in the case of Norwich Great Hospital, Flegg for St Benet's Abbey, Heacham and West Walton for Lewes Priory, and Thorpe Abbotts and Tivetshall for the abbey of Bury St Edmunds.[26] Yet although ecclesiastically owned manors dominate the record, accounting for 70 per cent of all surviving *compoti*, lay demesnes outnumbered them on the ground by four to one.[27]

Slightly over half of all recorded Norfolk manors were, in fact, in lay hands. A striking feature of even the best documented of these – the various Norfolk properties of the Bigod, earls of Norfolk, the manor at Bircham belonging to the Clare, earls of Gloucester and Hertford, that at Gressenhall, a possession of the Hastings, earls of Pembroke, or the manor of the L'Estrange family at Hunstanton – is that the accounts that survive mostly derive from the incumbency of a single lord of the estate.[28] The explanation lies in the periodic discontinuities of inheritance and management to which all privately rather than institutionally owned estates were prone. Short rather than long runs of accounts are consequently the norm of lay manors: Bircham, with the fullest series, has thirty-two accounts for a fifty-two-year period.[29] Only one in five lay manors retains at least ten accounts compared with one in three ecclesiastical manors, while for two out of every three recorded lay manors there are fewer than four accounts (see Figure 2.01). It follows that at an aggregate level a comparatively small number of well-recorded manors, and an even smaller number of estates (and those mostly in ecclesiastical ownership), contribute a disproportionate share of available accounts. In fact, 12 per cent of recorded manors furnish 55 per cent of surviving Norfolk accounts. The remaining 88 per cent of manors, many of them in lay hands, are represented by just a few stray *compoti*.[30]

A systematic search for all manors with extant accounts for the periods 1288–1315 and 1375–1400 in a ten-county area around London (comprising Bedfordshire, Berkshire, Buckinghamshire, Essex, Hertfordshire, Kent,

[25] The grouped accounts in question relate to the years 1300–1, 1307–8 and 1309–10: K. Biddick, *The other economy: pastoral husbandry on a medieval estate* (Berkeley and Los Angeles, 1989).

[26] NRO, Case 24, Shelf C; Diocesan Est/9, 10, 58; L'Estrange DG 1–6; Hare 210×2, 210×3; WAL 274×3, 274×6, 288×2.

[27] Blake, 'Norfolk manorial lords'; B. M. S. Campbell, 'Medieval manorial structure', in P. Wade-Martins (ed.), *An historical atlas of Norfolk* (Norwich, 1993), pp. 52–3.

[28] PRO, SC 6/929–38, 943, 944. NRO, ING 245×5/186; L'Estrange G1–6, BG 1–19.

[29] PRO, SC 6/930/1–31.

[30] Studies that demonstrate what can be made of such stray accounts include: R. C. Shaw, *The Royal Forest of Lancaster* (Preston, 1956), pp. 353–96; I. S. W. Blanchard, 'Economic change in Derbyshire in the late Middle Ages, 1272–1540', PhD thesis, University of London (1967), pp. 16–45, 164–92; R. H. Britnell, 'Minor landlords in England and medieval agrarian capitalism', *PP* 89 (1980), 3–22; Campbell, 'Arable productivity', pp. 379–404; Campbell and others, *Medieval capital*.

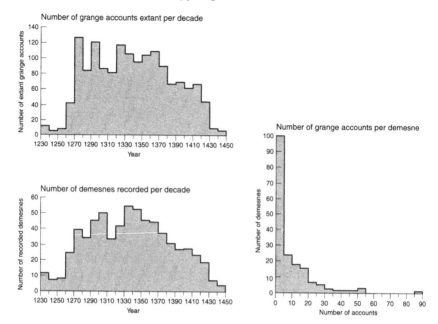

Fig. 2.01. Chronology of surviving manorial accounts (demesnes under direct management only): Norfolk, 1230–1450 (source: Appendix 2).

Middlesex, Northamptonshire, Oxfordshire, and Surrey) confirms both the exceptional fullness of Norfolk's coverage and the survival of significant numbers of accounts for other parts of the country.[31] A few fortunate counties combine both quantity and quality of coverage: Hampshire, with no fewer than twenty-three demesnes of the bishop of Winchester, is an obvious example; several of the Bury St Edmunds demesnes in Suffolk also retain superb series of accounts.[32] Not all counties are as well favoured, however, and in the remote north-west and south-west of the country it takes much searching in the archives to turn up any manorial accounts at all (Table 2.01 and Figures 2.03 and 2.04).[33] According to the Manorial Documents Register there is only one manor with extant accounts in Westmorland, six in Cumberland, six in Northumberland, seven in Lancashire, and five in Cheshire, and in most cases the number of accounts per manor is equally small

---

[31] Campbell and others, *Medieval capital*, pp. 18–22, 184–90; Galloway and others, *Kentish demesne accounts*; D. J. Keene, B. M. S. Campbell, J. A. Galloway, and M. Murphy, 'Feeding the city 2: London and its hinterland c. 1300–1400', Economic and Social Research Council, unpublished End of Award Report (1994).

[32] *The archives of the abbey of Bury St Edmunds*, ed. R. M. Thomson (Woodbridge, 1980).

[33] For productive gleanings, see H. S. A. Fox, 'Farming practice and techniques: Devon and Cornwall', in Miller (ed.), *AHEW*, vol. III, pp. 303–23.

(nor does it necessarily follow that these are accounts of manors whose demesnes were kept in hand). In these northern and north-western counties the scarcity of accounts is probably an indication that direct demesne management was never very important. Counties slightly further south, such as Staffordshire, Shropshire, and Herefordshire, are rather better served and appear to have modest numbers of documented demesnes. It is, however, in the heartland of seigniorial agriculture, in the closely settled arable counties of central, southern, and eastern England, that the documentary survival is greatest, its bulk obscured by its wide archival dispersal. Collectively these accounts comprise a data source of tremendous and, as yet, under-exploited potential. For his national study of the technology of horses and oxen J. L. Langdon recently assembled a sample of 1,565 accounts for some 874 demesnes; this probably represents less than a fifth of all the material that is potentially available.[34]

Rich and abundant though the accounts are as a data source, they do have certain drawbacks, of which the most conspicuous is their failure to be truly representative of the seigniorial sector at large. N. J. Mayhew, reworking data assembled by C. C. Dyer, has estimated the relative incomes of different propertied groups in 1300. He reckons that the landed classes enjoyed between them an annual income of approximately £640,000, of which a little under 5 per cent was received by the crown, 66 per cent by lay landlords (with the lesser gentry receiving the lion's share), 4 per cent by episcopal landlords, and 25 per cent by conventual and collegiate landlords.[35] Yet the ownership of manors represented by extant accounts is almost the inverse of this. In Norfolk episcopal, conventual, and collegiate manors make up 52 per cent of the total and lay manors of various types the remainder (crown manors whose accounts are enrolled in the annual Pipe Rolls are excluded). In the ten-county area around London the bias towards ecclesiastical manors of one sort or another is even greater, with three-quarters of all documented demesnes both in the period 1290–1315 and 1375–1400 belonging to this category of landlord.[36] The problem of such a pronounced ecclesiastical bias is two-fold. First, the

---

[34] These accounts are mostly listed in Appendix C of J. L. Langdon, 'Horses, oxen and technological innovation: the use of draught animals in English farming from 1066 to 1500', PhD thesis, University of Birmingham (1983), pp. 416–56.

[35] Mayhew, 'Modelling medieval monetisation', pp. 58–60. The Hundred Rolls indicate that just under a third of all arable belonged to church manors: Kosminsky, *Studies*, p. 109.

[36] For the mismatch between the social distribution of landed incomes and the institutional distribution of documented demesnes in the FTC1 accounts database (see below, n. 81), see B. M. S. Campbell, 'Measuring the commercialisation of seigneurial agriculture *c*. 1300', in Britnell and Campbell (eds.), *Commercialising economy*, p. 140. In Kent lay manors are particularly badly represented. Before 1350 only 6 per cent of documented manors were in lay hands, compared with 22 per cent in royal and 70 per cent in ecclesiastical ownership: Galloway and others, *Kentish demesne accounts*, p. vii. It follows that no sample of accounts can ever be a true random sample of demesnes; hence the application of inferential statistics which depend upon the assumption of randomness is inappropriate.

Table 2.01. *Geographical coverage of principal databases, 1250–1349 and 1350–1449*

County	National IPM database 1300–49 Total number of:			National accounts database 1250–1349 Total number of demesnes with complete data on:			FTC1 accounts database 1288–1315 and Norfolk accounts database 1250–1349 Total number of demesnes with complete data on:			FTC2 IPM database 1375–1400 Total number of:	National accounts database 1350–1449 Total number of demesnes with complete data on:			FTC2 accounts database 1375–1400 and Norfolk accounts database 1350–1449 Total number of demesnes with complete data on:		
	A	B	C	D	E	F	D	E	F	A and C	D	E	F	D	E	F
Bedfordshire	91	113	179	5	10	5	3	3	3	15	1	5	1	3	2	2
Berkshire	122	106	184	12	17	12	17	16	16	5	10	11	10	15	15	15
Berwickshire	0	0	0	1	1	1					1	1	1			
Buckinghamshire	148	163	239	7	15	7	14	15	14	30	8	10	7	17	17	17
Cambridgeshire	92	95	147	6	12	6					9	10	9			
Cheshire	27	15	35	1	1	0					2	2	2			
Cornwall	130	83	198	0	8	0					2	2	2			
Cumberland	121	54	177	3	3	3					0	0	0			
Derbyshire	98	63	127	3	3	3					0	0	0			
Devon	194	172	263	12	16	9					5	6	3			
Dorset	132	124	196	17	8	8					5	5	5			
Durham	22	8	28	10	26	10					10	18	10			
Essex	283	336	473	14	28	12	24	24	24	35	13	18	12	33	33	33
Gloucestershire	142	160	232	11	17	9					11	11	10			
Hampshire	185	175	272	25	36	24					30	31	30			
Hampshire (IOW)	43	30	50	7	8	7					1	1	1			
Herefordshire	112	92	160	0	4	0					3	7	3			
Hertfordshire	86	96	141	13	16	11	12	12	11	23	8	10	8	10	10	10

	A	B	C	D	E	F	C	D	E	F	D	E	F	D	E	F
Huntingdonshire	40	40	70	0	8	0	40	38	38	25	8	10	8			
Kent	235	245	349	25	49	25					24	27	23	26	26	26
Lancashire	144	50	168	3	6	3					1	1	1			
Leicestershire	144	100	219	10	9	8					2	4	2			
Lincolnshire	361	320	589	19	31	18					4	7	4			
Middlesex	31	33	58	5	15	5	14	13	13	4	3	5	3	9	9	9
Monmouthshire				7	6	0					0	0	0			
Norfolk	240	268	381	54	55	53	125	125	125	114	45	46	45	107	102	101
Northamptonshire	150	154	241	21	26	21	29	29	29	9	7	9	7	9	9	9
Northumberland	190	91	270	2	4	2					1	1	1			
Nottinghamshire	106	61	145	3	5	3					1	3	1			
Oxfordshire	120	120	189	12	29	7	27	30	27	11	9	11	9	7	7	7
Rutland	25	22	36	4	4	4					2	2	2			
Shropshire	196	121	280	1	2	1					2	2	2			
Somerset	213	214	328	60	43	41					10	11	10			
Staffordshire	102	71	151	5	9	5					0	1	0			
Suffolk	204	225	302	30	33	23					17	21	17			
Surrey	106	97	182	9	18	8	17	17	17	11	8	14	8	12	11	11
Sussex	224	179	309	14	22	13					23	20	17			
Warwickshire	90	90	155	17	20	16					3	4	3			
Westmorland	65	33	109	0	0	0					0	1	0			
Wiltshire	221	229	325	33	29	24					9	10	9			
Worcestershire	71	78	116	1	6	1					6	8	6			
Yorkshire ER	260	160	426	8	11	8					7	8	7			
Yorkshire NR	133	79	167	3	6	3					2	2	2			
Yorkshire WR	189	113	249	21	27	21					3	6	3			
Total	2,660			512	702	438	*197	*197	*192	168	316	382	304	*141	*139	*139

Notes:

A Places with *IPM* extents; B *IPM* extents for complete manors; C *IPM* extents; D Crops; E Livestock; F Crops and livestock;

*Totals for the 10 FTC counties only (omitting Norfolk); IOW = Isle of Wight; ER = East Riding; NR = North Riding;

WR = West Riding

Sources: National *IPM* database; National accounts database; FTC1 and FTC2 accounts and *IPM* databases; Norfolk accounts database (see Appendices 1–3).

management objectives of perpetual institutions with large permanent households to provision naturally differed from those of most other landlords with
more mobile and peripatetic lifestyles and very different households and
needs.[37] Second, there was a greater proportion of large manors on ecclesiastical estates and, on the evidence of the Hundred Rolls, the demesne was much
more linked up with villein land, and occupied a less independent position in
the manorial economy on large manors than on small.[38] Accordingly, customary labour tended to assume greater significance in the cultivation of the
demesne.

The bias towards ecclesiastical estates with their larger than average manors
is compounded by a general bias towards the manors of large estates in
general. In the ten-county area around London over two out of three documented manors belonged to what may be defined as great estates, notably
those of the crown, the knights templar (or at least their confiscated lands
under crown management), the earldoms of Cornwall, Lincoln, and Norfolk,
the archbishopric of Canterbury and bishoprics of Winchester and London,
plus the cathedral priories of Canterbury and Winchester and the abbeys of
Crowland, Peterborough, Ramsey, Waltham, and Westminster. Evidence of
how the several thousands of minor landlords, with perhaps only a single
manor to their name, worked their demesnes is therefore sparse in the
extreme.[39] This is particularly unfortunate, for these, like their glebe counterparts, provide the best clue to the management strategies of the next tier of
landholdings, those of substantial free tenants.

Geographically, too, the accounts are distinctly uneven in their coverage,
with areas of lowland arable farming far better served than the pastoral
uplands (Figures 2.03 and 2.04). Hence while much can be learnt about the
great arable demesnes of the south and east the management of the many specialist studs, vaccaries and bercaries in the north and west remains largely
masked from view.[40] In fact, pastoral land-use in general receives short
measure. Grassland, whether temporary or permanent, is rarely accounted for
directly in the accounts. The same is usually true of the unsown arable. This is
a major deficiency, since the ideal would be to analyse land-use, stocking den-

---

[37] B. F. Harvey, 'The aristocratic consumer in England in the long thirteenth century', in
M. Prestwich, R. H. Britnell, and R. Frame (eds.), *Thirteenth century England* (Woodbridge,
1997), vol. VI, pp. 19–37. For differences in the disposal of agricultural produce between
different types of estate, see Campbell, 'Measuring commercialisation', pp. 144–7, 156–63,
165–70, 174–5, 186–9.   [38] Kosminsky, *Studies*, p. 101.

[39] But see Britnell, 'Minor landlords'; Campbell, 'Complexity of manorial structure'.

[40] On studs, vaccaries, and bercaries see Blanchard, 'Economic change in Derbyshire', pp.
164–92; E. Miller, 'Farming in northern England', *Northern History* 11 (1975), 1–16;
R. A. Donkin, *The Cistercians: studies in the geography of medieval England and Wales*
(Toronto, 1978), pp. 68–102; M. A. Atkin, 'Land use and management in the upland demesne
of the de Lacy estate of Blackburnshire c. 1300', *AHR* 42 (1994), 1–19; B. Waites, *Monasteries
and landscape in north east England* (Oakham, 1997), pp. 117–45.

sities, labour inputs, productivity, and the like in terms of the total farm area or, at the very least, that of the total arable. To redress these imbalances and fill these gaps the accounts need to be supplemented with information from other sources, of which by far the most useful are manorial extents.

2.12 IPM *extents*

Manorial accounts record the use made of landed resources rather than the resources themselves. The latter are the explicit subject of manorial extents, which generally specify the principal categories of land-use on the demesne – arable, meadow, pasture, heath, marsh, turbary, and wood – their respective values, and, less consistently, their areas. Extents are a relatively common type of manorial record but the greatest and most concentrated collection of them occurs in conjunction with the *inquisitiones post mortem* (*IPM*s) preserved in the Public Record Office, London (Table 2.01).

Whereas accounts are the product of local manorial administration, the *IPM* extents emanate from central government and relate to the properties of tenants-in-chief of the crown. Since these were exclusively lay and included many lords of comparatively humble status, a large body of data is provided for precisely that class of estate which is least well served by extant accounts. Chronologically, the *IPM*s are available for much the same period as the accounts: they commence in the mid-thirteenth century, become increasingly numerous during the last quarter of that century and throughout the first half of the fourteenth century provide remarkably comprehensive and detailed coverage of the country as a whole (Figures 2.02 and 2.03). Thereafter, although *IPM*s continued as an important instrument of royal administration, far fewer of them include detailed extents, so that by the final quarter of the fourteenth century they are of greatly diminished use as a source of land-use information. At no stage are they as reliable as the accounts and they are far less consistent in the range and detail of the information they provide.[41]

IPM extents were created at the instigation of royal escheators following the death of a tenant-in-chief of the crown.[42] The source of the information contained in each extent was a sworn jury which was not necessarily closely acquainted with the detailed attributes of the demesne in question and may conceivably have had a vested interest in misrepresenting the true value of land and assets. How individual escheators and sub-escheators and the juries they empanelled interpreted their respective remits clearly varied, for the scope and

---

[41] Kosminsky, *Studies*, pp. 46–67; R. F. Hunnisett, 'The reliability of inquisitions as historical evidence', in D. A. Bullough and R. L. Storey (eds.), *The study of medieval records* (Oxford, 1971), pp. 206–35; J. A. Raftis, *Assart data and land values* (Toronto, 1974), pp. 12–18.

[42] On the work and organisation of the medieval escheators, see E. R. Stevenson, 'The escheator', in W. A. Morris and J. R. Strayer (eds.), *The English government at work 1327–36* (Cambridge, Mass., 1947), vol. II, pp. 109–67.

(a) 15 counties: annual totals of *IPMs*

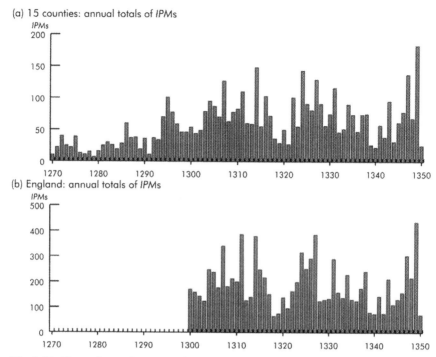

(b) England: annual totals of *IPMs*

Fig. 2.02. Chronology of *IPM* survival: England, 1300–49, and fifteen eastern counties, 1270–1349 (source: National and FTC1 *IPM* databases).

format of the extents is far from uniform. In some the amounts and values of the different land-uses are given in very rounded figures and summary form, in others they either incorporate information taken directly from manorial accounts and other documentation available at the time or are the product of an exact survey and evaluation.[43] The whole exercise was subject to periodic administrative reform and reorganisation, which affected the numbers of estates and manors netted and the amount of detailed information contained in individual extents. The most precise and informative extents of all were drawn up following the reform of 1323, when the jurisdictions of the various escheators were redrawn and escheators were clearly instructed to provide more precise information.[44] From then until 1342, when recording practice was relaxed somewhat, the extents often provide useful incidental information

[43] Kosminsky's own cross-check between the figures of demesne arable contained in the 1269–70 *IPM* for Roger Bigod, earl of Norfolk, and near contemporary manorial accounts shows a close and sometimes exact correspondence between the two. On this estate the most conspicuous discrepancies between the *IPM* and the accounts concern the revenues from extensive marshland grazings. Kosminsky, *Studies*, pp. 58–62.

[44] B. M. S. Campbell and K. C. Bartley, *Lay lordship, land, and wealth* (Manchester, forthcoming).

concerning such matters as the duration of common grazing rights, the frequency of fallowing, size of customary measures, and duration of coppicing cycles.[45]

Ambiguity, however, is more prevalent than precision. For instance, it is often unclear whether stated arable acreages include or exclude land lying fallow. The value of common rights also poses particular problems: these might be omitted altogether, subsumed into valuations of the pasture, or valued as an appurtenance of the arable. Much hinged upon the recording policy pursued by individual escheators. Here the escheators responsible for England north of the Trent appear to have been a law unto themselves, for the extents which they produced tend to be less comprehensive in their survey of demesne land-uses and more idiosyncratic in their language and format than those produced further south. Large customary measures of uncertain size also pose far greater problems in the northern escheatry than most other parts of the country. It was where pressure on all types of land-use was greatest, property rights were most clearly demarcated, and individual land-uses were most closely managed and profitable that the precision of recording is greatest, as in much of East Anglia and the south-east.[46]

These kinds of inconsistencies and uncertainties qualify the utility of this source and greatly complicate its analysis. Nevertheless, even though the absolute figures are not always to be trusted much can be learnt from the presence or absence of particular resources and their relative amounts and unit values. It was H. L. Gray who first appreciated the potential of this land-use information and used it to considerable effect in his seminal study of English field systems.[47] He also made extensive use of the *IPM*s in a more controversial study of the commutation of labour services.[48] Gray's work subsequently attracted strong criticism from E. A. Kosminsky, who is largely responsible for giving the *IPM*s their reputation for unreliability. In part this stems from Kosminsky's own exclusive reliance upon the thirteenth-century *IPM*s, produced before the whole operation was put on a tighter administrative footing in the fourteenth century. He failed to appreciate that Gray, with unerring historical instinct, had focused almost exclusively upon the very cream of *IPM*s from the years 1323–42 when many of the most detailed extents were produced.

Kosminsky nevertheless conceded that 'a comparison of the figures they contain, carried out over wide area, enables us to capture certain characteristic traits, certain peculiarities which, though vague, are vital. And the results thus obtained assume considerable weight, when confirmed by other sources,

[45] This feature was observed by H. L. Gray, who makes much use of *IPM* data from this period in his classic study, *English field systems* (Cambridge, Mass., 1915), and in his essay 'Commutation of villein services'.
[46] This especially applies to woodland: Galloway and others, 'Fuelling the city'.
[47] Gray, *English field systems*.   [48] Gray, 'Commutation of villein services'.

and by conclusions independently arrived at.'[49] This is the line followed by J. A. Raftis in his 1974 study of land values in eight counties in the east midlands over the period 1272–1350, and it is one with quite a long historical tradition.[50] Thus, *IPM*s were used in several of the county descriptions of medieval agriculture given in the early volumes of the *Victoria county history*.[51] Subsequently they have been employed by I. S. W. Blanchard in an analysis of the changing regional economy of medieval Derbyshire, by H. S. A. Fox in work on both the ecology of commonfield systems in the midlands and land-use variations in medieval Devon, and by L. R. Poos to chart the changing balance of land-use in medieval Essex.[52] Used both in aggregate and with caution there is therefore much that these extents alone can reveal.

2.13 Other sources

Accounts and extents by no means exhaust the range of sources capable of casting light upon the seigniorial sector, although they dominate by their sheer bulk and potential for both spatial and temporal analysis. Central estate accounts are particularly useful for the information they provide concerning that portion of production sent for consumption by the household.[53] They also record capital investments made from central funds. Charters, deeds, and leases – often enrolled in cartularies – may also record aspects of property ownership and land-use not otherwise covered by the accounts or extents, while occasional inventories of grain, stock, and implements reveal the technical side of seigniorial agriculture in especially rich detail.[54] Many of the sources which cast incidental light on peasant agriculture also illuminate aspects of the seigniorial sector. The more detailed of the local tax assessments, when these survive, are particularly useful here since they sometimes

---

[49] Kosminsky, *Studies*, p. 63.   [50] Raftis, *Assart data*.

[51] E.g. the *Victoria county history* volumes for Sussex (1907, vol. II, p. 169), Rutland (1908, vol. I, pp. 216–20), Oxfordshire (1907, vol. II, p. 181), and Nottinghamshire (1910, vol. II, p. 275). *IPM*s have since been used by L. F. Salzman, 'Social and economic history: medieval Cambridgeshire', in Salzman (ed.), *The Victoria history of the counties of England: Cambridgeshire and the Isle of Ely* (London, 1948), vol. II, pp. 65–6; by R. H. Hilton, 'Medieval agrarian history', in W. G. Hoskins (ed.), *The Victoria history of the counties of England: a history of Leicestershire* (London, 1954), vol. II, pp. 159–65; and by R. Scott, 'Medieval agriculture', in E. Crittal (ed.), *The Victoria history of the counties of England: a history of Wiltshire* (London, 1959), vol. IV, p. 15.

[52] Blanchard, 'Economic change in Derbyshire'; H. S. A. Fox, 'Some ecological dimensions of medieval field systems', in K. Biddick (ed.), *Archaeological approaches to medieval Europe* (Kalamazoo, 1984), pp. 119–58; Fox, 'Occupation of land: Devon and Cornwall', pp. 152–63; Poos, *Rural society*, pp. 44–9. See also B. M. S. Campbell, 'Medieval land use and land values', in Wade-Martins (ed.), *Historical atlas of Norfolk*, pp. 48–9.

[53] Campbell and others, *Medieval capital*, p. 22.

[54] E.g. *The prior's manor-houses*, trans. and ed. D. Yaxley (Dereham, 1988). Inventories are particularly common on manors belonging to Canterbury Cathedral Priory: Galloway and others, *Kentish demesne accounts*.

allow direct comparison of the seigniorial and peasant sectors.[55] Mention, too, should be made of the returns which resulted from the third commission of the *Nonarum inquisitiones* of 1341, for these often provide a detailed extent of the glebe, an important class of quasi-seigniorial holding otherwise poorly represented by available documentation.[56] The various contributions to Volumes II and III of *The agrarian history of England and Wales* illustrate the ingenuity with which these and other sources may be used to reconstruct the character of seigniorial agriculture in all its diversity.[57]

2.2 Methods and databases

2.21 *Approaches to the analysis of demesnes, estates, and regions*

Current knowledge of seigniorial agriculture springs from two complementary traditions, one emphasising the institutional and the other the geographical context within which management decisions were taken and husbandry systems developed. Thus, manorial- and estate-focused case studies have revealed much about how and why individual landlords managed their estates while incidentally yielding valuable insights into the husbandry of particular localities and regions. Since the prerequisite of such studies is a well-preserved estate archive this has, however, resulted in an almost exclusive emphasis upon the properties of perpetual ecclesiastical institutions.[58] Although this has not been without benefit, it has hardly yielded a balanced or comprehensive view

[55] E.g. *A Suffolk hundred in the year 1283*, ed. E. Powell (Cambridge, 1910).
[56] *Nonarum inquisitiones in curia scaccarii*, ed. Record Commissioners (London, 1807). Many unpublished returns are preserved in class E179 at the PRO.
[57] H. E. Hallam (ed.), 'Farming techniques', in Hallam (ed.), *AHEW*, vol. II, pp. 272–496; E. Miller (ed.), 'Farming practice and techniques', in Miller (ed.), *AHEW*, vol. III, pp. 175–323.
[58] F. R. H. Du Boulay, *The lordship of Canterbury* (London, 1966); Titow, 'Land and population'; D. L. Farmer, 'Grain yields on the Winchester manors in the later Middle Ages', *EcHR* 30 (1977), 555–66; E. Searle, *Lordship and community: Battle Abbey and its banlieu, 1066–1538* (Toronto, 1974); P. F. Brandon, 'Cereal yields on the Sussex estates of Battle Abbey during the later Middle Ages', *EcHR* 25 (1972), 403–29; I. Keil, 'The estates of Glastonbury Abbey in the later Middle Ages: a study in administration and economic change', PhD thesis, University of Bristol (1964); K. Biddick, 'Animal husbandry and pastoral land-use on the fen-edge, Peterborough, England: an archaeological and historical reconstruction, 2500 BC–1350 AD', PhD thesis, University of Toronto (1982); J. A. Raftis, *The estates of Ramsey Abbey* (Toronto, 1957); B. F. Harvey, *Westminster Abbey and its estates in the Middle Ages* (Oxford, 1977); Farmer, 'Grain yields on Westminster Abbey manors'; R. A. L. Smith, *Canterbury Cathedral Priory* (Cambridge, 1943); E. M. Halcrow, 'The administration and agrarian policy of the manors of Durham Cathedral Priory', BLitt thesis, University of Oxford (1959); R. B. Dobson, *Durham Priory, 1400–1450* (Cambridge, 1973); Saunders, *Norwich Cathedral Priory*; Stone, 'Estates of Norwich Cathedral Priory'; Dyer, *Lords and peasants*. For studies of other classes of estate see Davenport, *Norfolk manor*; Harvey, *Oxfordshire village*; K. Ugawa, *Lay estates in medieval England* (Tokyo, 1966); J. Hatcher, *Rural economy and society in the Duchy of Cornwall, 1300–1500* (Cambridge, 1970); Britnell, 'Minor landlords'; M. Mate, 'Profit and productivity on the estates of Isabella de Forz (1260–92)', *EcHR* 33 (1980), 326–34.

of medieval estate management. While the approach remains as valid as ever it needs to be broadened to encompass the full range of estates – large and small; royal, lay, and ecclesiastical – no matter how slight or fragmentary their documentation.[59]

It has been partly with the object of redressing this institutional bias that regionally focused case studies have come increasingly to the fore. Such studies are also better able to accommodate the many stray manorial accounts.[60] Both medieval volumes of *The agrarian history of England and Wales*, for example (like both early modern volumes), adopt an explicitly regional approach to the description and analysis of farming practice.[61] Such an approach acknowledges the fact that estates operated within a wider environmental and economic context and greatly sharpens the focus on those geographically specific factors which informed the production decisions of those charged with the management of individual demesne enterprises. Yet, whereas each of these separate regional studies has yielded important insights into its respective region, stitching them together has failed to produce a wholly satisfactory national picture; the seams are too conspicuous and too many gaps and holes remain. Most regional studies are perforce county based even though farming practice was no respecter of administrative boundaries. This problem is compounded when separate regional studies employ different methodologies. Small wonder, therefore, that national studies thus constructed reveal 'an economy and society split into many and various sub-economies and sub-societies': such a conclusion tends to be predicated by the approach.[62]

To be convincing a national picture requires a national scale of analysis.[63] It is this which this book aims to provide through the derivation of results from a set of nested databases capable of yielding both general and specific insights into many aspects of farming practice. The macro and micro scales of analysis are reconciled through the application of a consistent methodology to spatial and temporal analysis at a range of institutional and geographical scales. Here, the basic building block is the individual demesne represented by a set of meaned variables generated from however many sampled accounts contain usable data. In most cases all percentages and ratios are calculated at

[59] E.g. Campbell, 'Arable productivity'; Campbell, 'Measuring commercialisation'.

[60] Regional studies include the following: W. Rees, *South Wales and the March 1284–1415* (Oxford, 1924); H. J. Hewitt, *Mediaeval Cheshire* (Manchester, 1929); D. Roden, 'Demesne farming in the Chiltern Hills', *AHR* 17 (1969), 9–23; P. F. Brandon, 'Demesne arable farming in coastal Sussex during the later Middle Ages', *AHR* 19 (1971), 113–34; Davies, *Lordship and society*, pp. 110–19; Campbell, 'Agricultural progress'; M. Bailey, *A marginal economy? East-Anglian Breckland in the later Middle Ages* (Cambridge, 1989); Campbell and others, *Medieval capital*.

[61] Hallam (ed.), 'Farming techniques'; Miller (ed.), 'Farming practice and techniques'. On the initial proposal to inaugurate this series see H. P. R. Finberg, 'An agrarian history of England', *AHR* 4 (1956), 2–3.   [62] 'Flyleaf', Hallam (ed.), *AHEW*, vol. II.

[63] E.g. Langdon, *Horses*; B. M. S. Campbell, 'Towards an agricultural geography of medieval England', *AHR* 36 (1988), 87–98.

account level before being meaned at demesne level; in this way all years are weighted equally. The exceptions are where small numbers are a potential problem, as often applies to livestock. In such cases percentages and ratios are either calculated from account-level data aggregated at demesne level or from individual demesne means aggregated to some larger unit of analysis. The results are thereby weighted towards those years or demesnes with the largest figures.[64]

Any extension of analysis from individual variables to combinations of variables requires the development of typologies.[65] These are most useful if generated at a national level since this ensures that the focus is consistently upon similarities and differences which hold valid at a national rather than merely a sub-national or local scale. For the purposes of this study six specific typologies have been developed, as summarised in Table 2.02. Some are methodologically more refined than others but all share a common set of principles. First, their aim has been to classify the 'fundamental units of agricultural production', either the individual demesne farms or, in the case of land-use and land values, small regular hexagons containing 2.6 square kilometres.[66] In this way analysis of farming practice is liberated from the tyranny of administrative boundaries. Second, since the concern is with the nature of farm enterprise rather than the determinants of that enterprise, each typology has been based on a set of measurable criteria which reflect the inherent properties of those farms or land-use units. Such external attributes as soils, climate, distance from the market, field systems, or type of estate, which may have influenced the type of farming or land-use, are excluded. In this way the possibility of circularity is eliminated from explanations of observed agricultural differences.[67] Whether or not specific farming types were associated with, or exclusive to, particular soil types, regions, locations, field systems or estates can then be established independently by mapping or correlating the resultant typology.[68] Third, because differences and similarities of national significance are here of paramount interest, intermediate local and regional solutions have been rejected; thus, in each case the criteria of classification have been applied at a national or, in the case of the *IPM*s, a near-national scale.[69] Fourth, investigation of temporal continuities and discontinuities in farming systems and

---

[64] E.g. Campbell, 'Measuring commercialisation'.

[65] The literature on the classification of farming systems is substantial. For an introduction see D. B. Grigg, *The agricultural systems of the world* (Cambridge, 1974), pp. 2–4.

[66] Grigg, *Agricultural systems*, pp. 2–3.

[67] M. Overton, 'Agricultural regions in early modern England: an example from East Anglia', University of Newcastle upon Tyne, Department of Geography, Seminar Paper 42 (1985), p. 3.

[68] E.g. J. P. Power and B. M. S. Campbell, 'Cluster analysis and the classification of medieval demesne-farming systems', *TIBG*, new series, 17 (1992).

[69] For an illustration of the advantages of national over regional classifications see Campbell and others, 'Demesne-farming', pp. 143–54.

their geography has been undertaken by applying a common method of classification to similarly structured sets of data for the consecutive time periods 1250–1349 and 1350–1449.[70]

Each of the six typologies has been derived quantitatively rather than qualitatively using cluster analysis as the key statistical technique.[71] The appeal of cluster analysis for the classification of farming and land-use systems lies in its capacity to subdivide a dataset into groups based upon the degree of similarity or dissimilarity between cases measured across all the individual variables (e.g. the crops sown, livestock kept, or portfolio of land-uses) of which they are composed.[72] Classifications based on cluster analysis are most robust when consistent results are obtained from the application of several different clustering methods to the same dataset. Applying more than one method can also help to establish the natural number of groups present and focus attention on the characteristics of those 'core' cases common to all solutions and therefore most typical of the groups identified. Cases outside of these cores, which do not fall decisively into any one category – a common problem in classifications of land-use and farm enterprise – can then be assigned probabilities of group membership and classified accordingly using discriminant functions calculated on the core cases.[73] These discriminant functions can serve three further useful purposes. First, they allow the next nearest classification of each farm enterprise or land-use unit to be determined. Not only does this bring out much secondary patterning within the data but it is also helpful in establishing the degree of similarity or dissimilarity between farming and land-use types. Second, direct comparison can be made between the pre-1350 and post-1349 classifications of farm enterprise. Third, they may be used to classify additional cases as more data become available.[74] Here they have been used to extend national classifications of cropping and mixed-farming systems to both the 'Feeding the city' and Norfolk samples of demesnes and to apply classifications of

---

[70] E.g. Campbell and others, 'Demesne-farming'.

[71] For critiques of the qualitative approach, see B. M. S. Campbell, 'Laying foundations: the agrarian history of England and Wales, 1042–1350', *AHR* 37 (1989), 190–1; Campbell, 'Fair field', pp. 62–3; E. L. Jones, 'The condition of English agriculture, 1500–1640', *EcHR* 21 (1968), 615–16; M. Overton, 'Depression or revolution? English agriculture, 1640–1750', *Journal of British Studies* 25 (1986), 345–7.

[72] The use of cluster analysis to classify farming types was pioneered by M. Overton, 'Agricultural change in Norfolk and Suffolk, 1580–1740', PhD thesis, University of Cambridge (1981); Overton, 'Agricultural regions'. It was developed further by P. Glennie, 'Continuity and change in Hertfordshire agriculture 1550–1700: i – patterns of agricultural production', *AHR* 36 (1988), 55–75.

[73] Such an approach is similar to that utilised for the handling of poorly defined sets within regional geography: A. C. Gatrell, *Distance and space: a geographical perspective* (Oxford, 1983), pp. 11–13.

[74] K. C. Bartley, 'Classifying the past: discriminant analysis and its application to medieval farming systems', *History and Computing* 8.1 (1996), 1–10.

Table 2.02. *Principal agricultural and land-use typologies and their derivation*

Typology	Period	Databases included in cluster analysis	Additional databases classified using discriminant functions	No. of clusters	Clustering methods	Method	Tables	Figures
Cropping types	1250–1349	National acnts.	FTC1 acnts.; Norfolk acnts.	7	Relocation (and Friedman and Rubins)	A	6.01	6.01–6.07
Cropping types	1350–1449	National acnts.	FTC2 acnts.; Norfolk acnts.	6	Relocation (and Friedman and Rubins)	A	6.02	6.13–6.18
Pastoral types	1250–1349	National acnts.; FTC acnts.; Norfolk acnts.		6	Relocation (and Friedman and Rubins)	A	4.01	4.01–4.11
Pastoral types	1350–1449			5		A	4.01	
Mixed-farming types	1250–1349	National acnts.	FTC1 acnts.; Norfolk acnts.	8	Ward's, K-means, Normix	B	4.08	
Mixed-farming types	1350–1449	National acnts.	FTC2 acnts.; Norfolk acnts.	7		C	4.08	
Land types by unit value	1300–1349	National *IPM* south of the Trent and east of the Tamar	National *IPM* north of the Trent and west of the Tamar	4	Ward's, K-means, Within group linkage	D	9.01	9.01
Types of demesne land-use	1300–1349			6		D	3.05	3.14

Methods:

A B. M. S. Campbell and J. P. Power, 'Mapping the agricultural geography of medieval England', *JHG* 15 (1989), 24–39.

B J. P. Power and B. M. S. Campbell, 'Cluster analysis and the classification of medieval demesne-farming systems', *TIBG*, new series, 17 (1992), 227–45.

C B. M. S. Campbell, K. C. Bartley, and J. P. Power, 'The demesne-farming systems of post Black Death England: a classification', *AHR* 44 (1996), 131–79.

D K. C. Bartley and B. M. S. Campbell, '*Inquisitiones post mortem*, G.I.S., and the creation of a land-use map of pre Black Death England', *Transactions in G.I.S.* 2 (1997), 333–46.

land-use and unit land values to *IPM* data for the northern and Cornish escheatries.

All classifications are artificial constructs. Their purpose is to reduce the kaleidoscopic complexity of reality into something simpler and more comprehensible. They are useful only if they help to identify and make sense of genuine trends and patterns within the data. All those made use of in this volume are only as good as the methods on which they are based and the datasets from which they have been derived (Table 2.02). Different choice, specification, and weighting of the component variables, alternative cluster methods, and a larger and more comprehensive dataset would in each case have yielded a more refined typology. In that sense each should be regarded as no more than provisional. What each does offer, however, is a genuinely national classification which as far as possible expresses differences and similarities inherent within the data. None is either a local or an imposed solution. They therefore serve as convenient analytical tools for exploring the factors and influences which shaped seigniorial land-use and husbandry systems over the period 1250 to 1450 and should be regarded as no more than that.

2.22 The databases

The four cornerstones of the book are its four databases: a national *IPM* database, a national accounts database, a pair of accounts databases relating to the ten FTC counties around London, and a comprehensive accounts database for Norfolk. Their scale and scope has been determined as much by the time and costs of constructing and analysing them as by the survival and availability of the actual data. The national and Norfolk accounts databases were created independently of one another.[75] Both evolved by trial and error and were created by essentially manual methods over periods of several years' duration. The 'Feeding the city 1' accounts database built on the experience of these two earlier databases and, in turn, was improved upon by 'Feeding the city 2'. Both FTC accounts databases were mechanised from the outset. Evolution of the *IPM* database was more complex. Early regional and national samples of the source established its utility and suggested a methodology.[76] A mechanised database of *IPM*s was then created for the period 1270–1339 as part of the 'Feeding the city 1' project. Apart from the *IPM*s for the years 1270–99, however, this has been entirely superseded by the current *IPM* database which built upon all previous experience and was again mechanised from the outset. Over the long period that these databases were being

---

[75] The national accounts database is largely the work of John Langdon, the Norfolk accounts database of Bruce Campbell (with the financial support of the Social Science Research Council).

[76] Some of the resultant data for Lincolnshire, Huntingdonshire, Cambridgeshire, Norfolk, and Suffolk are included in the sample for the years 1270–99.

created computer hardware and software have both advanced considerably, with considerable benefits for data collection and processing. Insofar as a great deal more evidence remains to be collected, these databases and the results they have yielded must nevertheless be regarded as provisional. In particular, with further work in the archives the national accounts database could be greatly enhanced.

1. The national IPM *database*

The national *IPM* database comprises essential land-use information drawn from a total of almost 9,000 *IPM* extents for the half-century 1300–49 (Table 2.01 and Figure 2.02).[77] It was during this fifty-year period that employment of the *IPM* as a means of raising royal revenues attained its fullest development, as expressed both in terms of the number and range of estates netted and the amount of detailed information which the extents provide. This half-century also coincides with the climax of direct management when manorial accounts are also at their most abundant. Geographically, the extent and density of coverage of the *IPM*s is far superior to that of the accounts. For 4,600 unique locations there is at least some land-use information, and for over half of those locations there is a reasonably full extent of a complete manor. So efficient was central government by the early fourteenth century that the king's writ extended to the furthest reaches of the realm, with the result there is no sizeable area of the country without at least some extents (Figure 2.03). Even the palatine counties of Durham and Cheshire with Lancashire (which were technically exempt) retain a few *IPM* extents for the 1330s and 1340s. Nevertheless, these remain the least well represented counties, closely followed by Cornwall. At the opposite extreme, the density of documentation is at its greatest in an arc of country extending westwards from Essex, and south Suffolk through Hertfordshire, southern Cambridgeshire, Bedfordshire, and Buckinghamshire into Oxfordshire. For Essex alone there are 473 usable *IPM* extents, giving information for 219 unique locations.[78]

The *IPM* database serves four main functions. First, it helps to establish the size range of lay demesnes, manors and estates. Second, it enables broad spatial trends in seigniorial land-use to be reconstructed. Third, the unit values of arable and meadow serve as surrogate measures of land productivity. And fourth, it can be used to cast light upon temporal trends in land-use and land values. The chronological range of this last exercise has been extended back

[77] PRO C133–5. The National *IPM* database was created in conjunction with the project 'The geography of seigniorial land-ownership and land-use, 1270–1349', based at The Queen's University of Belfast, co-directed by John Power and Bruce Campbell, and funded by the Leverhulme Trust. The archival work was undertaken by Marilyn Livingstone and Roger Dickinson. The database was developed and analysed by Ken Bartley.

[78] The totals for Lincolnshire and Yorkshire (all three Ridings) are higher but as these are larger counties the density of coverage is lower.

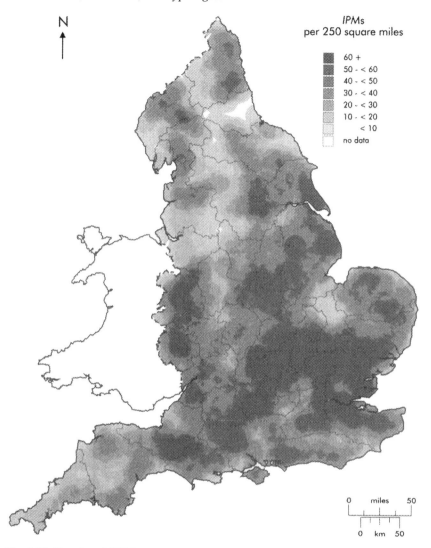

Fig. 2.03. Density of *IPM* coverage: England, 1300–49 (source: National *IPM* database).

to 1270 using additional *IPM* data for a fifteen-county area in eastern and south-eastern England comprising Bedfordshire, Berkshire, Buckinghamshire, Cambridgeshire, Essex, Hertfordshire, Huntingdonshire, Kent, Lincolnshire, Middlesex, Norfolk, Northamptonshire, Oxfordshire, Suffolk, and Surrey.[79]

[79] For ten of these counties the *IPM* data were collected as part of the FTC1 project (see below, n. 81). PRO C132–3.

2. *The national accounts database*

Creating and analysing a national dataset from manorial accounts is logistically a far greater task than doing so from the *IPM*s. Not only are the individual documents more information rich but the accounts themselves are dispersed among a large number of public and private archives. The core of the sample used here is therefore that assembled by Langdon for his study of horses and oxen.[80] This comprises information on the crops sown and the livestock kept on a national sample of several hundred demesnes. To this has been added further material culled from a variety of published and unpublished sources. The combined sample contains information from 1,904 accounts representing some 873 demesnes (most demesnes being represented by just one or two accounts but a few by longer runs) (Table 2.01). Out of this total, 792 demesnes have complete data on livestock and 601 demesnes have complete data on crops, but only 520 have data on both crops and livestock (the discrepancy being partly a function of the fact that Langdon was primarily concerned with livestock, but also of the failure of some accounts to specify the acreage as opposed to the seed sown). Not only are there more ecclesiastical than lay demesnes in the sample (56 per cent compared with 39 per cent), but the ecclesiastical demesnes are also on average better documented.

Every English county except Norfolk (represented by the 'Norfolk accounts database') features in this national sample, plus Monmouthshire and Berwickshire (Table 2.01). For much of northern and north-western England the density of coverage is sparse. The north-eastern counties of Durham and Yorkshire are somewhat better served; otherwise the sampled manors are strongly biased towards the lowland arable-farming counties of the south and east where direct seigniorial management of demesnes was most widely practised. This spatial bias in the coverage of the sample echoes the corresponding bias in the distribution of extant manorial accounts. Further assiduous work in the archives could nevertheless do much to fill out this picture. In the meantime it is sufficient as it stands to bring out the broad national picture, albeit sketched in simple but bold strokes.

Chronologically, the national accounts sample is drawn from the entire span of years for which accounts are extant – i.e. 1250 to 1450 – hence it lacks the narrower, temporal focus of the *IPM* sample. Since every decade between 1250 and 1450 has furnished at least some information, and between 1270 and 1400 there is only one decade with information for less than fifty demesnes, the sample can also be used to identify aggregate trends in various aspects of demesne production. Nevertheless, for the purpose of identifying broad regional variations in agriculture and investigating how these changed over time it is more effective to split the sample into two, taking the Black Death of 1348–9 as a convenient and far from inappropriate dividing line. The pre-1350

---

[80] See Appendix C, in Langdon, *Horses*, pp. 416–56.

sub-sample pivots around the decade 1300–9, which is the single best documented decade in the entire sample, closely followed by the 1280s and 1290s.
Because of the post-Black Death retreat of many landlords from direct
demesne management the post-1349 sub-sample is 25–40 per cent smaller and
thinner. The picture that emerges – which pivots around the decade 1380–9 –
is therefore marred by the patchiness and thinness of the data across much of
the country.

3. The FTC accounts databases

The patterns and trends that emerge from the national accounts database may
be clarified and tested by comparison with the two 'Feeding the city' databases
(FTC1 and FTC2).[81] These contain a much wider array of more precisely
specified information and thereby greatly extend the range and precision of
possible analyses. They are also chronologically more tightly focused, each
spanning a twenty-five- rather than hundred-year period, and their respective
densities of coverage are the best that available documentation permits. Both
encompass a common ten-county area (Table 2.01) very roughly coincident
with the provisioning hinterland of London, medieval England's single largest
and most lucrative domestic market, within which were to be found a wide
array of different agricultural opportunities.[82] Although all known manors
with accounts are included the number of sampled accounts per manor is
limited to a maximum of three.

As with the national accounts database, the earlier of the two FTC accounts
databases contains data for the greater number of manors (204 manors drawn
from 460 manorial accounts), as is consistent with the far higher proportion
of demesnes that were in hand at the opening compared with the close of the
fourteenth century.[83] The FTC2 accounts database actually contains more
data, since the later accounts are more detailed and informative and a wider
range of variables was collected, but from 1375 the leasing of demesnes
became so prevalent that there are significantly fewer available accounts of
demesnes still kept in hand. It consequently comprises data for 141 manors
drawn from 360 manorial accounts.[84] Both databases are decidedly patchy in

---

[81] Created in conjunction with the projects 'Feeding the city 1' and 'Feeding the city 2', both of
which were based at the Centre for Metropolitan History, University of London and undertaken in collaboration with The Queen's University of Belfast. 'Feeding the city 1' was funded
by the Leverhulme Trust and co-directed by Derek Keene and Bruce Campbell. 'Feeding the
city 2' was funded by the Economic and Social Research Council (Award No. R000233157)
and co-directed by Derek Keene, Bruce Campbell, Jim Galloway, and Margaret Murphy. The
archival work on both projects was undertaken by Jim Galloway and Margaret Murphy.
[82] Campbell and others, *Medieval capital.*
[83] For fifty-eight manors there is only one account, for thirty-nine there are two, and for 107 three
or more.
[84] For thirty-eight manors there is only one account, for thirty-four there are two, and for sixty-
nine three or more.

their spatial coverage and both are dominated by manors in ecclesiastical ownership of one sort or another. In some localities ecclesiastical manors are the only ones documented, hence the results that emerge are strained through a strong institutional filter. These spatial and institutional shortcomings are, however, those of the available documentation.

4. The Norfolk accounts database
Both the national accounts database and the two FTC accounts databases are samples. The Norfolk accounts database, in contrast (the first to be created), makes use of all known extant manorial accounts for the county spanning the entire period 1238 (the date of the earliest Norfolk account) to 1450. With the exception of accounts enrolled in the exchequer pipe rolls it therefore purports to be complete. For this one county 220 demesnes are documented by approximately 2,000 separate manorial accounts. The nature and scope of these accounts has already been described so little more need be added here. Their temporal range is slightly longer than that of the national sample, but their general chronological distribution is much the same, as is the balance between ecclesiastical and lay manors (Figure 2.01). Adding corresponding data for nine demesnes just across the county boundary into Suffolk, five of which have the merit of possessing substantial runs of accounts (that for Hinderclay being exceptionally complete), raises to 229 the total number of demesnes and to 2,211 the total number of accounts represented.[85] This supremely rich documentation makes Norfolk an ideal case study. The chronological range of surviving accounts means that it is possible to chart the development of demesne husbandry within the county over a considerable length of time. Likewise, the sheer density of documented demesnes enables the county's own agricultural geography to be reconstructed in quite exceptional detail.

Its rich documentation apart, there are several good reasons for making a special case study of Norfolk. First, on the evidence of the 1327–32 lay subsidies and 1377 poll tax returns, it was fourteenth-century England's most densely populated county.[86] Second, the county stood in the van of several technological innovations of the age and is known to have been home to one of the most distinctive and productive of medieval husbandry systems.[87] Third, strong internal institutional, economic, and environmental contrasts within the county lend considerable intrinsic interest to a detailed reconstruction of its husbandry. It serves, in fact, as a microcosm of the country as a whole.

---

[85] The demesnes concerned are Brandon, Bungay, Denham, Hinderclay, Hoxne, Redgrave, Rickinghall, Wattisfield and one that is unidentified.

[86] Campbell and Bartley, *Lay lordship, land, and wealth*; Baker, 'Changes', pp. 190–1.

[87] Campbell, 'Agricultural progress'; Power and Campbell, 'Cluster analysis', pp. 232–42; Campbell and others, 'Demesne-farming', pp. 143–54.

Fig. 2.04. Distribution of all accounts used, 1250–1349: England, Norfolk, FTC counties (source: National, Norfolk, and FTC1 accounts databases).

In the Middle Ages Norfolk, like England, contained coastal districts and inland districts, fenlands and broadlands with rich alluvial pastures and meadows, extensive windswept sandy heaths, areas of ancient hedgerows interspersed with coppiced woodland, and great tracts of open country with scarcely a hedgerow or enclosure to be seen. Pedologically the county contains some of the best and some of the worst agricultural land in Britain. Its soils

Fig. 2.05. Distribution of all accounts used, 1350–1449: England, Norfolk, FTC counties (source: National, Norfolk, and FTC2 accounts databases).

include light, drought-prone, and sterile sands, deep, warm, and fertile loams, stiff, cold, and heavy clays, and rich fen peat and alluvial silts in need of extensive drainage and flood protection before they could be brought into cultivation. Via differential access to coastal ports and the navigable rivers which selectively penetrated the county some localities always enjoyed readier access to urban markets than others, whether within or beyond the county.

Deep-rooted variations in field-systems, manorial structures, and socio-economic attitudes further heightened internal regional differences.[88] For all these reasons, Norfolk agriculture has always exhibited a decidedly varied aspect, with strong contrasts in production, technology, and productivity occurring over comparatively short distances.[89] It therefore serves as a particularly instructive and sharply focused case study.

Each of the three manorial accounts databases differs in its contents and composition. Each therefore fulfils a different purpose in the chapters that follow. For certain analyses, however, they can fruitfully be combined. This maximises geographical coverage, as illustrated in Figures 2.04 and 2.05, at the price of reinforcing the strong bias of the national accounts database to the south and east.

---

[88] B. Dodwell, 'The free peasantry of East Anglia in Domesday', *NA* 27 (1939), 145–57; B. M. S. Campbell, 'The regional uniqueness of English field systems? Some evidence from eastern Norfolk', *AHR* 29 (1981), 16–28; Campbell, 'Medieval manorial structure'.

[89] Campbell, 'Arable productivity'; Overton, 'Agricultural regions'; Campbell and Overton, 'New perspective'.

3

The scale and composition of the seigniorial sector

3.1 The seigniorial share of agricultural output and land-use

Lords differed from other agricultural producers in the scale and relative factor endowments of their respective production units. The feudal system ensured that lords generally enjoyed privileged access to land, coercive powers over labour and often a superior command over capital.[1] Lords held their land on more generous terms than other producers, especially unfree tenants, and their individual demesne farms were commonly operated as components of federated estates. The greatest of these estates comprised dozens of individual demesnes, thousands of acres of land, and encompassed a wide range of agricultural environments.[2] As 'firms' many were therefore both horizontally and vertically integrated and consequently enjoyed significant scale economies.[3] Concomitant management structures, especially on the greatest and most far-flung estates, could nevertheless be excessively bureaucratic, to the detriment of both efficiency and enterprise.[4] Low work motivation and high policing costs also went hand in hand with a traditional reliance upon forced servile labour. That was why the more enlightened and progressive lords increasingly commuted labour services and substituted hired labour, both permanent and

---

[1] Postan, 'Agrarian society', p. 602.
[2] For an analysis of a production system organised at estate level, see Biddick, *The other economy*.
[3] Sheep farming, for instance, was often organised on an estate basis, with centralised sales of wool: F. M. Page, '"Bidentes Hoylandie": a medieval sheep farm', *Economic History* 1 (1929), 603–13; D. L. Farmer, 'Marketing the produce of the countryside, 1200–1500', in Miller (ed.), *AHEW*, vol. III, pp. 395–8.
[4] For a case study of one demesne and its managerial superstructure see Thornton, 'Determinants of productivity', pp. 201–7. On the conservativeness of the seigniorial sector see J. L. Langdon, 'Was England a technological backwater in the Middle Ages?', in Astill and Langdon (eds.), *Medieval farming*, pp. 275–92. For a contrasting view, see S. Fenoaltea, 'Authority, efficiency and agricultural organization in medieval England and beyond: a hypothesis', *JEH* 35 (1975), 693–718.

casual.[5] The latter was nevertheless inferior in motivation, application and enterprise to the family labour employed on most peasant holdings.

Lords produced both for consumption and exchange. They used their estates to provision their households and to generate a cash income from the sale of surplus produce, of which there could be a great deal since the scale of many demesnes and estates was more than equal to the consumption requirements of the seigniorial households which they were intended to support.[6] Indeed, by the late thirteenth century most lords were in constant and pressing need of cash to maintain an increasingly sophisticated and costly life-style and to satisfy the mounting tax demands of the crown.[7] Whether, as a result, the seigniorial sector was by 1300 more commercialised than the peasant is a moot point, for peasants, in their battle to survive, increasingly resorted to the market to sell goods and labour and purchase food, foodstuffs, fuel, raw materials and basic consumption goods.[8] Peasants, in fact, were less able than lords to insulate themselves from the market and satisfy the bulk of their needs from their own resources.

Whether measured by its share of the total agricultural area, the value of its production, the size of the labour force employed, or the proportion of the total population directly dependent upon it for subsistence, the seigniorial sector was smaller and less important than the peasant. On the evidence of Domesday Book G. D. Snooks has estimated that demesnes and free peasants together contributed approximately 58 per cent of the rural component of national income in 1086.[9] Excluding consumption by free peasants reduces this proportion to 45 per cent and excluding the traded surplus of free peasants would reduce it further.[10] Mayhew, however, argues that Snooks exaggerates the demesne share of non-urban national income and estimates that by 1300 this amounted to only 20 per cent of the total.[11] Although both estimates rest on debatable assumptions and are mutually incompatible, they do suggest that the seigniorial sector produced perhaps 20–40 per cent of agricultural output. This can be compared with the seigniorial share of agricul-

---

[5] Fenoaltea, 'Authority', 695–6; M. M. Postan, *The famulus: the estate labourer in the XIIth and XIIIth centuries* (Cambridge, 1954); Stone, 'Hired and customary labour'.

[6] Campbell and others, *Medieval capital*, pp. 145–56; Campbell, 'Measuring commercialisation', pp. 174–6, 186–91.

[7] C. C. Dyer, *Standards of living in the later Middle Ages* (Cambridge, 1989), pp. 71–85; Biddick, *The other economy*, pp. 50–61; K. Biddick (with C. C. J. H. Bijleveld), 'Agrarian productivity on the estates of the bishopric of Winchester in the early thirteenth century: a managerial perspective', in Campbell and Overton (eds.), *Land, labour and livestock*, pp. 98–104; R. H. Britnell, *The commercialisation of English society 1000–1500* (Cambridge, 1993), pp. 105–8.

[8] C. C. Dyer, 'The hidden trade of the Middle Ages: evidence from the west midlands of England', *JHG* 18 (1992), 142; Masschaele, *Peasants, merchants, and markets*, pp. 33–54.

[9] G. D. Snooks, 'The dynamic role of the market in the Anglo-Norman economy and beyond, 1086–1300', in Britnell and Campbell (eds.), *Commercialising economy*, p. 32.

[10] Snooks, 'Dynamic role', p. 38.   [11] Mayhew, 'Modelling medieval monetisation', p. 58.

tural land-use, for which the evidence is firmer but nevertheless still far from secure.

The Hundred Rolls of 1279 – 'a cadastral survey far superior to Domesday Book in detail and accuracy' – provide the single best guide to the seigniorial share of agricultural land-use.[12] Returns are extant for the greater part of Cambridgeshire, Huntingdonshire, and Oxfordshire, plus portions of Bedfordshire, Buckinghamshire, and Warwickshire, and a small part of Norfolk. In addition, copies of portions of the original rolls are available for parts of Suffolk and Leicestershire.[13] Information is thus extant for over 800 vills, a number equivalent to perhaps 6 per cent of those in the country as a whole. This is a tantalisingly small sample, and not even a very representative one at that. It is spatially biased towards the old-settled and heavily manorialised champion country of central England and that part of the country where large arable demesnes were particularly prominent (Figure 3.05). As with most medieval sources, there is little overall consistency in the manner in which the size of demesnes is recorded; sometimes no area is given at all, often it is stated only approximately, and invariably the arable and meadow are the only land-uses consistently extended. A modern re-analysis of this major source is long overdue, but in the meanwhile reliance has to be placed on the figures produced by Kosminsky over sixty years ago.[14]

Kosminsky was unaware of the handful of Norfolk returns (which have only recently come to light) and excluded from his analysis the materials for Suffolk and Leicestershire which survive only as later copies. His statistics of the relative proportions of demesne and peasant arable thus derive from four hundreds in Huntingdonshire, six hundreds in Cambridgeshire (for five additional Cambridgeshire hundreds he encountered difficulties in calculating demesne land), two hundreds in Bedfordshire, four in Buckinghamshire, eleven in Oxfordshire, and two in Warwickshire. He found that out of a total arable area of over half a million acres, 31.8 per cent was in demesne, 40.5 per

---

[12] Kosminsky, *Studies*, p. 3.

[13] On the coverage of the Hundred Rolls see E. A. Kosminsky, 'The Hundred Rolls of 1279–80 as a source for English agrarian history', *Economic History Review* 3 (1931–2), 16–44; J. B. Harley, 'The Hundred Rolls of 1279', *Amateur Historian* 5 (1961), 9–16; T. John (ed.), *The Warwickshire Hundred Rolls of 1279–80* (Oxford, 1992), pp. 1–16. For the Norfolk rolls see D. E. Greenway, 'A newly discovered fragment of the Hundred Rolls of 1279–80', *Journal of the Society of Archivists* 7 (1982), 73–7; Campbell, 'Complexity of manorial structure'; *Lordship and landscape in Norfolk 1250–1350*, ed. W. Hassall and J. Beauroy (Oxford, 1993), pp. 22, 26–36, 215–30. The principal editions of the Hundred Rolls are: J. Nichols, *The history and antiquities of the county of Leicester*, 4 vols. (London, 1795, reprinted Wakefield, 1971); *Rotuli Hundredorum*, ed. Record Commissioners, 2 vols. (London, 1812, 1818); *The Pinchbeck Register of the abbey of Bury St Edmunds etc.*, ed. F. Hervey (Brighton, 1925), vol. II, pp. 30–282; John, *Warwickshire Hundred Rolls*, pp. 25–332.

[14] Kosminsky's researches into the Hundred Rolls were originally published in Russian in 1935, revised in 1947; an English-language edition – *Studies in the agrarian history of England in the thirteenth century* – only appeared in 1956.

cent was villein land, and 27.7 per cent was free land.[15] Equivalent figures cannot be produced for the other main categories of land-use – grassland, woodland, private hunting grounds etc. – since they are recorded with even less consistency and precision. Kosminsky believed that the figures for arable tend, if anything, to understate the amount of agricultural land in demesne. He also noted that there was a tendency for the proportion of demesne arable to decline from east to west, an observation which accords with regional variations in demesne size observable from account rolls and *IPMs* (Figure 3.05).

If a third of all arable within these old-settled and heavily manorialised counties was in demesne in 1279 the proportion within the country as a whole is likely to have been significantly smaller. Nationally, a third of all plough-teams were already in demesne in 1086, yet over the next two centuries it was tenants rather than lords who were most active in adding to the area under tillage, especially in those weakly manorialised areas with greatest scope for reclamation.[16] Even within the already densely populated Deanery of Waxham in eastern Norfolk there was a quadrupling in the number of tenants' teams during the 150 years after Domesday.[17] In woodland and marshland areas all over England and those extensive parts of the north exposed to planned resettlement tenant gains must have been at least as great.[18] Nationally, therefore, the proportion of the arable in demesne is likely to have been reduced over the twelfth and thirteenth centuries from a third to nearer a quarter, and if lords made only modest arable gains between 1086 and 1300 it is conceivable that it could have fallen to as little as a fifth of the total.

Not all demesne land was necessarily kept and managed in hand. Piecemeal leasing of small portions of the arable and herbage was always widespread.[19] Moreover, even when direct management was at its height some landlords preferred to lease rather than manage their demesnes directly, thereby transfer-

[15] Kosminsky, *Studies*, pp. 87–95.

[16] S. P. J. Harvey, 'The extent and profitability of demesne agriculture in England in the later eleventh century', in Aston and others (eds.), *Social relations*, p. 53, reckons that the average ratio of demesne plough-teams to tenant plough-teams was 1:2. There are many revealing case studies of the contribution made by tenants to the post-Domesday expansion of the tillage, e.g. P. F. Brandon, 'Medieval clearances in the East Sussex Weald', *TIBG* 48 (1969), 135–53; E. C. Vollans, 'The evolution of farm-lands in the central Chilterns in the twelfth and thirteenth centuries', *TIBG* 26 (1959), 197–235; J. B. Harley, 'Population and agriculture from the Warwickshire Hundred Rolls of 1279', *EcHR* 11 (1958), 8–18; B. K. Roberts, 'A study of medieval colonization in the Forest of Arden, Warwickshire', *AHR* 16 (1968), 101–13; G. H. Tupling, *The economic history of Rossendale* (Manchester, 1927); E. M. Yates, 'Dark Age and medieval settlement on the edge of wastes and forests', *Field Studies* 2 (1965), 133–53; J. A. Sheppard, 'Pre-enclosure field and settlement patterns in an English township: Wheldrake, near York', *Geografiska Annaler* 48B (1966), 59–77; J. S. Moore, *Laughton: a study in the evolution of the Wealden landscape* (Leicester, 1965).

[17] Campbell, 'Commonfields in eastern Norfolk', pp. 18–20.

[18] H. C. Darby, 'The changing English landscape', *Geographical Journal* 117 (1951), 377–94; Darby and others, 'Distribution of wealth', pp. 249–56.

[19] E.g. Davenport, *Norfolk manor*, pp. 31–2; Biddick, *The other economy*, pp. 57, 96–7.

ring responsibility for production to their tenant farmers. According to Kosminsky, leasing was of limited importance at the time that the Hundred Rolls were drawn up, but *c.* 1300 it was probably sufficient to reduce the share of arable in the direct control of landlords to nearer 20 than 25–30 per cent (a figure which corresponds with Mayhew's estimate that the seigniorial sector contributed approximately 20 per cent of the rural component of national income at that time).[20] Subsequently, as more and more landlords found it expedient to lease all or part of their estates, the proportion of land in demesne contracted progressively and with it the seigniorial contribution to agricultural production. In fact, Titow has shown that the bishops of Winchester were reducing the amount of land in hand from as early as the 1270s, and shortly afterwards, according to Farmer, the abbots of Westminster were doing likewise.[21] In contrast, the prior of Norwich, with a much smaller acreage in demesne, kept expanding the amount of arable in hand until the early 1300s (Figure. 5.03).[22] Depending upon which experience was the more prevalent, the opening of the fourteenth century may possibly represent the high-tide of direct demesne production.

On the Norwich Cathedral Priory estate, as on many others, it was the exceptionally low prices of the third decade of the fourteenth century that initiated the first real retreat from direct management. Thenceforth landlords resorted more and more to the wholesale or piecemeal leasing of demesnes. The Black Death precipitated a whole spate of leasings as, in its aftermath, real wages commenced their inexorable climb and when, in turn, grain prices finally collapsed in the late 1370s most landlords realised that leasing represented a more reliable method of generating an income. From that date it was more common for demesnes to be leased than managed directly and those that remained in hand were mostly restricted to the home farms of monastic and noble households.[23] By the middle of the fifteenth century direct management had been abandoned on virtually all estates. Throughout the fourteenth and fifteenth centuries, therefore, the seigniorial sector was a contracting sector.

[20] Cf. the estimate of Masschaele, *Peasants, merchants, and markets*, pp. 53–4, that 'Peasants held at least two-thirds of the country's assessable surplus, possibly significantly more.'

[21] Titow, 'Land and population', p. 15; Farmer, 'Grain yields on Westminster Abbey manors', p. 339.

[22] NRO, DCN 40/13; 60/4, 8, 10, 13–15, 18, 20, 23, 26, 28, 29, 33, 35, 37; L'Estrange IB 4/4. See also R. Virgoe, 'The estates of Norwich Cathedral Priory, 1101–1538', in I. Atherton, E. Fernie, C. Harper-Bill, and H. Smith (eds.), *Norwich Cathedral* (London, 1996), p. 352.

[23] On the leasing of demesnes, see E. M. Halcrow, 'The decline of demesne farming on the estates of Durham Cathedral Priory', *EcHR* 7 (1954), 345–56; F. R. H. Du Boulay, 'Who were farming the English demesnes at the end of the Middle Ages?', *EcHR* 17 (1964–5), 443–55; B. F. Harvey, 'The leasing of the abbot of Westminster's demesnes in the later Middle Ages', *EcHR* 22 (1969), 17–27; R. A. Lomas, 'The priory of Durham and its demesnes in the fourteenth and fifteenth centuries', *EcHR* 31 (1978), 339–53; J. N. Hare, 'The demesne lessees of fifteenth-century Wiltshire', *AHR* 29 (1981), 1–15; M. Mate, 'The farming out of manors: a new look at the evidence from Canterbury Cathedral Priory', *Journal of Medieval History* 9 (1983), 331–44.

On the Norwich Cathedral Priory estate the amount of arable in hand fell to 90 per cent of its 1300 level in the 1330s, 74 per cent in the 1350s, 46 per cent in the 1380s, recovered briefly to 71 per cent in the 1400s, and then fell to 38 per cent in the 1420s (Figure 5.03). In 1432 the last of its demesnes was finally leased out.[24] On this estate, like most others, the demesnes that were retained in hand until the bitter end tended to be that distinctive minority most directly involved in the provisioning of the household. From *c.* 1330, therefore, the selective impact of leasing changed the composition as well as the scale of the seigniorial sector.

How much more or less the seigniorial sector contributed to agricultural production than its share of arable land-use depended upon the respective land productivities and stocking densities of the seigniorial and peasant sectors. These are matters about which there has been much speculation but for which there is little direct evidence.[25] If the seigniorial sector in both its arable and pastoral husbandry was the more productive it perhaps contributed between a quarter and a third of agricultural production *c.* 1300. Alternatively, if the peasant sector was the more productive the corresponding proportion is unlikely to have been much more than a fifth (i.e. the proportion proposed by Mayhew). Whichever the proportion, it was probably reduced by a half or more over the course of the next century. Beyond this, on present knowledge, it is not possible to go.

3.2   The scale and composition of estates

Lords, except those with only one manor, organised and undertook agricultural production on the basis of federated estates. To them it was the size of their estate that was most material; the size and location of its constituent demesnes were secondary considerations. Few demesnes were managed in complete isolation of other properties belonging to the same estate, with livestock husbandry in particular being amenable to inter-manorial forms of organisation.[26] Intra-estate transfers of livestock sometimes took place over impressive distances.[27] Estates were administered centrally, although on the largest estates of all sub-groupings of manors known as bailiwicks constituted an intermediate administrative tier.[28] Key decisions concerning the acquisition or disposal of land were commonly taken centrally as were those concerning

[24] NRO, DCN 40/13; 60/4, 8, 10, 13–15, 18, 20, 23, 26, 28–30, 33, 35, 37; 61/35–6; 62/1–2; L'Estrange IB 1/4, 3/4; NNAS 20 D1–3; Raynham Hall, Norfolk, Townshend MSS; BLO, MS rolls Norfolk 29–45.

[25] Postan, 'Agrarian society', p. 602; Postan, 'Village livestock'; Campbell, 'Agricultural progress', pp. 39–41; Masschaele, *Peasants, merchants, and markets*, pp. 42–7, 52–3, 158–9; Chapter 8, pp. 404–5.   [26] Biddick, *The other economy*; Chapter 4.

[27] E.g. Davies, *Lordship and society*, p. 116; Farmer, 'Marketing', pp. 378–88.

[28] E.g. *The Pipe Roll of the bishopric of Winchester 1301–2*, ed. M. Page (Winchester, 1996), pp. xvii–xviii.

major items of capital investment such as the construction of new mills and barns, which were sometimes financed from central funds.[29] Central rather than local officials usually decided when paid should be substituted for servile labour. Whether demesnes sold all, some or none of their produce was also normally a matter of estate policy, different policies typically applying to wool, livestock, and grain.[30]

The larger and more far-flung the estate the more elaborate the administrative superstructure and the more detached the lord and his leading officials from the practical business of farming.[31] The greatest lords, both lay and ecclesiastical, commanded large numbers of manors often scattered over a wide geographical area. For instance, the estate of the bishops of Winchester – the mightiest episcopal landowner in England – comprised up to sixty functioning demesnes spread across a seven-county area in southern England.[32] At its peak between *c.* 1225 and 1270 it had a total cropped area of 13,000 acres, although by 1325 this had declined to 8,200 acres.[33] This compares with the thirty-four demesnes and 8,400 cropped acres of Canterbury Cathedral Priory in the 1320s and 14,500 acres of arable demesne land (sown and unsown) possessed by the abbot and convent of Westminster at about the same time.[34] The leading lay estates were even larger and geographically much more widely dispersed. Gilbert de Clare, earl of Gloucester and Hertford, held over 160 manors in 1314 with a total demesne arable acreage in excess of 18,800 acres.[35] The following year Guy de Beauchamp, earl of Warwick, died possessed of almost 15,000 arable acres in demesne scattered across a hundred different manors.[36] The earldom of Cornwall comprised forty-six manors at the end of the thirteenth century, which for administrative purposes were divided into nine groups each under its own steward.[37] The earldom of Norfolk was even more extensive: the 1306 *IPM* of Roger Bigod, earl of Norfolk, lists almost a hundred English manors scattered through eighteen counties plus manors, castles, and boroughs in Wales and Ireland.[38] Such estates are conspicuous by dint of their sheer size and the legacy of extant documentation which they

---

[29] On the estate of the priors of Norwich, for instance, piecemeal land purchases which enhanced the size of individual demesnes were recorded centrally in the rolls of the Master of the Cellar (i.e. the prior): NRO, DCN 1/1/1–30. The costs of maintaining the various boats which linked the priory with several of its manors and the port of Yarmouth were accounted for similarly: L. F. Salzman, *Building in England down to 1540* (Oxford, 1952), pp. 392–3.

[30] This was conspicuously the case on the estate of Peterborough Abbey: Biddick, *The other economy*; Farmer, 'Marketing'.

[31] Denholm-Young, *Seignorial administration*, pp. 6–31. For a case study of burgeoning bureaucracy at a local level see Thornton, 'Determinants of productivity', p. 202.

[32] *Pipe Roll*, pp. xii–xv.   [33] Titow, 'Land and population', pp. 21–2.

[34] Smith, *Canterbury Cathedral Priory*, p. 141. Smith's figures omit the 55–75 acres sown on the priory's small demesne at Deopham, Norfolk: Harvey, *Westminster Abbey*, p. 127.

[35] PRO, C134 Files 42–4.   [36] PRO, C134 Files 49–50.

[37] *Ministers' accounts of the earldom of Cornwall 1296–7*, ed. M. Midgley (London, 1942), vol. I, pp. xvii–xxiv.   [38] PRO, C133 File 127.

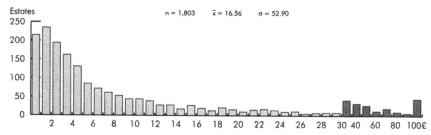

Fig. 3.01. Value of demesne land per lay estate: England, 1300–49 (source: National *IPM* database).

have bequeathed. Yet, of 1,511 lay lords represented by *IPM*s in the period 1300–49 only 1 per cent had estates containing 5,000 acres of arable or more. Even estates of 1,000 arable acres or more accounted for only 9 per cent of the total. If great estates impress by their size small estates impress by their numbers. Over 90 per cent of lay lords held less than 1,000 arable acres and 80 per cent held less than 500 acres. Whereas the broad demesne acres of a handful of mighty magnates yielded £100 or more per year, the average lay lord held only three demesnes from which he derived an annual revenue of £16 11s. 2d. Such lords were comparatively well off for they were far outnumbered by those with just a single demesne worth no more than a few pounds per year (Figure 3.01). Because these single-manor estates are poorly represented by extant records – of whose creation they, of course, had less need – they have failed to receive the historical attention that their numerical importance merits.

Ecclesiastical estates could not match the mightiest lay estates for size, nor were they geographically as dispersed in location. Otherwise the size variation in ecclesiastical estates echoed that of lay estates. Late-thirteenth-century England contained approximately a thousand religious houses, with the number of poor houses far outnumbering the rich.[39] The many single-manor lay estates also found their counterparts in the numerous glebe holdings of rectors, most of which were modest but a few of which were as large and valuable as a good-sized demesne. As a class of producers rectors have largely been overlooked in accounts of the period. Like other single-enterprise producers, their priorities of production must have differed in many significant ways from those of the great multi-manorial estate complexes which have attracted a disproportionate amount of historical attention. Redressing this historiographic imbalance is likely to revise or at least qualify the verdict returned upon lords as a class of producer.

[39] D. Knowles and R. N. Hadcock, *Medieval religious houses* (London, 1953), p. 364.

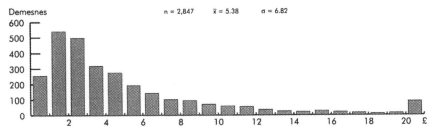

Fig. 3.02. Value of land-use per demesne: England, 1300–49 (source: National *IPM* database).

3.3   The scale and composition of demesnes

3.31   Aggregate value

According to the *IPM*s, the average lay demesne was worth £5 7s. 7d. in the first half of the fourteenth century. As a source the *IPM*s are prone to under-valuation but against this must be set their failure to net many of the lesser gentry and thus the smallest demesnes of all. Within the three east midland counties of Bedfordshire, Cambridgeshire, and Huntingdonshire, for instance, small arable demesnes of less than 50 acres comprise 17.2 per cent of those recorded by the 1279 Hundred Rolls but only 8.4 per cent of those recorded by the *IPM*s.[40] Many ecclesiastical demesnes, as the Hundred Rolls again make clear, were larger and more valuable than their lay counterparts, although these, too were counterbalanced by the innumerable small glebe properties of rectors. A mean annual net value somewhere in the range £5–£6 thus seems most plausible. The mean nevertheless masks a modal value of less than £2, indicative of the fact that whereas a small minority of demesnes were both large and valuable the vast majority were neither (Figure 3.02). The typical as opposed to the average demesne was modest in both scale and value.

Humble demesnes might be found almost anywhere and must often have been virtually indistinguishable from the more substantial free holdings. Blurring the distinction further must have been the reliance of many of the smallest demesnes upon hired rather than customary labour (with which they were ill endowed).[41] Substantial demesnes of conspicuously high value were altogether more circumscribed in their distribution. They were apparently well represented in those parts of Yorkshire which had been the subject of much

[40] Hundred Roll information from Allen, *Enclosure and the yeoman*, p. 64. *IPM* information from national *IPM* database. The 290 demesnes recorded by the Hundred Rolls have a mean of 165 arable acres; the corresponding mean of the 167 demesnes recorded by the *IPM*s is 176 acres. Excluding demesnes smaller than 50 acres yields means of 196 acres and 188 acres respectively (n = 240 and n = 153). For these three counties the Hundred Rolls thus largely endorse the validity of the *IPM* evidence.   [41] Kosminsky, *Studies*, pp. 268–78.

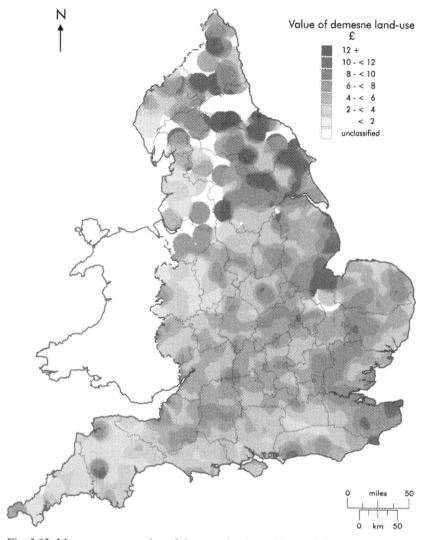

Fig. 3.03. Mean aggregate value of demesne land-use: England, 1300–49 (source: National *IPM* database).

rationalisation of settlement following the harrying of the north by William the Conqueror (Figure 3.03).[42] They also show up strongly in the Lincolnshire fens where reclamation had brought much highly fertile land into production: here, in much of Kent, and in several other parts of the south and east it was as much the unit value as the quantity of land which made demesnes so valu-

---

[42] T. A. M. Bishop, 'The distribution of manorial demesne in the vale of Yorkshire', *EHR* 49 (1934), 386–406.

able. In other localities demesnes owed their superior value to their great size.

For historical reasons to do with the origins and evolution of manors and estates demesnes were simply larger in some parts of England than others. The heavy boulder-clay country of Suffolk, west Essex, east Hertfordshire, and east Middlesex, for instance, was characterised by consistently large demesnes, as was much of the Oolitic belt which ran diagonally from Lincolnshire to south-east Somerset. Adding in the ecclesiastical demesnes would obviously qualify the pattern revealed by Figure 3.03 a good deal. In west Suffolk and central Somerset the many substantial demesnes of the abbeys of Bury St Edmunds and Glastonbury would have reinforced the strong lay bias towards large demesnes of above average aggregate value. In Hampshire, on the other hand, the modest value of most lay demesnes seems to have been a function of the immodest size and value of those belonging to the Church which had long been entrenched as the major landowner within the county. There is, however, good reason to believe that in much of the remote north-west of England, the west and north-west midlands, and the south-western counties of Devon and Cornwall the low aggregate value of most lay demesnes is representative of the demesne sector as a whole.

Demesnes could potentially comprise a wide range of different land-uses with the result that some were territorially very extensive enterprises. Some stood out in scale and value because of a particularly favourable endowment with meadow, marsh or wood. In the vast majority of cases, however, it was the arable that was the single most valuable component of land-use, accounting for just over 60 per cent of all lay land-use by value in the first half of the fourteenth century (Table 3.01). Indeed, it was a rare demesne that was entirely lacking in arable. Tillage lay at the heart of the seigniorial sector and, after the meadow, was the most carefully and consistently recorded of all the land-uses; but tillage was unsustainable without at least some pasturage. The latter came in several different guises, which collectively accounted for 37 per cent of all land-use by value (Table 3.01). Grassland was by far the most widely recorded and highly valued of these sources of pasturage. Mowable and unmowable grassland together accounted for 88 per cent of all demesne pasturage by value, with 72 per cent of that value coming from the meadow and 28 per cent from pasture and herbage.[43] The remaining 12 per cent of pasturage were contributed in roughly equal measure by non-grassland of various usually environment-specific types – 'waste', moor, heath, and marsh – and the browsing and pannage afforded by woods and private hunting grounds. The final 5 per cent of aggregate demesne land-use was contributed by woodland, private hunting grounds (mostly of greater recreational than financial value), and an assortment of minor land-uses. Their limited aggregate importance should not obscure the fact that at a local scale their

---

[43] The terms 'pasture' and 'herbage' are used more or less interchangeably in the extents, with the latter having greatest currency in the north.

Table 3.01. *Mean land-use composition of lay demesnes: England, 1300–49 (excluding gardens and orchards etc.)*

Land-use type	IPMs recording land-use type		Mean value of land-use type (excluding 0) £		Mean value of land-use type (including 0) £		
	n	%	Mean	Std	Mean	Std	%
Arable	3,906	97.1	3.41	3.79	3.27	3.77	60.8
Pasturage:	3,933	97.8	2.06	3.88	1.98	3.83	36.8
Grassland:	3,891	96.8	1.84	3.57	1.74	3.49	32.3
Mowable	3,575	88.9	1.44	2.99	1.25	2.83	23.2
Unmowable	3,157	78.5	0.91	2.05	0.62	1.74	11.5
Non-grassland	624	15.5	1.26	3.37	0.13	1.14	2.4
Wood-pasture	1,071	26.6	1.17	2.31	0.11	0.80	2.0
Wood, forest, chase, park*	2,180	54.2	0.80	2.39	0.23	1.34	4.3
Misc.	68	1.7	1.95	5.57	0.03	0.73	0.6
All land-use	4,019	100.0	5.39	6.82	5.38	6.82	100.0

Note:
* Includes wood-pasture.
Source: National *IPM* database.

significance could be considerable and sometimes gave rise to distinctive local economies.

3.32 Arable land-use

As arable farms many demesnes ranked as very substantial enterprises. The single greatest sown acreage recorded in the national, FTC and Norfolk accounts databases is the 1,004 acres cropped in 1282–3 on the demesne belonging to Roger Bigod, earl of Norfolk at Bosham in Sussex.[44] A few lay demesnes recorded by the *IPMs* were as large or larger. For instance, at Westbourne-and-Stansted, also in Sussex, Edmund of Woodstock, earl of Kent, died possessed of an arable demesne of 1,436 acres in 1330, while at Rothwell on the Lincolnshire Wolds Robert de Rowelle's arable demesne was extended at 1,655 acres in 1304.[45] Such extensive 'prairie' farms were very much the exception rather than the rule. On the evidence of manorial accounts, the average Norfolk demesne sowed 152 acres in the period 1250–1349, compared with 200 acres in the country as a whole and 224 acres in the FTC counties.[46] Very likely these averages are inflated by the over-representation of ecclesiastical demesnes, which tended to be larger than their lay counterparts.[47] Thus, the mean arable acreage of 3,883 lay demesnes recorded by the *IPMs* is 151 acres.

Most *IPM* extents fail to distinguish between the sown and unsown arable. In commonfield areas where the unsown demesne arable was subject to common grazing and was therefore of no direct value to the lord there was a tendency to extend the sown arable only since this alone yielded income. A mere 138 *IPM* extents record both the sown and the unsown arable; these demesnes had an average arable area of 181 acres, of which 59 per cent was sown and the remaining 41 per cent was fallowed.[48] A total of 652 *IPM* extents provide explicit information of the proportions but not necessarily the areas sown and unsown; these confirm that between half and two-thirds of the arable was generally sown, with a mean of 62 per cent.

---

[44] PRO, SC 6/1020/14.   [45] PRO, C 135 File 24, C133 File 113(6).

[46] These national figures exclude Cornwall and Devon in the south-west and Cumberland, Westmorland, Lancashire, Cheshire, Staffordshire, Derbyshire, Nottinghamshire, Yorkshire, Durham, and Northumberland in the north, where outsize customary acres present problems.

[47] Kosminsky, *Studies*, pp. 109–12.

[48] For instance, an extent of 1334 for Monewden in Suffolk specifies that out of a total of 140 acres, 50 acres were (winter) sown with wheat and worth 4d. an acre, a further 40 acres were worth only 2d. an acre because they lay in common from 1 August to 25 March (presumably because they were spring sown), and the remaining 50 acres were worth nothing because they lay in the commonfield and were fallowed every third year: PRO, C135 File 41(19). Likewise, at Sherringham in Norfolk in 1335 60 acres of arable were valued at 4d. an acre when sown, and nothing when unsown because they then lay in common throughout the year. The same extent goes on to specify that 20 acres should be sown in the winter, and 20 acres in the spring, implying that the remaining 20 acres were fallowed: PRO, C135 File 43(12).

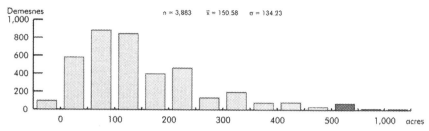

Fig. 3.04. Total acreage of arable per demesne: England, 1300–49 (source: National *IPM* database).

It follows that even on the most generous of allowances the average demesne is unlikely to have contained more than 250 acres of arable (both sown and unsown). Indeed, given that many of the smallest demesnes of all escaped the net of the *IPM* escheators, a figure nearer 150 acres seems more probable. This is the top end of the modal range of 50–150 acres indicated by the *IPMs* (Figure 3.04). Any demesne which regularly cropped 250 acres or more therefore ranked as a very major enterprise. After 1349 the amounts of arable kept in hand were generally somewhat smaller as some arable was converted back to pasture, outfield cultivation contracted, and portions of demesne were leased piecemeal to tenants. Norfolk lords sowed on average an eighth less, lords within the FTC counties sowed a fifth less, and nationally lords sowed almost a quarter less land than they had done in the heady days of arable production at the opening of the fourteenth century (Table 5.06).

Table 3.02 gives a breakdown of the size distribution of demesnes according to the various accounts and *IPM* databases, from which it will be seen that the mean and modal sown acreages both before and after 1349 were consistently in the range 100–200 sown acres. Many lay and glebe demesnes were, however, significantly smaller than this (Figure 3.04) and are particularly poorly served by extant accounts even though they might be encountered almost anywhere.[49] In contrast, large demesnes of 300 or more sown acres, although far less widely distributed (Figure 3.05), are considerably better documented.[50] For instance, Cambridgeshire, Essex, Hertfordshire, and Suffolk are counties in which demesnes of at least 300 sown acres were particularly common, in contrast to adjacent Middlesex and Norfolk which emerge as having been characterised by much smaller demesnes.

---

[49] On glebe demesnes, see D. Postles, 'The acquisition and administration of spiritualities by Oseney Abbey', *Oxoniensia* 51 (1986), 69–77. The single greatest source of information on the size and composition of glebe demesnes is undoubtedly the *Nonae* Rolls of 1340–1, a substantial portion of which are available in print: *Nonarum inquisitiones in curia scaccarii.*

[50] On the oft-cited estates of the bishops of Winchester, by contrast, the mean sown area in the period 1300–24 was some 224 acres (i.e. well above the national average), and a fifth of these demesnes had sown acreages in excess of 300 acres: Titow, *Winchester yields*, pp. 136–9.

Table 3.02. *Size distribution of arable demesnes: England, Norfolk, and FTC counties, pre- and post-1350*

	England[a]				FTC counties			Norfolk	
	Undifferentiated arable	Sown arable			Sown arable		Undifferentiated arable	Sown arable	
	IPMs	IPMs	Manorial accounts		Manorial accounts		IPMs	Manorial accounts	
Acreage	1300–1349 %	1300–1349 %	1250–1349 %	1350–1449 %	1288–1315 %	1375–1400 %	1375–1400 %	1250–1349 %	1350–1449 %
<50	*17.6*	*22.5*	*4.2*	*7.3*	*3.5*	*3.6*	*16.4*	*9.2*	*8.1*
50 – <100	**22.3**	**27.7**	*15.7*	*21.8*	*11.6*	*13.0*	**19.2**	*22.3*	*28.8*
100 – <150	*21.6*	*24.0*	**19.4**	**25.6**	*14.7*	**26.1**	**19.2**	**23.1**	**32.4**
150 – <200	*10.2*	*6.5*	*15.2*	*19.1*	**23.2**	**26.1**	*12.3*	*20.8*	*15.3*
200 – <250	*11.7*	*11.7*	*9.3*	*10.3*	*13.6*	*13.8*	*11.6*	*12.3*	*8.1*
250 – <300	*3.7*	*2.8*	*6.1*	*8.0*	*13.6*	*8.7*	*2.7*	*3.1*	*1.8*
300 – <350	*5.3*	*1.8*	*4.2*	*3.4*	*3.0*	*4.4*	*7.5*	*3.9*	*3.6*
350 – <400	*2.1*	*0.9*	*2.5*	*1.2*	*6.6*	*0.7*	*1.4*	*3.9*	*1.8*
400 – <450	*2.3*	*1.2*	*1.2*	*0.8*	*2.5*	*1.5*	*3.4*	*1.0*	*0.0*
450 – <500	*0.8*	*0.3*	*2.5*	*1.2*	*3.5*	*0.7*	*0.0*	*0.0*	*0.0*
500 – <750	*1.9*	*0.6*	*0.5*	*0.0*	*3.5*	*1.5*	*6.2*	*1.0*	*0.0*
750 – <1,000	*0.4*	*0.0*	*0.0*	*0.0*	*0.5*	*0.0*	*0.0*	*0.0*	*0.0*
1,000+	*0.3*	*0.0*	*0.0*	*0.0*	*0.0*	*0.0*	*0.0*	*0.0*	*0.0*
n	3,669	325	407	262	198	138	146	130	111
Mean acreage	153.2	117.5	199.5	155.8	223.7	178.4	172.2	152.3	131.8
Std	136.9	95.3	122.5	87.4	133.7	89.5	141.6	92.2	72.5

Notes:
Roman = median; **bold** = modal class
[a] Excluding Cornwall, Devon, Northumberland, Durham, Yorkshire, Nottinghamshire, Derbyshire, Cheshire, Lancashire, Westmorland, and Cumberland owing to the predominance of outsize customary acres.
Sources: National *IPM* database; National accounts database; FTC1 and FTC2 accounts and *IPM* databases; Norfolk accounts database.

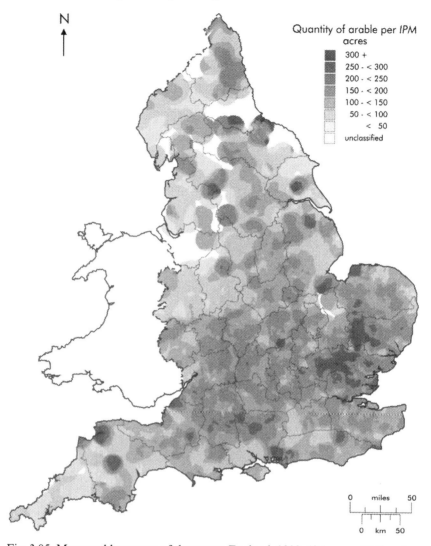

Fig. 3.05. Mean arable acreage of demesnes: England, 1300–49 (source: National *IPM* database).

These spatial variations in the size of arable demesnes were of considerable managerial significance, not least because, on the evidence of the Hundred Rolls, large demesnes often formed part of a much more fully articulated manorial economy, in which villein land and customary labour played an especially crucial role.[51] Large demesnes are therefore likely to have had greater

---

[51] 'We are left with a very definite impression that on small manors, the demesne was much less linked up with villein land, and occupied a more independent position in the manorial economy, than on large manors': Kosminsky, *Studies*, p. 101.

resources of servile labour to draw upon, with all that this implies in terms of poor work motivation and high supervision costs. Among the most crucial and onerous of servile obligations was the performance of ploughing services. The more that lords relied upon these to work their demesnes the less they were obliged to maintain their own plough-teams and employ *famuli* to do the ploughing. Significantly, the counties where mean arable acreages were greatest – Cambridgeshire, Essex, Hertfordshire, and Suffolk – were also those which before 1350 had fewest demesne ploughs per 100 sown acres, the deficiency presumably being made good by customary ploughings.[52] After 1349 as commutation and a general decline in the supply of servile labour forced landlords back on their own draught resources all four of these counties registered a substantial improvement in the ratio of demesne ploughs to sown acres. Nationally, a 25 per cent fall in the mean number of sown acres per demesne plough over the period 1250–1450 implies that the demesne sector as a whole was becoming increasingly self-reliant in matters of ploughing (Table 4.02).

The size of arable demesne had two further important implications for the conduct of husbandry. First, as cropped acreages rose so stocking densities, at least within the FTC counties, tended to fall (Table 4.09). In other words, the greater the arable acreage the stronger the arable bias of production. Second, the larger the arable area the higher in all probability the ratio of land to labour. Large arable demesnes therefore tended to be extensively rather than intensively cultivated. The productivities of land and labour on such large demesnes are thus hardly representative of those within the seigniorial sector as a whole let alone of those pertaining on the far smaller holdings of the peasantry.[53]

3.33 Pastoral land-uses

Grassland: meadow

Several grassland was the most typical type of demesne pasturage. Only 3 per cent of demesnes were without some. It was prized because it was generally capable of supporting the highest stocking densities. Oxen in particular – the

---

[52] Demesne ploughing was a common customary service owed by villeins. The more villeins and the more onerous their services the smaller was the number of ploughs which the lord needed to maintain on the demesne (unless he chose to commute those services and substitute a permanent force of *famuli* for the wage-less labour of his villeins): Postan, *The famulus*, pp. 3–4.

[53] See, for example, E. Van Cauwenberghe and H. Van der Wee, 'Productivity, evolution of rents and farm size in the southern Netherlands agriculture from the fourteenth to the seventeenth century', in H. Van der Wee and E. Van Cauwenberghe (eds.), *Productivity of land and agricultural innovation in the Low Countries (1250–1800)* (Leuven, 1978), p. 135, in which they claim: 'The theory that the intensity of agriculture increases as the area diminishes seems to be proved beyond all doubt for 14th century Flanders: as farms were split into even smaller units, there was a manifest intensification of cultivation techniques and an appreciable increase in land productivity.'

Table 3.03. *Composition of pasturage on lay demesnes: England, 1300–49*

		Cumulative % of total lay demesnes:				
% of total value	Pasturage as % of (arable + pasturage)	Mowable grassland as % of pasturage	Unmowable grassland as % of pasturage	All grassland as % of pasturage	Non grassland as % of pasturage	Wood-pasture as % of pasturage
0	3.4	13.2	28.9	5.1	88.6	90.5
<10	13.4	15.1	36.1	5.7	90.8	92.9
<20	31.6	18.3	48.3	6.4	92.6	95.1
<30	48.6	22.1	60.9	6.9	93.7	96.3
<40	65.0	25.8	70.2	7.4	94.8	97.1
<50	79.0	31.3	77.4	8.1	95.8	97.9
<60	87.8	38.6	83.0	9.1	96.6	98.4
<70	92.5	49.6	86.5	10.3	97.2	98.9
<80	95.0	60.4	89.3	11.3	97.5	99.4
<90	96.5	72.6	91.9	13.2	98.4	99.7
<100	97.0	79.1	93.0	15.4	99.0	99.9
≤100	100.0	100.0	100.0	100.0	100.0	100.0
n	2,501	2,527	2,496	2,493	2,507	2,549
Mean	34.2	61.6	29.3	91.2	4.9	3.0
Std	22.7	34.5	30.8	25.5	17.5	12.4

Source: National *IPM* database.

most widely employed demesne draught animal – were largely grass fed.[54] Because hay was required in quantity to keep them and other core livestock through the winter the most highly valued grassland was almost invariably the meadow. It was also the most labour-intensive since it alone could be mown to yield a hay crop. The preconditions for hay production were fertile and well-watered grassland combined with good drying conditions in early summer when the meadows were mown. Only after the hay had been cut were meadows generally used for grazing, when, like the arable, they might be pastured in common.

In much of lowland England meadowland was an improved land-use. Typically it occupied land with a naturally high water-table, drained and maintained by careful ditching, embanked where necessary as a protection from seasonal flooding, regularly mown and closely supervised.[55] Its artificial nature is well exemplified by the Lea Valley of Middlesex, Essex, and

---

[54] J. L. Langdon, 'The economics of horses and oxen in medieval England', *AHR* 30 (1982), 32–5.
[55] For examples of the use of ditching and drainage to upgrade marshland into meadow, see Davenport, *Norfolk manor*, p. 31; H. S. A. Fox, 'The alleged transformation from two-field to three-field systems in medieval England', *EcHR* 39 (1986), 544–5.

Hertfordshire, where extensive meadows were developed in response to the concentrated demand for hay from London. Not only did the River Lea help keep the meadows well watered in what was one of the drier parts of the country; it also provided a cheap means of boating the bulky hay crop to market.[56]

If the Lea Valley owed the extent of its developed meadows to economic incentives, Lincolnshire and Yorkshire, the country's most meadow-rich counties (Figure 3.06), owed theirs to the exceptional abundance of suitable environmental opportunities. Both possessed quantities of fertile, low-lying, well-watered, alluvial land coupled with the right seasonal balance of rain and sun. Only human labour was wanted to make such land productive. In Lincolnshire the process of meadow-creation was well advanced by 1086, when the Domesday Survey shows that this already populous county was better endowed with meadow than any other. At this stage, Yorkshire, recently devastated by the Conqueror, lagged far behind.[57] Two centuries later Lincolnshire still retained its lead over the rest of the country but on the evidence of the *IPM*s now shared it with Yorkshire. Patently, land reclamation and improvement had made striking headway in the latter county during the twelfth and thirteenth centuries. By the early fourteenth century it was quite usual in both counties for there to be at least 1 acre of meadow for every 5 acres of arable. Many of Lincolnshire's fen-edge demesnes were even more richly endowed, with at least 1 acre of meadow for every 2 acres of arable. In this low-lying area quality grassland had gained more from drainage and reclamation than tillage, providing an extreme example of the expansion of pasture farming so often overlooked in accounts of the period.[58]

Opposite circumstances pertained on the Lincolnshire Wolds. On these elevated limestone uplands significant numbers of demesnes lacked meadow altogether.[59] They either had to procure their hay from the adjacent meadow-rich lowlands or rely upon straw and fodder crops to see their livestock through the winter. This was a common predicament in many areas of wold-land and downland.[60] North-west Norfolk, for instance, was one of the most meadow-deficient localities in the country. With a low rainfall, sandy soils

---

[56] J. F. Edwards and B. P. Hindle, 'The transportation system of medieval England and Wales', *JHG* 17 (1991), 130–2; J. L. Langdon, 'Inland water transport in medieval England', *JHG* 19 (1993), 4–5; Campbell and others, *Medieval capital*, pp. 194, 196–7.

[57] H. C. Darby, *Domesday England* (Cambridge, 1977), pp. 142–8, 248–52.

[58] H. C. Darby, *The medieval Fenland* (Cambridge, 1940), pp. 48–52, 67–82; H. E. Hallam, *Settlement and society* (Cambridge, 1965), pp. 174–96.

[59] G. Platts, *Land and people in medieval Lincolnshire* (Lincoln, 1985), pp. 103–10, 157–62.

[60] H. S. A. Fox, 'The people of the wolds in English settlement history', in M. Aston, D. Austin, and C. C. Dyer (eds.), *The rural settlements of medieval England* (Oxford, 1989), pp. 85–9, 94–6; B. Harrison, 'Demesne husbandry and field systems on the north Hampshire estates of Saint Swithun's Priory, Winchester, 1248–1340', unpublished paper. Hay was regularly purchased for use on the demesne at Cuxham in Oxfordshire: Harvey, *Oxfordshire village*, pp. 101–2.

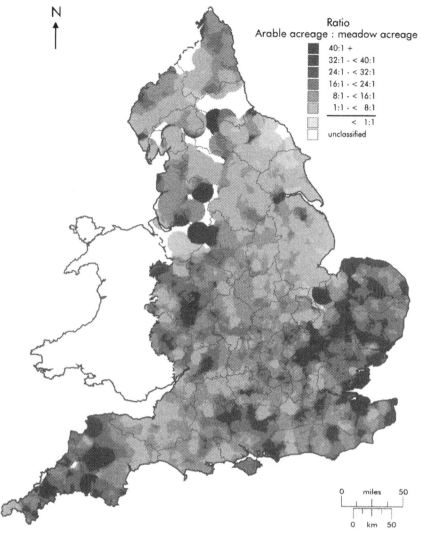

Fig. 3.06. Ratio of arable acreage to meadow acreage: England, 1300–49 (source: National *IPM* database).

overlying chalk, and no substantial rivers, opportunities for the establishment of hay meadows were mostly non-existent.[61] Thus circumstanced, demesne managers in this area early substituted the grain-fed horse for the grass-fed ox (Figure 4.13). Elsewhere in East Anglia, on the heavy boulder-clay soils that predominated in south-east Norfolk and most of Suffolk and Essex, environmental circumstances were less constraining. Even so, there was rarely more than 1 acre of meadow for every 20 acres of arable. Faced with such a scarcity of meadow it therefore made good pastoral sense to replace oxen partially or wholly with horses and thereby reserve the decidedly limited supplies of hay to the more or less exclusive use of the intensive dairy herds which were a conspicuous specialism of this closely settled region.[62] Demesne managers in fertile but meadow-scarce northern and eastern Kent adopted a similar expedient with the difference that they grew vetches in quantity to augment their limited supplies of hay.[63] There were other options. On the meadow-deficient chalklands of Hampshire and Wiltshire, for example, demesne managers reserved their hay to their essential draught oxen, kept few other cattle (buying in replacement oxen from elsewhere), and concentrated their remaining pastoral resources on the rearing of sheep, an activity for which supplies of hay were less of a limiting factor. Demesnes on the Gloucestershire Cotswolds adopted a similar strategy. For them replacement oxen were readily obtained from Welsh rearers across the River Severn to the west.[64]

Some localities otherwise well endowed with pastoral resources were also lacking in meadow. In this respect the Essex marshes, exposed to periodic salt-water inundation, contrast strikingly with the Lincolnshire Fenland. Meadow was also conspicuously scarce in much of the extreme north-west and south-west of the country. In both cases the high rainfall that promoted grass growth hindered production of hay. This was less of a problem in Cornwall and Devon where winters were short and comparatively mild than it was in Cumberland and Lancashire where they were longer and harsher. In both cases it encouraged an emphasis upon hardy breeds of animals and led cattle farmers to concentrate upon breeding, since young stock could be sold off at the end of the year thereby minimising the number of animals which had to be kept through the winter.

Meadow scarcity was thus greatest in the east and west of England. In

---

[61] There is no mention of meadow at Hunstanton in 1275–6, Ringstead and Great Bircham in 1295–6, Southmere in 1310–11, Broomsthorpe in 1304–5, Stanhoe in 1308–9, Docking in 1310–11, Creake and Hillington in 1323–4, Ingoldisthorpe in 1328–9, Roydon in 1329–30, and Syderstone and Helhoughton in 1337–8: PRO, C133 Files 14 (4), 76 (5), 77 (3), 119 (2), C134 Files 8 (9), 20 (14), 81 (19), C135 Files 11 (2), 15 (23), 51 (10).

[62] Campbell, ' Chaucer's reeve', pp. 271–96; Chapter 4, pp. 148–51.

[63] Campbell, 'Diffusion of vetches'.

[64] H. P. R. Finberg, 'An early reference to the Welsh cattle trade', *AHR* 2 (1954), 12–14; C. Skeel, 'The cattle trade between Wales and England from the fifteenth to the nineteenth century', *TRHS*, 4th series, 9 (1926), 137–8.

between lay the midland plain. From Somerset and east Devon in the south-west to the Vale of Pickering in Yorkshire's North Riding in the north-east, it was in the clay vales of this broad diagonal band of country that meadowland was most consistently well represented (Figure 3.06). Except on the wolds, few demesnes were without at least some meadow and many possessed little other pasturage. Yet, paradoxically, it is here that some historians have questioned the completeness of the meadow information contained in the *IPM*s, claiming that the jurors only returned what existed over and above the amount of meadow needed to sustain ox-teams.[65] Such a formula is most likely to have been employed in those extents which use the generic term 'land' rather than 'arable' and measure it in carucates and bovates rather than acres. These are most common in northern England and the north-east midlands, where much meadow is independently documented in the extents, and in the west midlands – in Worcestershire, Herefordshire, Shropshire, and Staffordshire – where meadow is far less plentifully recorded (Figure 3.06). Nevertheless, in East Anglia and the south-east most extents make a clear distinction between the arable and meadow and record both in acres. Here meadow bears every semblance of being the most accurately and consistently recorded category of land-use, as vital and valuable – with a unit value commonly three to four times that of arable – as it was scarce.

Grassland: pasture and herbage

If grassland is under-recorded in the *IPM* extents it is the pasture rather than the meadow entries that are likely to be most defective. Pasture is a far less specialised category of land-use than meadow; nevertheless, it is recorded by fewer *IPM*s (79 per cent compared with 89 per cent) and rarely with as much precision. Often, especially in the north and west of the country, a lump-sum valuation is all that is given. In many cases this must have been because the physical extent of unenclosed pastures had never been accurately extended, possibly because their unit value was too low to warrant the effort. Enclosed pasture was another matter. It had scarcity value, was finite in amount, could be more effectively managed and hence is normally recorded in some detail in the extents. Such pasture closes are, however, far from representative of pasture as a whole.

In principle what distinguished pasture as grassland from meadow was that it could not be mown. In practice that distinction was far from clear cut. Instances of unmowable meadow do occur and many pastures must, on occasion, have yielded a hay crop.[66] What is described as pasture in some localities

---

[65] 'The amount of land in demesne was under-estimated, and, naturally enough, the most valuable land – the meadow – was most undervalued': Hilton, 'Medieval agrarian history', pp. 161–2.

[66] Meadow at Kirksanton, Cumberland was unmowable in 1319 owing to the depredations of the Scots: PRO, C134 File 64(9). That at Hambledon, Buckinghamshire was unmowable in 1338 because it was worn out: PRO, C135 File 56.

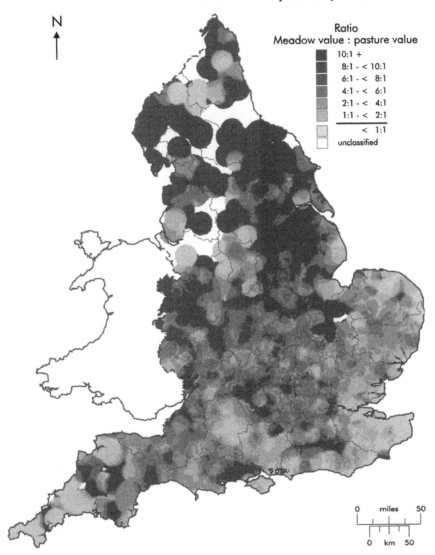

Fig. 3.07. Ratio of meadow value to pasture value: England, 1300–49 (source: National *IPM* database).

sometimes commanded a unit value significantly above grassland described as meadow in others.[67] Jurors reported land-use as they perceived it, and perceptions and uses varied both from place to place and over time. Meadow and pasture existed on a continuum and in practice the distinction between them is unlikely to have been as consistently drawn as the precise terminology of the documents might imply.

Although in reality there was undoubtedly more pasture than meadow, as Tables 3.01 and 3.03 demonstrate, pasture generally comprised a smaller component of demesne land-use. Under-recording apart, this was primarily because most extents recorded only that pasture in the exclusive possession of the lord. Except in a group of counties focusing upon Wiltshire the value of any common pasture to which demesnes were entitled rarely appears in the extents.[68] Consequently, pasture tends to feature most prominently as a demesne land-use in localities where common pastures were scarce or non-existent. This is most conspicuously the case in Kent and neighbouring East Sussex, one of the few parts of England where common rights either early disappeared or were never established.[69] Here, on average, recorded acreages of demesne pasture either equalled or exceeded those of meadow (Figure 3.07). Considerable quantities of demesne pasture are also recorded in Cornwall and Devon, another area where most demesnes were held and worked in severalty. Here, too, demesne pasture was all the more important because meadow was so scarce. In fact, right the way across southern England from Kent to Cornwall, in commonfield and non-commonfield districts alike, pasture was a more regular component of recorded demesne land-use than in most other parts of the country.

Pasture was also well recorded in East Anglia, but rarely in quantity. Demesnes in this region ranked among the least grassy in the country. Where several grassland was so scarce communal grassland consequently assumed disproportionate importance. In much of north and west Norfolk and in the Norfolk and Suffolk Breckland, for instance, extensive common pastures and heaths were a prominent component of overall land-use at this time, even though there is little to intimate this in the *IPM* extents.[70] This was, of course, even more the case in the hill country of the north and west of England where

---

[67] Pasture valuations as high as 24d., 30d., and even 36d. an acre sometimes occur, e.g. Sherington, Bucks., Dunmow, Essex, Ifield, Kent, Westleton, Suffolk, and Great Wishford, Wilts. (all 24d.); Little Linford, Bucks., and Saltfleetby, Lincs. (both 30d.); and Great Munden, Herts. (36d.): PRO, C133 File 100(10), C135 File 48(2), C133 Files 123(10), 108(1), C135 File 64(17), C133 File 106(8), C135 File 43(10), C133 File 105(1).

[68] M. R. Livingstone and K. C. Bartley, 'Historical problems, GIS solutions? Spatio-temporal patterns in medieval data', in A. Barnual (ed.), *Mapping historical data* (Moscow, forthcoming).

[69] A. R. H. Baker, 'Field systems of south-east England', in A. R. H. Baker and R. A. Butlin (eds.), *Studies of field systems in the British Isles* (Cambridge, 1973), pp. 393–419.

[70] Bailey, *A marginal economy?*, pp. 25–36; Campbell, 'Land use and values'.

there is little in the extents to hint at the abundance of pasture which prevailed throughout most parts of this extensive upland region (Figure 3.07). Probably it is here that the grassland resources of demesnes are most under-recorded. Certainly, few *IPM* extents produced within the Northern Escheatry (which encompassed the whole of England north of the Trent) record demesne land-use in the detail which is so characteristic of most East Anglian *IPM*s. But it was also the case that within this region several pasture was vastly exceeded in area by common pasture which, because it was not in the sole ownership of the demesne, does not appear in the extents.[71] The strong emphasis upon arable in so many of these northern extents thus belies the overwhelming bias of land-use within the region towards pasture.[72]

In much of the commonfield country of the midlands, however, and in the east midlands above all, the limited appearance of pasture in the extents is symptomatic of the genuine deficiency of pasture in the region. In the east midland counties of Leicestershire, Rutland, Northamptonshire, Hunting-donshire, Cambridgeshire, Bedfordshire, and north-east Hertfordshire, the extents either record very small amounts of pasture or none at all. Almost invariably, less pasture is recorded than meadow (Figure 3.07). Such pasture as they possessed was mostly communal, but so closely settled was the region that almost everywhere this was severely restricted in amount. In fact, Fox believes that much of the countryside within the territory of the midland commonfield system was experiencing a crisis in the provision of pasture in the thirteenth century.[73] Raftis concurs: 'Whereas meadow was an integral part of the arable economy in the East Midlands, appearing in most extents of the *Inquisitiones Post Mortem*, and valued high because it supported animals required for work on the arable in both winter and summer, pasture was a much more marginal resource.'[74] Not even in arable East Anglia was several pasture as scarce. Hence the premium which all regular commonfield townships placed upon the temporary pasturage afforded by the fallow arable. Since demesne and tenant land usually lay intermixed, more often than not unsown demesne arable is reported as worth nothing to the lord because it was grazed in common.[75]

Common pasture
What was the value to demesne lords of their own rights of common pasture, on the fallow and on land set aside permanently as pasture? A few *IPM* extents

---

[71] These common pastures show up most clearly during the era of parliamentary enclosure in the eighteenth and nineteenth centuries: A. Harris, 'Changes in the early railway age: 1800–1850', in Darby (ed.), *New historical geography*, pp. 478–86; J. R. Walton, 'Agriculture and rural society 1730–1914', in R. A. Dodgshon and R. A. Butlin (eds.), *An historical geography of England and Wales*, 1st edition, London, 1978, pp. 239–65.
[72] E. Miller, 'Farming techniques: northern England', in Hallam (ed.), *AHEW*, vol. II, pp. 408–11.   [73] Fox, 'Ecological dimensions', p. 123.
[74] Raftis, *Assart data*, p. 73.   [75] Gray, *English field systems*, pp. 450–509.

provide an answer. In the case of ninety-two demesnes south of the Trent and east of the Exe the value of the common pasture is recorded along with the value of all other pasturage and the value of the arable. The bulk come from the broad midland zone of regular commonfield systems, by far the greater number hailing from a group of six counties in the south-west midlands focusing upon Wiltshire. As such they probably reflect the activities of a particular sub-escheator. On average, the value to these demesnes of their common pasture was 4 shillings; equivalent in worth to little more than a couple of acres of good meadow. In a clear majority of cases it was worth even less, and in only half a dozen cases was it worth £1 or more. As a component of total demesne land-use it was small. Common pasture accounted for half by value of all pasture, an eighth of all demesne pasturage, and a mere twentieth of the combined value of demesne arable and pasturage.

On these ninety-two demesnes, costing in the common pasture raises the value of all pasturage from 36 to 39 per cent of the combined value of arable and pasturage. On this evidence common pasture hardly rated as a major component of demesne land-use. Presumably, this was because even when common pastures were physically extensive they were rarely capable of supporting a high stocking density and, of course, rights in them were divided among many different commoners. There were exceptions. At Haversham in north Buckinghamshire and Warminster and Westbury in Wiltshire on the western edge of Salisbury Plain common pasture accounted for an eighth of demesne land-use. At Headborne Worthy in mid-Hampshire and Stoughton in West Sussex on the edge of the South Downs the corresponding proportions were respectively a fifth and a quarter, and in all five of these cases the common pasture in question was worth a respectable 20 to 25 shillings.[76] Nevertheless, the single most striking exception occurs not in southern England but on the Welsh border. At Trelleck in Monmouthshire it was reported that the common pasture was worth £2 in 1314, thereby exceeding in value all other components of land-use on this modest demesne.[77] This illustrates the far greater quantities of common pasture often available on England's upland margins compared with the predominantly arable countryside of the lowlands.

Waste, moor, heath, and marsh

Some demesnes possessed quantities of rough pasture of their own. In the south-west and at a scattering of locations north of the Trent there was a good deal of what the extents misleadingly describe as 'waste'. This was unimproved pasture, often of the roughest sorts, which yielded rough grazing and a variety of natural products. Other demesnes possessed quantities of moorland, whose waterlogged soils likewise yielded poor-quality grazing. Instances occur in

---

[76] PRO, C133 Files 122 (4), 94 (7), 114 (8), 127.   [77] PRO, C134 File 43.

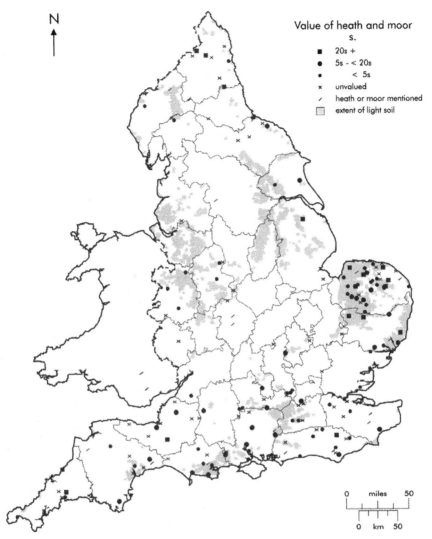

Fig. 3.08. *IPMs* recording heath and moor: England, 1300–49 (source: National *IPM* database).

both upland and lowland locations. Heath, in contrast, was mostly a land-use of arid lowland settings, particularly in southern England and East Anglia where it was almost invariably associated with light acidic soils. Heather rather than grass was thus the predominant vegetation, which was more effectively grazed by sheep than cattle. Many demesnes in north and west Norfolk possessed tracts of heath which were an essential adjunct of the foldcourses operated on most manors in the region (Figure 3.08).[78] This was a system whereby provision was made for fertilising the arable by feeding manorial flocks on permanent sheepwalks during the day and then folding them on the arable by night. Since this systematically robbed the permanent sheepwalks of nutrients, heathland may to some extent represent degraded vegetation. Coastal and freshwater marshland, in contrast, was exceptionally rich in nutrients. It yielded rich seasonal grazings along with abundant opportunities for fishing and wildfowling, the harvest of reeds, digging of turves, and even the manufacture of salt. Such marshes could be a lucrative demesne asset and show up most prominently in and around the East Anglian Fenland and on opposite sides of the Thames estuary, as well as at a scattering of locations elsewhere in East Anglia and the south-east (Figure 3.09).[79] It is in these counties that the *IPM*s differentiate in most detail between the different categories of pasturage and here, too, that market demand at home and abroad was strongest for the range of pastoral products produced from these marshes.

On individual manors these alternative sources of pasturage could sometimes underpin distinctive rural economies which had evolved in response to the peculiar opportunities which waste, moor, heath, and marsh variously presented and the unique management problems which went with them.[80] Sometimes the availability of these land-uses in abundance stimulated particular pastoral specialisms and lent a strong pastoral cast to the overall agrarian economy. Often, too, they compensated for shortages of more conventional sources of pasturage. At a national scale, however, their contribution to demesne pasturage was small, as Tables 3.01 and 3.03 clearly demonstrate. Only one in seven of all demesnes had access to at least one of these types of pasturage, which collectively contributed less than 7 per cent of total demesne pasturage. This is undoubtedly a minimum figure. It would be surprising if much moor, heath, and marsh was not subsumed in many extents under the generic term 'pasture'. Probably, too, the jurors may have masked the true value of these resources, knowing that without detailed local knowledge they were difficult to value and quantify.[81] Nevertheless, even on the most generous allowances these were minority land-uses.

Wood-pasture

More frequently recorded, although collectively far less valuable, was the pasture and pannage available in seigniorial woods and private hunting

[78] K. J. Allison, 'The sheep-corn husbandry of Norfolk in the sixteenth and seventeenth centuries', *AHR* 5 (1957), 12–30; Campbell, 'Regional uniqueness', pp. 17–18; M. Bailey, 'Sand into gold: the evolution of the foldcourse system in west Suffolk, 1200–1600', *AHR* 38 (1990), 40–57.

[79] B. M. S. Campbell, J. A. Galloway, and M. Murphy, 'Rural landuse in the metropolitan hinterland, 1270–1339: the evidence of *inquisitiones post mortem*', *AHR* 40 (1992), 10–12.

[80] E.g. Darby, *Medieval Fenland*, pp. 42–85; Hallam, *Settlement and society*, pp. 162–73; Bailey, *A marginal economy?*, pp. 40–96.   [81] Kosminsky, *Studies*, pp. 58–63.

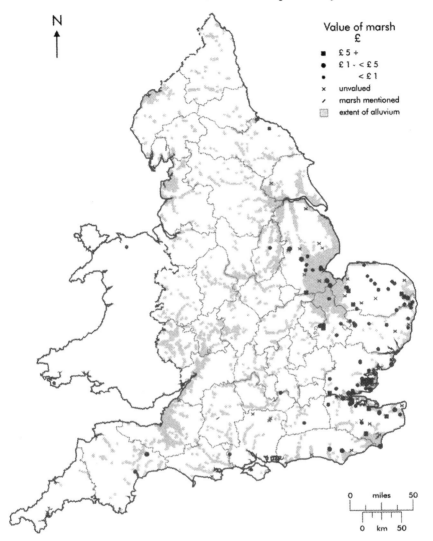

Fig. 3.09. *IPM*s recording marsh: England, 1300–49 (source: National *IPM* database).

grounds (Figure 3.10). As primary land-uses woods, private forests, chases, and parks were themselves far from ubiquitous. Just over half of all *IPM*s record them. The proportion is lower in north-west Norfolk, most of Cambridgeshire and Fenland Lincolnshire, and a swathe of open-field country running diagonally from the Yorkshire Wolds in the north-east to Salisbury Plain in the south, where woods and parks were genuinely thin on the ground. Such an explanation, however, hardly accounts for the paucity of

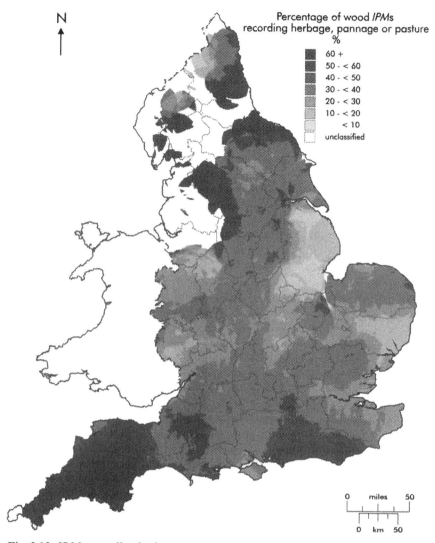

N

Fig. 3.10. *IPM*s recording herbage, pannage or pasture as a percentage of those recording wood: England, 1300–49 (source: National *IPM* database).

wood entries throughout most of north-west England. Here, under-recording is more likely to be a factor.

As with other land-uses, wood was most scrupulously recorded where it was both scarce and valuable, hence the care with which even small amounts of wood are recorded in East Anglia and the south-east. The most detailed of these entries – roughly half of the total – provide a separate valuation of any pasture, herbage, or pannage available in the wood or park. Only in

Lincolnshire and Suffolk do the sub-escheators in question appear to have omitted such secondary information on a systematic basis. Elsewhere its incidence is a function of recording convention combined with genuine availability. Thus, in Sussex, Devon, and Cornwall clear majorities of wood, forest, chase, and park entries provide a separate valuation of any pasture or pannage the particular land-use afforded. Indeed, in most of the English counties south of the Thames (Kent and Hampshire are the exceptions) wood-pasture – like pasture – was clearly recognised as a valuable component of demesne land-use (Figure 3.10). The value of such pasturage is also more often noted than not in those northern *IPMs* which record woods, forests, chases or parks. Wood-pasture was less regularly recorded in the midlands and East Anglia where there may have been a more exclusive emphasis upon the management of woodland to yield a regular crop of underwood. Browsing, after all, was harmful to the regeneration of trees and damaging to intensively managed coppice woods.[82]

Overall there can be little doubt that more demesnes profited from the pastoral opportunities available in their woods, forests, chases, and parks than the one in four which are recorded as having done so. Nevertheless, as a source of pasturage it was rarely of more than subsidiary importance (Table 3.03). Nationally, wood-pasture in its various forms accounted for barely 6 per cent by value of all demesne pasturage. Allowance for under-recording is unlikely to raise this proportion above 10 per cent. As with waste, moor, heath, and marsh, it was at a local scale that the availability of wood-pasture had greatest impact. This was most especially the case in localities such as the West Sussex Weald where an abundance of pannage promoted something of a specialist interest in the extensive production of swine.[83]

3.34 Arable versus pasturage

Demesne pasturage was therefore made up of many different elements whose distinctive land-use characteristics often shaped the pastoral systems which evolved in response to the specific opportunities they afforded. On average each lay demesne had pasturage of one sort or another worth just under £2 in the first half of the fourteenth century (Table 3.01). This was equivalent to only 60 per cent of the average value of the arable. Probably, for the reasons given above, it is an under-estimate. Some meadowland may be subsumed under the category 'land'; unenclosed pasture with a low unit value may be under-recorded; exceptional and extensive land-uses such as wastes, moors, heaths, and marshes may be under-valued and in some locations even omitted

---

[82] O. Rackham, *Ancient woodland* (London, 1980), pp. 22, 47–8, 59, 136; K. Witney, 'The woodland economy of Kent, 1066–1348', *AHR* 38 (1990), 27.

[83] P. F. Brandon, 'Farming techniques: south-eastern England', in Hallam (ed.), *AHEW*, vol. II, pp. 315–16.

altogether; and there was undoubtedly more wood-pasture and pannage than the *IPM*s admit. At local and even regional scales these omissions may have a seriously distorting effect – much of England north of the Trent is a case in point – but nationally even the most generous allowances are unlikely to raise the value of the pasturage relative to the arable by more than a few percentage points. Costing in the common pasture would raise it yet further, but not, on the available evidence, by very much. Overall there can be little doubt that the value of demesne pasturage was inferior to that of demesne arable and decisively so in much of the country.

The gap in value between the recorded quantities of arable and pasturage is widest in many parts of the north of England where the descriptions of demesne land-use contained in the *IPM*s are at their most laconic (Figure 3.11). The scant attention paid to pasturage in such a conspicuously pastoral region can only mean two things. Either it was so common-place and low in value that the escheators did not deem it worth recording, or, more probably, the bulk of it was common pasture and therefore technically ineligible for inclusion as a demesne asset. The latter is certainly true of some demesne properties on the fringes of Dartmoor in Devon, which convey the same misleading impression that they were deficient in pasturage when, in fact, the moor on their doorstep furnished common pasturage in abundance.[84] Some other localities almost as poorly endowed with demesne pasturage were also able to survive because they too had access to major reservoirs of pasturage not very far away. For the populous Isle of Thanet and stretch of coast immediately to the south in eastern Kent, for instance, the intervening marshes of the Ash Levels must have provided a vital but slender pastoral lifeline. Similarly, the crowded and almost treeless countryside of eastern Norfolk took full advantage of the reed beds, turbaries, and grazings available on the neighbouring Broadland marshes, within which most vills in the district had some territorial stake.[85] Demesnes deficient in pasturage on and near the fen edge in central Cambridgeshire and northern Huntingdonshire probably made similar resort to the diverse pastoral resources of the peat fens.[86]

Other localities were less fortunately placed. Demesne pasturage was conspicuously scarce in south-central, northern, and western Norfolk. Here all demesnes had to fall back on were the scanty common grazings available on the region's heaths and commons. Much the same appears to have been true on the downland of northern Hampshire. Worst placed of all were demesnes in a belt of closely settled commonfield country, which stretched from western

---

[84] H. S. A. Fox, 'Medieval Dartmoor as seen through its account rolls', in *The archaeology of Dartmoor* (Exeter, 1994), pp. 156–62.

[85] J. M. Lambert, J. N. Jennings, C. T. Smith, C. Green, and J. N. Hutchinson, *The making of the Broads* (London, 1960), pp. 82–99; D. Dymond, *The Norfolk landscape* (London, 1985), p. 115.

[86] J. R. Ravensdale, *Liable to floods: village landscape on the edge of the Fens, 450–1850* (Cambridge, 1974), pp. 41–69.

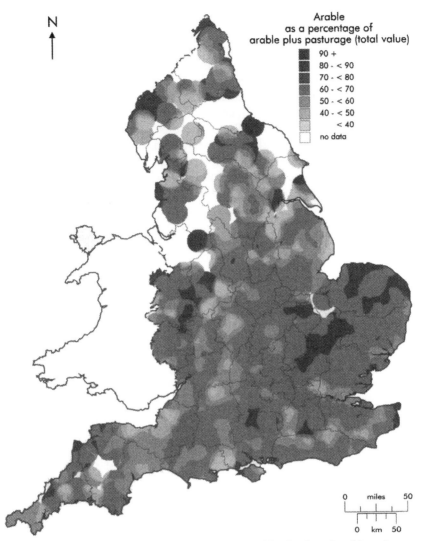

N

Arable
as a percentage of
arable plus pasturage (total value)

90 +
80 - < 90
70 - < 80
60 - < 70
50 - < 60
40 - < 50
< 40
no data

0   miles   50

0   km   50

Fig. 3.11. Value of arable as a percentage of the combined value of arable and
pasturage: England, 1300–49 (source: National *IPM* database).

Suffolk through southern Cambridgeshire into north-eastern Hertfordshire
and neighbouring portions of Bedfordshire. Here, on average, for every £1 of
pasturage there were at least £4-worth of arable (Figure 3.11); nor was there
any convenient local reservoir of pasturage to which they could turn.
Demesnes therefore had to rely upon such precious pasturage as they possessed combined with any common pasturage to which they were entitled on
the common pastures and fallow arable. Here more than anywhere, on the

IPM evidence, the frontier between grass and grain had shifted most danger-ously grainwards by the beginning of the fourteenth century.

Significantly, it is in Bedfordshire and Cambridgeshire, along with Buckinghamshire and adjacent portions of Hertfordshire, that the *Nonae* Rolls afford greatest evidence of an early fourteenth-century contraction in cultivation. In 1341 land lay untilled in scores of parishes in these counties because of declining village populations, the impoverishment of the tenants, and a shortage of seed-corn.[87] Postan believed that in many corn-growing parts of the country the frontier between grass and grain had 'crossed its limits of safety' by the beginning of the fourteenth century, reducing pasture and animal populations to levels 'incompatible with the conduct of mixed farming itself'.[88] If that holds true for any part of the country it is surely here.

The problem was only marginally less acute in much of East Anglia and the east midlands, where the ratio of arable to pasturage was rarely less than two to one. There was no other single extensive region in England where demesne pasturage resources were consistently so limited. Only pockets and ribbons of marginally superior provision in north-central Norfolk, Breckland, the Sandlings of east Suffolk, and the Stour Valley of Suffolk and Essex relieved it. Elsewhere shortages of pasturage were equally real but more circumscribed in geographical extent. The more notable include central Kesteven in Lincolnshire; a band of country stretching south-westwards from the wold-lands of east Leicestershire into south-east Warwickshire; the vale country of central Buckinghamshire and Oxfordshire; the Kennet Valley of south-west Berkshire and adjoining north Hampshire Downs; most of eastern Surrey lying between the Thames and the Weald; portions of northern Kent includ-ing the Isle of Sheppey; the south-westernmost portion of the Sussex coastal plain around Chichester; and south Devon.

In about a tenth of England south of the Trent supplies of demesne pastur-age were seriously constrained and in over 40 per cent of southern England there was only half as much pasturage, by value, as there was demesne arable. At the opposite extreme, only one in five of all demesnes had pasturage worth at least as much as their arable (Figure 3.11). South of the Trent localities with amounts of pasturage equal to or greater than amounts of arable occupied a mere eighth of the total area. None were at all extensive and almost all lay to the north and west of a line from the Wash to the Solent. The sole conspicu-ous exceptions are the lower Lea Valley of Middlesex and Essex (whose valu-able meadows were exactly the kind of land-use specialism which the Thünen land-use model would predict of a location so close to a great city); the Wey Valley of west Surrey which was an important supplier of fat animals to the capital; the Isle of Oxney in the Rother Valley of south-east Kent which was well placed to export pastoral products via the port of Rye; and Ashdown and

---

[87] Baker, 'Contracting arable lands'.   [88] Postan, *Economy and society*, pp. 58–9.

Horsham Forests deep in the wooded, heavy-clayland core of the Weald of Sussex.[89]

Outside of this predominantly arable eastern zone there were many more localities where demesnes were at least as well provided with pasturage as they were with arable (Figure 3.11). Conspicuous among them are the lower Trent Valley and Isle of Axholme in north-west Lincolnshire; the low-lying alluvial coastlands of eastern Lincolnshire; the vicinity of Cannock Chase in southern Staffordshire; the hill country of south-west Shropshire; the wooded Arden district of central Warwickshire; the traditional 'cheese' country of north Wiltshire; the country lying immediately to the west of the New Forest in sandy and infertile south-west Hampshire and eastern Dorset where arable husbandry was particularly unrewarding; most of Somerset immediately inland from the Severn estuary (a land-use specialism mirrored on the opposite south Welsh side of the estuary); south-east Devon; and Dartmoor and Bodmin Moor.[90] These were all localities where environmental factors promoted a strong predisposition towards pastoral and woodland land-uses. The one reasonably extensive tract of country to show up with a consistently ample provision of pasturage straddled the Severn estuary and on the English side stretched in a widening wedge from Gloucester in the north to Devon's lush Exe Valley in the south. From a demesne perspective this appears to have been England's single grassiest region. North of the Trent, where there was also undoubtedly much pasturage, the thinner coverage of the evidence and its greater inconsistency produces a far less clearly focused picture. Nevertheless, the vales of both York and Pickering show up as well furnished with pasturage as, more fitfully, do portions of the eastern and western flanks of the Pennines and the Eden Valley of Cumberland and Westmorland.

3.35 The relative unit values of grassland and arable

Judging the relative importance of the arable and pastoral sectors on the basis of the value of their respective shares of demesne land-use is, however, less clear cut than might at first appear (measures of relative area are not a viable alternative option since comprehensive areal information is rarely provided by the extents and pannage and wood-pasture cannot be adequately quantified in that way). Demesnes held arable and pasturage in combination rather than separately; the value of one was therefore determined in part with reference to the other. Postan, for instance, believed that it was relative scarcity which drove up the value of grassland during the course of the thirteenth century and rendered it disproportionately more valuable than arable. Such an

---

[89] Campbell and others, *Medieval capital*, p. 5.

[90] For an examination of the dual character of medieval Wiltshire agriculture see J. N. Hare, 'Lord and tenant in Wiltshire, *c.* 1380–*c.* 1520, with particular reference to regional and seigneurial variations', PhD thesis, University of London (1976).

inversion, in his view, meant that conversion of pasture to arable use would have made no sense since it could not offer any economic inducement to the converter.[91] Yet, as Adam Smith explained, there are sound economic reasons why limited amounts of several grassland should command higher rents than the neighbouring, and much more extensive, arable:

in an open country . . . of which the principal produce is corn, a well-inclosed piece of grass will frequently rent higher than any corn field in its neighbourhood. It is convenient for the maintenance of the cattle employed in the cultivation of the corn, and its high rent is, in this case, not so properly paid from the values of its own produce, as from that of the corn lands which are cultivated by means of it.[92]

Where arable rather than grassland was in strictly limited supply the transference of value could also work in reverse, the rental value of the arable being enhanced by its vital role in supplying the larger pastoral sector with the fodder crops and straw necessary to sustain animals through the winter. In such contexts the value of arable holdings might be further inflated by the rights of common pasture appurtenant to them. This may be why arable was valued so high relative to grassland in so much of the north and especially north-west of England and why pasture *per se* appears so fitfully in these northern extents (Figure 3.12).

Examination of the relative unit values of arable and grassland and the associated ratios of arable-to-meadow and meadow-to-pasture highlights the complexity of the relationships between them. Nationally, grassland had on average a unit value 4.1 times that of arable while meadow had a unit value 4.7 times that of arable and 6.2 times that of pasture. These ratios, however, were subject to considerable spatial variation (Figures 3.12 and 3.13). For instance, in several distinctive localities within eastern and south-eastern England – the East Anglian silt fens, the Flegg district of eastern Norfolk, northern and eastern Kent including the Isle of Thanet, and the Walland and Romney Marshes in south-east Kent – the unit values of arable and grassland approached parity. Indeed, in a few extreme instances the arable actually exceeded the meadow in unit value.[93] In all of these locations this exceptional state of affairs arose from the conjunction of very highly valued arable with more modestly valued grassland. The high arable values reflected its quality,

[91] Postan, *Economy and society*, pp. 59–60.
[92] Adam Smith, *The wealth of nations, Book I*, new edition (Edinburgh, 1872), p. 69.
[93] Kosminsky, *Studies*, p. 50, believed that the value of meadow land 'is always higher than that of the arable', yet instances of arable land being assessed more highly than meadow in the same vicinity include Burnham, Halvergate, Hickling, the Hundred of Flegg, Scratby, South Walsham, and Upton (all Norfolk), Appleby, Frampton, and Risgate in Lincolnshire, and Great Sutton in Essex: PRO, C133 File 80 (1), C132 File 38 (17), C134 Files 75 (22), 77 (5), C133 File 48 (10), C132 File 38 (17), C134 File 76 (1), C133 File 42 (5), C134 Files 34 (4), 59 (12), C133 File 70 (17). The bulk of these examples come from areas of rich coastal alluvium where highly fertile arable land existed in close juxtaposition with a relative abundance of meadow and pasture.

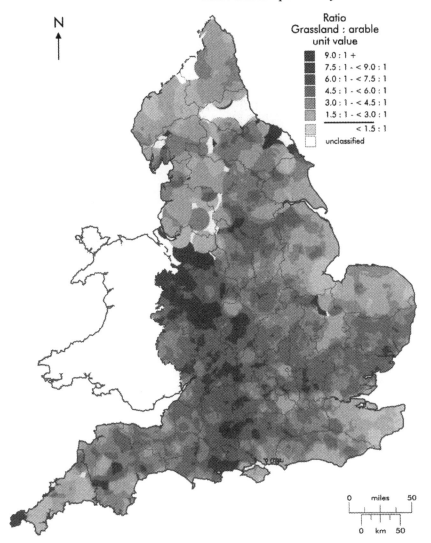

Fig. 3.12. Ratio of the unit value of grassland to the unit value of arable: England, 1300–49 (source: National *IPM* database).

productivity, and profitability; the modest grassland values, either its unusual abundance or inferior quality.

Inland from each of these distinctive coastal localities lay several more extensive areas within which the differential between the unit values of grassland and arable was narrower than that within the country as a whole. Arable of above average value was the common denominator of each. In most of Kent, outside the coastal marshlands, highly valued arable co-existed with

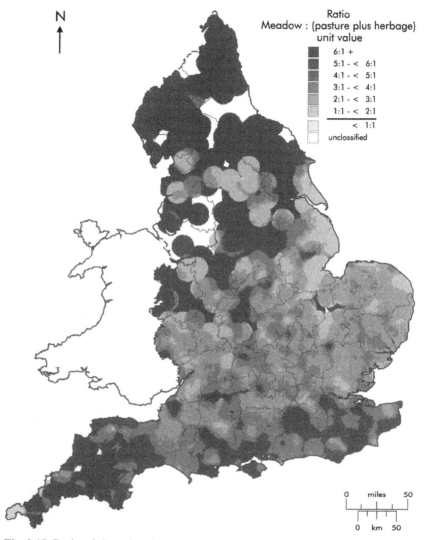

Fig. 3.13. Ratio of the unit value of meadow to the unit value of pasture/herbage:
England, 1300–49 (source: National *IPM* database).

grassland which, although available in relative abundance, was poor in quality
and low in unit value. A similar relative abundance of grassland in conjunction with arable of above-average value also pertained in much of
Lincolnshire but with the difference that here the grassland was generally of
much better quality. In most of Norfolk and Cambridgeshire, however, grassland was in exceptionally limited supply. Its quality, moreover, was often
indifferent. Scarcity of grassland, therefore, did not translate into high unit

values. Arable, in contrast, was generally of superior productivity; hence the differential in unit value between the arable and grassland remained a relatively narrow one. Indeed, but for the scarcity of the grassland it would doubtless have been narrower still. On the boulder-clay soils of neighbouring Suffolk, Essex, and Hertfordshire, in contrast, this qualitative differential between arable and grassland tended if anything to be reversed. Arable existed in quantity but was of indifferent quality; grassland was much more circumscribed in supply, but evidently highly productive and correspondingly highly valued. Suffolk boasted the highest-valued meadowland in the country (Figure 4.15), indicative of a highly developed pastoral sector.[94] Here, therefore, a wide differential existed between the unit values of arable and grassland.

Much the same applied on heavy land throughout the interior counties of lowland England. High costs of cultivation, in conjunction with prices that were low by comparison with other parts of England, kept arable unit values down; at the same time, for the reasons stated by Adam Smith, scarce grassland of quality commanded a disproportionately high rent. Nowhere was this differential wider than on the poorer soils in the land-locked, ox-ploughing, and open-field west midlands. Here was some of the lowest-valued arable land in the country. To the immediate north-west, however, the differential between the unit values of arable and grassland narrowed once again, as befitted a predominantly pastoral region where scarce but relatively valuable arable co-existed with abundant but relatively poor grassland. A similar explanation would appear to fit south Wales, parts of Devon, and much of Cornwall. Here, but for very different reasons, the narrow differential between the unit values of arable and grassland echoes that of the profitable arable-farming districts on the opposite side of the country.

To suppose that scarcity alone drove up unit land-values is consequently far too crude. In practice, abundance was more likely to depress the value of arable, meadow, or pasture than scarcity was to raise it. This was because landuses were often poor in quality as well as scarce in supply. The inherent quality and productivity of any given land-use was therefore an important determinant of its unit value. So, too, were the uses to which it was put and the productivity and profitability of those uses. Where husbandry systems were intensive and productive and their products commanded a high price, unit land-values, whether of arable or grassland, tended to be comparatively high.[95] Variations in unit land-values were thus in part a function of variations

---

[94] Campbell, 'Chaucer's reeve', pp. 284–8.

[95] As Raftis, *Assart data*, p. 73, observed of pasture values in Leicestershire and Rutland, 'the future bias towards a pastoral economy in this part of the East Midlands is already apparent'. Mavis Mate, 'Profit and productivity', p. 329, has also speculated that 'the shortage of pasture was not caused, as Sir Michael Postan thought, primarily by the encroachment of grain upon grass, but [by] the continued expansion of the animal population'.

in farm enterprise and thereby of the economic factors by which they in their turn were in part determined. Unit land-values were also, to a lesser extent, a function of property rights. Producers could not realise the full value of their land when others were entitled to rights of common grazing on it. Other things being equal, land held in severalty was always worth more to the owner than land held in common, except when rights of common grazing appurtenant to that land were incorporated in its valuation.

3.4   Demesne land-use combinations and their geography

There was no standard portfolio of demesne land-uses. The ways in which arable, pastoral, and woodland land-uses were combined varied a good deal from place to place, as did the relative unit values of those land-uses. Using value to quantify different land-uses reveals much about their relative importance but can be distorted by the tendency for scarce land-uses to acquire borrowed value from those that were in abundance. Quantifying land-use in terms of area reveals another aspect of the equation, but is dependent upon the availability of appropriate information. One way of summarising so much diversity is in terms of a land-use classification which takes into account the presence or absence of the principal land-uses and the relative areas, values, and unit values of the various sub-components of the arable/pastoral equation.[96]

The classification of demesne land-use employed here and summarised in Table 3.04 and Figure 3.14 is based upon the twelve components of land-use set out in Table 3.05. The way in which each is specified is determined by the manner in which the relevant information is recorded in the *IPM*s. Four components measure the relative composition of arable and pastoral land-use in terms of some specified combination of either area or value. Since these require comparable information for more than one major category of land-use they are based upon a sub-set of the most complete *IPM*s relating to properties with the territorial and jurisdictional attributes of full manors. Three further variables relate to the presence or absence of woods and private hunting grounds and a fourth to the presence or absence of such minor categories of land-use as bogs, turbaries, bracken, rushes, reeds, and warrens. In their case analysis is extended to all manors and all relevant *IPM*s. The same is true of the four last variables, which encapsulate different aspects of the relative and absolute unit values of arable and pastoral land-uses. Here the value per acre of meadow is included as a control on the three ratios, meadow being chosen for this purpose because of the superior accuracy and consistency with which it is recorded. These twelve variables have been differentially weighted,

[96] K. C. Bartley and B. M. S. Campbell, '*Inquisitiones post mortem*, G.I.S., and the creation of a land-use map of pre Black Death England', *Transactions in G.I.S.* 2 (1997).

Table 3.04. *Component variables used for the classification of demesne land-use*

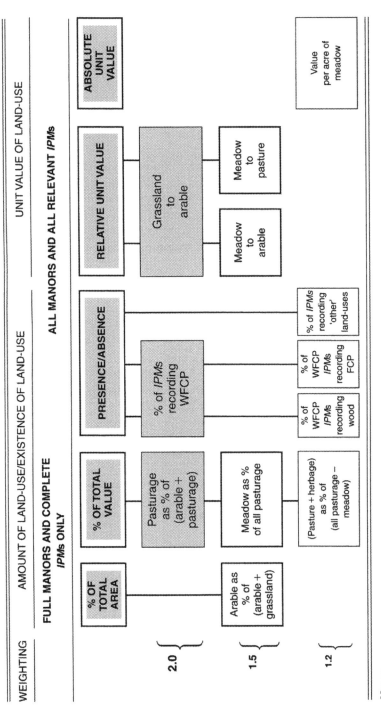

Notes:
Pasturage = grassland, 'land', forelands, verges, herbage, waste, heath, broom, moor, marsh; grassland = meadow, pasture; 'other' land-uses = bog, turbary, bracken, reeds, rushes, warrens, heronries; WFCP = wood, forest, chase, park; FCP = forest, chase, park.

as indicated in Table 3.04, so that those deemed to be of greatest diagnostic significance exercise the greatest influence upon the final land-use classification.

The classification itself has been derived using cluster analysis since this technique provides a means of generating a solution that reflects genuine differences inherent within the data.[97] In this case the optimum solution is a six-fold classification of demesne land-use types which when mapped in Figure 3.14 reveals considerable spatial differentiation of land-use. Since location *per se* was not a component of the classification there is a clear implication that this pattern is genuine. Spatial differentiation shows up both at a broad regional level and more locally in the many sub-gradations of land-use. Land-use variation, it will be noted, was greater around the coast than it was within the land-locked interior of England. This was partly because of the natural diversity of coastal and estuarine environments but also because coastal areas were more fully exposed to the differentiating effects of market forces. The six basic demesne land-use types may be broadly defined as follows:

Type 1– Poor land with low unit values
This land-use type was limited to a few highly specific but quite widely separate locations, notably south-western Surrey, the New Forest in southern Hampshire, the Isle of Purbeck in Dorset, the Brendon Hills in Somerset, south-eastern and south-western Devon, the upper Severn Valley in Shropshire, and the Huntingdonshire fen edge. A version of the same land-use type also prevailed in the East and North Ridings of Yorkshire and several other localities in the north of England. Unit values of meadow, pasture, and arable well below the national average are its most distinguishing feature, with, nevertheless, an exceptionally wide differential between the unit values of meadow and arable possibly betokening some transference of value to the former from the latter. These characteristics are consistent with the known physical limitations of most of these areas, especially with respect to arable husbandry.

Type 2 – Open arable country with limited differentiation of unit land-values
This land-use type was most widely represented in Lincolnshire, the Isle of Ely, and southern Cambridgeshire. Other occurrences include the limestone country of central Northamptonshire and mid-Buckinghamshire, the Vale of the White Horse in Berkshire, the Dorset Downs, and the immediate environs

---

[97] The initial classification has been restricted to England south of the Trent and east of the Tamar since this is the area for which the *IPM*s give fullest and most consistent coverage. The north of England and Cornwall have then been classified using discriminant functions derived from the original classification. See Bartley and Campbell, 'Creation of a land-use map', 44; Chapter 2, p. 342.

Table 3.05. *Core demesne land-use types, 1300–49: variable means and standard deviations*

| Land-use variable | Weighting | Land-use type | | | | | | All land-use |
		1	2	3	4	5	6	
Value of pasturage as % of (arable + pasturage)	2.0	32.0	36.7	31.3	34.5	38.9	22.3	32.4
		13.2	*12.6*	*8.4*	*8.9*	*9.5*	*7.3*	*10.8*
% of *IPMs* recording WFCP	2.0	36.8	13.9	45.8	55.8	30.9	25.8	36.9
		12.6	*8.2*	*14.0*	*10.8*	*12.2*	*14.1*	*18.0*
Unit value of grassland:unit value of arable	2.0	6.4	2.7	5.6	1.8	4.0	2.6	4.3
		2.8	*1.0*	*1.5*	*0.8*	*1.2*	*0.8*	*2.1*
Area of arable as % of (arable + grassland)	1.5	85.1	81.6	93.2	69.8	86.3	91.5	87.1
		16.2	*15.7*	*6.8*	*8.5*	*11.9*	*7.8*	*12.9*
Value of meadow as % of all pasturage	1.5	61.2	76.5	65.1	28.6	67.2	44.3	60.3
		24.2	*11.9*	*14.6*	*13.6*	*15.5*	*15.2*	*20.8*
Unit value of meadow:unit value of arable	1.5	7.0	3.1	6.7	3.7	4.5	3.3	5.2
		2.6	*1.1*	*1.5*	*1.3*	*1.4*	*1.2*	*2.2*
Unit value of meadow:unit value of pasture	1.5	12.3	3.3	2.9	5.9	4.2	2.5	4.3
		3.0	*3.0*	*1.0*	*2.9*	*2.4*	*0.8*	*3.4*
Value of (pasture + herbage) as % of (all pasturage – meadow)	1.2	64.4	74.3	92.9	67.8	90.6	58.0	80.3
		32.9	*28.3*	*11.7*	*29.5*	*13.7*	*25.0*	*25.5*
% of WFCP *IPMs* recording wood	1.2	93.7	98.6	95.3	92.4	73.3	98.6	92.5
		7.8	*4.4*	*4.9*	*8.4*	*14.6*	*3.0*	*11.2*
% of WFCP *IPMs* recording forest, chase or park	1.2	15.9	4.0	24.0	21.2	60.0	22.2	25.2
		12.5	*6.6*	*13.9*	*13.3*	*14.9*	*17.2*	*20.9*
% of *IPMs* recording other land-uses	1.2	2.4	2.3	2.0	2.2	2.0	19.3	4.2
		2.7	*2.3*	*1.9*	*1.8*	*2.1*	*6.6*	*6.4*
Value per acre of meadow (d.)	1.2	13.6	15.9	22.1	19.4	15.3	15.5	18.3
		3.4	*3.2*	*3.4*	*5.7*	*3.3*	*2.8*	*4.9*
No. of cases		302	447	1,243	332	468	395	3,187

Notes:
Roman = mean; *italics* = standard deviation. Pasturage = grassland, 'land', forelands, verges, herbage, waste, heath, broom, moor, marsh; grassland = meadow and pasture; other land-uses = bog, turbary, bracken, reeds, rushes, warrens, heronries. Land-use types are explained on pp. 96–100. WFCP = wood, forest, chase or park.
Source: National *IPM* database.

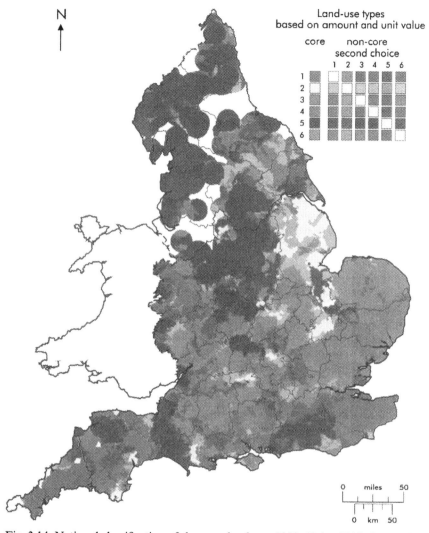

Fig. 3.14. National classification of demesne land-use, 1300–49 (see Table 3.05 and pp. 96–100 for explanation of land-use types; p. 44 for 'core' and 'non-core' classifications) (source: National *IPM* database).

of Exeter in Devon. The most striking features of this land-use type are the scarcity of woods and private hunting grounds, the relatively narrow differential between the unit values of meadow and arable (indicative of the abundance or limited quality of one and profitability of the other), and the dominance of pasturage by grassland and particularly by meadow.

Type 3 – Arable country with limited but valuable grassland
This was the characteristic land-use type of much of the heavier land of lowland England, especially inland and away from the coast. No other land-use type was as well or as widely represented. It was the predominant land-use of the closely settled East Anglian boulder-clay plateau stretching in a broad swathe from south-east Norfolk across Suffolk and Essex into eastern Hertfordshire. Other notable occurrences included the lower Welland Valley of Northamptonshire and adjacent Rutland, northern Bedfordshire, north-eastern Buckinghamshire and south-western Northamptonshire, the Avon Valley of south Warwickshire and Worcestershire, the mid-Severn Valley, eastern Shropshire, the Vale of Oxford, and substantial portions of the Hampshire and Wiltshire Downs. It was the typical land-use type of much of commonfield England as well as of several extensive areas outside that zone. Arable was everywhere present in above average quantities while grassland, partly because it was scarce and carefully managed, was valuable, especially the meadow. A higher than average number of demesnes also possessed some woodland, although as with the grassland the quantities were mostly small.

Type 4 – Superior arable with several pasture and wood
This minority land-use type is associated almost exclusively with central and eastern Kent and neighbouring portions of Sussex. Elsewhere there are hints of it in south-eastern Hampshire and parts of Devon and Cornwall. These are all areas where, 'waste' apart, the bulk of all land was held in severalty. In the absence of much or any common pasture, demesnes in these areas were endowed with well above average quantities of several pasture. No doubt for similar reasons, woods and private hunting grounds were also exceptionally well represented. Equally distinctive was the high unit value of arable, resulting in a narrower differential between the unit values of arable and grassland than in any other land-use type (Table 3.05).

Type 5 – Inferior arable and pasturage with private hunting grounds
This land-use type is well represented in several widely separate areas: the West Sussex Weald, a belt of country extending from the Mendip Hills southwards across mid-Somerset into west and south Dorset, the hilly country of north Devon, north Oxfordshire and the Stour Valley of south Warwickshire, a scattering of locations along the Welsh border, and much of the north midlands. Unit values of arable, meadow, and pasture all tended to be below average. Grassland was present in above-average quantities and was complemented by several other forms of pasturage. Private hunting grounds, enclosed from the waste and wood, were also quite a typical feature of these areas. A version of this demesne land-use type was also widespread throughout much of the north of England.

Type 6 – Open arable country with assorted lesser land-uses
This was the predominant land-use type of Norfolk and was scarcely ever encountered outside that county, the only notable exceptions being the immediately neighbouring portion of the Suffolk Breckland and the Sandlings of south-east Suffolk which shared many of the same physical characteristics (Figure 3.14). Within Norfolk this land-use type was pre-eminent everywhere apart from on the stiff boulder-clay soils of the south-east, the Norfolk Fenland in the extreme west, and along the boulder-clay watershed which separated east from west Norfolk. It was therefore most characteristic of soils which were either light or medium and free draining. In both parts of the county demesne land-use was overwhelmingly dominated by arable. Several grassland was present in below-average quantities and on the evidence of mean land-values was often of inferior quality. The differential between the unit values of meadow and pasture was smaller than in any other land-use type. Woodland was also under-represented by national standards and it is clear that much of the countryside wore a very open aspect. In this respect, together with the level and general relationship of unit land values, there are strong affinities with land-use type 2. Where these two land-use types differ is in the balance struck between arable and grassland, the quantity and quality of their meadowland, and, above all, in the relative importance of such minor land-use resources as warrens, turbaries, and rushes. The last feature more prominently in land-use type 6 than in any other. It is the prominence of such miscellaneous land uses that explains why an analogous land-use type shows up in Cornwall, at the opposite end of the country.

The common denominator of each of these major land-use types is their domination by arable. Everywhere demesne grassland was greatly inferior to the arable in physical extent while the combined value of all sources of pasturage was rarely more than half that of the arable. North, south, east, or west, the typical demesne was an arable concern. It is small wonder, therefore, that customary labour services almost everywhere were directed towards performance of the key arable tasks of ploughing, sowing, harrowing, weeding, harvesting, and carting. Such services, until commuted, reinforced the existing land-use bias towards arable. Of course, the omission of common pastures means that this is only a partial picture, although only in certain specific localities and regions would the inclusion of common pasture have tipped the land-use balance in favour of the pastoral sector. Specialist seigniorial pastoral farms are certainly known to have existed but vaccaries and bercaries are surprisingly poorly represented among the class of demesne recorded by the *IPM*s.[98]

[98] Tupling, *Rossendale*, pp. 17–33; Shaw, *Royal Forest*, pp. 353–80; Blanchard, 'Economic change in Derbyshire', pp. 164–91; Miller, 'Farming: northern England', pp. 409–11; Atkin, 'Land use and management'.

Many Cistercian granges were of this type, their exclusively pastoral function often constituting the most rational use of their specific endowments of land and labour. Again, however, they are poorly served by surviving accounts.[99]

The consistency with which the arable dominated all six basic demesne land-use types is one of the most striking features of this classification. Insofar as land-use varied locally and regionally, the key differences derived in the main from the precise combination of pastoral land-uses, the relative unit values of the arable, meadow and pasture, and the presence or absence of woods and private hunting grounds. Such differences as existed were consequently more a matter of degree than of kind with the result that across wide areas of lowland England demesne land-use was often more remarkable for its sameness than its variety. This is hardly surprising given that neither arable nor pastoral husbandry was easily conducted in complete isolation from the other. Draught animals and animal manure were vital to arable husbandry just as pastoral enterprises relied upon fodder crops and straw to help see their animals through the winter and restore them to strength in the spring. Virtually all demesnes therefore found it necessary to combine arable and pastoral land-uses in some measure. How these land-uses were then exploited was another matter. Pastoral production could serve as a mere servant of the arable or it could be developed as an important enterprise in its own right.

---

[99] Donkin, *The Cistercians*, pp. 51–82; C. Platt, *The monastic grange in medieval England* (London, 1969), pp. 183–245.

4

Seigniorial pastoral production

Lords, their immediate household members, and their estate officials were medieval England's greatest *per capita* consumers of livestock and their products.[1] First and foremost, draught animals were required in quantity to work the land and provide transport. Many conventual households maintained expensive cart-horses on their demesnes so that provisions could be delivered in bulk to the central household and cash crops – typically grain and wool – carried to market.[2] Lay lords with substantial itinerant households also required significant numbers of cart and pack animals for their periodic removals from manor to manor. Moreover, it was lords who kept and rode the most prestigious and expensive horses, spending, on Dyer's reckoning, about a tenth of all expenditure on the marshalsea or stable department alone.[3] Good riding horses were costly both to buy and maintain and purpose-bred destriers or war-horses even more so.

Some lords spent as much on textiles – silks, linens, worsteds and, above all, woollens – as they did on transport.[4] What they wore was as much a badge of status as what they rode. Accordingly, they bought woollen cloth in quantity to provide raiment for themselves and their families, liveries for their servants and followers, and a variety of hangings and coverings for their residences. Leather, from hides, was another important article of aristocratic dress, an essential component of all types of armour, and the raw material out of which the finest saddlery and harness were fashioned. It was animal skins, too, which furnished the parchment and vellum consumed in quantity by a class increasingly dependent upon written records.

From extant household and other accounts it is plain that aristocratic diets were dominated by meat, substituted with fish on holy days and during Lent.[5]

---

[1] Dyer, *Standards of living*, pp. 55–8; *Household accounts from medieval England*, ed. C. M. Woolgar, 2 vols. (Oxford, 1992, 1993).

[2] E.g. Biddick, *The other economy*; Campbell and others, *Medieval capital*, pp. 56–8.

[3] Dyer, *Standards of living*, p. 71.   [4] Dyer, *Standards of living*, p. 78.

[5] Dyer, *Standards of living*, pp. 58–62; Grant, 'Animal resources', pp. 161–75.

Beef was evidently the prime meat eaten, followed by pork, often purpose-produced for the table. Mutton was not much eaten by the aristocracy, but game – produced in the many private hunting grounds and warrens – was.[6] Many religious households also found means of incorporating more meat into their diets than was strictly permissible under the rule of St Benedict. They supplemented it with a higher *per capita* intake of lard and dairy produce than was normal in equivalent lay households where such foodstuffs were not held in high regard.[7] Above all, lords were great consumers of cash. Sales of livestock and livestock products produced on their estates were consequently a potentially lucrative source of revenue.[8]

Seigniorial producers therefore managed their pastoral resources to produce a range of commodities. Horses and oxen supplied draught power to be used both on and off the farm. Cattle, sheep, and swine produced meat, although they reproduced it at different rates. Mature female cattle and sheep yielded milk, which could be consumed either unprocessed or processed into butter and cheese. From cattle came tallow and from swine lard. Sheep alone produced wool. All animals produced dung while alive and a skin or hide once dead, from which a variety of products could be manufactured. Finally, poultry of various sorts were kept for eggs, meat, and feathers. Cattle and sheep were the most versatile of animals since they could be managed to yield the greatest variety of products. The effectiveness with which they did so nevertheless depended upon the breed, different breeds having been developed for different purposes.[9] Because the types and breeds of livestock varied in the commodities they produced, their hardiness, their feed requirements, and their suitability to different terrains, it was a rare farm which did not stock a range of animals. Only the most intensive, however, stocked the horses, dairy, neat and store cattle, sheep, swine, and poultry which Chaucer describes as in the care of Oswald, the reeve of Bawdeswell in Norfolk.[10]

4.1   Types of pastoral husbandry

The principal types of pastoral husbandry practised on demesnes before and after 1350 can be established by applying cluster analysis (Relocation Method) to the livestock statistics collectively contained in the national, FTC, and Norfolk samples of manorial accounts.[11] These statistics enumerate the types and numbers of animals on the demesne at the end of the farming and

---

[6] Campbell, 'Measuring commercialisation', pp. 164, 168; Grant, 'Animal resources', pp. 150–5, 164–7.

[7] Biddick, *The other economy*, pp. 38–40, 137–9; Dyer, *Standards of living*, pp. 56, 63–4.

[8] Campbell, 'Measuring commercialisation', pp. 147–53; below, pp. 183–7.

[9] R. Trow-Smith, *A history of British livestock husbandry to 1700* (London, 1957), pp. 160–6; M. L. Ryder, 'Medieval sheep and wool types', *AHR* 32 (1984), 14–28.

[10] Campbell, 'Chaucer's reeve', pp. 271–84.

[11] Chapter 2; Campbell and Power, 'Mapping'; cf. cropping classification, Tables 6.01, 6.02.

financial year, which in the vast majority of cases was Michaelmas (29th September). They omit, therefore, any animals which may have been on the demesne at other times of the year, especially those moved from manor to manor as part of an inter-manorial system of management.[12] Unless accounted for independently, these inter-manorial flocks and herds are accredited to whichever manor they were on at Michaelmas, even though they were supported by more than that one manor's resources. Sheep are most likely to be affected in these ways, especially on those estates which managed and accounted for them centrally. On the Fenland properties of Crowland Abbey central sheep accounting was initiated as early as 1276, it was adopted by Norwich Cathedral Priory on its Norfolk manors in 1392, and was firmly in place on the estates of Pershore and Winchcombe abbeys by 1415 and 1435 respectively.[13] A few of these central accounts do survive but none are included in this sample.[14] Asset stripping by the crown during a minority, wardship or ecclesiastical vacancy could also make the pastoral husbandry of a demesne appear more rudimentary than was normally the case.[15] Yet although the pastoral husbandry of some individual demesnes may as a result be mis-characterised, the overall picture of the seigniorial sector as a whole is unlikely to be seriously distorted. The numbers of demesnes in both combined samples are sufficiently great – 836 before 1350, 519 after 1349 – and, although spatially biased towards East Anglia and the south-east, they are wide enough in geographical coverage to ensure reliability.

For the purposes of classification and comparison it is necessary to convert the raw numbers of livestock of each type into standardised 'livestock units'. The units employed here – horses $\times 1.0$; bulls, oxen and cows $\times 1.2$; immature cattle $\times 0.8$; sheep $\times 0.1$; swine $\times 0.1$ – have been adapted from those employed by historians of seventeenth-century agriculture to investigate the livestock statistics contained in probate inventories. J. A. Yelling, for instance, used conversion ratios of 1.0 for horses and colts; 1.2 for oxen, bullocks, and steers; 0.8 for all other cattle; 0.1 for sheep and lambs; and 0.1 for swine, in his analysis of agriculture in Worcestershire between 1540 and 1750.[16] These are derived from modern scientific ratios based upon feed equivalents. R. C. Allen has

---

[12] A classic example of this is provided by the prior of Norwich's small demesne at Thornham in north-west Norfolk where sheep cease to appear in the accounts after 1317. In fact, sheep continued to be kept on the manor but were accounted for on the neighbouring manor of Sedgeford, which was the prior's principal sheep manor: NRO, DCN 60/37/1–21, 60/33/1–30; L'Estrange IB 1/4.

[13] Page, 'A medieval sheep farm', pp. 603–5; NRO, L'Estrange IB 3/4; R. A. L. Smith, 'The estates of Pershore Abbey', MA thesis, University of London (1939), pp. 215–16; R. H. Hilton, 'Winchcombe Abbey and the manor of Sherborne', *University of Birmingham Historical Journal* 2 (1949–50), 50–2.

[14] Allison, 'Sheep-corn husbandry'; NRO, DCN 62/5, 12, 16, 17, 19, 22, 23, 25, 28, 29; 64/1–12.

[15] E.g. Titow, 'Land and population', pp. 45–6; Biddick, 'Agrarian productivity', pp. 98–104.

[16] J. A. Yelling, 'Probate inventories and the geography of livestock farming: a study of east Worcestershire, 1540–1750', *TIBG* 51 (1970), 115.

used identical ratios in his analysis of seventeenth-century livestock in the south midlands.[17] They contrast with ratios of 1.0 for horses, oxen, and cattle, and 0.25 for sheep, initially adopted by Titow in his study of the Winchester estates and subsequently employed by Farmer.[18] Not only do these omit swine, they are also needlessly crude given the detail with which livestock are actually recorded in manorial accounts, and almost certainly attach too much weight to sheep.[19] Thus, a comparative analysis of flock and herd sizes indicates a generally ten-fold differential between them in the period 1250–1349.[20] After 1350, when sheep gained in importance relative to cattle, flocks were on average eleven times larger than herds.[21] These figures suggest that a ratio of approximately ten sheep per head of cattle is of the right general order of magnitude.[22]

Some indication of how contemporaries viewed the relative feed requirements of cattle and sheep is provided by rates of payment for agistments. At Halvergate, Cawston, Forncett, South Lopham, and Eccles in Norfolk in the late thirteenth and early fourteenth centuries such payments suggest that oxen, cows, and plough-horses were expected to eat between five and eight times as much as a ewe and at least ten times as much as a lamb.[23] Similarly, on the estates of northern Cistercian abbeys cattle were reckoned to have a grazing requirement four to eight times that of sheep.[24]

Another approach altogether is to weight animals by type, age, and sex according to their relative sale prices as recorded in the accounts. For the FTC counties this yields the following detailed weightings:[25]

---

[17] Allen, *Enclosure and the yeoman*, p. 194; R. C. Allen, *The 'capital intensive farmer' and the English agricultural revolution* (Vancouver, 1987), pp. 27–33.

[18] Titow, *Winchester yields*, pp. 136–9; Farmer, 'Grain yields on Winchester manors', pp. 563–4; Campbell, 'Agricultural progress', pp. 29–31.

[19] For similar ratios, see M. E. Turner, 'Livestock in the agrarian economy of Counties Down and Antrim from 1803 to the Famine', *Irish Social and Economic History* 11 (1984), 29–30.

[20] Mean of 31.3 cattle per herd (oxen omitted); mean of 304.1 sheep per flock: 95 per cent of herds contained fewer than ninety animals and 95 per cent of flocks fewer than 900 animals: National accounts database.

[21] Mean of 35.1 cattle per herd (oxen omitted); mean of 397.3 sheep per flock: 95 per cent of herds contained fewer than ninety animals and 93 per cent of flocks fewer than 900 animals.

[22] Clark, 'Labour productivity', p. 217, estimates that *c.* 1300 'cattle had a carcass weight about 8.7 times that of a sheep'.

[23] PRO, SC 6/936/17, 6/1090/4; Davenport, *Norfolk manor*, p. xxxi; PRO, SC 6/937/29; Raynham Hall, Norfolk, Townshend MSS; PRO, SC 6/934/14.

[24] Donkin, *The Cistercians*, p. 71, n. 4.

[25] These are 1.2 times greater than the weightings given in Campbell, 'Measuring commercialisation', p. 163, n. 84. For the mean sale price of animals in the London region, see M. Murphy and J. A. Galloway, 'Marketing animals and animal products in London's hinterland *circa* 1300', *Anthropozoologica* 16 (1992), 97. Corresponding modern weightings based on feed-equivalents are: horses × 1.00; cows, bulls and other cattle two years old and over × 1.00; other cattle one year old × 0.67; other cattle under one year old × 0.33; rams and wethers × 0.10; ewes × 0.20; other sheep × 0.067; boars × 0.25; sows × 0.50; other pigs × 0.14: T. R. Coppock, *An agricultural atlas of England and Wales* (London, 1964), p. 213.

horses and cart-horses	×1.20
stots	×0.68
affers, plough-horses, and mares	×0.60
young horses	×0.40
oxen	×1.20
bulls	×1.12
cows	×0.86
bovecti and *juvencule*	×0.68
boviculi	×0.56
juvence	×0.68
yearling cattle	×0.30
calves	×0.11
rams and wethers	×0.13
ewes	×0.12
hoggets	×0.11
gimmers	×0.10
lambs	×0.06
goats	×0.10
boars	×0.46
sows	×0.50
pigs	×0.26
hogs	×0.20
young pigs	×0.13
piglets	×0.11

These confirm a wider differential between cattle and sheep than that envisaged by Titow and, interestingly, suggest that, far from being omitted, swine should be given twice the weighting of sheep.[26] Unfortunately, neither the national nor the Norfolk livestock data were collected in sufficient detail to permit the application of such a detailed set of ratios. Moreover, these ratios are better suited to a more fine-tuned analysis of the pastoral sector than that for the most part offered here.[27]

Separate classification of the two composite samples identifies six basic pastoral types (Table 4.01). Five are common to both periods, while one is specific to the earlier. The first and most conspicuous source of differentiation between pastoral types is the relative importance of working and non-working animals. In pastoral types 1 to 4, comprising over three-quarters of all sampled demesnes in the period 1250–1349 and more than four-fifths of demesnes in the period 1350–1449, non-working animals account for between a half and four-fifths of all livestock units. In pastoral types 5 and 6,

[26] At Forncett in Norfolk in 1272–3 agistment payments for swine were ten times those for sheep: Davenport, *Norfolk manor*, p. xxxi.   [27] But see Tables 4.05, 4.06, and 4.09.

Table 4.01. *National classification of pastoral-husbandry types, 1250–1349 and 1350–1449*

Variable		Pastoral type					All[6]
	1	2	3	4	5	6	
1250–1349:							
Working animals[1] as % of livestock units[2]	*20.9*	*45.6*	*35.9*	*54.3*	*90.6*	*98.1*	*53.4*
Oxen per 100 horses	*<1*	*286*	*360*	*524*	*508*	*698*	*399*
Oxen as % of total cattle units[3]	*0.1*	*38.1*	*47.3*	*95.5*	*98.2*	*97.7*	*58.5*
Adults as % of non-working cattle[4]	*56.6*	*56.2*	*61.7*	*7.1*	*9.7*	*1.4*	*40.8*
Cattle as % non-working units[5]	*55.6*	*85.8*	*42.6*	*3.4*	*9.5*	*99.6*	*60.0*
Sheep as % non-working units[5]	*38.2*	*7.9*	*51.9*	*88.5*	*1.8*	*0.0*	*29.8*
Swine as % non-working units[5]	*6.2*	*6.4*	*5.5*	*8.1*	*88.7*	*0.4*	*10.1*
No. of demesnes	*37*	*286*	*236*	*101*	*48*	*128*	*836*
% of total classified	*4*	*34*	*28*	*12*	*6*	*15*	*100*
1350–1449:							
Working animals[1] as % of livestock units[2]	*16.7*		*29.4*	*38.0*	*83.3*	*93.2*	*38.2*
Oxen per 100 horses	*<1*		*268*	*425*	*386*	*494*	*286*
Oxen as % of total cattle units[3]	*0.1*		*31.9*	*93.9*	*66.1*	*78.3*	*44.6*
Adults as % of non-working cattle[4]	*59.3*		*64.8*	*6.3*	*7.2*	*8.2*	*45.2*
Cattle as % non-working units[5]	*51.1*		*61.1*	*3.5*	*3.8*	*98.7*	*49.8*
Sheep as % non-working units[5]	*44.3*		*33.6*	*90.6*	*1.5*	*0.0*	*40.0*
Swine as % non-working units[5]	*4.5*		*5.3*	*5.9*	*94.7*	*1.3*	*10.3*
No. of demesnes	*72*		*278*	*90*	*31*	*48*	*519*
% of total classified	*14*		*54*	*17*	*6*	*9*	*100*

Method:
Cluster analysis (Relocation method) applied to the following variables: working animals as % of livestock units (×3); oxen per 100 horses (×2); sheep as % of (sheep + cattle) units (×2); swine as % of non-working units (×2); oxen as % of (oxen + cattle) units; adults as % of non-working cattle (omitting oxen).

Table 4.01. (*cont.*)

Notes:
[1] All horses plus oxen
[2] (Horses $\times 1.0$) + (oxen, cows, and bulls $\times 1.2$) + (immature cattle $\times 0.8$) + (sheep and swine $\times 0.1$)
[3] Cows, bulls, and immature cattle
[4] Oxen omitted (raw numbers, not units)
[5] Cows, bulls, immature cattle, sheep, and swine
[6] Unweighted by region
(The demesnes are listed in Appendix 1)
Sources: National accounts database; Norfolk accounts database; FTC1 and FTC2 accounts databases.

Fig. 4.01. Demesnes practising pastoral type 1, 1250–1349 (see Table 4.01 and pp. 106–20 for explanation of pastoral types) (source: National, Norfolk, and FTC1 accounts databases).

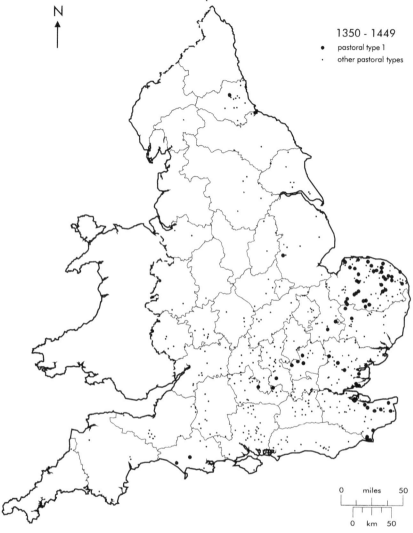

Fig. 4.02. Demesnes practising pastoral type 1, 1350–1449 (see Table 4.01 and pp. 106–20 for explanation of pastoral types) (source: National, Norfolk, and FTC2 accounts databases).

Fig. 4.03. Demesnes practising pastoral type 2, 1250–1349 (see Table 4.01 and pp. 106–20 for explanation of pastoral types) (source: National, Norfolk, and FTC1 accounts databases).

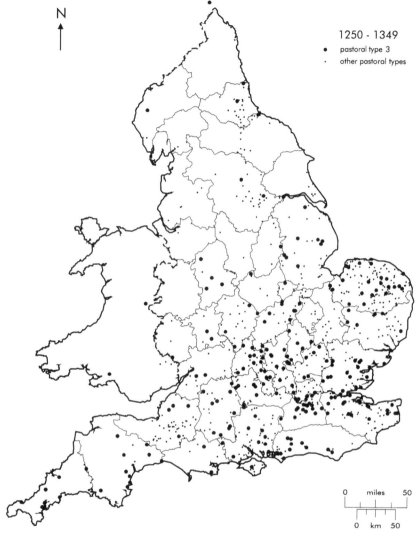

Fig. 4.04. Demesnes practising pastoral type 3, 1250–1349 (see Table 4.01 and pp. 106–20 for explanation of pastoral types) (source: National, Norfolk, and FTC1 accounts databases).

Fig. 4.05. Demesnes practising pastoral type 3, 1350–1449 (see Table 4.01 and pp. 106–20 for explanation of pastoral types) (source: National, Norfolk, and FTC2 accounts databases).

Fig. 4.06. Demesnes practising pastoral type 4, 1250–1349 (see Table 4.01 and pp. 106–20 for explanation of pastoral types) (source: National, Norfolk, and FTC1 accounts databases).

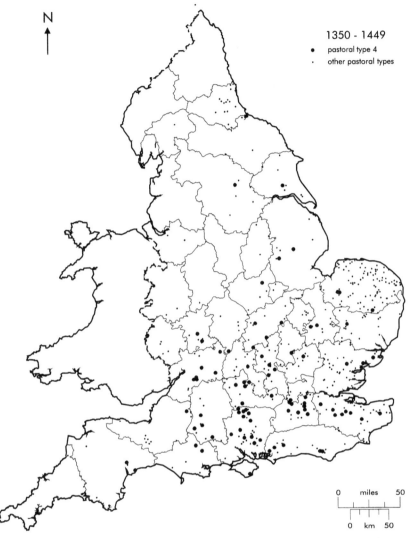

Fig. 4.07. Demesnes practising pastoral type 4, 1350–1449 (see Table 4.01 and pp. 106–20 for explanation of pastoral types) (source: National, Norfolk, and FTC2 accounts databases).

Fig. 4.08. Demesnes practising pastoral type 5, 1250–1349 (see Table 4.01 and pp. 106–20 for explanation of pastoral types) (source: National, Norfolk, and FTC1 accounts databases).

Fig. 4.09. Demesnes practising pastoral type 5, 1350–1449 (see Table 4.01 and pp. 106–20 for explanation of pastoral types) (source: National, Norfolk, and FTC2 accounts databases).

Fig. 4.10. Demesnes practising pastoral type 6, 1250–1349 (see Table 4.01 and pp. 106–20 for explanation of pastoral types) (source: National, Norfolk, and FTC1 accounts databases).

Fig. 4.11. Demesnes practising pastoral type 6, 1350–1449 (see Table 4.01 and pp. 106–20 for explanation of pastoral types) (source: National, Norfolk, and FTC2 accounts databases).

in contrast, working animals (horses and oxen) predominate, sometimes, as in the case of pastoral type 6, to the virtual exclusion of all others. Further differentiation arises from the composition of the working and non-working sectors. On the working front, there were demesnes which relied more or less exclusively upon oxen (pastoral types 4, 5, and 6), others which employed only horses (pastoral type 1), and by far the greater number which used varying combinations of the two (pastoral types 2 and 3). On the non-working front, there were demesnes which concentrated upon cattle, usually for breeding and/or dairying (pastoral type 2), others which specialised in sheep (pastoral type 4), a good number which combined cattle with sheep (pastoral types 1 and 3), and some, even, whose prime interest was in the production of swine (pastoral type 5).

4.2   Working animals

The dictates of arable husbandry in a horse- and ox-propelled age meant that working animals always occupied pride of place within the pastoral sector. They were the single common denominator of all pastoral types (Tables 4.01 and 4.02). Oxen and horses were required for ploughing, harrowing, carting, and a variety of other draught tasks and no farm could manage without them. They were the most valuable livestock and the most expensive to feed since the amount of work energy they produced was a direct function of the amount of food energy they consumed. Grain was consequently an essential component of their diets.[28] These working animals were invariably oxen or horses. Occasionally other categories of cattle might be pressed into service, and mules and donkeys do sometimes feature in grange accounts, but their overall contribution to demesne motive power was so small that it can safely be disregarded.[29]

Obviously, not all oxen and horses drew ploughs and hauled carts. Younger horses were not put regularly to work until they had reached maturity, usually in their fourth year. Additional horses might be kept as riding animals, a function which was not strictly justified by the needs of agriculture. Some demesnes with horses or oxen were rearing them for sale rather than maintaining them for work; they stand out in the record because the numbers stocked are in excess of the draught requirements of the area under crops.[30] When oxen reached the end of their working lives they were usually fattened up for meat but some were purpose-bred for the table and never put to the plough. This became increasingly the case after 1375, as tillage contracted, *per capita* incomes rose, and meat consumption expanded. With these mostly minor exceptions the bulk of the oxen and horses enumerated in the accounts owed

---

[28] Langdon, 'Economics of horses', pp. 32–5.   [29] Langdon, *Horses*, p. 87.
[30] Ormesby in east Norfolk, endowed with rich marshland grazings, became a significant producer of oxen in the 1430s: PRO, SC 6/939/11–13, 6/940/1–8.

their existence to the work requirements of their respective demesnes. In this most fundamental respect and to this extent livestock were subservient to tillage.

Of all the tasks which draught animals performed, ploughing was by far the most demanding. It was undertaken in both the autumn and the spring. Repeated summer ploughing was also the most effective method of cleansing intensively cropped land of weed growth.[31] Within lowland England as a whole there were on average 78.5 sown acres per demesne plough in the period 1250–1349, which declined by 15 per cent to 66.6 sown acres per plough in the period 1350–1449 (Table 4.02). When fallows are taken into account this is roughly equivalent to 110 to 130 arable acres per plough (i.e. remarkably close to a conventional carucate of 120 acres). An average demesne with at least 150 acres under crop would therefore have needed at least two and possibly even three plough-teams, depending upon the soil and terrain, the type of plough and composition of team, and any contribution made to demesne ploughing by tenant teams.[32] The course of cropping also made a difference. Two-course cropping and multi-course cropping tended to result in a marked seasonal asymmetry of ploughing need. Three-course cropping, on the other hand, divided ploughing requirements more equally between autumn and spring. This may be why three-course demesnes in Cambridgeshire, Suffolk, Essex, and Hertfordshire may have been able to get away with more sown acres per demesne plough than was the norm elsewhere in the country. On the other hand, on the evidence of the *IPMs* this was the area where customary plough-ings probably made their greatest contribution.[33] The 15 per cent reduction overall in the mean number of sown acres per demesne plough over the course of the fourteenth century may represent the progressive removal by commu-tation of that hidden subsidy. It can be hardly a coincidence that the reduc-tion in sown acres per plough was most marked in that wedge of eastern and midland counties where the *IPMs* reveal the performance of customary ser-vices to have been most prevalent.

For each plough there were on average eight to ten working animals (Table 4.02), at least one or two of which would almost always have been reserved for harrowing and carting rather than ploughing. Eliminating young horses and riding horses from the calculation reduces the mean number of plough animals (affers, stots, and oxen only) per plough to 9.0 in the period 1250–1349 and 7.8 in the period 1350–1449, a 13 per cent reduction. This shrinkage in mean plough-team sizes was most pronounced in the counties of eastern and central England where oxen were increasingly being replaced with horses over that period. Substitution of the horse for the ox was invariably undertaken with the aim of raising ploughing speeds and reducing team sizes. All the main

---

[31] Campbell, 'Agricultural progress', p. 29.   [32] Langdon, *Horses*, pp. 118–41.
[33] Campbell and Bartley, *Lay lordship, land, and wealth.*

areas of mixed or all-horse ploughing either had fewer teams or smaller team sizes than was normal in much of the rest of the country. Nowhere did this process proceed further than in Norfolk. Here, where horses earliest made a significant contribution to ploughing, mean team sizes shrank by a fifth from 4.9 animals in the period 1250–1349 to 3.9 animals in the period 1350–1449. By the close of the fourteenth century the number of sown acres per plough within the county was 10 per cent above the national average and the first one-man and two-horse ploughs had made their appearance on the county's lighter soil demesnes (Figures 4.13 and 4.14).[34]

Accelerating ploughing speeds facilitated bringing more land into cultivation and cropping it more frequently. That usually meant converting pasture to arable and reducing the frequency of fallowing to the minimum compatible with effective weed control. Since pasture was scarce and fallows now provided little or no forage the cultivation of fodder crops became inevitable. Because this imposed a significantly increased workload on the labour force it became important to convert that fodder into traction with the maximum degree of efficiency; hence the partial or complete substitution of the horse for the ox.[35] Such a changeover was further encouraged by the fact that the greater intensity of cropping entailed a much more demanding ploughing schedule with, often, a major seasonal imbalance between autumn and spring. This pattern of development proceeded furthest in east Norfolk and north-eastern Kent and in both cases demesnes eventually converted to all-horse ploughing (Figures 4.01 and 4.02).[36]

Speeding up ploughs and reducing team sizes also economised upon the share of pastoral resources which it was necessary to dedicate to the provision of draught power. Here, substituting the partially grain-fed horse for the largely grass-fed ox yielded a double bonus: not only were hay and grass released to the benefit of other categories of livestock, but fewer back-up animals were required for the reproduction of replacement draught beasts. Norfolk spearheaded the introduction of mixed- and all-horse ploughing. It was also one of England's most arable counties (Figure 3.11). Yet, paradoxically, working animals consistently accounted for a smaller share of demesne livestock than in any other part of the country. Whereas, nationally, working animals accounted for 60 per cent of demesne livestock units in the second half of the

---

[34] These small plough-teams feature in a number of contemporary Flemish illustrations of rural life, e.g. Holkham Hall, Norfolk, MS 311 fol. 41 verso; BL, Add. MS 24098 fol. 26b. In England they had undoubtedly long been a feature on peasant holdings, but as wage rates escalated towards the end of the fourteenth century even the demesne managers had to abandon their traditional preference for two-man ploughs. This can be seen on the abbey of St Benet at Holme's demesne at Flegg in Norfolk, where the changeover took place some time between 1380 and 1407: NRO, Diocesan Est/9. Whether the change to one-man ploughing required further refinements in plough design and harness is not as yet clear.

[35] In this context see the arguments advanced by E. Boserup, *The conditions of agricultural growth* (Chicago, 1965), pp. 35–9.

[36] Campbell, 'Agricultural progress'; Mate, 'Agrarian practices'.

thirteenth century, in Norfolk the equivalent proportion was only 35 per cent. By the first half of the fifteenth century the demands of tillage had everywhere contracted considerably releasing resources to the pastoral sector. Cart and plough-animals now accounted for 42 per cent of all demesne livestock nationally and a mere 20 per cent in Norfolk (Table 4.02). This is an impressively small proportion for such an arable county which compares favourably with the modest proportions of working animals maintained by Norfolk farmers throughout the seventeenth, eighteenth, and nineteenth centuries when the country was undergoing a fodder-based agricultural transformation.[37]

4.21 Draught horses

In the thirteenth century the horse's advent as a draught animal was still a comparatively recent phenomenon. Langdon has recently researched the history of its introduction to demesne husbandry with great thoroughness.[38] According to his estimates, at the time of Domesday horses accounted for little more than 5 per cent of total animal draught force on the demesne, and no more than 10 per cent in any of the regions for which there are figures (there are signs that the level of horses was already higher among peasant draught stock, but this cannot be quantified). In contrast, by the beginning of the fourteenth century horses accounted for at least 20 per cent of the animal draught force on demesnes and almost 50 per cent on peasant farms, and these figures exceeded 50 per cent and 75 per cent respectively in certain regions.[39] Moreover, only 5 per cent of demesnes kept no horses at all. The ubiquity of the horse derived from the fact that it came in a variety of types suited to different purposes. Along with the affers and stots and associated mares and younger animals which collectively belonged to the category of work horse (mares offered demesne managers the double advantage that they could both work and reproduce), there were the more expensive and powerful cart-horses, plus pack horses and riding horses. Demesnes might therefore keep horses for haulage but not for traction, or harrow with a horse while ploughing with oxen, or keep no working horses at all and merely maintain a riding horse or two for the convenience of the reeve and other estate officials. The reeve of Cuxham in Oxfordshire, for instance, used one of the demesne horses to make regular visits to the important river entrepot of Henley-on-Thames.[40] Chaucer's Norfolk reeve travelled further, setting off to Canterbury on pilgrimage riding what may have been one of the demesne stots.[41] The latter is

[37] Overton and Campbell, 'Norfolk livestock farming'.   [38] Langdon, *Horses*, pp. 80–171.
[39] Langdon, *Horses*, pp. 86–94, places the contribution of horses to demesne draught power rather higher at 25 per cent of the total: the difference is possibly to be explained by his use of different regional weightings.   [40] Harvey, *Oxfordshire village*, pp. 66, 103.
[41] Campbell, 'Chaucer's reeve', pp. 303–5. Oswald, reeve of Bawdeswell in Norfolk, is described as mounted upon 'a ful good stot, that was all pomely grey and highte Scot': *The general prologue to the Canterbury tales*, ed. J. Winny (Cambridge, 1966), p. 70, lines 617–18.

Table 4.02. *Pastoral trends within the working sector: England, Norfolk, and the FTC counties, 1250–1449 (demesne means)*

Years	Sown acres	Plough beasts[1]	Working animals[2] as % of livestock units[3]	Working animals[2] Horses	Oxen	Cart-horses	Oxen	Immature cattle	Cattle units[4]
	per plough			per 100 sown acres		per 1,000 horses	per 100 horses	per 100 oxen	per 100 ox units[5]
England:									
1250–1299	78	9.3	59.8	4.2	10.4	105	498	44	178
1275–1324	74	8.8	54.1	4.4	11.5	131	448	41	180
1300–1349	70	8.8	53.0	4.8	13.0	137	404	41	216
1325–1374	70	7.9	50.7	4.3	9.9	200	396	48	253
1350–1399	65	7.8	44.8	4.4	10.4	208	435	55	320
1375–1424	64	7.9	43.6	4.7	11.3	198	420	60	360
1400–1449	60	8.1	41.6	5.0	12.2	132	489	74	236
Norfolk:									
1250–1299			35.3*	9.3	3.9		84	125*	252*
1275–1324	74	4.9	31.8*	9.1	3.9	34	88	141*	279*
1300–1349			26.4*	8.3	3.0		64	187*	404*
1325–1374			22.5*	8.0	2.3		45	279*	639*
1350–1399			20.1*	7.7	2.1		41	310*	784*
1375–1424	73	3.9	19.9*	6.3	1.5	29	34	273*	814*
1400–1449			20.2*	5.5	1.1		27	236*	773*

FTC counties:

1288–1315	*46.9*	*12.6*	*4.7*	*6.6*	*235*	*191*	*140*	*375*
1375–1400	33.7	13.6	5.9	6.4	265	146	221	669

Notes:

[1] Affers, stots, and oxen

[2] All horses + oxen

[3] (Horses ×1.0) + (oxen and adult cattle ×1.2) + (immature cattle ×0.8) + (sheep and swine ×0.1)

[4] (Adult cattle ×1.2) + (immature cattle ×0.8)

[5] (Oxen ×1.2)

* Calculated at aggregate not demesne level

Method:

All Norfolk means are the product of four regional sub-means weighted equally. All national means are the weighted product of six regional means, comprising the weighted mean for Norfolk ×0.081; the eastern counties (Cambs., Essex, Herts., Hunts., Lincs., Middx., Suffolk) ×0.214; the south-east (Hants., Kent, Surrey, Sussex) ×0.12; the midlands (Beds., Berks., Bucks., Leics., N'hants., Oxon., Rut., Warks.) ×0.164; the south-west (Devon, Dorset, Cornwall, Gloucs., Heref., Mon., Somerset, Wilts., Worcs.) ×0.209; and the north (Berwick., Ches., Cumb., Derbs., Durham, Lancs., Northumb., Notts., Salop, Staffs., Westmor., Yorks.) ×0.213. The weightings are based on each region's share of assessed lay wealth in 1334 and poll tax population in 1377.

Sources: National accounts database; Norfolk accounts database; FTC1 and FTC2 accounts databases.

the term by which plough-horses are commonly described in East Anglia and adjacent parts of southern and eastern England, whereas in the rest of the country they were known as affers.

Overall, it was in the counties of eastern and south-eastern England that the horse became most widely used for all aspects of farm work during this period (Figures 4.01 and 4.02). Although some specialist cart-horses were kept, the all-purpose work horse predominated; so much so, that in a county such as Norfolk oxen had effectively been eliminated from all but a minority of demesnes by the close of the Middle Ages (Table 4.02 and Figure 4.13).[42] In the period 1250–1349 horses accounted, on average, for a third or more of demesne draught animals in Northamptonshire, Huntingdonshire, Cambridgeshire, Norfolk, Suffolk, Essex, Hertfordshire, Middlesex, and Kent.[43] North and west of this core zone – in the middle Thames Valley, the east midlands, parts of the lower Trent Valley, and Co. Durham – horses were present in rather smaller numbers and there was a heavier emphasis upon horse-haulage rather than horse-traction. In most of the rest of the country horses made little contribution to either ploughing or carting. Horses accounted for fewer than one in seven of all demesne draught animals in Sussex and the Isle of Wight, along with most of the south-west, north-west and north of England. Paradoxically, economically peripheral upland areas of relative land abundance were more notable for their horse-breeding than their horse-power, as is well attested by the substantial stud farms which have been documented at Ightenhill in Lancashire, Macclesfield in Cheshire, Blansby Park outside Pickering in Yorkshire, and Woodstock in the Derbyshire High Peak.[44] Wales, too, was a significant supplier of horses and Farmer has drawn attention to the tendency for horses to be traded over greater distances than any other category of livestock.[45] As a demesne activity horse breeding was, nevertheless, a relatively unusual specialism.[46]

---

[42] Campbell, 'Towards an agricultural geography', pp. 92–3; Overton and Campbell, 'Norfolk livestock farming', pp. 382–4.

[43] Horses were also remarkably common in Cornwall: Hatcher, *Duchy of Cornwall*, p. 16.

[44] Shaw, *Royal Forest*, pp. 381–91; P. H. W. Booth, *The financial administration of the lordship and county of Chester 1272–1377* (Manchester, 1981), pp. 93–5; Hewitt, *Mediaeval Cheshire*, p. 56; Waites, *Monasteries and landscape*, pp. 130–2; Blanchard, 'Economic change in Derbyshire', pp. 164–8.

[45] Farmer, 'Marketing', pp. 378–85. Little is as yet known about the medieval horse trade. As far as the supply of riding animals to the crown was concerned, a great inter-manorial chain of stud farms existed linking horse-breeding establishments in England and Wales: Blanchard, 'Economic change in Derbyshire', pp. 165–8. Whence the peasantry and demesne managers obtained their working horses is, however, unknown: Langdon, *Horses*, pp. 272–3, 287–8. For the horse trade's structure and organisation in later centuries see P. Edwards, 'The horse trade in the midlands in the seventeenth century', *AHR* 27 (1979), 90–100.

[46] Possible examples, taken from the national accounts database and all from the West Riding of Yorkshire, are: Cowick with Snaith, 1330, six affers and twenty-nine other horses; Methley, 1435, no affers or stotts but twenty-one other horses; Paddockthorpe, 1354, four affers and eight other horses. For known royal studs used for breeding riding horses and war horses, see R. H. C. Davis, *The medieval warhorse* (London, 1989), pp. 86–97.

Differential trends in the ratio of oxen to horses highlight the fundamental spatial dichotomy between the innermost and outermost of these three zones. In the former, the horse continued to consolidate its position after 1349; in much of the latter (notably southern and south-western England and the west midlands) it lost some of the ground it had already gained. The upshot was that, nationally, horses only seem to have increased at the expense of oxen down to the middle years of the fourteenth century (Table 4.02). Moreover, outside the four East Anglian counties of Norfolk, Suffolk, Essex, and Hertfordshire the ox always remained numerically the more important draught animal. The histories of haulage and traction were nevertheless somewhat different in this respect and it is instructive to consider them separately.

Cart-horses

Of the two categories of draught horse it was the more expensive and powerful cart-horse which followed the most dynamic diffusion path down to the end of the fourteenth century (Table 4.02). Langdon has argued that by the close of the thirteenth century horses already dominated the carriage of goods by vehicle.[47] At that point one in every eight horses appears to have been maintained primarily for carting and there was approximately one cart-horse to every 250 sown acres. Great though the increase in horse haulage may have been during the thirteenth century, it was during the middle decades of the following century that it reached its medieval zenith. By the close of the fourteenth century fully one in five horses was used for carting and there was now one cart-horse to every 110 sown acres.

On the face of it the economic benefits of this massive application of horse power to haulage seem clear enough. Other things being equal, the more rapid transit of goods thereby facilitated should have reduced transportation costs, extended the sphere of the market, and increased the rate of circulation. The only problem is that the heavy oat-fed cart-horse favoured on many demesnes was extremely costly to maintain. For instance, Biddick has calculated that Peterborough Abbey invested more in cart-horses and transport (fodder, shoeing, maintenance of carts, wages of carters, etc.) than it made in wool sales in the opening decade of the fourteenth century.[48] Of course, on many demesnes and all peasant holdings ordinary working horses served the dual function of traction and haulage, and this would have helped to keep transport costs down, but the fact remains that improved haulage was frequently only obtained at a high financial price. It is therefore no surprise to find that adoption of the cart-horse proceeded furthest in those parts of the country which most stood to gain from a closer involvement with the market.

Throughout the period 1250–1449 cart-horses were mainly restricted to central and eastern England (Figure 4.12). Within that area they were

---

[47] J. L. Langdon, 'Horse hauling: a revolution in vehicle transport in twelfth- and thirteenth-century England?', *PP* 103 (1984), 37–66.   [48] Biddick, *The other economy*, p. 120.

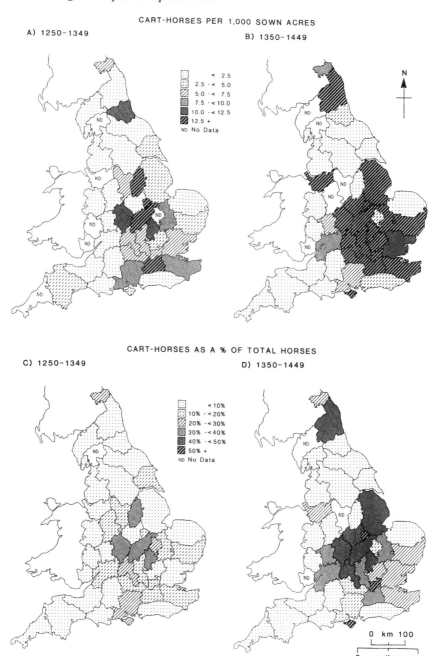

Fig. 4.12. Types of horsepower employed on demesnes: England, 1250–1449 (source: National accounts database).

originally concentrated in two specific geographical contexts: the immediate hinterland of London and the valleys of the rivers Trent, Nene, Welland, and Ouse in the east midlands. These were the parts of the country most deeply penetrated by metropolitan, national, and international demand for grain, wool, and other bulky agricultural products. For demesne managers the key to supplying that demand lay in getting their produce either to river ports, coastal ports (particularly King's Lynn, Boston, and Hull), or directly to London itself. To do so they needed to invest in road transport, both carts and other vehicles and cart-horses.[49] After 1350, as more pastoral resources became available, the cart-horse consolidated its position as an essential draught animal on demesnes throughout these areas. In virtually every county south of the Humber, east of the Warwickshire Avon, and north of the Weald, there was now one cart-horse to every 80 sown acres and on many individual demesnes one cart-horse to every 50 sown acres. Nowhere was the cart-horse more popular than in the midland counties where, on demesne after demesne, virtually every other horse appears to have belonged to this category. North and west of this core area of horse haulage the fall-off in the utilisation of cart-horses was very abrupt, a circumstance which implies a more limited participation in wider orbits of agricultural exchange. Evidently, Langdon's medieval revolution in road haulage was less than a national phenomenon.

Plough-horses
The adoption of horses for ploughing was altogether more circumscribed in its distribution since it was subject to greater environmental and economic constraints and was conditional upon a much larger supply of animals. Nor was it a decision which could sensibly be taken without reference to the rest of the husbandry system. Because horses consumed more grain and less grass than oxen their adoption had consequences for both the arable and pastoral sectors. It was in East Anglia that horse ploughing was introduced earliest and proceeded furthest. A shift towards a greater use of horses was already taking place in the first half of the twelfth century, ahead of both the Home Counties and the east midlands, neither of which followed East Anglia's lead until the end of the century. Within these areas the spread of horses was very much a diffusion process and in Norfolk the documentation is sufficiently complete to trace the stages by which horses eventually supplanted oxen as the sole beasts of traction on most demesnes in the county (Table 4.02 and Figure 4.13).[50]

Horses seem first to have been utilised for ploughing on the light, dry soils of the meadow-deficient north-west of Norfolk, where in the reign of Henry I they are already recorded in significant numbers on the Ramsey Abbey

---

[49] Campbell and others, *Medieval capital*, pp. 56–60. There was also much private investment in bridge building: D. Harrison, 'Bridges and economic development, 1300–1800', *EcHR* 45 (1992), 240–61; Britnell, 'Commercialisation and economic development', pp. 17–18.

[50] Campbell, 'Towards an agricultural geography', pp. 91–3.

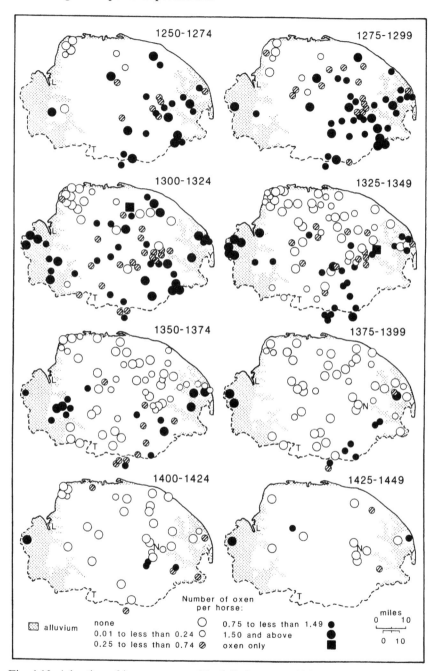

Fig. 4.13. Adoption of horsepower on Norfolk demesnes, 1250–1449 (L = Lynn; N = Norfolk; T = Thetford; Y = Yarmouth) (source: Norfolk accounts database).

demesnes of Brancaster-with-Deepdale, Ringstead, and Holme-next-the-Sea.[51] By the mid-thirteenth century demesnes in this locality had converted to all-horse teams and in the county as a whole horses already significantly outnumbered oxen. In this respect Norfolk was far ahead of the rest of the country, where horses were still outnumbered over four to one by oxen. Moreover, in Norfolk once the process of substitution began it proceeded without reversal. Demesnes converted first to mixed teams and then to all-horse teams. By the middle of the fifteenth century it was horses which outnumbered oxen by four to one, the inverse of the situation still prevailing nationally.

Even in Norfolk, though, there were localities where horses were slow to make headway. Typically, these were at some distance from the initial source of innovation, were reasonably well provided with grass, and had heavy soils. On the boulder-clay soils of the south-east of the county mixed teams of horses and oxen remained the norm for as long as demesnes remained in hand. In the grass-rich Norfolk Fenland oxen dominated the draught sector for even longer: it was large-scale drainage in the seventeenth century which transformed the fens from a stronghold of ox-ploughing to one of horse power.[52] Elsewhere in the county the horse had long reigned more or less supreme and as its use spread so plough-teams got smaller (Figure 4.14), although whether this was contingent upon improvements in harnessing and plough construction remains to be established. Such full conversion by demesnes to all-horse teams occurred comparatively rarely during the Middle Ages, the only other areas to convert being northern and eastern Kent (which had much environmentally, economically and institutionally in common with Norfolk), the Chiltern Hills (whose flinty soils were treacherous to the tread of oxen), and the Yorkshire Wolds (which shared a strong environmental resemblance to north-west Norfolk) (Figures 4.01 and 4.02).[53] In most other parts of the country, when demesnes employed plough-horses it was invariably in combination with oxen.

4.22 Draught-oxen

Striking as were these developments in the use of horses, it was upon the ox that the bulk of demesnes relied for their draught power (Table 4.01). As Table 4.02 demonstrates, oxen consistently outnumbered horses by at least four to one. They may not have been able to work as fast or for as long as horses, but they were steadier of pull, less demanding in their feed requirements and therefore cheaper to maintain, and at the end of their working lives they could be fattened and sold off for meat.[54] Pedologically the ox was at greatest

[51] Langdon, *Horses*, pp. 43, 50–4. [52] Campbell and Overton, 'New perspective', p. 80.
[53] Langdon, *Horses*, pp. 100–5; Roden, 'Demesne farming'.
[54] *Walter of Henley and other treatises on estate management and accounting*, ed. D. Oschinsky (Oxford, 1971), pp. 160–4; Langdon, 'Economics of horses'.

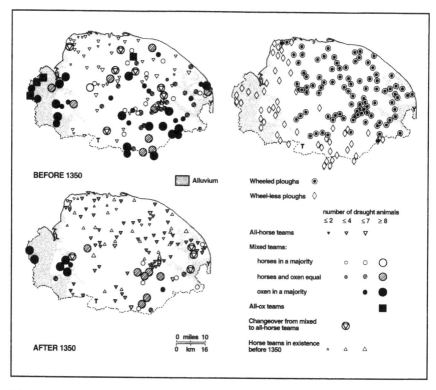

Fig. 4.14. Plough-teams and plough-types in Norfolk, 1250–1449 (L = Lynn;
N = Norfolk; T = Thetford; Y = Yarmouth) (source: Norfolk, accounts database).

advantage on heavy soils, where large teams were unavoidable and speed
scarcely an option; ecologically it scored in situations where grass and hay
were plentiful and cheap. It was a more land-extensive animal than the horse
and consequently enjoyed greatest comparative advantage in areas of medium
to low economic rent.

For all these reasons the relatively remote and economically isolated counties of the north and west and, to a lesser extent, those of the remote south-
west, were the greatest bastions of both ox-ploughing and ox-hauling in the
Middle Ages. It is here that the highest ratios of oxen to sown acres are registered. Indeed, so many oxen are often recorded that either plough-teams were
of above average size or surplus beasts were being produced for transfer or
sale.[55] Even though horses were bred in these regions, few demesnes showed
any inclination to substitute them for oxen with the result that throughout the
period horses were massively outnumbered by oxen. To the immediate south

---

[55] Langdon, *Horses*, pp. 122–3.

and east lay a broad arc of counties where substantial all-ox teams were the norm for ploughing but where few surplus oxen were carried, replacement animals often being obtained from elsewhere. Often ox-ploughing was accompanied by horse-hauling. Mixed plough-teams were few and also usually contained more oxen than horses. Only in a contiguous block of counties in East Anglia and the south-east – Huntingdonshire, Cambridgeshire, Norfolk, Suffolk, Essex, Hertfordshire, Middlesex, and Kent – did oxen regularly yield to horses as the most important demesne draught animal (Figure 4.12). Without horses, recorded numbers of oxen on most demesnes in these counties would have been wholly inadequate to the draught requirements of arable husbandry.

4.23 Working animals in perspective

Throughout the period 1250–1450 demesnes stocked roughly one working animal for every 6 acres sown, at an average cost, on Langdon's estimate, of 20d. per beast and 18–20d. per sown acre.[56] Without these animals, and the pastoral resources which in large part supported them, arable production could not have been sustained. Of course, there were spin-offs for the pastoral sector of maintaining so many draught beasts – there was hay to be carted and wool to be transported – but for the most part working animals were there to service the arable sector. As such they necessarily had first claim upon available pastoral resources.

   Circa 1300, before the substitution of horses for oxen had reached its medieval climax, horses and oxen together accounted for two-fifths of all demesne livestock units. This figure includes young horses – between a fifth and a tenth of the total – and some oxen which likewise rarely worked. Even so, well over a third of all recorded livestock units at the beginning of the fourteenth century were dedicated primarily to work. A century later the working component of the pastoral sector had shrunk considerably. Demesnes now had less land under cultivation and so, on average, had less need of working animals. Moreover, the pastoral sector had expanded, as resources had been diverted from producing crops to producing animals. Here non-working animals were the prime beneficiaries, doubly so because on-going substitution of horses for oxen in eastern and south-eastern England improved the productivity of draught animals thereby facilitating some reduction in the number of beasts per plough (Table 4.02). Spurring this development was the dwindling contribution which customary ploughing and carrying services were now making to demesne production, as customary services either lapsed or were commuted. By c. 1400, therefore, although the number of working animals per sown acre was little altered their share of total livestock units had contracted

[56] Langdon, 'Economics of horses', p. 37.

to barely 30 per cent of the total (less when allowance is made for young horses and non-working oxen).

This relative contraction in the importance of working animals is apparent whether comparison is made between the two national, the two FTC, or the Norfolk samples of demesnes (Table 4.02). It also shows up in a more controlled comparison between demesnes practising the same pastoral type (Table 4.01). Although it was the specialist sheep-farming demesnes practising pastoral type 4 which expanded their non-working sectors most, it was the combined cattle- and sheep-farming demesnes of pastoral type 1 which consistently kept non-working animals in greatest proportion (Table 4.01). The key to the latter's success lay in the wholesale substitution of horses for oxen. Using small, fast, all-horse plough-teams such light-soil demesnes as Brancaster, Burnham Thorpe, Gimingham, Hevingham-with-Marsham, and Sedgeford (all Norfolk) required only a tenth of all livestock units to satisfy their draught requirements. To have squeezed this proportion further would have been incompatible with the continued conduct of arable cultivation. At the opposite extreme, there were always some demesnes – pastoral types 5 and 6 – which stocked little else than working animals (Figures 4.08, 4.09, 4.10, and 4.11). About one in five of all sampled demesnes were of this type (Table 4.01). Many were demesnes, which, for one reason or another, had been temporarily asset-stripped of their non-working animals, but there were also some narrowly preoccupied with the more extensive forms of arable production (pastoral type 6). They were as exclusively arable as was possible in this animate age and, significantly, were more conspicuous at the opening of the fourteenth century than they were to be at its end.

4.3 Non-working animals

Only after draught requirements had been satisfied could demesne managers devote any pastoral resources that remained to animals managed primarily for their meat, milk, wool or breeding capacity. These might be cattle other than oxen, sheep, and/or swine. Over time all categories of non-working animal gained steadily in importance (Table 4.03). Before 1325 they benefited from productivity gains within the draught sector coupled, on some demesnes, with a greater emphasis upon fodder crops, which together helped release scarce resources to other pastoral activities. Thereafter they benefited from a general production shift from crops to livestock and further expansion in the fodder cropping of legumes (Table 5.08). Change, although slow, was far reaching. In relative terms non-working animals increased their share of total livestock units by 25 per cent over the period 1250–1450 (Table 4.03). In absolute terms, demesnes were stocking 70 per cent more non-working animals in the early fifteenth century than they had done in the late thirteenth, at a density per 100 grain acres that was 150 per cent higher (Tables 4.02 and 4.03). These trends denote a pastoral sector characterised by considerable dynamism; a sector, moreover, which gained steadily in importance relative to the arable.

4.31 Breeding replacement draught beasts

Just as arable production had first call on the pastoral sector so, too, at least in principle, it had first claim upon the non-working component of that sector. Draught animals had only a finite working life and ultimately had to be replaced. Complete self-sufficiency in draught power thus required maintaining enough breeding animals to ensure an adequate supply of replacement horses and oxen. Often this was more viable at the level of the estate than the individual demesne, especially when estates comprised quite widely flung manors which encompassed a diverse range of resources. Peterborough Abbey, for instance, organised pastoral husbandry on its twenty-three manors at an estate rather than demesne level and used inter-manorial transfers of stock to maintain its plough-teams.[57] So, too, did the abbot (but not the convent) of Westminster, Bicester Priory, and Oseney Abbey. Such a strategy, however, was very much the exception rather than the rule and was impractical on the smallest estates. When estates and demesnes could not breed sufficient replacement animals they had no other recourse but to buy them. In fact, it was often cheaper to do so.[58] Rather than pursue a policy of pastoral self-sufficiency it frequently made better economic sense to specialise according to comparative advantage and purchase whatever replacement draught beasts were required from those best placed to supply them.

The recurrent need for replacement working animals could be considerable. On Langdon's estimation draught oxen, plough-horses, and cart-horses had average working lives of 5.1, 5.5 and 7.0 years respectively.[59] A typical mid-fourteenth-century demesne with two ploughs and sixteen to eighteen draught animals (four-fifths of them oxen and one-fifth horses) would thus have required at least three replacement animals per year. Since oxen and plough-horses both took three years to reach maturity, at least nine immature beasts would need to have been stocked to meet this requirement. Breeding, in turn, would have been impossible without a stallion and mares and a bull and cows. Moreover, whereas fillies could be bred up to make good workhorses only steers were eligible to become oxen. Since cows calved but once a year and produced males and females in equal proportion, an adequate supply of steers could only be guaranteed if at least twice as many cows were kept as the number of steers that were required. Merely to produce two replacement oxen a year would thus have required a herd comprising a bull, four cows, and nine immatures (three of them females ultimately intended as replacement cows).

---

[57] Campbell, 'Measuring commercialisation', p. 170; Biddick, *The other economy*, pp. 81–90, 117–18.   [58] Farmer, 'Prices and wages', p. 747.

[59] Langdon, 'Economics of horses', p. 36.

Table 4.03. *Pastoral trends within the non-working sector: England, Norfolk, and the FTC counties, 1250–1449 (demesne means)*

Years	Non-working livestock units[1]	Adult cattle	Immature cattle	All cattle	Sheep	Swine	Cattle	Sheep	Swine	Sheep per 10 cattle	Immature cattle per 100 adults[2]
		as % of non-working livestock units[1]					per 100 sown acres				
England:											
1250–1299	34.4	33.0	22.1	55.1	32.5	12.5	11.2	63.3	7.3	71	119
1275–1324	38.9	30.3	18.1	48.5	37.7	13.9	10.8	87.4	11.9	83	110
1300–1349	40.2	30.8	16.2	47.1	36.4	16.6	12.5	107.0	16.1	86	102
1325–1374	40.5	29.1	16.9	46.0	37.7	16.3	12.8	107.1	13.9	90	95
1350–1399	48.9	29.7	15.4	45.2	41.2	13.7	15.7	156.7	14.0	102	102
1375–1424	50.6	27.2	14.9	42.0	38.6	19.5	16.5	164.3	15.7	98	122
1400–1449	57.9	31.1	16.4	47.5	37.3	15.3	20.0	190.8	17.5	100	110
Norfolk:											
1250–1299	29.5*	46.1*	22.7*	68.8*	26.1*	5.1*	11.4	44.5	8.9	39	85
1275–1324	31.7*	44.2*	22.4*	66.6*	29.3*	4.1*	12.6	56.7	8.0	45	86
1300–1349	33.8*	43.8*	19.5*	63.3*	32.3*	4.4*	14.1	74.4	10.0	53	79
1325–1374	36.6*	45.9*	18.9*	64.8*	30.9*	4.4*	17.0	84.9	12.3	50	66
1350–1399	39.4*	47.5*	17.0*	64.5*	31.2*	4.3*	18.9	96.9	13.6	51	53
1375–1424	34.7*	46.1*	13.3*	59.4*	36.3*	4.3*	13.9	92.1	10.9	66	53
1400–1449	34.7	38.3*	9.8*	48.1*	47.3*	4.6*	9.7	103.1	10.3	106	52
FTC counties:											
1288–1315	40.4	35.5	15.2	50.7	36.6	12.7	11.5	94.9	12.5	483	77
1375–1400	58.3	33.2	10.0	43.3	46.3	10.5	19.7	204.6	18.9	810	50

Notes:

[1] (Adult cattle ×1.2) + (immature cattle ×0.8) + (sheep and swine ×0.1)

[2] Minimum herd size of 10; imposed maximum of 400

* Calculated at aggregate not demesne level

Method:

All Norfolk means are the product of four regional sub-means weighted equally. All national means are the weighted product of six regional means, comprising the weighted mean for Norfolk ×0.081; the eastern counties (Cambs., Essex, Herts., Hunts., Lincs., Middx., Suffolk) ×0.214; the south-east (Hants., Kent, Surrey, Sussex) ×0.12; the midlands (Beds., Berks., Bucks., Leics., N'hants., Oxon., Rut., Warks.) ×0.164; the south-west (Devon, Dorset, Cornwall, Gloucs., Heref., Mon., Somerset, Wilts., Worcs.) ×0.209; and the north (Berwick., Ches., Cumb., Derbs., Durham, Lancs., Northumb., Notts., Salop., Staffs., Westmor., Yorks.) ×0.213. The weightings are based on each region's share of assessed lay wealth in 1334 and poll tax population in 1377.

Sources: National accounts database; Norfolk accounts database; FTC1 and FTC2 accounts databases.

Table 4.04. *Sources of replacement horses and oxen on demesnes in the FTC counties, 1288–1315 and 1375–1400*

	% of demesnes which added beasts			
	Horses		Oxen	
Method of addition	1288–1315	1375–1400	1288–1315	1375–1400
Graduated in	26.4	26.4	48.9	39.5
Transferred in	25.4	21.4	20.7	14.3
Bought	68.5	71.4	56.4	53.8
Heriot		21.4		23.5
Other	32.5	20.0	18.1	5.0
Total demesnes	197	140	188	119

	% of beasts added			
	Horses		Oxen	
	1288–1315	1375–1400	1288–1315	1375–1400
Graduated in	12.1	11.4	30.2	18.7
Transferred in	20.4	12.5	24.7	15.0
Bought	52.3	63.2	35.2	45.5
Heriot		6.4		5.5
Other	15.2	6.5	9.9	2.5
Total beasts	416	262	565	234

	Immature beasts per 100 adults			
	Horses		Cattle	
Type of beast	1288–1315	1375–1400	1288–1315	1375–1400
All immatures	12	8	75	79
Male immatures	6	5	38	38

Source: FTC1 and FTC2 accounts databases.

In sum, this amounts to almost as many livestock units as those already engaged in draught work. It is small wonder, therefore, that few demesnes aspired to complete self-sufficiency in draught power.

Analysis of the cattle stocked on the sampled demesnes demonstrates that the demesne sector as a whole, at least in lowland England, was incapable of satisfying its own demand for replacement oxen. As Table 4.02 demonstrates, far too few immature cattle were stocked. Nationally, there were fewer than fifty immature cattle per 100 oxen before 1350 and never more than seventy-five thereafter. These figures understate the shortfall in steers, since at least half of these immatures would have been heifers, especially in the many herds with a dedicated dairying function. Even in the FTC counties, where the ratio of immature cattle to oxen was well above the national average, the number of male immatures – at less than forty per 100 oxen – was too small to ensure replacement (Table 4.04).

The recorded provenance of replacement oxen in these counties confirms this point. Less than half of all sampled FTC demesnes before 1350 and fewer than 40 per cent of demesnes thereafter bred their own replacement oxen. Moreover, demesne-bred animals constituted barely 30 per cent of replacement animals at the beginning of the fourteenth century and less than 20 per cent at the end. Estate-bred replacements were of even smaller importance and declined in significance over the course of the fourteenth century. At the beginning of the century one in five demesnes obtained oxen by transfer from elsewhere on the estate, such transfers contributing one in four of all replacements. By the end of the century only one in seven demesnes were still obtaining oxen by this means, which now accounted for just one in seven of all replacements. Some additional animals were received as heriot, a few more were commandeered or confiscated from the tenantry. Otherwise by far the greatest single source of replacement oxen was the market: 35 per cent of all replacement beasts were bought *c.* 1300, rising to 45 per cent *c.* 1400. In both periods over half of all demesnes resorted to the market to obtain oxen. The market was even more important as a source of replacement horses, 50 per cent of which were bought *c.* 1300 and 60 per cent *c.* 1400. In contrast, demesne-bred horses provided only one in eight of all replacements. Estate-bred animals, in this instance, were of marginally greater importance, although as with oxen that importance diminished as the fourteenth century drew to a close (Table 4.04). As estate- and demesne-produced animals waned in significance, so the market gained in prominence as a source of supply.

Whence came such a substantial market supply of young draught animals? Specialist stock farms in regions of surplus pastoral resources were one potential source. The many seigniorial vaccaries in the north and north-west of England are a case in point. Here was a region with limited draught requirements of its own, abundant upland grazings, and a ready market for finished animals in the ox-dependent counties immediately to the south and east where self-sufficiency in oxen was more the exception than the norm.[60] Although horse studs and substantial sheep flocks were both to be found in this region,

---

[60] In the period 1250–1349 six out of nine sampled demesnes in Leicestershire, eight out of twenty in Warwickshire and thirteen out of twenty-one in Somerset kept oxen but no cattle. Yet within the country as a whole the equivalent was true of only one demesne in four, and in Devon and Cornwall and parts of the Home Counties – Hertfordshire, Middlesex, and Surrey – of only one demesne in ten: National accounts database.

cattle rearing was the pre-eminent pastoral activity, more so than in any other part of the country. The Pennine dales, flanks of the hills, and various forests of the region were studded with demesne vaccaries.[61] A major function of these vaccaries was to supply associated lowland demesnes with draught animals. Hence the fact that over half of all enumerated cattle were oxen.

The dependent relationship between lowland demesnes and upland stock farms has been described most clearly by Blanchard in Derbyshire.[62] Here, few of the lowland manors bred their own oxen and, instead, replacement animals were obtained from central stock-breeding establishments within the estates, most of them established on the uplands of the Middle and High Peak and on the pastures of the Derbyshire forests. Dairying was a secondary activity of these upland vaccaries, producing cheese and butter and also milk for personal consumption, whereas production of meat and hides for the market was entirely unspecialised and dependent upon the vagaries of cattle disease. A similar emphasis upon the production of oxen can be observed on the vaccaries of the de Lacys, earls of Lincoln, in Lancashire.[63] Initially developed to satisfy an entirely local demand, by at least the middle of the thirteenth century the production of oxen had grown to a scale where it was yielding large numbers of surplus animals for sale mostly in the markets of Pontefract and Bolton. Animals were sold both for draught and for meat. In 1258 oxen and cows from Lancashire were sent to the royal larder at Westminster and over the next half-century the number of vaccaries and their output appear to have risen significantly. The revenues thus realised might be considerable. For instance, during the accounting year 1304–5 income received by the de Lacys from the sale of 213 oxen, five bulls, 168 cows, and two calves totalled £173 1s. 6d.; at the end of that year there were 2,518 cattle on their vaccaries in the chase of Blackburnshire alone.[64] As in Derbyshire, dairying was practised as a sideline and sales of butter and cheese yielded a valuable subsidiary income.

The typical northern vaccary comprised a combination of good-quality and sheltered grazing land in the river valleys draining the uplands and a wide area of hill pasture for summer grazing. On the de Lacy estate each vaccary possessed an average of thirty milk cows and a roughly equivalent number of followers up to three years: in 1297 the average size of some twenty vaccaries in Wyresdale was some fifty-four animals. The stock and land of each vaccary were leased to a keeper for a rent of about £3 0s. 0d. He was responsible for the maintenance in good condition of the entire stock of the farm and answer-

---

[61] Kershaw, *Bolton Priory*, pp. 97–103; Donkin, *The Cistercians*, pp. 68–82; Miller, 'Farming: northern England', pp. 409–11; E. Miller, 'Farming practice and techniques: Yorkshire and Lancashire', in Miller (ed.), *AHEW*, vol. III, pp. 188–9.

[62] Blanchard, 'Economic change in Derbyshire', pp. 168–74.

[63] Tupling, *Rossendale*, p. 25; Shaw, *Royal Forest*, p. 354; Atkin, 'Land use and management'.

[64] Shaw, *Royal Forest*, pp. 359–60.

able to the sub-stockmaster of the particular bailiwick where the vaccary was situated. In Blackburnshire there was a sub-stockmaster for each of the forests of Pendle, Trawden, Accrington, and Rossendale, all of them subject to a chief stockmaster of the chase of Blackburnshire, assisted by clerks who drew up the accounts for submission to the steward. Each vaccary under the control of its keeper was run by the aid of a group of herdsmen. They were mostly paid in kind and lived in bothies or cottages, either near the principal grange or farmstead or in more distant parts of the forest where the cattle were grazed.

Such vaccaries, although well represented in the region, are under-represented in the national samples of demesnes.[65] Nor were they confined to the north of England. In Monmouthshire R. R. Davies has shown that the lord of Brecon had a full-time stock-keeper in charge of a vaccary of some 300 cows and calves in Fforest Fach throughout most of the fourteenth century.[66] So profitable did this enterprise prove that during the first half of the century three further vaccaries were established. In this case many of the livestock went to cater for the needs of the lord's larder and for the restocking of his English lands. In 1349, for instance, twenty drovers accompanied over 400 head of cattle to the Bohun household in Essex and in the following year an almost equal number was driven to Kimbolton in Huntingdonshire and Oaksey in Wiltshire. These particular long-distance stock movements took place within the context of a specific estate but there is evidence of other contemporary droves of Welsh cattle to England which were not so constrained.[67] Demesne-produced animals probably constituted only a minority of those regularly driven across the Welsh border in this way.

Wales was potentially a major reservoir of replacement animals (as, until the Wars of Independence, may have been Scotland).[68] Environmentally and locationally it was ideally placed to specialise in breeding and rearing. The same applied to much of Devon and Cornwall in the extreme south-west. Here available manorial accounts indicate high stocking densities of cattle, with oxen again comprising slightly over half of the total.[69] The one difference from the situation in the north of England was that cattle had to vie with sheep for

[65] Booth, *Financial administration*, pp. 86–97; Donkin, *The Cistercians*, pp. 68–79; Waites, *Monasteries and landscape*, pp. 117–45.   [66] Davies, *Lordship and society*, pp. 115–16.

[67] Davies, *Lordship and society*, pp. 115–16; Finberg, 'Welsh cattle trade'; Skeel, 'Cattle trade', pp. 137–8.

[68] C. Thomas, 'Thirteenth-century farm economies in North Wales', *AHR* 16 (1968), 1–14; Jack, 'Farming: Wales and the Marches', pp. 482–96.

[69] Finberg, *Tavistock Abbey*, pp. 131–43; Hatcher, *Duchy of Cornwall*, p. 16; N. W. Alcock, 'An east Devon manor in the later Middle Ages. Part I: 1374–1420. The manor farm', *Reports and Transactions of the Devonshire Association* 102 (1970), 141–87. Yet the role of livestock within the economy should not be exaggerated. Important as was the sale of livestock and their produce on the estates of the earls of Devon in 1286–7, there is no doubt that the profits of arable husbandry predominated: K. Ugawa, 'The economic development of some Devon manors in the thirteenth century', *Reports and Transactions of the Devonshire Association* 94 (1962), 652. See also Hatcher, 'Farming: south-western England', pp. 395–7.

importance. Outside of these locationally peripheral and traditionally pastoral counties of the west and north with their natural excess of pasture so invisible in the sources, the only other major area to have evolved an equivalent economy was the East Anglian Fenland.[70] Here was a rich reservoir of pasture in the arable heartland of the country whose economic potential was not lost on those producers who had access to it.[71]

These areas of pastoral surplus were vital to arable production throughout the greater part of lowland England. This functional inter-dependence was articulated through a network of markets and fairs which facilitated the substantial inter-regional transfer of animals. As Farmer points out, 'most fairs were in the summer and early autumn. Those in the early summer offered the chance to buy young stock, or to dispose of sheep after shearing; those in September supplied cattle to drovers and graziers for fattening before slaughter; those in November a last opportunity to sell the culls which were not to be kept over the winter.'[72] As a supply system it attained its fullest development before 1350. Thereafter, as stocking densities rose on many lowland demesnes, especially in the midlands, a healthier balance was established between oxen and other classes of cattle. Some lowland demesnes now had surplus animals of their own to dispose of, so that the market for northern and Welsh animals must have changed accordingly.[73] Only in the north-east – in Yorkshire and Durham – does the old upland–lowland dependency appear to have persisted, with the maintenance of specialist cattle farms on the upland fringes and extensive arable demesnes on the neighbouring lowlands.

Nevertheless, not all replacement oxen were necessarily bred and reared at a distance. Surplus male calves were one of the by-products of the specialist dairying practised on many of the more intensively managed lowland demesnes. Dairying often went hand in hand with the substitution of horses for oxen, which both reduced the demand for replacement oxen and released the hay and grassland necessary to support a dairy herd. Keeping cows in milk required keeping them in calf. Enough female calves would have been retained to maintain the herd at full milking strength; the rest would have been sold, including most males. Who bought these males the accounts do not record, but since lords were sellers rather than buyers of calves it can rarely have been other demesnes. The vast majority were presumably purchased by other cate-

---

[70] On stock farming in the Fens see Raftis, *Estates of Ramsey Abbey*, pp. 129–57; Hallam, *Settlement and society*, p. 220; Donkin, *The Cistercians*, pp. 69–70; Biddick, *The other economy*, pp. 17–19, 84–6, 114, 118, 119. Examples of Fenland stock farms included within the national accounts database are 'Munkelode' (1258), Hildick (1295), and Nomansland (1258), all in Lincolnshire. At Bolingbroke on the Fen edge a small arable demesne coexisted with very extensive marshland grazings and livestock husbandry consequently predominated: PRO, C134 File 22 (1).

[71] Platts, *Land and people*, pp. 103–11; H. E. Hallam, 'Farming techniques: eastern England', in Hallam (ed.), *AHEW*, vol. II, p. 309.   [72] Farmer, 'Marketing', pp. 339–40.

[73] Dyer, *Warwickshire farming*, pp. 20–2.

gories of producer either within or outside the region and reared up for subsequent sale as either work or meat animals. For small producers, without breeding stock of their own, it could be quite a lucrative proposition. In the early fourteenth century male calves could be bought for 15d. or less but three years later, when reared up and finished as oxen, they sold for over 12 shillings. It was thus that many undoubtedly re-entered the demesne sector. Some must have left it again at the end of their working lives, when they were sold off for fattening and slaughter.

A single ox could thus pass through at least four different hands over the course of its lifetime.[74] In this case the key functional relationship was not so much that between lowland and upland but between large producer and small. Possibly, too, it was smaller producers who were most active in meeting the demand for replacement horses. After all, peasants rather than lords were most active in the changeover to horsepower, which would have been difficult to achieve without an adequate supply of peasant-bred animals. Such few seigniorial studs as are known were more concerned with breeding the higher-status riding animals than the lowly affers and stots which drew the plough and hauled the cart.[75]

4.32   Cattle-based dairying

Cattle other than oxen consistently accounted for about half or more of all non-working animals (Table 4.03).[76] Initially lords favoured cattle over other categories of non-working animal because of the constant need for replacement oxen. Over time, however, the development of an active trade in draught animals released many of them from that obligation. As economic rent rose, so it made progressively better economic sense to concentrate upon milk production since it gave a far better financial and food return per unit of land and per unit of available solar radiation.[77] Dairying was, of course, an inevitable adjunct of all forms of cattle rearing. Nevertheless, as an object in itself, it tended to be restricted to the more populous and commercialised districts since it was only here that its not inconsiderable overhead costs were justified. Within such localities dairying often became the dominant pastoral component of manure-intensive mixed-farming systems. In such systems dairying was often valued less for its commercial potential than as a source of both the cheese consumed in quantity by estate workers and the farmyard manure so necessary to the maintenance of a demanding cropping regime.

---

[74] Cf. D. Dickson, *New foundations: Ireland 1660–1800* (Dublin, 1987), pp. 112–13.

[75] E.g. Hewitt, *Mediaeval Cheshire*, pp. 56–8.

[76] Case studies of conventual cattle husbandry include Smith, *Canterbury Cathedral Priory*, pp. 157–65; Finberg, *Tavistock Abbey*, pp. 133–44; Biddick, *The other economy*, pp. 81–99.

[77] On the food productivity of different pastoral systems see I. G. Simmons, *The ecology of natural resources* (London, 1974), pp. 201–6.

Breeding has greatly transformed the size and milk yield of cows since the Middle Ages. At the beginning of the fourteenth century Clark reckons net milk production to have been approximately 100 gallons per year whereas by the mid-nineteenth century it had risen to 450 gallons.[78] The anonymous author of the *Husbandry* expected cows kept 'in good pasture' to yield 98 gallons between May and Michaelmas and at least a further 14 gallons during the remaining months of the year (less could be expected from heifers 'in their first year of bearing'). This was enough to produce 8 stones of cheese and 1 stone 2 lbs. of butter (made from whey rather than whole milk) at a ratio of cheese to butter of seven to one.[79] Walter of Henley expected roughly similar yields but also stressed the adverse impact upon milk output of inferior pasture; he expected cows fed on the pasture of salt marshes to yield 50 per cent more milk than those fed in wood-pasture or on the stubble and aftermath of fallowed arable. Moreover, on his reckoning a cow should have yielded ten times as much milk as a ewe.[80] These predicted yields are broadly in line with those calculated from manorial accounts. R. Trow-Smith concluded 'that from a sampling of the records of manors in Kent, Devon, Northamptonshire and Sussex annual yields of between 120 and 150 gallons, according to the quality of the pasture available for summer grazing, are the most that can be credited to the late medieval cow'.[81] The herd of 230 to 256 cows stocked on the estates of Peterborough Abbey yielded at a rate of 100 to 130 gallons per cow during the summer milking season at the opening of the fourteenth century.[82] At Wootton St Lawrence in Hampshire, Enford in Wiltshire, and Hallow and Grimley in Worcestershire milk yields were somewhat lower, they were lower still on the Somersetshire estates of Glastonbury Abbey and on the Yorkshire vaccaries of Bolton Priory and Fountains Abbey, and fell to only 40 gallons per cow on the manors of Clyst, Hurdwick, and Werrington in Devon.[83] These low south-western yields contrast with the 150 to 225 gallons per cow which Ramsey Abbey obtained on its more closely managed east midland and East Anglian manors and imply a yield gradient which declined with the intensity of husbandry from east to west.[84]

No branch of cattle husbandry required closer management than dairying. Maximising the proportion of milk animals within the herd and keeping milk production at a high level demanded constant attention. Care needed to be taken in putting each cow to the bull, since cows needed to be kept in calf if they were to continue to produce milk and heifers only yielded milk once they

[78] Clark, 'Labour productivity', pp. 214–15, 218.
[79] *Walter of Henley*, p. 431.   [80] *Walter of Henley*, pp. 333–5.
[81] Trow-Smith, *Livestock husbandry*, p. 122.   [82] Biddick, *The other economy*, pp. 94–5.
[83] Miller (ed.), *AHEW*, vol. III, pp. 192, 235, 300; Hallam (ed.), *AHEW*, vol. II, pp. 397–8; M. Ecclestone, 'Dairy production on the Glastonbury Abbey demesnes 1258–1334', MA thesis, University of Bristol (1996).   [84] Miller (ed.), *AHEW*, vol. III, p. 220.

had calved.[85] An annual rate of reproduction of one calf per cow was the ideal. Yet, as Biddick points out, with a forty-week gestation period for calves and an interval of at least three to four weeks from calving to first heat, cows had only three mating opportunities per year to maintain yearly production of a calf.[86] Within the FTC counties in the period 1288–1315 cows achieved a calving rate of between 60 and 73 per cent, according to whether it is calculated on the gross number of cows present during the year or the net number enumerated at the start of the year. Each year herds had to be culled of any sterile and decrepit animals and disposal made of all calves surplus to the maintenance of the herd. Unlike herds producing animals for meat and draught there would have been few intermediate sales or transfers. Equal attention also had to be paid to grassland management, since rates of milk production were also a function of diet. Milk output rose rapidly with the resumption of grass growth from late March, peaked in June, and thereafter subsided as grass growth abated until it fell off dramatically in October with the cessation of grass growth.[87] Over the winter months cows had to get by on a diet of hay, straw, pulses and whatever meagre pasturage might be available. Often those lords most interested in developing dairying found it expedient to substitute horses for oxen, since this helped economise on the share of grassland resources which it was necessary to reserve exclusively to the support of the plough.

The kind of large-scale dairying undertaken by demesnes tended to require a considerable capital outlay upon housing for the cows. Stables, byres, and cowhouses were features of many manorial complexes and imply that animals must have been stall-fed for at least part of the year. Some lords went further and invested in purpose-built and equipped dairy houses. The prior of Norwich maintained dairies on his manors of Plumstead, Newton-by-Norwich, and Hemsby. A detailed inventory of 1352 records their contents. That at Hemsby contained: one bench, five Eastland tables, one table with two trestles, one table for drying cheese, five cheese vats, two pressing-boards, one stoup (i.e. wooden bucket), one churn, nine dishes, nine plates, twelve saucers, two hanging tables, one press, one jug, and one broken tong.[88] The purchase of salt, cheese cloths, and replacement items of equipment involved in the cheese- and butter-making processes show up as regular items of expense in Hemsby's manorial accounts, as they do on most dairying demesnes.[89] Costessey in Norfolk, for example, spent 3s. 8¾d. in 1278–9 on stoups, a board, buckets, a press, plates, a bench, a churn, and sundry other items for the dairy.[90]

Getting the most out of a dairy herd, in the form of milk, butter, and cheese, surplus calves, and the occasional mature animal which could be fattened for

[85] *Walter of Henley*, p. 431.   [86] Biddick, *The other economy*, p. 90.
[87] Ecclestone, 'Dairy production'.   [88] *Prior's manor houses*, pp. 14–15.
[89] NRO, DCN 60/15.   [90] PRO, SC 6/933/13.

slaughter, entailed matching substantial capital investment with experience and expertise. Properly managed, a single cow, costing in the first half of the fourteenth century between 8 and 12 shillings to purchase, could be as profitable as several acres of prime arable.[91] Auditors sometimes calculated the annual income per cow and appended a note to that effect on the account. For instance, on the prior of Norwich's Norfolk demesnes of Plumstead, Martham, and North Elmham, it was calculated that the lactage of each cow was worth, respectively, 2s. 7d., 4s. 7d., and an impressive 6s. 0d. in 1326–7.[92] From that time on, cows were increasingly farmed out to a lessee in return for an annual rent, thereby anticipating the more general farming of demesne lands which followed later in the century.

Sometimes cows were farmed for their milk only, with the lord retaining their issue. Within the FTC counties *c.* 1300 lessees usually paid between 3s. 0d. and 4s. 6d. per cow for lactage only (mean of 4s. 0d.).[93] Equivalent rates could also prevail when the lessee was entitled to both the lactage and the calves.[94] Usually, however, lords were able to demand a higher rent when lessees retained both lactage and calves, since the latter were worth anything between 8d. and 15d. each. In Norfolk recorded payments per cow for the farm of both milk and calves mostly fall within the range 4s. 9d. to 6s. 8d.[95] Corresponding rates in the FTC counties are marginally lower, at 4s. 6d. to 5s. 6d. per cow, with a mean rate of 5s. 1½d.[96] These are impressive rental levels, given that the lessees also had to make a livelihood as well as cover their expenses, and imply higher levels of efficiency outside rather than within the demesne sector. The arrangement seems to have been that the herd continued to be managed using the grassland resources of the demesne but that the lessee was responsible for calving, milking, and the manufacture of butter and cheese using the dairying equipment of the demesne, along with the marketing of those products. As an arrangement it must have operated to the benefit of both parties for it endured for many years on significant numbers of demesnes.

The intensive and profitable management of grassland which these rental rates imply is reflected in the high unit value generally placed upon grassland in the main areas of commercial dairying (Figure 4.15). Within Norfolk unit valuations of meadowland, as recorded by the *IPM*s, are highest in precisely that part of the county where seigniorial dairying was most fully developed. Exceptionally high meadow valuations in Suffolk – some as high as 5 shillings an acre – imply

---

[91] Farmer, 'Prices and wages', p. 748.
[92] B. M. S. Campbell, 'Commercial dairy production on medieval English demesnes: the case of Norfolk', *Anthropozoologica* 16 (1992), 113.
[93] Campbell, 'Measuring commercialisation', p. 173.
[94] E.g. at Wroxham in 1342–3, Hainford in 1363–4 and Haveringland in 1356–7 and 1376–7: Campbell, 'Commercial dairy production', p. 113.
[95] Campbell, 'Commercial dairy production', p. 113.
[96] Campbell, 'Measuring commercialisation', p. 173.

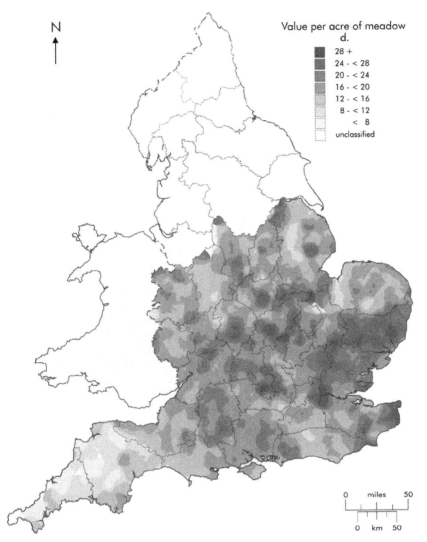

Fig. 4.15. Unit value of meadowland: England south of the Trent, 1300–49 (source: National *IPM* database).

an even more developed interest in dairying.[97] Neither scarcity nor quality can alone explain these remarkable valuations. Instead, like high arable valuations, they are better interpreted as evidence of the high 'rental' income to be obtained from the more specialised and productive forms of husbandry.[98]

[97] Unit valuations of meadowland reached a national peak in high Suffolk: Figure 4.15.
[98] Chapter 7, pp. 345–52.

One of the surest signs of a specialist interest in dairying is the inclusion of a separate dairy account within the annual manorial account, recording the cheeses and butters made, their methods of disposal, and any income that was realised from cash sales. For the royal manor of Costessey, situated in the lush valley of the River Wensum four miles north-west of Norwich, a series of detailed though damaged dairy accounts survive from the 1270s.[99] These reveal a demesne actively engaged in the large-scale commercial production of butter and cheese presumably for the Norwich market. A full-time cowman was employed to tend the herd of twenty-five to thirty milking cows. This bore fruit, on the three occasions for which there are legible figures of the number of cows kept and calves born, in a calving rate of 100 per cent. In order to maximise the amount of milk available for butter and cheese production, two out of three calves born were subsequently sold. Careful culling of aged and sterile females also helped maintain the herd at full milking strength. Manufacture of butter and cheese was placed in the charge of a permanent dairymaid. On at least two occasions the sale of cheese, butter, milk, and calves produced by the herd of twenty-five cows yielded an income of just over 5 shillings per cow. On both occasions well over 90 per cent of cheeses and butters were sold.

The hallmarks of dairying demesnes such as Costessey were several. First, and most obviously, cattle herds were demographically dominated by mature females, usually with a single accompanying bull. In Norfolk, where demesne dairying was well developed, herds generally contained between five and twenty-five cows, the number tending to be higher after 1350 than before. There appears to have been a natural upper limit of thirty-five to forty cows per herd, with, on average, one bull to every thirteen and a half cows. Second, the number of followers – mostly heifers – was kept to the minimum compatible with maintaining the population of dairy cows at full strength (unlike herds geared towards rearing in which immatures often outnumbered adults). In many specialised dairy herds adults outnumbered immatures by two to one. For the same reasons, dairying demesnes also endeavoured to minimise the number of oxen that they stocked, since oxen competed directly with cows for scarce hay and grassland. One of the most effective ways of achieving this was to substitute oxen with horses. Fewer horses would perform the same amount of draught work and do so on a diet that made smaller demands on available grassland. It is therefore no coincidence that the most specialised dairy herds were generally to be found on demesnes employing mixed- or all-horse teams. On such demesnes dairying typically comprised one element within a generally intensive pastoral regime, which, in addition to horses, might include sty-fed swine and a range of farmyard poultry (this confirms the view that adoption of the horse is often best interpreted with reference to developments

---

[99] PRO, SC 6/933/13.

taking place within the pastoral sector as a whole). Sheep were less commonly a component. They were an essentially extensive animal and tended only to be kept on dairying demesnes when either wool prices were exceptionally high or pastoral opportunities were available which only sheep could effectively exploit.

Dairying was the dominant pastoral activity on all demesnes classified as pastoral type 2 (itself the single most prominent pastoral type before 1350) (Figure 4.03). It was also a prominent activity on many demesnes classified as pastoral type 1 and an important subsidiary activity on demesnes classified as pastoral type 3 (Table 4.01). Collectively, the national, FTC, and Norfolk samples of demesnes furnish many examples of these three pastoral types, thereby testifying to a widespread demesne involvement in dairying both after and, especially, before 1350 (Figures 4.01, 4.02, 4.03, 4.04 and 4.05).[100] Within a lowland context they show up in several areas where an abundance of grassland resources encouraged a specialist interest in cattle. Examples include the East Anglian Fen edge, the Rother Valley and Walland and Romney Marshes in south-east Kent, the Somerset Levels and east Devon.[101] More remarkable are the far greater numbers of demesnes which specialised in cattle-based dairying in localities lacking any obvious environmental advantages for pastoral husbandry. Central and south-eastern Norfolk stand out in this respect, as do High Suffolk, eastern Hertfordshire, the immediate environs of London, and southern Hampshire. For the most part these localities owed their specialised pastoral regimes, distinguished by impressively high proportions of non-working animals, to economic and institutional advantages rather than any superior endowment of grassland.

Already in the thirteenth century East Anglia and the Home Counties were relatively highly commercialised. Norfolk, for example, boasted a dense network of over 120 rural markets. These counties were also deeply penetrated by the demand of major urban food markets, both domestic and overseas. Proximity to both Norwich, a city of approximately 25,000 inhabitants in the 1330s, and the densely populated rural textile-producing district to its north together undoubtedly explain why demesne dairying developed further in east-central Norfolk than in any other part of the county.[102] The influence

---

[100] Whether medieval cattle were much valued for their milk has been questioned by Trow-Smith, *Livestock husbandry*, pp. 122–3, and Grant, 'Animal resources', pp. 156–7.

[101] For the Fen edge see Raftis, *Estates of Ramsey Abbey*, pp. 129–58; Biddick, *The other economy*, pp. 84–6. For Romney Marsh and east Kent see Smith, *Canterbury Cathedral Priory*, pp. 146–65. For east Devon see Alcock, 'East Devon manor'; Ugawa, 'Economic development', pp. 646–52; Hatcher, 'Farming: south-western England', pp. 395–8; H. S. A. Fox, 'Peasant farmers, patterns of settlement and pays: transformations in the landscapes of Devon and Cornwall during the later Middle Ages', in R. Higham (ed.), *Landscape and townscape in the south-west* (Exeter, 1990), pp. 57–64.

[102] Campbell, 'Commercial dairy production', pp. 111–13; Campbell, 'Chaucer's reeve', pp. 294–6.

of London, a city three or four times larger than Norwich, was felt across a far wider area. Via a dense network of subordinate trade centres the capital drew upon an extensive hinterland for its provisions. Milk, in its preserved forms of cheese and butter, was capable of transportation over a considerable distance and was high in value relative to its bulk. Much may therefore have been sent to London from a distance. Specialisation in dairying would have been further encouraged by the city's appetite for meat, since this would have provided a ready market for many of the surplus calves and sterile cows – suitably reared and/or fattened – which were such characteristic by-products of dairying.[103]

Nevertheless, the proportion of dairy produce that entered these wider orbits of exchange must always have been small. In the fourteenth century the bulk of all seigniorial dairy production was still undoubtedly intended for local consumption. At the beginning of the century demesnes within the FTC counties sold only half of their dairy produce. The rest was retained for consumption by seigniorial households and estate workforces. Cheese, in particular, long formed a staple of the food liveries paid by lords to their workers. Dyer reckons that dairy produce accounted for a fifth of the value of all foodstuffs consumed by harvest workers on the prior of Norwich's substantial demesne at Sedgeford in north-west Norfolk between 1256 and 1341.[104] On the same estate, manorial consumption alone accounted for half of all cheeses produced on the seven demesnes of Gnatingdon, Thornham, North Elmham, Taverham, Monks Granges, Plumstead, and Martham in 1326–7, a proportion which rose to over three-quarters at Plumstead and Monks Granges. For many lords what the market offered, therefore, was an opportunity to dispose of the surplus which remained after the combined demands of household and estate had been satisfied, coupled with an incentive to specialise and develop production further than would otherwise have been possible.

Over time, as will be seen from Tables 4.02 and 4.03, non-working cattle – many of them dedicated to milk production – gained steadily in relative importance. From the second quarter of the fourteenth century the numbers stocked per 100 sown acres began to edge upwards, and from the last quarter of the fourteenth century, as arable husbandry went into retreat, that rise became pronounced. It was from mid-century, too, that oxen and ox production began to wane as conspicuous components of cattle husbandry. Until 1375, both nationally and in Norfolk, the rise in the stocking density of cattle was accompanied by a progressively greater emphasis upon adult animals, as dairying gained relative to rearing. Thereafter, developments in Norfolk diverge from those within the country at large. In Norfolk the growing numerical imbalance between mature and immature cattle persisted whereas nation-

---

[103] Langdon, *Horses*, pp. 261–2.   [104] Dyer, 'Changes in diet', p. 25.

ally there was an abrupt return to the status quo of the late thirteenth century, with immatures outnumbering adults (Table 4.03).[105]

It was during the third quarter of the fourteenth century that commercial dairying as a demesne activity attained its peak of development. In the immediate aftermath of the Black Death more resources became available for pastoral production and *per capita* demand for dairy produce rose. These favourable circumstances proved transitory. As wage rates rose so many lords found it increasingly cost effective to farm out their dairy operations. Dairying thus became a 'peasant' rather than seigniorial activity, albeit employing working capital provided by lords. Second, continued population decline, rising living standards, and further changes in relative prices collectively eroded the market for dairy produce. A smaller and increasingly well-paid workforce could now afford to replace cheese with meat. By the early fifteenth century, for instance, dairy produce accounted for barely 10 per cent by value of the foodstuffs consumed by harvest workers at Sedgeford, as workers consumed less bread and cheese than in the thirteenth century and more ale and meat.[106] Third, rising costs, especially of labour, and falling prices encouraged a shift to cheaper and more extensive forms of livestock enterprise.

By the close of the fourteenth century the intensive arable-based pastoral husbandry in which Norfolk and much of the rest of East Anglia and the Home Counties had excelled, and of which dairying was one manifestation, had become too expensive to remain economic. In most cases dairying lapsed from being an object in itself and once more became a by-product of stock raising. As economic rent fell so lords turned increasingly to sheep farming. The latter not only had lower labour and feeding costs than cattle, it also produced a variety of products – wool, milk, and meat – which provided a hedge against uncertain markets. By the close of the fourteenth century the number of sheep per head of non-working cattle was 25 per cent higher than it had been at the close of the thirteenth century (Table 4.03). Not until the sixteenth century, when population growth and associated price and wage trends encouraged a return to more intensive forms of husbandry, would specialised dairy farming again become a conspicuous feature of the east and south-east of England.[107]

4.33 Sheep

Sheep are the least satisfactorily recorded of demesne livestock. From the early fourteenth century great estates increasingly managed and accounted for

---

[105] Methods of accounting for dairy herds let at farm undoubtedly mask many immature beasts from view, since demesne managers had only to account for as many animals as were held directly at lease.   [106] Dyer, *Standards of living*, pp. 157–60.

[107] J. Thirsk, 'The farming regions of England', in J. Thirsk (ed.), *AHEW*, vol. IV, *1500–1640* (Cambridge, 1967), pp. 40–9.

Table 4.05. *Composition of livestock gains on demesnes in the FTC counties,*
1288–1315 and 1375–1400

Livestock type	No. of demesnes gaining	% of livestock units gained from:				Total livestock units gained
		Birth	Transfer	Purchase	Other	
1288–1315:						
Equines	153	7	22	55	16	562
Bovines	161	17	30	38	16	1,082
Ovines	129	44	20	33	2	1,824
Porcines	116	74	14	11	1	446
All livestock	165	35	23	35	8	3,914
1375–1400:						
Equines	113	5	11	71	13	365
Bovines	111	10	16	64	11	731
Ovines	108	46	17	34	3	1,746
Porcines	107	90	1	6	3	464
All livestock	127	40	14	41	6	3,305

Note:
The livestock units used are those given on p. 106. They are based on relative
purchase and sale prices within the FTC counties and are specific by age and sex.
Source: FTC1 and FTC2 accounts databases.

them centrally. Sometimes as a result the combined flock of several demesnes
is credited to a single manor; at others, sheep fail to show up in a manor's
accounts even though they were present on the demesne. Nor, with inter-
manorial methods of flock management, does non-appearance at Michaelmas
necessarily mean that sheep were absent throughout the year. Sheep, therefore,
are prone to under-recording. Sporadic clerical use of the long-hundred
further compounds the problem. Sheep are the one category of livestock regularly to have been reckoned in hundreds. Unfortunately, it is frequently
unclear whether those hundreds were long or short. Consequently, counting
sheep from manorial accounts cannot be a precise science. Nor are manorial
accounts themselves adequately representative of seigniorial sheep farming in
general. Too few accounts survive from the upland zone of the north and west,
where the potential for sheep farming was considerable. The many Cistercian
communities of the region, which were among the most active and organised
of early-fourteenth-century wool producers and collectively one of the leading
suppliers of wool to the foreign market, are particularly serious lacunae.[108]

[108] Thirsk, 'Farming regions', pp. 100–2.

Table 4.06. *Composition of livestock losses on demesnes in the FTC counties, 1288–1315 and 1375–1400*

	No. of demesnes losing	Deaths as % of gross livestock units lost	% of livestock units lost (net of deaths) by:					Gross livestock units lost
			Slaughter	Transfer	Transfer ad hospicium	Sale	Other	
Livestock type								
1288–1315:								
Equines	156	38	0	23	0	70	7	373
Bovines	169	15	10	25	3	59	3	1,571
Ovines	134	35	3	31	1	59	5	1,846
Porcines	135	13	19	24	8	44	4	919
All livestock	173	24	9	27	3	56	4	4,709
1375–1400:								
Equines	109	53	0	10	2	85	4	212
Bovines	114	24	4	20	5	70	1	1,048
Ovines	110	40	3	29	6	58	4	1,807
Porcines	106	22	6	13	26	51	5	780
All livestock	127	33	4	22	10	61	3	3,855

Note:
The livestock units used are those given on p. 106. They are based on relative purchase and sale prices within the FTC counties and are specific by age and sex.
Source: FTC1 and FTC2 accounts databases.

Beaulieu Abbey in southern England is one of the very few Cistercian estates for which there are extant accounts.[109]

Sheep, unlike cattle, were kept exclusively for non-draught purposes. Insofar as they were kept to service the arable, it was as walking dung machines on light-land demesnes. This required a sturdy breed which could thrive on the meagre grazing of heaths, sheepwalks, and degraded common pastures. Such arable sheep had to be good walkers because each evening they would be moved onto the arable to be penned in folds, returning to their sheep-walks the following morning (the arable, in the interim, having been 'tathed' with their treading, dung, and urine). In downland and heathland areas the distances involved could be considerable. In west Norfolk and the Suffolk Breckland, where the seigniorially organised foldcourse system was instrumental in keeping much light land under cultivation, the local breed yielded only a relatively coarse fleece of low value. It was the corn which until *c.* 1375 was the most profitable component of this region's sheep-corn husbandry, not the sheep. Without the sheep, however, it would have been difficult to maintain yields and keep so much land under cultivation.

As a food animal sheep were kept much more for their milk than their meat.[110] Within the FTC counties only horses were less likely to be slaughtered for their meat (Table 4.06). It was in the fifteenth century, with the glutting of the market for wool and higher *per capita* meat consumption, that sheep first appear to have been reared for their mutton.[111] Their use as a dairy animal was of much longer standing. Trow-Smith calculated that medieval cows yielded between ten and twenty times more milk than ewes.[112] Within the FTC counties the lactage of a ewe was valued at between 1½d. and 2½d., whereas that of a cow was over twenty times greater at 3s. 0d. to 4s. 6d. The Luttrell Psalter produced in East Anglia in the early fourteenth century contains a celebrated illustration of women milking folded ewes.[113] In the early thirteenth century sales of sheep's cheese provided Peter des Roches, bishop of Winchester, with a lucrative source of income. Biddick reckons that several of the bishop's demesnes produced more cheese per cultivated acre than wool. She calculates that 'dairy income per ewe varied between 66 and 100 per cent of the wool income per ewe and wether of the sheep flock'.[114] Sheep's cheese was also a notable product of the extensive flocks maintained on the Essex salt marshes.[115] At the opening of the fourteenth century demesnes within the

---

[109] *The account-book of Beaulieu Abbey*, ed. S. F. Hockey (London, 1975); Donkin, *The Cistercians*, pp. 83–102.   [110] Cf. Grant, 'Animal resources', pp. 154–5.

[111] Mate, 'Pastoral farming', pp. 535–6.   [112] Trow-Smith, *Livestock husbandry*, p. 122.

[113] BL, Add. MS 42130 fol. 163 verso, reproduced in J. Backhouse, *The Luttrell Psalter* (London, 1989), p. 17.   [114] Biddick, 'Agrarian productivity', p. 116.

[115] On the importance of the Essex marshlands for sheep farming see H. C. Darby, *The Domesday geography of eastern England* (Cambridge, 1971), pp. 241–4. The distribution of sheep in 1341 can be reconstructed with considerable detail for the parts of the country covered by those extant returns of the *Nonarum inquisitiones* which distinguish between the respective tax

FTC counties stocked almost as many ewes as wethers. Relatively favourable lambing rates of 64–88 per cent (the lower calculated on the gross number of ewes present during the year, the higher on the net number of ewes recorded at the start of the year) served to keep the ewes in milk and were underpinned by a mean ratio of one ram to 30 ewes.[116] Nevertheless, on soils and in situations where alternative forms of pastoral husbandry are possible, sheep farming is one of the least productive methods of producing human food both per unit of land and per unit of solar energy.

It was as a source of textile fibre rather than food that sheep tended to give their best returns.[117] Wool was the raw material of the largest and most specialised of medieval industries. Since wethers produced the heaviest fleeces – twice those of hoggs (two-year-olds) and a third heavier than those of ewes – they tended to dominate the most specialised wool-producing flocks.[118] The weight and quality of the fleece were also affected by the breed, the environment, and the quality and intensity of management.[119] For example, the Cotswold sheep of north Oxfordshire bore fleeces twice as heavy as medieval Welsh hill sheep.[120] Wool yields on the estates of the bishopric of Winchester, the abbeys of Crowland, Peterborough, and Ramsey, Merton College Oxford, the countess of Aumale, and Adam de Stratton's manor of Sevenhampton in Wiltshire have been the subject of detailed investigation by M. J. Stephenson.[121] Wool yields reported in Volume III of the *Agrarian history* basically confirm Stephenson's findings, as do those calculated from the two FTC databases.[122] On the extensive estates of the bishops of Winchester which stocked up to 35,000 sheep between 1210 and 1454 mean fleece weights ranged from a minimum of 0.87 lbs. at Bishopstoke to a maximum of 1.70 lbs. at Upton, with fleeces in excess of 4 lbs. sometimes being produced at Adderbury in north Oxfordshire (which seems to have stocked sheep bearing middle- rather than short-staple wool). Across the Winchester estates as a whole fleece weights averaged 1.35 lbs., although mean fleece weights rose to a temporal maximum of 1.77 lbs. in the early 1320s and fell to a minimum of 1.04 lbs. in

receipts of corn and of wool and lambs: e.g. R. A. Pelham, 'The distribution of sheep in Sussex in the early fourteenth century', *Sussex Archaeological Collections* 125 (1934), 128–35. The broader national distribution of sheep can be reconstructed from the near-contemporary wool tax: Ormrod, 'Crown and economy', pp. 178–9; Campbell, 'Tradable surpluses'.

[116] These rates apply to the periods 1288–1315 and 1375–1400 and are calculated from the FTC1 and FTC2 accounts databases.

[117] J. H. von Thünen, *Der isolierte Staat* (Hamburg, 1826), trans. C. M. Wartenberg, *Von Thünen's isolated state*, ed. P. Hall (Oxford, 1966), pp. 177–81.

[118] Trow-Smith, *Livestock husbandry*, p. 149; M. J. Stephenson, 'Wool yields in the medieval economy', *EcHR* 41 (1988), 373.

[119] J. P. Bischoff, '"I cannot do't without counters": fleece weights and sheep breeds in late thirteenth and early fourteenth century England', *Agricultural History* 57 (1983), 142–60.

[120] Trow-Smith, *Livestock husbandry*, pp. 167–8.   [121] Stephenson, 'Wool yields'.

[122] Miller (ed.), *AHEW*, vol. III, pp. 192, 209, 220, 235, 281, 296–7, 320.

the late 1440s.[123] The great bulk of all fleeces shorn on the Winchester estates weighed 1.25–1.75 lbs. and this seems to have been the normal weight range of the predominantly short-woolled medieval fleece. Mean fleece weights were significantly lower in the counties of the extreme south-west and higher in the East Riding of Yorkshire but across lowland arable England as a whole seem not to have deviated far from the 1.4 lbs. per fleece obtained by demesnes in the FTC counties *c.* 1300. When wool prices were at their peak in the first quarter of the fourteenth century such a fleece would have fetched just over 6d.; at that time a dozen fleeces were roughly equivalent in value to a single quarter of wheat. Sheep farming therefore had to be conducted on a large scale before it became a major gross earner. At the height of the early-fourteenth-century wool export boom, wool accounted for just over a quarter of the gross sales revenue of the pastoral sector and less than a tenth of gross agricultural sales income on demesnes within the FTC counties.[124] By the close of the fourteenth century, although the proportion of wool-producing manors had risen from 70 per cent to 83 per cent, wool was still contributing only 10 per cent of gross agricultural revenues while its share of gross pastoral revenues had declined from 27 per cent to 22 per cent as sales of live animals and the farming of demesne dairies both gained in importance.

Seigniorial wool production almost everywhere was an intrinsically land-extensive activity, reliant upon extensive tracts of permanent pasture. Supplementary feed, winter shelter in purpose-built cotes, and closer supervision could help maximise lambing rates and minimise mortality. Such measures, or the want of them, also undoubtedly impacted upon fleece weights.[125] Careful sorting, grading and packing of wool could also improve its marketability.[126] Otherwise sheep-farming responded less well than most other branches of pastoral husbandry to intensification of capital and labour inputs. For seigniorial producers sheep-farming consequently offered its greatest economic advantages at the extensive rather than intensive margin of agriculture: hence its close association with regions of low population density, cheap land, and low economic rent outside of England's demographic and economic core. Remoteness from markets was not an insurmountable problem to wool producers because their product was high in value relative to its bulk and if properly stored could be kept for periods at a time without

[123] Stephenson, 'Wool yields', pp. 370–81.
[124] Campbell, 'Measuring commercialisation', pp. 147–53.
[125] Trow-Smith, *Livestock husbandry*, pp. 148–69; P. D. A. Harvey, 'Farming practice and techniques: the Home Counties', in Miller (ed.), *AHEW*, vol. III, pp. 265–6; H. E. J. Le Patourel, 'Rural building in England and Wales', in Miller (ed.), *AHEW*, vol. III, pp. 878–81; Dyer, 'Sheepcotes'; C. Thornton, 'Efficiency in thirteenth-century livestock farming: the fertility and mortality of herds and flocks at Rimpton, Somerset, 1208–1349', in P. R. Coss and S. D. Lloyd (eds.), *Thirteenth century England* (Woodbridge, 1992), vol. IV, pp. 25–46.
[126] Lloyd, *English wool trade*, pp. 288–98.

serious deterioration. Wool, alone among agricultural products, was purchased in bulk by international dealers.

As a food animal, sheep were most likely to be preferred over cattle at locations where economic rent was comparatively low and in periods when it was falling. On this reasoning, there should have been a shift in lowland England from sheep to cattle during the thirteenth century and a shift in the opposite direction during the late fourteenth and the fifteenth centuries. Yet, paradoxically, it was in the late thirteenth and early fourteenth centuries, when the urban textile industries of Italy and Flanders were at their height, that wool production gave its best returns. Moreover, the strength of this overseas demand meant that wool early became the single most valuable component of English exports, rendering it unique among agricultural products from the late thirteenth century in the degree of government interference to which it became subject.[127] Producers were thus exposed to a variety of at times countervailing influences, rendering sheep one of the most complex components of medieval agriculture.[128] This complexity is manifest in the tendency for wool prices to behave differently from prices of other agricultural products.

At an aggregate level sheep consistently took second place to cattle (Table 4.03). Fewer demesnes kept sheep and in aggregate sheep accounted for a smaller share of livestock units. This holds true nationally, within the FTC counties, and in Norfolk. That being said, sheep gained consistently in relative importance over the period under consideration. Notwithstanding the growing tendency for great estates to account for sheep separately, a higher proportion of demesnes stocked sheep at the close of the fourteenth century than had done so at its opening (three out of four rather than two out of three). They also stocked them in larger numbers. Flocks grew from a mean of just over 400 to almost 500 animals and those within the FTC counties contained a higher proportion of wool-producing wethers.[129] The growth of seigniorial sheep farming was the single most dynamic element within the non-working sector. Over the course of the fourteenth century sheep gained in importance relative to cattle by 18 per cent nationally, 47 per cent in Norfolk, and 68 per cent in the FTC counties. They gained even more relative to the sown area, for the number of sheep stocked per 100 sown acres rose by 60 per cent in Norfolk, 90 per cent nationally, and 115 per cent in the FTC counties.

Before 1325 the relative rise in sheep numbers reflects the powerful

[127] Lloyd, *English wool trade*; W. M. Ormrod, 'The English crown and the customs, 1349–63', *EcHR* 40 (1987), 27–40.

[128] J. H. Munro, 'Environment, land management, and the changing qualities of English wools in the later Middle Ages', paper presented at the 20th International Congress on Medieval Studies, Kalamazoo, May 1985.

[129] By the close of the fourteenth century wethers outnumbered ewes by three to two, whereas at the century's opening their numbers had been roughly equal: FTC1 and FTC2 accounts databases.

production incentive provided by buoyant and rising prices for English wools on the international market. Lords built up their flocks mainly by natural increase – where necessary drafting animals from one manor to another – augmented by purchase (Table 4.05). E. Power believed that a demand-stimulated rise in the medieval sheep population had been taking place progressively throughout the twelfth and thirteenth centuries, notwithstanding the concurrent expansion of tillage.[130] Since sheep are a mainly grass-fed animal, this would only have been possible if there was a concomitant expansion in the amount of pasture. Nowhere is the expansion of the pastoral area more conspicuous than on the estates of the many new Cistercian monasteries, whose ethos and land endowments predisposed them towards a predominantly pastoral economy. The Cistercians were active reclaimers of the waste and thereby created numerous upland granges staffed by lay brothers. By the close of the thirteenth century Cistercian estates had become some of the largest individual suppliers of wool to the export trade.[131] The Yorkshire houses alone exported the clip of 70,000 sheep, 20,000 of them belonging to Fountains, the greatest sheep-farmer of them all.[132]

Wool prices peaked in the opening decades of the fourteenth century and it was then, in Power's view, that the national flock attained its medieval maximum.[133] The quality of English wool made it particularly sought after. In the first decade of the fourteenth century annual exports were worth in excess of £¼m.; equivalent in value to the fleeces of almost 10 million animals.[134] Yet more sheep were involved in producing the wool exported as cloth. When allowance is made for the fact that domestic is likely to have matched if not exceeded overseas consumption of wool, it follows that the national sheep population *c.* 1310 can scarcely have been less than 20 million and is likely to have been far greater.[135] This implies a national stocking density of almost 300

---

[130] Power, *Wool trade*, pp. 31–5.

[131] Power, *Wool trade*, pp. 22–3; Donkin, *The Cistercians*, pp. 85–91.

[132] Trow-Smith, *Livestock husbandry*, p. 139.

[133] Power, *Wool trade*, pp. 34–5; R. A. Pelham, 'Fourteenth-century England', in H. C. Darby (ed.), *An historical geography of England before A.D. 1800* (Cambridge, 1936), p. 240.

[134] Estimated for 1300–9 as follows: mean annual value of wool exports £258,188 (Miller and Hatcher, *Towns, commerce and crafts*, p. 213); mean wool price 4.83 shillings per stone; mean value per fleece 0.53 shillings (Farmer, 'Prices and wages', p. 757); mean fleece weight 1.5 lbs. (Stephenson, 'Wool yields', p. 377; within the FTC counties mean fleece weight was 1.4 lbs. *c.* 1300: FTC1 accounts database). If the 41,310 sacks of wool exported annually during the years 1304–9 (Carus-Wilson and Coleman, *Export trade*, p. 41) each contained the fleeces of at least 240 sheep, it would have required at least 10m. sheep to supply the export trade alone. In 1341, when the crown was granted one-fifteenth of all the wool produced in the kingdom, over 25,000 sacks of wool were collected, which was equivalent to the product of over 6m. sheep: Ormrod, 'Crown and economy', pp. 176–81.

[135] Trow-Smith, *Livestock husbandry*, p. 140, estimates a national wool-producing flock of about 12m. animals via a slightly different method. In comparison, Gregory King reckoned that there were 11m. sheep in England and Wales in 1688–95 and by 1741 the equivalent number, based on the estimate made by the anonymous author of *A short essay on trade*, had risen to

sheep per 100 sown acres. Such a density is roughly thrice that pertaining on sampled demesnes both nationally and within the FTC counties in the first half of the fourteenth century and four times that on Norfolk demesnes in the same period (Table 4.03). So great a discrepancy cannot be accounted for by under-recording and under-representation alone.[136] Rather, the demesne share of the national sheep flock must have been substantially smaller than its share of the national arable area. Here, the Norfolk evidence, because it is the fullest, is the most telling. Throughout the period 1250–1450 Norfolk demesnes consistently stocked sheep at a density per sown acre well below the national average; yet in 1341–2 the county contributed over an eighth of the entire national wool tax, paying two-and-a-half times more than any other county.[137] Such a massive contribution would only have been possible if sheep in Norfolk (and, by implication, within the country as a whole) were primarily non-demesne animals.[138] The existence of large numbers of peasant sheep would certainly help to explain why it was so much in the interests of Norfolk lords to assert a prerogative right of foldcourse over their tenants' animals.

In the second quarter of the fourteenth century wool prices slumped as the crown for financial and military reasons began to interfere in the export trade. This culminated in the imposition of a heavy customs duty on exported wool in the 1330s. Producers also had to contend with repeated taxation, purveyancing, adverse weather, and recurrent outbreaks of murrain. Such setbacks were, however, uneven in their impact and essentially transitory in their effects. Those lords most readily deterred were 'fair-weather' sheep-farmers, lured into sheep-keeping by apparently easy profits and quick to abandon it when prices slumped and costs rose.[139] More committed sheep-farmers persevered. Within the demesne sector as a whole sheep continued to make advances.

More dramatic gains accrued from the economic transformation triggered by the Black Death and the new demographic era which it initiated. Suddenly labour rather than land was the scarce factor of production. Relative prices and production costs shifted increasingly in favour of pastoral production, with extensive rather than intensive methods proving the most cost effective. These developments favoured sheep even more than cattle. Overseas demand for wool was nevertheless weaker than formerly and domestic demand, although strengthening, was not yet substantial enough to provide adequate

about 13m. P. J. Bowden, 'Agricultural prices, wages, farm profits, and rents', pp. 1–118 in J. Thirsk (ed.), *AHEW*, vol. V.ii, *1640–1750, Agrarian change* (Cambridge, 1985), p. 11.
[136] Adding in the ¼m. sheep reckoned by Trow-Smith, *Livestock husbandry*, p. 139, to have been in monastic ownership in Yorkshire would do little to bridge the gap.
[137] Ormrod, 'Crown and economy', pp. 178–9.
[138] Masschaele, *Peasants, merchants, and markets*, pp. 52–3.
[139] E.g. sheep disappear from the accounts of the manor belonging to the de Ingaldesthorp family at Wimbotsham on the Norfolk Fen edge after 1325; hitherto a small flock of forty to 100 sheep had been kept: NRO, Hare 213 × 1/4272, 213 × 4.

compensation. Wool prices, therefore, registered a general decline.[140] Notwithstanding Power's view that these lower prices discouraged any recovery of sheep numbers to their early fourteenth-century peak, seigniorial interest in sheep-farming patently grew significantly over this period (Table 4.03).[141] It did so not because of any sustained price incentive but because sheep farming incurred lower costs than most alternative activities.

Nor was sheep-farming itself immune from the general de-intensification of husbandry methods which followed in the wake of the Black Death. As with husbandry in general it became a more land-extensive activity. On the estates of the bishops of Winchester fleece weights suffered as, from the 1370s, methods of management became more lax.[142] Increasingly, flocks were maximised and labour inputs minimised. In some parts of the country unscrupulous lords expropriated fold rights and expanded their own sheep numbers at the expense of their tenants'. Nowhere did this process proceed further than in Norfolk. Whereas in the early fourteenth century the county's sheep were in disproportionately peasant ownership, by the close of the sixteenth century comparatively few sheep remained in the hands of tenant farmers; instead the majority of recorded sheep were concentrated in the hands of a few substantial flockmasters, mostly scattered through the west and south-west of Norfolk.[143] Initially, this progressive transformation of ownership arose from the realisation that sheep rather than corn had become the more profitable enterprise. Accordingly, arable demesnes were leased out, flocks retained in hand, and foldcourse rights extended and exploited to the benefit of the landlord and detriment of the tenantry.[144] By 1500 the prior of Norwich had long since leased out all his demesne arable but was nevertheless managing four times as many sheep as in 1300.[145] Resentment at this engrossing of fold rights was an important source of the mounting rural tension within the county which culminated in 1546 in the outbreak of Kett's Rebellion.[146] Yet the ascendancy of seigniorial sheep farming was not to be reversed, as is apparent from the highly skewed distribution of flock sizes recorded in seventeenth-century probate inventories.[147]

---

[140] Lloyd, *Wool prices*, pp. 24–30; Farmer, 'Prices and wages, 1350–1500', pp. 461–4.

[141] Power, *Wool trade*, p. 35. On the estates of the bishops of Winchester sheep numbers peaked in the second rather than the first half of the fourteenth century: Stephenson, 'Wool yields', pp. 385–6.   [142] Stephenson, 'Wool yields', pp. 378–81.

[143] Some of the county's greatest flockmasters managed as many as 15,000 animals: Allison, 'Flock management', pp. 99–101; A. Simpson, *The wealth of the gentry, 1540–1660* (Cambridge, 1963), pp. 179–216.   [144] Bailey, 'Sand into gold', pp. 43–51.

[145] NRO, L'Estrange IB 4/4; DCN 62, 64.

[146] K. J. Allison, 'Flock management in the sixteenth and seventeenth centuries', *EcHR* 11 (1958), 98–112; D. MacCulloch, 'Kett's Rebellion in context', *PP* 84 (1979), 51–3. Paradoxically, contemporary estimates suggest that by the 1540s the national flock contained 3.0–8.4m. animals and hence was substantially smaller than 200 years earlier: M. W. Beresford, 'The poll tax and the census of sheep, 1549', *AHR* 1 (1953), 12.

[147] 70 per cent of extant Norfolk probate inventories of the sixteenth and seventeenth centuries record no sheep: Overton and Campbell, 'Norfolk livestock farming', pp. 383, 386.

For all wool's importance to the national economy sheep farming was always an essentially regional activity. This is most immediately self-evident in the wide variations in wool quality to which the pronounced geographical variations in price levels bear testimony. At the beginning of the fourteenth century the very finest wools were grown in the Welsh Marcher counties of Monmouthshire, Herefordshire, and Shropshire, and the next finest in the Kesteven and Lindsey divisions of Lincolnshire, narrowly followed by those of the Cotswolds. Whereas the Marcher wools retained their pre-eminent position throughout the Middle Ages, by the fifteenth century the rank order of the Lincolnshire and Cotswold wools had been reversed. The sheep which produced these finer wools were less productive and required more nourishing feed and greater care than those which bore the coarser wools. With the exception of Lincolnshire, areas specialising in fine wools therefore tended to lie in the land-locked western interior of England. They were well provided with suitable pasturage, had below average population densities, and were characterised by generally low levels of economic rent. In such regions only a high-valued, non-perishable product could provide a lifeline to the wider commercial world.

By the early fourteenth century some high-quality wools were also being produced by the better organised northern abbeys: Holmcultram in Cumberland, Furness in Lancashire, Stanlaw and Combermere in Cheshire, Fountains and Kirkstall in Yorkshire, and Newminster in Northumberland. Otherwise northern wools were among the poorest in quality. As, after 1350, methods of flock management deteriorated, so there was a concomitant deterioration in the quality of these northern wools. By the early fifteenth century all save those of Yorkshire were exempt from the Staple of Calais. Much the same applied to wool from the extreme south-west. Wools produced in Devon and Cornwall were of such poor quality – the latter were derisively known as 'Cornish hair' – that they scarcely appeared at all in the various wool-price schedules of the later Middle Ages. Production of lower-quality wools also predominated in the south-east and in Norfolk and Suffolk. These wools occupied the bottom-most ranks of the price schedules and seem, like the northern wools, to have suffered some deterioration in quality during the fourteenth and early fifteenth centuries. The best of them were produced in Hampshire and the adjoining counties of south-central England.[148] To producers in these lowland arable counties coarse wools had the merit that per unit area they returned a higher yield at less cost than their finer and higher-priced alternatives.

Nationally, over a third of all sampled demesnes had a developed interest in sheep-farming in the period 1250–1349, and over a half in the period 1350–1449. Generally, the greater their specialisation, the more extensive their

---

[148] J. H. Munro, 'Wool-price schedules and the qualities of English wool in the later Middle Ages, circa 1270–1499', *Textile History* 9 (1978), 118–69.

pastoral husbandry. Those classified as pastoral type 4, for example, employed mainly oxen for draught work and kept mostly sheep for profit (Table 4.01 and Figures 4.06 and 4.07). Both were essentially grass-fed animals with a heavy reliance upon natural rather than produced fodder. Such an exclusive concentration on sheep was always a minority specialism. Far more common both before and after 1350 were demesnes classified as pastoral type 3. These made fuller use of horses for draught purposes and thereby supported a marginally larger non-working sector within which they combined sheep with cattle. After 1350 these became the single most common pastoral type (Table 4.01). Seemingly, it was the more extensive forms of seigniorial sheep-farming which gained most during the 100 years which followed the Black Death.

Seigniorial sheep-farming was an exceptionally widely distributed pastoral activity (Figures 4.04, 4.05, 4.06, and 4.07). As a primary or secondary pastoral specialism it could be found on demesnes in almost any part of the country throughout the period 1250–1450. Lords kept sheep on upland and lowland, on light-land and heavy-land, on wolds and downs and in fens and marshes. Only a minority of demesnes, however, took specialisation in sheep to the extreme. Before 1350 demesnes with a more or less exclusive interest in sheep-farming appear to have been absent from the north-west and south-west. In most other parts of the country demesnes with this extreme form of specialisation generally only occur as relatively isolated examples. The exception was on the chalk downlands of central-southern England and along the Oolitic limestone belt which runs diagonally across the country from Gloucestershire to Lincolnshire; here sheep-farming comprised the dominant pastoral type.

Permanent pastures in these downland and woldland environments produced a sweet, short turf upon which sheep thrived. In the absence of much surface water and meadowland cattle were not a viable alternative. The scale of seigniorial sheep-farming in these areas could be considerable. The bishops of Winchester, for example, ran flocks in excess of 1,500 animals on their main downland manors of Knoyle and Downton in Wiltshire and Twyford and East Meon in Hampshire. After 1350 it is in the chalk and limestone country of southern England that the growth of specialist sheep-farming was most marked, a system of sheep-corn husbandry enabling it to dovetail very effectively with extensive cultivation of grains. On the pastoral side, more and more demesnes in Dorset, Wiltshire, Hampshire, Berkshire, Surrey, and Sussex went over exclusively to sheep-farming. Nowhere did this development go further than on Salisbury Plain where the pastoral husbandry practised by most demesnes amounted, in effect, to sheep monoculture.[149] Where seigniorial sheep-farming developed so strongly there was not always the same scope for other classes of producer. It should therefore be no surprise to find many tenants without sheep in these classic sheep-farming areas.[150]

[149] Scott, 'Medieval agriculture', pp. 19–21; Hare, 'Wiltshire agriculture', pp. 4–9.
[150] Postan, 'Village livestock'.

Elsewhere in lowland England sheep tended to be kept to complement cattle and thereby make the most of available environmental and economic opportunities. The precise balance struck between sheep and cattle varied from place to place and over time. In the south-west, outside of the cattle country of east Devon, most demesnes kept both since this was the most effective way of exploiting the full range of pastoral resources available within this diverse region. In contrast, within the Thames basin it was the range of strong market opportunities for sheep and cattle products which explains why all but the heaviest-soiled demesnes and those within the immediate environs of the capital stocked both. London itself was at once the country's greatest single food consumer and leading wool-exporting port. Within the closely settled countryside of Kent, East Anglia, and the east midlands, however, demesnes generally only combined sheep with cattle when this was the only effective method of exploiting available pastoral resources to the full. This applied on the lighter soils of the north-west and south-west of Norfolk but not on the heavier and richer soils of the centre and east. In Suffolk, too, demesne sheep were rarely encountered in number on the heavy boulder clay soils which dominated so much of the county. Similarly, within the commonfield country of the east midlands it was more usual for demesnes to keep cattle alone than sheep and cattle.

If sheep were rarely kept by demesnes in these counties the same cannot be true of other classes of producer, for Norfolk, Suffolk, Cambridgeshire, Huntingdonshire, Bedfordshire, Northamptonshire, and Rutland all contributed to the national wool tax of 1341 at above average rates per unit area.[151] Perhaps sheep were better suited than cattle to the more limited pastoral and capital resources available to peasant producers. Possibly, too, the absence of much seigniorial interest in sheep-farming left greater opportunities for others to exploit. Certainly, the wool tax confirms Trow-Smith's observation that 'hurdled arable flocks' were 'the true basis of medieval sheep husbandry', while comparison with the demesne evidence leaves little doubt that these arable flocks were largely non-seigniorial in ownership.[152] After 1350 lords throughout East Anglia and the south-east took a more active interest in sheep farming and it became the norm rather than the exception for demesnes to stock sheep as well as cattle. Only in the east midlands did sheep fail to make much progress. The pastoral resources available within its commonfield townships were still too limited to allow effective development of seigniorial sheep-farming on any very significant scale. It would take the depopulation movement of the fifteenth century to transform this situation.[153]

The one major exception to the post-1350 expansion of seigniorial sheep-farming was the north of England. The imposition of direct taxes on wool and

[151] Trow-Smith, *Livestock husbandry*, p. 141.
[152] Trow-Smith, *Livestock husbandry*, p. 142.
[153] M. W. Beresford and J. G. Hurst, *Deserted medieval villages* (London, 1971), pp. 11–17.

establishment of the Staple of Calais evidently had a disproportionately adverse effect upon wool producers in the north and from the second quarter of the fourteenth century sheep numbers progressively dwindled.[154] It had always required capital and entrepreneurship to overcome the region's natural disadvantages for wool production. Adverse environmental conditions were a major reason for the coarseness of much northern wool. Many northern flocks suffered high mortality levels and the various forests and chases still harboured several natural predators. Nor were sheep integral to the mixed husbandry of the region in the way that they were on the downs and wolds of southern England. Demographic and economic change therefore further marginalised the north as a major sheep-farming region. Once the major monastic houses began leasing their estates the fate of large-scale seigniorial sheep-farming within the region was sealed. With the abandonment of their active interest in wool production a major source of quality control was removed. The north's established emphasis upon cattle therefore became yet further pronounced as the later Middle Ages progressed. By the time of the Dissolution there were almost twice as many cattle as sheep on the estates of Fountains Abbey, the greatest of the northern Cistercian houses.[155] From *c.* 1450 a clear spatial dichotomy is apparent between a cattle-dominated north of England and an increasingly sheep-dominated south.

The development of seigniorial sheep-farming over the period 1250–1450 thus displays several paradoxical features. First, notwithstanding powerful incentives to develop more food-productive forms of pastoralism, sheep numbers rose in the late thirteenth and early fourteenth centuries in response to strong international demand for English wool. Nor did that expansion of seigniorial sheep numbers cease after 1349, even though government policies had wrecked established wool-marketing structures and wool prices underwent a long process of secular decline. The incentive now was not higher prices but the lower production costs that animals offered compared with crops and that sheep in particular offered compared with other categories of livestock. Second, the regions of highest wool prices were not necessarily those with the heaviest seigniorial commitment to sheep-farming, nor were they the regions within which the post-Black Death swing to sheep farming was most marked. To a disproportionate extent the fourteenth- and fifteenth-century expansion of seigniorial sheep-farming took place in the areas of lower-priced wools in the arable south and east of England, with the result that an increasing proportion of the national wool supply must have been made up of inferior-quality wools. Such wools underpinned the contemporary growth of those branches of textile manufacture which utilised the cheaper long-staple

---

[154] Munro, 'Changing qualities', pp. 8–10.
[155] R. A. Donkin, 'Cattle on the estates of medieval Cistercian monasteries in England and Wales', *EcHR* 15 (1962), 31–53.

wools.[156] The expansion of seigniorial sheep numbers may, however, have been at the expense of those kept by other producers with the result that the national sheep population may actually have declined. A double dichotomy therefore existed between regions that specialised in sheep-farming and those that did not and producers that specialised in sheep-farming and those that did not. Within areas such as the Hampshire and Wiltshire chalklands sheep-farming was dominated by lords; within others, such as Norfolk, it was initially more in the hands of peasants. Viewed in isolation the geography of seigniorial sheep-farming is therefore an imperfect guide to the geography of sheep-farming in general.

4.34 Swine

Swine were the one category of demesne livestock kept solely for their food value as meat and lard.[157] Game and poultry apart, they were the single most productive source of meat available to medieval farmers, with an output of human food per acre and per unit of solar energy which compared favourably with that of dairying.[158] One source of this productivity was their naturally high rate of reproduction; within the FTC counties sows produced, on average, nine to twelve piglets per year.[159] Swine also matured faster than other livestock. Pigs could be slaughtered for consumption at almost any age, but made the best baconers and porkers when they were rising two (less than half the age at which oxen reached their full carcass weight).[160] They were rarely allowed to live long enough to die a natural death. Instead, they were more likely to be slaughtered – on the manor or elsewhere on the estate – than any other category of livestock (Table 4.06). Certainly, most excavated pig bones are of slaughtered juvenile animals.[161]

Swine can be managed either extensively or intensively. Browsing and foraging in woodlands and rough pastures represents the most typical medieval form of extensive management, with the natural pannage afforded by beech mast, acorns, and the like providing a seasonal opportunity to fatten the animals.[162] This was the prevalent form of swine management at the time of

[156] J. H. Munro, 'Structural changes in late medieval textile manufacturing: the Flemish response to market adversities, 1300–1500', public lecture delivered at the Katholieke Universiteit Leuven on 5 November 1986; A. F. Sutton, 'The early linen and worsted industry of Norfolk and the evolution of the London Mercers' Company', *NA* 40 (1989), 201–25.

[157] Biddick, *The other economy*, p. 40; Campbell, 'Measuring commercialisation', pp. 164, 168; Grant, 'Animal resources', pp. 157–9.   [158] Simmons, *Ecology*, p. 203.

[159] 1288–1315, 9.7 piglets per sow were produced; 1375–1400, 12.0 piglets per sow were produced: FTC1 and FTC2 accounts databases.

[160] Trow-Smith, *Livestock husbandry*, p. 128.   [161] Grant, 'Animal resources', p. 158.

[162] Biddick, *The other economy*, p. 45; Miller (ed.), *AHEW*, vol. III, pp. 190, 243, 267, 417–18; D. L. Farmer, 'Woodland and pasture sales on the Winchester manors in the thirteenth century: disposing of a surplus or producing for the market?', in Britnell and Campbell (eds.), *Commercialising economy*, pp. 105–7, 112–13.

Domesday – for much of the woodland in that great survey is measured in terms of the number of swine it fed or rendered – and was still probably the method in use at Monks Risborough, on the edge of the Chiltern woodlands, and at West Tanfield in the North Riding of Yorkshire over two centuries later.[163] Over the intervening period, however, colonisation and assarting had greatly eroded the opportunities for this type of pig-keeping. In the half-century after 1300 only a minority of *IPM*s identify pannage as an independent source of income from demesne woodlands. References to pannage are most common in the West Sussex Weald (where it had long been important and significant stands of thick woodland remained), in the forests and chases of the far north-west of England, and in parts of Devon and Cornwall (Figure 3.10).[164] These must have been among the few areas where pig management was still largely conducted on more or less exclusively extensive lines. In most other parts of the country, except where woodland was very extensive, pannage alone offered little more than a seasonal bonus to other sources of feed.[165] Where this was the case, pig-keeping, in common with most other aspects of husbandry, responded by becoming more intensive.

In its most intensive medieval form pig-keeping involved sty-feeding animals on legumes, poor-quality grains, the by-products of dairying and brewing, and household waste.[166] This was the method of pig husbandry followed on the home manors of Peterborough Abbey at the start of the fourteenth century. According to Biddick, these piggeries yielded sufficient dressed meat to feed the community of 140 men for 157 days at 2,500 calories a day.[167] Sty husbandry was similarly employed on the abbey of St Benet at Holme's home demesnes of Potter Heigham and Hoveton in Norfolk. Both combined pig-keeping on a substantial scale with the large-scale cultivation of legumes.[168] In fact, in their most intensive, legume- and grain-based, form swine represent the most extreme example of a pastoral activity underpinned by arable production. Swine were therefore a characteristic component of those intensive mixed-farming systems which relied upon substantial sowings of legumes to replenish nitrogen supplies within the soil. After 1349, when legumes were partially substituted for grain within regular two- and three-course rotations, swine likewise tended to increase in numbers.

The association between pigs and legumes shows up in several parts of the

[163] Darby, *Domesday England*, pp. 171–8.

[164] National *IPM* database; Hewitt, *Mediaeval Cheshire*, p. 58. On woodland and forest economies see J. R. Birrell, 'The forest economy of the Honour of Tutbury in the fourteenth and fifteenth centuries', *University of Birmingham Historical Journal* 8 (1962), 114–34.

[165] Farmer, 'Woodland and pasture sales', pp. 106, 120.

[166] Grant, 'Animal resources', p. 158.

[167] K. Biddick, 'Pig husbandry on the Peterborough Abbey estate from the twelfth to the fourteenth century A.D.', in J. Clutton-Brock and C. Grigson (eds.), *Animals and archaeology* (Oxford, 1985), vol. IV, pp. 161–77; Biddick, *The other economy*, pp. 121–5.

[168] NRO, Diocesan Est/11; Church Commissioners 101426 7/13, 2/13, 11/13.

country but rarely more strikingly than on Westminster Abbey's demesne of Hardwicke in Gloucestershire. Here, in the second half of the fourteenth century, legumes occupied a third of the arable area and the pigs which they fattened were periodically sent to the abbey's own larders 100 miles away at Westminster, using the intermediate manors of Islip and Denham as staging posts.[169] Swine were more likely to be disposed of by intra-estate transfer than other livestock (Table 4.06). In most cases, like the Hardwicke pigs, their fate upon reaching their destination was to be slaughtered for pork and bacon, many animals being sent to seigniorial households explicitly *ad hospicium* of the lord. Although sale was another even more important method of disposal, swine were less likely to be sold than horses, cattle or sheep.

Lords used the market to sell rather than buy pigs. By implication, it was finished animals ready for almost immediate butchering that were mostly being sold. Only a few can have been bought to replenish other herds. Unlike horses, cattle, and sheep, demesne pig production was rooted in self-sufficiency. Reproduction on the manor was the overwhelming source of all replacement animals (Table 4.05). Except on a few big estates the swine husbandry practised on most demesnes was a largely self-contained affair.[170] There was next to no specialisation by age, sex, or product. Most animals must therefore have been destined for local consumption. Insofar as they were traded over long distances, it was as bacon rather than as live animals. Nevertheless, little of the bacon that was traded appears to have been demesne-produced, at least within the FTC counties. Instead, peasant producers were almost certainly the more active suppliers of bacon to the market.[171]

Most swine were kept as part of a composite pastoral enterprise. Hogs combined well with horses and cattle, hence few demesnes were wholly without them (Table 4.01).[172] There was even a distinctive handful of demesnes – pastoral type 5 – on which they were the single most important category of non-working animal kept (Table 4.01 and Figures 4.08 and 4.09). Some may conceivably have been demesnes whose cattle were at farm and whose sheep were managed inter-manorially and accounted for centrally. Others, for a variety of reasons, may have been temporarily shorn of their major livestock. When cattle and sheep were either asset-stripped or devastated by disease swine were the quickest and cheapest means of redressing the loss. For example, when the dairy herds painstakingly built up over a period of thirty years on the prior of Norwich's demesnes of Martham and Hemsby

---

[169] Farmer, 'Marketing', p. 387.   [170] Biddick, *The other economy*, pp. 121–5, 132.

[171] Bacon is conspicuous by its absence from sales of demesne produce: Campbell, 'Measuring commercialisation', p. 169, n. 89. The large quantities of bacon purchased by purveyors to provision the royal armies appear to have been obtained from numerous small producers: PRO, E101/574/8, 24, 30; E101/575/5.

[172] Legumes were frequently fed to both horses and hogs: Biddick, *The other economy*, pp. 122–3, 132, 202; Miller (ed.), *AHEW*, vol. III, pp. 191, 270.

were decimated by cattle plague in 1319, swine numbers were temporarily boosted to bridge the gap until the herds could be rebuilt.[173] For other demesnes with severely limited pastoral resources, intensive, arable-based pig husbandry may have been the only viable pastoral option. This may account for the superior importance of pig-keeping in the champion country of the east midlands, as also in densely populated, arable eastern Kent.[174] Above all, demesnes specialised in swine when they were charged with provisioning a major household with fresh meat.[175] Monkwearmouth in Co. Durham, Oakham in Rutland, and Exning in Suffolk all kept pigs in substantial numbers for this reason. It is therefore local and institutional rather than regional factors which largely account for this unusual pastoral specialism.

More generally, swine were of greatest importance as a demesne animal in three main areas. First, they show up strongly on demesnes in northern England and the north-west midlands, a circumstance which can be directly related to the browsing and pannage available in the forests and chases of the region. Second, they were a prominent feature of pastoral husbandry in Kent and Sussex, where swine herding had long been associated with the woods and forests of the Weald. By the late thirteenth century, however, this traditional, extensive mode of management was increasingly being combined with fodder-based systems of husbandry. Intensive rather than extensive methods must certainly have prevailed in the third main area of pig-rearing, the east midlands. Counties such as Rutland, Huntingdonshire, Bedfordshire, and Buckinghamshire supported substantial numbers of swine without the advantage of extensive areas of woodland or 'waste'. Elsewhere, swine were generally the least important category of demesne livestock and were probably kept in greater numbers by other classes of producer.

As an animal which responded well to more intensive methods of production, swine appear to have gained steadily in relative numbers and importance in most parts of the country over the period 1250–1325. In the south-east and the midlands this upward trend persisted into the middle decades of the fourteenth century, but in the south-west and the north numbers apparently fell back sharply. Thereafter, in the ensuing period of contracting demand and declining land-values, it was evidently in areas where swine were husbanded extensively that they fared best, notably in parts of the midlands and north.[176]

---

[173] B. M. S. Campbell, 'Field systems in eastern Norfolk during the Middle Ages: a study with particular reference to the demographic and agrarian changes of the fourteenth century', PhD thesis, University of Cambridge (1975), pp. 94, 138.

[174] Grant, 'Animal resources', p. 159.

[175] Biddick, *The other economy*, pp. 37–40, 121; Dyer, *Standards of living*, pp. 59–60.

[176] See Grant, 'Animal resources', p. 159, for archaeological evidence of a relative decline in pig-keeping in the fourteenth and fifteenth centuries.

4.35 Non-working animals in perspective

Demesnes varied a good deal in the scale and composition of their non-working sectors. Many factors contributed to this. Land-use variations were one obvious source of pastoral differences. On wolds, downs, and sandy heaths sheep enjoyed decided advantages over cattle, since their lower nutritional requirements enabled them to make more effective use of poorer land.[177] Forests and woods afforded excellent sustenance to swine and could be grazed to advantage by cattle. Meadowland was nevertheless indispensable to dairy cattle. Where pasture was abundant there was scope for free-range herding of cattle and sheep. Areas with most pasturage were generally those which specialised in the more land-extensive forms of pastoralism: stock-rearing, fine wool production, and pannage-fed pigs.

Nevertheless, scarcity of pasture did not preclude development of a non-working pastoral sector. Indeed, contrary to all ecological expectations, it was in the arable counties of East Anglia and the south-east rather than the grassier counties of the north and west that seigniorial pastoral husbandry assumed its most intensive and developed forms. Here were concentrated many of the demesnes with specialist dairy herds, and here, too, those which sty-fed their pigs. When sheep were kept their milk could be valued almost as much as their wool, which was coarse rather than fine. Underpinning these activities were fuller use of labour, capital, and enterprise, exemplified by the winter housing of livestock, intensive management of hay meadows, and supplementation of natural pasture with produced fodder. By such means it was possible to support surprisingly large proportions of non-working livestock, as in the case of pastoral types 1, 2, and 3, all of which were exceptionally well represented in 'arable' East Anglia (Figures 4.01–4.05).

These intensive and highly developed pastoral regimes were a response to the higher levels of economic rent generated by relative land scarcity and proximity to substantial markets, both rural and urban. Through the partial or, occasionally, complete substitution of horses for oxen demesnes were able to maximise development of their non-working sectors. The pastoral products produced – cows' and ewes' milk, calves, coarse wool, legume-fed pigs, and fat animals – gave higher returns per unit of land than the replacement stock, lean animals, fine wool, and pannage-fed pigs associated with a greater abundance of pasturage. These contrasting forms of pastoralism were mutually interdependent, with the links between them articulated via commercial or intra-estate transfers of animals and their products.

Maintaining the specialised flocks and herds kept by most fourteenth-century demesnes invariably entailed the selective buying and selling of animals. Many flocks and herds were not fully self-replacing, hence fresh stock

---

[177] Grant, 'Animal resources', p. 156.

had constantly to be bought in from other producers. The cattle trade in particular appears to have involved substantial inter-regional movements of animals.[178] At the same time, there were always finished animals and surplus stock to be sold off. Typically, these were old and infirm beasts plus surplus calves and lambs. Each year many animals changed hands, creating the potential for smaller producers without breeding stock to obtain either old animals for fattening or young stock for rearing. As commercial products, live animals had the considerable merit that they could be walked to market.[179] Only the finishing of fat animals had to take place close to market and hence was an activity closely associated with urban butchers.[180] The potential range of live-stock movements is well attested by the far better documented intra-estate transfer of animals, which bears witness to some remarkably substantial droves. Livestock products, too, tended to have a high unit value relative to their bulk and so could bear the costs of transport over a considerable distance, as is well exemplified by the international trade in fine wools produced in land-locked and remote locations.

Since animals and their products were intended for local, regional, national, and international markets the commercial forces which partially shaped livestock production operated at a variety of overlapping spatial scales to create a kaleidoscopic pattern of pastoral types. Certain broad trends do, however, stand out. In the period 1250–1349 East Anglia, parts of the east midlands, and much of the south-east were dominated by the most intensive and developed pastoral regimes, exemplified by pastoral types 1, 2, and 3 (Figures 4.01, 4.03, and 4.04). Pastoral types 5 and 6 (Figures 4.08 and 4.10), with their weakly developed non-working sectors, are poorly represented in all these areas and mostly owe their occurrence to specific institutional circumstances. Flocks and herds were always vulnerable to disease. They were also capital assets liable to be sold off when necessity demanded.[181] More significant is the limited occurrence of a wholesale specialism in sheep-farming – pastoral type 4 – within these areas (Figure 4.06). This was one of the more land-extensive pastoral regimes and such scattered examples as occur are mostly explicable in terms of local environmental circumstances: the heathlands of west

---

[178] Farmer, 'Marketing', pp. 377–81; I. S. W. Blanchard, 'The continental European cattle trade, 1400–1600', *EcHR* 39 (1986), 427–60.

[179] For the scale of the late medieval cattle trade and evidence of long-distance droving in continental Europe, see Blanchard, 'European cattle trade'.

[180] Farmer, 'Marketing', pp. 387–92; Dyer, *Warwickshire farming*, pp. 20–1.

[181] For examples of disease mortality among livestock see H. Harrod, 'Some details of a murrain of the fourteenth century; from the court rolls of a Norfolk manor', *Archaeologia* 41 (1866), 1–14; Kershaw, 'The Great Famine', pp. 20–9. Titow, 'Land and population', pp. 44–7, has shown that on the estates of the bishops of Winchester large numbers of sheep and cattle were sold off by the bishop's executors or officers the moment that a vacancy occurred or it became obvious that one was imminent. This was to prevent appropriation of these capital assets by the crown. Consequently, there were periods of time when the non-working animal population was artificially low.

Norfolk and the marsh pastures of Broadland, the Fens, and Essex. This pastoral type was better represented away from the main rural and urban concentrations of population. It shows up particularly strongly on the downlands of Kent, Sussex, northern Hampshire and Wiltshire, on the Gloucestershire and Oxfordshire Cotswolds, and in the limestone country of south Yorkshire. It is further west and north again that pastoral types 5 and 6, with their limited non-working sectors and heavy emphasis upon cattle, are most prominent (Figures 4.08 and 4.10). They show up in areas both of pastoral shortage and pastoral abundance. In these remote upland areas the latter were invariably stock ranches, breeding and rearing animals for sale and transfer by the most land-extensive methods.

After 1350 differences between pastoral types are less sharply etched. On demesnes throughout the country, but especially in central and southern England, livestock numbers rose. Non-working animals almost everywhere gained in relative and absolute importance. Within the non-working sector sheep gained most from the growth of pastoral farming. Few pastoral types remained immune to the growth of sheep-farming (Tables 4.01 and 4.03). On many demesnes there was a reversion towards more land-extensive systems of production and in some cases this was reinforced by renewed reliance upon oxen within the working sector. The impression is of a pastoral sector in which land-use considerations were of enhanced, and economic considerations of diminished, importance in determining the range and purpose of animals stocked. Although demesnes relied even more upon the market to buy and sell animals, at the expense in many cases of intra-estate transfers (Tables 4.05 and 4.06), it is probable that the exchange of animals was increasingly within rather than between regions.

4.4   Animals versus crops

Pastoral husbandry was rarely conducted as a wholly independent enterprise. So long as agricultural technology remained predominantly organic and animate, pastoral and arable husbandry were perforce interdependent. Livestock relied upon tillage for fodder, bedding, and pasturage. Cropping needed animals for traction, haulage, and manure. Neither could easily exist without the other. Virtually all medieval husbandry systems were therefore to a degree mixed. Whether that also meant that arable and pastoral production were *integrated* was another matter. The hallmarks of integrated mixed-farming systems were permanently or temporarily housed livestock, fodder cropping, hay production, and the controlled grazing of fallow arable, temporary leys, and mown meadows: in short, the joint use of land to produce both crops and animals.[182] Alternatively, where there were adequate supplies of

[182] Campbell and Overton, 'New perspective', pp. 88–95.

arable and pasturage, crop and livestock production could co-exist as largely separate enterprises with the minimum of overlap between them.[183]

The price of integration was the substitution of labour, capital, and enterprise for land. It was therefore more likely to occur when and where land was scarce rather than abundant, labour and capital were cheap rather than dear, economic rent was high rather than low, and property rights were amenable to the alternation of land between tillage and temporary pasture. The rewards of integration were higher levels of agricultural output per unit of land. Paradoxically, therefore, the most intensive and land-productive pastoral systems tended to be found in arable or semi-arable contexts. This is because semi-fodder-dependent pastoral husbandry constituted a more energy-productive foodchain than pastoral systems based more or less exclusively upon natural grassland.[184]

4.41 Stocking densities and farming systems

Lack of pasturage was no barrier to the development of a significant economic interest in livestock. Via improvements to the efficiency of the draught sector, greater specialisation within the non-draught sector, closer management of such pastoral resources as were available, and more effective use of fodder crops, lords as a whole prevented any erosion of the numbers of livestock units maintained on their demesnes during the period 1250–1349 (Table 4.07). This was no mean achievement considering that until *c.* 1315 pastoral husbandry had to contend with a steadily expanding sown arable area. It testifies, as Biddick has observed, to 'a pastoral sector of some dynamism and complexity and dispels any notion of linear relations between animal and cereal husbandry'.[185] Perpetual institutions like Peterborough Abbey and Norwich Cathedral Priory were in the strongest position to undertake the simultaneous development of arable and pastoral husbandry since their flocks and herds were least at risk to the periodic asset-stripping that accompanied minorities, wardships, and episcopal vacancies. What is, perhaps, more remarkable is the patience with which many lay and episcopal lords repeatedly rebuilt flocks and herds after they had been asset-stripped either by executors or by the crown. Not all were so assiduous. Roger, the fifth and last Bigod earl of Norfolk, was more interested in living off rather than building up his vast inheritance. When he died in 1306 stocking levels on his Norfolk demesnes were little different from when he had inherited the estates in 1270.[186] Upon his death the estate reverted to the crown, which was an even less energetic

[183] E.g. Biddick, 'Agrarian productivity', p. 115.
[184] Simmons, *Ecology*, pp. 201–6; Grigg, *Dynamics*, p. 71; Campbell, 'Chaucer's reeve', pp. 289–91.   [185] Biddick, *The other economy*, p. 65.
[186] PRO, SC 6/932/11–26; 6/933/20–29; 6/934/1–39; 6/935/2–37; 6/936/2–32; 6/937/1–10, 27–33; 6/938/1–11; 6/944/1–10; 6/991/16–28.

landlord. Within the FTC counties royal demesnes and those in royal hands were the most seriously understocked of all.[187]

The net effect of the changes taking place within the pastoral sector over the period 1250–1349 was to maintain total livestock numbers in a state of dynamic equilibrium. Nationally, the numbers stocked by a typical demesne held remarkably constant at about sixty-five livestock units. The same trend prevailed in Norfolk, although the county's fragmented manorial structure and strong arable bias meant that demesnes stocked on average only about forty-six livestock units (Table 4.07). Even after *c.* 1325, when many lords reduced the cropped area of their demesnes, livestock numbers failed to register any perceptible increase. Either the arable was simply being cropped less intensively or it was being leased out to tenants. Certainly, the seigniorial pastoral sector does not appear to have benefited in any direct way. With less land under crop the ratio of livestock units to cropped acres did, nevertheless, register a modest improvement. Because of changes in the composition of both the arable and pastoral sectors there was a more marked improvement in the ratio of non-working livestock units to grain acres (Table 4.07). Norfolk was in the vanguard of these developments. Remarkably for so arable a county, its mean number of non-working livestock per 100 grain acres matched the national average.

Stocking densities are best expressed per unit of agricultural or, failing that, of arable land. Unfortunately, few manorial accounts provide the relevant information. For the majority of demesnes stocking densities can therefore only be expressed per sown acre or per grain acre. The latter is generally to be preferred since it takes account of the boost to stocking densities provided by the substitution of legumes (a fodder crop) for bare fallows and provides a better index of potential manure supplies. Land was rarely manured in preparation for legumes, which were themselves often sown for their nitrifying properties.

Poor medieval yields have often been blamed on inadequate supplies of manure. Those who have claimed that stocking densities were too low have cited the situation on the extensive estates of the bishops of Winchester scattered through southern England. Yet these were the estates of a capital-rich landlord, many of which were relatively well endowed with pasturage by seigniorial standards (Figure 3.11). Their pastoral profiles, too, tended to be dominated by oxen and sheep, the two most land-extensive and grass-dependent animals. Ignoring swine, weighting horses and cattle alike at 1.0 and sheep at 0.25, Titow calculated that demesnes on the Winchester estate had on average sixty-five livestock units per 100 sown acres during the final quarter of the thirteenth century. In his view such a stocking density was so low as to constitute a state of chronic under-manuring.[188] If that was the case the situation on the

[187] For the restocking of royal manors see Stacey, 'Agricultural investment'.
[188] Titow, *Winchester yields*, p. 30.

Table 4.07. *Trends in mean stocking densities: England, Norfolk, and the FTC counties, 1250–1449 (demesne means)*

Years	Mean cropped area		Mean livestock units		Mean stocking density	
	Sown acres	Grain acres	All livestock[1]	Non-working[2] only	Livestock units[1]	Non-working livestock units[2]
					per 100 grain acres	
England:						
1250–1299	189.2	176.7	64.2	36.2	40.6	21.8
1275–1324	193.4	176.7	67.7	39.9	52.7	30.7
1300–1349	172.1	155.7	64.8	39.0	59.0	34.6
1325–1374	156.4	134.7	63.8	40.9	59.0	35.1
1350–1399	147.1	124.9	75.0	51.4	63.7	42.2
1375–1424	144.7	123.9	78.6	53.4	63.3	42.6
1400–1449	142.8	117.4	89.3	62.8	92.1	66.6
Norfolk:						
1250–1299	172.9	149.2	45.6	29.5*	30.5*	19.8*
1275–1324	171.1	140.8	46.5	31.7*	33.0*	22.5*
1300–1349	146.0	126.6	45.9	33.8*	36.3*	26.7*
1325–1374	132.8	115.3	47.2	36.6*	41.0*	31.7*
1350–1399	126.8	110.6	49.3	39.4*	44.6*	35.6*
1375–1424	136.9	120.1	43.3	34.7*	36.1*	28.9*
1400–1449	158.6	140.7	43.5	34.7*	30.9	24.7*

FTC counties:

1288–1315	224.4	205.9	66.4	40.4	40.1	26.0
1375–1400	172.0	146.4	79.3	58.3	68.6	52.5

Notes:

[1] (Horses ×1.0) + (oxen and adult cattle ×1.2) + (immature cattle ×0.8) + (sheep and swine ×0.1)

[2] (Adult cattle ×1.2) + (immature cattle ×0.8) + (sheep and swine ×0.1)

* Calculated at aggregate not demesne level

Method:

All Norfolk means are the product of four regional sub-means weighted equally. All national means are the weighted product of six regional means, comprising the weighted mean for Norfolk ×0.081; the eastern counties (Cambs., Essex, Herts., Hunts., Lincs., Middx., Suffolk) ×0.214; the south-east (Hants., Kent, Surrey, Sussex) ×0.12; the midlands (Beds., Berks., Bucks., Leics., N'hants., Oxon., Rut., Warks.) ×0.164; the south-west (Devon, Dorset, Cornwall, Gloucs., Heref., Mon., Somerset, Wilts., Worcs.) ×0.209; and the north (Berwick., Ches., Cumb., Derbs., Durham, Lancs., Northumb., Notts., Salop, Staffs., Westmor., Yorks.) ×0.213. The weightings are based on each region's share of assessed lay wealth in 1334 and poll tax population in 1377.

Sources: National accounts database; Norfolk accounts database; FTC1 and FTC2 accounts databases.

estates of Westminster Abbey must have been far worse for, on the same method, Farmer reckoned that the Westminster demesnes had a stocking density some 50 per cent lower.[189] Like Titow, he draws attention to the poor yields which supposedly arose from these low animal ratios.[190]

Inconveniently for such an argument, stocking densities thus calculated were no better in eastern Norfolk, even though this was one of the most intensive and productive arable-farming districts in the country (Table 7.07 and Figure 7.09).[191] On the Titow method the mean for the country as a whole during the half-century 1250–99 was forty-five livestock units per 100 sown acres. By this yardstick stocking densities on the Winchester demesnes were actually 40 per cent above average whereas those on the Westminster demesnes and in east Norfolk were over 25 per cent below average. Incorporating swine, recalculating these stocking densities using the more refined set of weightings employed in Table 4.07, and expressing them per 100 grain acres qualifies but does not overturn the thrust of this basic comparison. By the standards of the time, the Winchester demesnes were relatively well stocked with livestock. It was elsewhere that the pastoral sector was most under duress, nowhere more so than in the closely settled arable countryside of Norfolk and the east midlands. Livestock units per 100 grain acres on Norfolk demesnes, however calculated, were consistently 25 per cent below average. Seemingly, it was here, if anywhere in late thirteenth-century England, that arable productivity was potentially most at risk from under-manuring. There is, however, little evidence to suggest that this risk ever materialised.

Maintaining sufficient livestock was one thing; ensuring that the principal grain crops benefited from their manure was another. It mattered not how many animals were stocked if arable and pastoral husbandry were conducted as largely separate enterprises, as appears to have been the case on the Winchester estates under bishop Peter des Roches in the early thirteenth century.[192] Lower stocking densities could be equally if not more effective, provided that arable and pastoral husbandry were integrated. On the downland demesnes of the bishops of Winchester that meant the perfection of a system of sheep-corn husbandry whereby pasture-fed sheep were nightly folded upon the arable. In that way arable soils were systematically improved in texture and enriched with nitrogen and other nutrients contained in the dung of the sheep. Excessive folding was, however, to be avoided since it encouraged parasitic infection of the sheep.

---

[189] Farmer, 'Grain yields on Westminster Abbey manors', pp. 340–2.
[190] 'Slightly more manure did not guarantee the Westminster manors better grain yields. Probably there was still too little': Farmer, 'Grain yields on Westminster Abbey manors', p. 342. On the relationship between crop yields and stocking densities in the seventeenth and eighteenth centuries, see Allen, *Enclosure and the yeoman*, p. 205.
[191] Campbell, 'Agricultural progress', pp. 29–31.
[192] Biddick, 'Agrarian productivity', p. 115.

'Sheep-corn husbandry' was one of eight basic seigniorial mixed-farming systems identifiable in the period 1250–1349 (Table 4.08).[193] It was one of the more extensive and depended upon the availability of substantial tracts of permanent pasture. 'Extensive mixed farming' and 'Oats and cattle' were similarly pasture-dependent. Both were characterised by essentially extensive forms of pastoralism and in location were strongly associated with pasture-rich locations, among them the East Anglian Fens, the Kent and Sussex coastal marshes, the Wiltshire Downs, south Devon, and a wide scatter of locations in northern England. These were the farming systems with least functional cross-over between the arable and pastoral sectors and minimal utilisation of fodder crops.

This lack of cross-over was also true of 'extensive arable husbandry'. It was practised by demesnes which stocked working animals only. With few livestock and little manure at their disposal they perforce relied upon regular fallowing as the principal means by which soil fertility was maintained. There is a strong northern, western, and south-western bias to the distribution of demesnes employing this farming type, i.e. regions of below-average population density mostly remote from major markets for arable produce but with easy local access to sources of livestock and their products.

The four other mixed-farming systems were all more intensive. In every case livestock were a vital component of the nitrogen cycle upon which the maintenance of soil fertility depended.[194] The livestock, in return, relied upon the arable for forage and fodder. Demesnes employing these mixed-farming systems utilised a variety of methods to maximise the potential and compensate for the often limited numbers of their livestock. On the pastoral side, hay meadows, fodder cropping, and the seasonal grazing of stubble, fallows, and annual leys all contributed to the sustenance of the animals. On the arable side, sheep folding, fallow grazing, systematic application of farmyard manure, marl and lime, multiple ploughings, flexible rotations which included courses of nitrifying legumes, and the periodic alternation of land between tillage and temporary pasturage variously helped conserve and maintain levels of soil nitrogen and other vital nutrients. Because the pastoral component of these mixed-farming systems had a strong arable base they were distributionally more closely associated with the 'arable' east than the 'pastoral' west.

It was on demesnes practising 'intensive mixed farming' that arable and pastoral husbandry were most closely integrated, to the mutual advantage of both sectors. Significantly, this was the preferred farming system in those localities where agriculture is known to have been most intensive and productive, notably eastern Norfolk, eastern Kent, and the Soke of Peterborough. Notwithstanding the strong arable bias to land-use in these localities, stocking densities per 100 grain acres compared favourably with those prevailing in

---

[193] Power and Campbell, 'Cluster analysis'.   [194] Shiel, 'Improving soil fertility', pp. 62–70.

Table 4.08. Mixed-farming systems and stocking densities, 1250–1349 and 1350–1449 (national, Norfolk, and FTC accounts databases combined)

Mixed-farming type	No. of demesnes	Mean livestock units per demesne	% non-working livestock units	Mean sown acres per demesne	% grain acres	Mean livestock units per 100 sown acres		Mean livestock units per 100 grain acres	
						Mean	Std	Mean	Std
1250–1349:									
1 Intensive mixed farming	105	78.7	65.1	188.8	81.0	45.4	35.0	57.3	48.0
2 Light-land intensive	61	54.2	72.8	181.3	91.4	34.1	19.4	37.4	21.7
3 Mixed farming with cattle	124	63.0	49.4	244.8	94.1	29.5	18.4	31.4	19.8
4 Arable husbandry with swine	24	26.8	31.5	171.0	85.0	18.6	10.3	22.2	13.5
5 Sheep-corn husbandry	74	47.0	54.6	167.6	90.8	35.1	39.5	39.0	45.6
6 Extensive mixed farming	100	107.9	61.5	226.6	93.6	59.6	63.9	63.9	65.8
7 Extensive arable husbandry	96	23.5	0.9	142.9	93.4	18.8	11.8	20.2	12.1
8 Oats and cattle	10	39.6	32.6	113.7	95.5	39.1	25.1	40.7	26.6
All	594	62.2	48.4	194.1	90.6	36.5	37.7	41.0	42.5
1350–1449:									
1 Intensive mixed farming	84	61.1	77.5	149.8	86.3	44.7	24.4	52.0	28.0
2 Light-land intensive	30	67.0	83.5	123.7	93.8	63.7	49.3	73.6	80.0
3 Mixed farming with sheep	105	101.4	62.2	168.2	81.1	65.8	56.9	81.9	68.7
4 Arable husbandry with swine	19	33.7	38.8	144.2	76.5	23.1	10.8	32.1	16.9
5 Sheep-corn husbandry	74	70.0	66.9	132.2	89.7	71.0	83.3	80.7	103.8
6 Extensive mixed farming	98	84.5	66.5	155.3	88.6	67.1	49.3	77.8	61.8
7 Extensive arable husbandry	29	24.2	1.9	148.1	85.3	18.7	9.0	22.9	12.5
All	439	74.2	63.4	150.3	86.1	57.8	55.0	68.4	68.9

Note:
Livestock units = [horses ×1.0] + [(oxen + adult cattle) ×1.2] + [immature cattle ×0.8] + [(sheep + swine) ×0.1]

Source: National accounts database; Norfolk accounts database; FTC1 and FTC2 accounts databases; J. P. Power and B. M. S. Campbell, 'Cluster analysis and the classification of medieval demesne-farming systems', *TIBG*, new series, 17 (1992), 227–45; B. M. S. Campbell, K. C. Bartley, and J. P. Power, 'The demesne-farming systems of post Black Death England: a classification', *AHR* 44 (1996), 131–79.

many regions better endowed with permanent pasturage. Here is proof positive that when and where environmental, economic, and institutional circumstances were right demesne managers were perfectly capable of developing the kind of mixed-farming systems in which the arable and pastoral sectors were complementary rather than competitive. What mattered most was the prevailing 'system of husbandry' or technological complex, not merely the physical amounts of arable and pasture or the relative numbers of animals (Table 4.08).

Institutional and structural factors also made a difference since they affected the capacity of such systems to develop. Regular commonfield systems inhibited adoption of the more flexible and intensive mixed-farming systems, especially those practising a form of convertible husbandry. 'Intensive mixed farming' and 'light-land intensive' husbandry were consequently mostly employed on demesnes outside and to the east of the midland zone of regular commonfields. Few commonfield demesnes progressed beyond 'mixed farming with cattle' and 'arable husbandry with swine'. Examples of both show up in the east midlands and, of the former, in Warwickshire and south Somerset, as well as in East Anglia and the Home Counties. Certain types of landlord were also more assiduous than others in improving the management of their demesnes and investing in the development of pastoral husbandry. Within the FTC counties the highest stocking densities tended to be encountered on episcopal demesnes and on those belonging to conventual and collegiate institutions. Mean stocking densities on lay demesnes, many of them the home farms of magnate households, were consistently inferior. Lower still were the stocking densities of demesnes under royal management: collectively, they tended to be 30–50 per cent below average.

The size of demesne also exercised a bearing upon the relative balance struck between crops and animals. In the FTC counties, as Table 4.09 demonstrates, there was a loose inverse relationship between sown acreage and stocking density, and much the same applied in Norfolk. Relatively small arable demesnes of less than 100 sown acres, and especially those with less than 50 sown acres, supported conspicuously higher stocking densities than large arable demesnes of 300 sown acres or more. Seemingly, large arable demesnes were genuinely more committed to arable production than their smaller counterparts. The sheer scale of the great 'prairie' farms operated by some landlords militated against adoption of the more intensive and integrated forms of mixed husbandry which were better suited to small production units. The size of most demesnes thus partially determined the form of their enterprise. Significantly, the same inverse relationship between farm size and stocking density prevailed in the ensuing early modern period.[195]

[195] Overton and Campbell, 'Norfolk livestock farming', pp. 388–90; Allen, *Enclosure and the yeoman*, pp. 204–5.

After 1349, and especially from the late 1370s, relative prices and production costs proved increasingly favourable to the pastoral sector. Released from the overriding imperative to produce grain, some demesnes expanded the area devoted to legumes (Table 6.02 and Figure 6.18), feeding them unthreshed to livestock. Others either reduced the frequency of cropping or took land out of tillage altogether (a process which was to gather considerable momentum in the fifteenth century).[196] By the close of the fourteenth century demesnes had on average 25 per cent less land under crop than they had done at the beginning of the century and 30 per cent less under grain. Some of this land was transferred by lease to the non-demesne sector but the bulk was released to pastoral use. Demesnes, accordingly, began to stock more animals. Livestock numbers started to rise during the third quarter of the fourteenth century and the rise was maintained thereafter (Table 4.07). Over the second half of the fourteenth century the mean number of livestock units stocked rose by 23 per cent. With a contracting requirement for draught power, the mean number of non-working units increased by 30 per cent over the same period. Stocking densities, consequently, were transformed. By *c.* 1400 demesnes were stocking 25 per cent more livestock units per 100 grain acres than they had done *c.* 1350 and over 40 per cent more non-working livestock units. Over the entire period 1250–1450 stocking densities per 100 grain acres rose by almost 90 per cent. The corresponding rise for non-working livestock was a remarkable 160 per cent (Table 4.07).

Not surprisingly, this relative swing from corn to horn was far from uniform throughout the country and took place within the context of an absolute contraction in the scale of the seigniorial sector. It is barely apparent in the counties north of the Trent, whose agrarian economy maintained its traditional division into complementary but largely separate arable and pastoral sectors. Nor is it strongly expressed in Norfolk and the other eastern counties, whose chief competitive advantage always lay in arable husbandry and arable-based pastoralism. Here, livestock husbandry remained as an adjunct of, rather than alternative to, tillage. Many Norfolk demesnes responded to the altered economic circumstances by leasing their herds rather than their arable and when they substituted sheep for cattle stocking densities tended to suffer because pastoral husbandry became less intensive. By *c.* 1400 they were little better endowed with livestock than they had been *c.* 1300. In the midlands, as in later centuries, the situation was entirely different. For environmental and economic reasons, land here converted well from arable to pasture. But for the prevalence of common property rights the retreat from tillage would undoubtedly have proceeded much faster. Commonfield arable which could not be taken out of cultivation tended to be sown instead with fodder crops. Stocking densities on many a midland demesne, hitherto often depressed by a scarcity

---

[196] Miller (ed.), *AHEW*, vol. III, pp. 214–18, 222–30; Dyer, *Warwickshire farming*, pp. 9–16.

Table 4.09. *Stocking density and grain acreage on demesnes in the FTC counties, 1288–1315 and 1375–1400*

Grain acreage	No. of demesnes	Aggregate method	Mean livestock units per 100 grain acres: Demesne-mean method		
			Minimum	Mean	Maximum
1288–1315:					
<50	9	180.5	61.4	191.2	508.0
<100	33	70.4	11.4	93.2	508.0
100 to <200	70	43.3	11.2	44.0	97.0
200 to <300	43	41.4	6.0	41.4	120.8
300 to <400	16	39.2	11.5	39.4	130.8
400 to <500	13	27.6	9.5	27.7	74.4
500+	4	32.0	28.6	31.7	37.9
1375–1400:					
<50	5	320.4	88.9	330.4	547.6
<100	29	116.0	24.0	142.6	547.6
100 to <200	79	70.0	7.6	70.1	213.1
200 to <300	19	51.6	11.3	52.2	118.4
300 to <400	2	27.2	27.2	27.2	27.2
400+	1	31.3	31.3	31.3	31.3

Note:
Livestock units are weighted according to relative prices; see text p. 106
Aggregate method = [(total livestock units of all demesnes in size range) ÷ (total grain acres of all demesnes in size range)] ×100
Demesne-mean method = mean of individual demesne means.
Source: FTC1 and FTC2 accounts databases.

of pasturage, now rose by more than the national average.[197] Nevertheless, it was the counties of southern England, and especially those of the south-west, which registered the greatest increases of all. The incentive to convert arable to pasture and invest more heavily in pastoral production was here provided by the vigorous expansion of seigniorial sheep-farming, which in turn was encouraged by the growing demand for wool from the expanding domestic cloth industry.[198]

These developments wrought far-reaching changes to established systems of mixed farming. Within the arable sector there was increased scope for

[197] Kussmaul, 'Agrarian change'.
[198] Hare, 'Wiltshire agriculture', pp. 6–8; Mate, 'Pastoral farming'; Stephenson, 'Wool yields', pp. 385–6.

flexibility and experimentation; within the pastoral sector there were mounting incentives to adopt more land-extensive forms of enterprise. The pastoral sector became less subservient to the arable and seigniorial agriculture in general became more mixed. The net effect was to render mixed-farming systems more alike one another. Seven rather than eight types of seigniorial mixed-farming system may be identified in the period 1350–1449, each statistically less sharply differentiated from the other (Table 4.08). No fundamentally new mixed-farming systems came into being; instead, those already in existence were modified and developed. 'Sheep-corn husbandry' and 'extensive arable husbandry' stand out as the most distinctive of these seven systems. The former is differentiated by the uniquely specialised character of its pastoral sector, which was dominated by oxen and sheep; the latter, by the virtual absence of non-working animals. 'Arable husbandry with swine' is almost as distinctive but very much a minority type, associated in the main with the home farms of permanent households. All other farming systems represent different gradations of the same basic mixed-farming type.

At the intensive extreme, 'intensive mixed farming' and 'light-land intensive' husbandry constitute examples of integrated and manure-intensive systems; one, because of a scarcity of pasturage, characterised by below-average stocking densities, the other, owing to an abundance of pasturage, with stocking densities well above average (Table 4.08). Examples of both are limited in number and circumscribed in distribution, showing up most strongly in Norfolk. 'Mixed farming with sheep' represents an altogether less intensive and less integrated type of mixed farming. It assumed its most intensive (i.e. most fodder-dependent) form in Norfolk and Suffolk and immediately adjacent parts of the east midlands and its most extensive (i.e. most grass-dependent) form in the south midlands and southern England. 'Extensive mixed farming' was by definition more extensive again and in distribution by far the most widely represented mixed-farming type. Examples are to be found in areas of both relatively intensive and extensive agriculture, and in some of the economically most developed and economically most remote parts of the country. It was the most grass-dependent mixed-farming system and was particularly characteristic of heavy soils.

As mixed-farming systems waned in intensity and land was converted from tillage to grass so stocking densities became increasingly a function of a demesne's relative endowment with arable and pasturage. Compared with the situation before 1350 it was the more extensive mixed-farming systems which both became more numerous and more mixed and which registered the greatest increases in stocking densities. By implication these were the systems within which the conversion to grass had proceeded furthest by *c.* 1400. 'Sheep-corn husbandry' – a particularly extensive system – gained most of all, becoming the dominant seigniorial husbandry type in much of central-southern England. It was here, in fact, that farming systems and stocking densities together changed most dramatically in the half-century following the Black Death.

When mixed-farming systems are viewed collectively rather than individually in the period 1350–1449 the picture that emerges is more heterogeneous than homogeneous. With some notable exceptions, few types of farm enterprise were unique to a single locality or region. It was the most intensive systems that were the most circumscribed in distribution; in this period the more extensive might be found in almost any part of the country, including populous areas of high economic rent. Norfolk, for example, although dominated by 'intensive mixed farming' and 'light-land intensive' husbandry, contained examples of the full range of national farming types. A wide range of mixed-farming types was also to be found within the immediate hinterland of London. No doubt it was the demand of that great city which engendered different farming specialisms among those demesnes most strongly exposed to it. In remoter and economically less developed and commercialised parts of the country the range of farming types appears to have been correspondingly narrower, with the most intensive systems either rare or absent. Compared with the situation in the period 1250–1349, over-arching centripetal influences appear to have been of less importance in structuring the overall distribution of farming systems. The very fact that demesne husbandry was becoming less sharply differentiated suggests that incentives to specialise and intensify had weakened. Individual regions and localities were increasingly self-sufficient in what they produced. The latent patterns of specialisation and intensification detectable at the climax of medieval economic expansion *c.* 1300 were falling into abeyance in an agrarian world now being restructured along more local and regional lines.

4.42   *The relative contributions of crops and livestock to gross revenues*

Another way of looking at the balance struck between animals and crops is in terms of their relative contributions to gross revenues, costing all intra-estate transfers as sales (Table 4.10). Commodities consumed directly on the manor either in the production process (as seed and fodder) or by the lord, his household, and his resident officials and servants, are excluded from the calculation. Within the FTC counties animals and their products contributed approximately 35–40 per cent of gross agricultural revenues at the beginning of the fourteenth century. By the close of the century, following a 70 per cent rise in mean stocking densities, revenues from animals and their products had grown to match those from crops and crop products. Costing in the revenues generated from pastoral land-uses leased rather than managed directly tilts the balance decisively in favour of the pastoral sector at the latter date.

The bulk of the enhanced pastoral revenues of the late fourteenth century came from the expanded output of animal products which the enlarged flocks

Table 4.10. *Relative contributions of crops and animals to estimated gross revenues from sales and transfers within the FTC counties, 1288–1315 and 1375–1400*

Sample of manors	No. of manors	Crops and crop products *as % gross revenues*[1]	Animals and animal products	Mean gross revenue[1] from crops and animals £	Animals, animal products, and herbage as % of gross revenues[1] (incl. herbage)	Mean gross revenue[1] from crops, animals, and herbage £
1288–1315:						
A) Any manors:						
All conventual and collegiate manors	111	*54.7*	*45.3*	13.23		
All episcopal manors	18	*71.3*	*28.7*	33.42		
All lay manors	31	*71.1*	*28.9*	29.99		
All royal manors	43	*66.7*	*33.3*	27.84		
All manors	201	*64.4*	*35.6*	20.70		
B) Manors common to both periods:						
Canterbury Cathedral Priory's manors	13	*50.1*	*49.9*	14.96		
Westminster Abbey's manors	12	*52.5*	*47.5*	16.39		
Bishopric of Winchester's manors	12	*73.7*	*26.3*	41.37		
All manors	60	*60.7*	*39.3*	22.91		

1375–1400:

A) Any manors:

All conventual and collegiate manors	88	*45.3*	*54.7*	27.09	*56.1*	27.94
All episcopal manors	13	*51.6*	*48.4*	42.24	*50.3*	43.77
All lay manors	38	*46.0*	*54.0*	29.03	*57.4*	31.29
All royal manors	2	*65.2*	*34.8*	14.72	*47.2*	18.00
All manors	142	*47.8*	*52.2*	28.87	*56.5*	30.27

B) Manors common to both periods:

Canterbury Cathedral Priory's manors	13	*54.3*	*45.7*	15.82	*49.4*	17.14
Westminster Abbey's manors	12	*47.7*	*52.3*	16.41	*53.1*	16.68
Bishopric of Winchester's manors	12	*51.9*	*48.1*	31.77	*48.6*	32.11
All manors	60	*54.4*	*45.6*	19.08	*47.0*	19.55

Method:
Based on accounts with no missing sales values; all percentages calculated at aggregate level.

Note:
[1] Actual revenue from sales plus estimated revenue from commodities transferred off the manor valued using mean prices calculated for the FTC counties.

Source: FTC1 and FTC2 accounts databases.

and herds made possible. The heady days of the international wool export trade may have been long past, but more manors now produced and sold wool than had done so at the height of the export boom. Income from wool sales was the component of pastoral revenue which grew most. Dairying, too, was a more lucrative source of revenue than hitherto, although this had more to do with the fact that the bulk of dairy herds were now farmed out for cash rather than managed directly, with much of the produce often being consumed directly on the manor. A modest rise in the revenues obtained from sales of live animals may reflect the strengthening demand for meat.

At both the opening and the close of the fourteenth century it was on demesnes in conventual and collegiate ownership that animals and animal products made their greatest relative contribution to gross revenues (Table 4.10). As perpetual institutions they were better placed to build up their flocks and herds than any other category of landlord. Conventual and collegiate lords were also smaller consumers of meat than their episcopal, lay, and royal counterparts and hence had greater numbers of surplus live animals available for disposal. Nevertheless, it was on episcopal and lay manors that livestock appear to have made their greatest relative revenue gains over the course of the century. By its close only the handful of sampled demesnes under royal management were failing to exploit the full revenue-generating potential of livestock. Royal demesnes were conspicuously under-stocked and it is no coincidence that it was on these manors that the leasing of herbage and sale of agistments made its greatest relative contribution to gross revenues. In fact, sales of herbage largely compensated for the limited direct income generated by animals (Table 4.10). The same held true in general. Across all ownership types animals, animal products, and herbage collectively generated 47–57 per cent of gross agricultural revenues within the FTC counties by the close of the fourteenth century. In every case this was a significantly greater share of revenues than that which had pertained at the century's opening.

Gross revenues should not, of course, be confused with net profits. The revenues generated by the disposal of crops and animals and their products had to be offset against the costs of their production, in terms of capital, labour, land, and enterprise. There was also the 'hidden' subsidy provided by customary labour services, which was always far greater for crops than for animals. Labour inputs in general were far greater per unit of arable than pastoral production and unit labour costs rose dramatically in the final decades of the fourteenth century. One of the attractions of pastoral husbandry after 1349, therefore, lay in its lower labour requirements and the superior profit margins which these increasingly delivered. All in all, therefore, it was pastoral production which, as the fourteenth century advanced, delivered the better returns.

The profits to be made from expanding pastoral production were nevertheless far from limitless. There was only a finite market for dairy produce and by the fifteenth century the domestic market for wool was becoming glutted, with the result that producers had to contend with depressed prices. Meanwhile, real labour costs continued to rise until the 1440s (Figure 1.01). Maintaining profit margins thus became increasingly contingent upon curbing labour inputs. This had potentially detrimental repercussions for flock and herd management. Fertility rates are likely to have suffered and mortality rates may well have risen, especially among sheep. Certainly, fleece weights on the Winchester estates displayed a long-term tendency to decline as management methods became increasingly extensive.[199] Nor does the archaeological evidence suggest much, if any, recovery in carcass weights, notwithstanding the greatly increased availability of pasturage. That had to await the effects of improved breeding in the sixteenth century and after.[200]

In fact, great as were the gains made by the pastoral sector during the hundred years or so following the Black Death, there was much ground still to be made up. Indeed, over the next four to five centuries it was to be the pastoral rather than the arable sector that constituted the most dynamic component of English agriculture. Whereas grain yields per unit area roughly doubled between *c.* 1300 and *c.* 1850, fleece weights increased by two-and-a-half fold, carcass weights of cattle and sheep trebled, and milk yields per cow quadrupled. Greater cultivation of a more productive range of fodder crops also supported stocking densities per cultivated acre which were 25 per cent higher in 1850 than those that had prevailed in 1300.[201]

These post-medieval gains in pastoral productivity do not necessarily imply that thirteenth-century pastoral husbandry was lacking in dynamism. On the contrary, during the 100 years or so before the Black Death pastoral husbandry, like arable husbandry, was undergoing a process of selective intensification. On many demesnes the most striking feature of that intensification was closer integration of the two sectors, hence the lack of much correlation between stocking densities and supplies of grassland *c.* 1300. Lords also invested much capital in the build-up of flocks and herds and in housing to shelter them. Above all, seigniorial pastoral husbandry became a great deal more specialised, as animals and their products were traded in ever-growing quantities over ever-greater distances. The pastoral sector was, in fact, more commercialised than the arable. Whereas lords were mainly sellers of grain they entered the market both to buy and sell animals. Moreover, cost-distance was less of a commercial handicap to them as pastoral than as arable producers. It was, after all, the pastoral sector which made the single greatest contribution to thirteenth-century English exports.

---

[199] Stephenson, 'Wool yields', pp. 376–89.
[200] Grant, 'Animal resources', pp. 176–8; Clark, 'Labour productivity', pp. 217–18; U. Albarella, 'Size, power, wool and veal: zooarchaeological evidence for late medieval innovations', in G. De Boe and F. Verhaeghe (eds.), *Environment and subsistence in medieval Europe: papers of the 'Medieval Europe Brugge 1997' conference*, I.A.P. Rapporten 9 (Zellik, 1997), pp. 19–21.
[201] Overton and Campbell, 'Norfolk livestock farming', pp. 387–8; Campbell and Overton, 'New perspective', pp. 83–8.

5

Seigniorial arable production

Carnivorous as lords and their immediate households may have been, the typical lowland demesne was a predominantly arable concern. Nationally, the arable comprised 60 per cent of all demesne land-use by value in the first half of the fourteenth century (Table 3.01) and on roughly one in three of all demesnes that proportion rose to three-quarters.[1] The arable bias to demesne land-use was particularly marked within the ten FTC counties, which had to feed both themselves and London.[2] Crops and crop products accounted for 84 per cent of gross agricultural output on demesnes within these counties c. 1300, a proportion which is consistent with Clark's estimate that within lowland England 'arable crops accounted for 80 per cent by value of total food output c. 1300'.[3] Underpinning that pronounced production bias were buoyant grain prices and an abundant cheap labour force.[4] By the close of the fourteenth century relative prices and factor costs were less favourable, yet although the arable sector's dominance weakened, it was by no means eclipsed. As a proportion of demesne land-use within the FTC counties the arable declined from 67 per cent to 60 per cent of the total.[5] Bolstered by the proximity of London, arable products continued to contribute half of all gross sales revenue compared with 60 per cent at the beginning of the century.[6]

In the vast majority of cases, therefore, crop production was, and long

[1] National *IPM* database. On the extensive estate of the abbot and convent of Westminster 'The acreage of demesne meadow, pasture, and woodland . . . was equivalent to c. 20 per cent of the total acreage of the arable demesnes': Harvey, *Westminster Abbey*, p. 127.

[2] Campbell and others, 'Rural landuse', p. 18; Campbell and others, *Medieval capital*, pp. 37–8.

[3] Campbell, 'Measuring commercialisation', p. 174; Clark, 'Labour productivity', p. 234. On the estates of the bishopric of Winchester grain production was by far the most important aspect of demesne farming right up to the Black Death, and the single most important source of manorial revenue almost to the very end of the period 1209–1350, when it was overtaken by rents. For the greater part of the thirteenth century well over 40 per cent of profits were contributed by grain production. Titow, 'Land and population', pp. 11–12.

[4] Farmer, 'Prices and wages', pp. 716–45, 760–72.

[5] National *IPM* database and FTC2 *IPM* database.

[6] FTC1 and FTC2 accounts databases.

remained, the first concern of demesne managers. To it they devoted the larger share of their resources and upon it they depended for the greater part of their gross income. There were exceptions, of course, and where circumstance and opportunity dictated the arable sector took second place to the pastoral. On the evidence of the *IPMs*, that was probably true of one out of three lay demesnes at the beginning of the fourteenth century and is likely to have become more rather than less common by that century's close. Nevertheless, in the thirteenth and fourteenth centuries an overwhelming emphasis upon pastoral rather than arable production was quite exceptional and almost always confined to certain specific parts of the country, types of environment and categories of estate.[7] Demesnes with small or non-existent arable sectors made up no more than a fifth of the total in the period 1300–49.[8]

5.1 The objectives of production

Lords used their arable to produce bread and brewing grains, and pottage and fodder crops. In order to do so they first had to ensure adequate inputs of seed, fertiliser, draught power, and labour.[9] The imperatives to consume and exchange thus had to be reconciled with the needs of sustainability. This determined both the choice of crops produced and the method of their production.

5.11 Sustainability

Under medieval yields the proportion of the harvest that had to be set aside as seed for the next year was substantial. On the sampled demesnes in the ten FTC counties seed accounted for 33 per cent of the total grain harvest by value net of tithe in the period 1288–1315 and 35 per cent in the period 1375–1400 (when yields were lower).[10] For individual crops the proportions ranged from less than 30 per cent (rye) to over 40 per cent (oats and legumes). Much here depended upon the place and role of crops in rotations. Oats, for instance, were often sown last in intensive rotations with the result that they had to be seeded thickly in order to compete with weed growth and consequently generally yielded poorly. With legumes, on the other hand, the high proportion of seed relative to yield often arose from the practice of feeding them unthreshed to livestock; few reeves followed the injunction of the anonymous author of the *Husbandry* to estimate and account for the portion of the harvest thereby consumed.[11]

Since a specific market in seed had yet to develop, many demesnes sowed their own home-grown grain. Walter of Henley advised against this, at least as far as winter-sown crops were concerned. He gave no reason but claimed that

---

[7] The Cistercians, for instance, were notable pastoral farmers: Donkin, *The Cistercians*, pp. 68–102.   [8] National *IPM* database.   [9] Pretty, 'Sustainable agriculture'.
[10] See Chapter 7, pp. 375–81.   [11] *Walter of Henley*, p. 439.

the advantages of sowing new and healthy seed were readily demonstrable by experiment. Later commentators have generally assumed that better yields were thereby obtained because of a diminished risk of crop disease.[12] By the opening of the fourteenth century increasing numbers of manors were following this recommendation. On the larger estates grain for seed was often exchanged between manors; on smaller estates grain was sometimes purchased for use as seed.[13] If, for whatever reason, there was a temporary cut-back in the supply of seed (as particularly tended to be the case following years of severe harvest shortfall and high prices), either land had to be sown more thinly – usually at some sacrifice to yields per acre – or less land could be sown.[14]

An additional proportion of the crop had to be set aside as fodder for the draught animals and other livestock. Without animals to draw the plough, haul the cart, and supply manure the arable could not be cultivated. Demesnes well endowed with labour services might shift part of the cost of maintaining plough-teams onto their tenants but it was a rare demesne that was entirely without working horses and oxen.[15] Plough-teams varied greatly in size and composition. Teams of eight oxen were common but far from universal. Langdon has noted larger teams of ten or more animals on the stiffest and heaviest soils but by the late thirteenth century on the lighter soils of the east smaller mixed teams of horses and oxen were increasingly making an appearance along with the first all-horse teams.[16] Draught animals were fed on a combination of grain (usually oats, but on occasion rye, dredge, or even barley, plus legumes – especially vetches), straw, chaff, hay, and pasture. Cart-horses were by far the greediest consumers of fodder crops, consuming on average over three times as much grain as plough-horses which, in turn, consumed almost six times as much grain as the predominantly straw- and grass-fed ox.[17] Nor did fodder requirements end here, for most farm animals were potential consumers of grain, legumes, straw, stubble, and other by-products of arable cultivation. All the more intensive arable regimes were components of integrated agro-systems in which the arable supplied fodder and temporary grazing to the animals and the animals supplied draught power and manure to the arable. Thus, it was standard practice for livestock to graze on the stubble and aftermath of the harvest, together with the temporary pasture

---

[12] *Walter of Henley*, pp. 174–5.   [13] E.g. Smith, *Canterbury Cathedral Priory*, pp. 134–5.

[14] The harvest of 1294 was particularly bad, occasioning the highest grain prices of the thirteenth century. The resultant scarcity of grain may explain a 10 per cent reduction in the seeding rates of wheat and oats and 4 per cent reduction in the seeding rate of barley on the earl of Norfolk's demesne at Suffield in Norfolk in 1294–5. The next year on the earl's demesne of Caister-cum-Markshall, also in Norfolk, seeding rates of wheat and barley were 5 per cent below their 1292–3 level and 13 per cent less land was sown: PRO, SC 6/944/7–8, 6/932/11–26. For a case study of one East Anglian community's response to this harvest crisis see Schofield, 'Dearth, debt'.   [15] See Chapter 4, p. 120.

[16] Langdon, *Horses*, pp. 118–27; Chapter 4, pp. 121–2.

[17] Langdon, 'Economics of horses', p. 33.

afforded by annual fallows and leys. In addition, pigs were sometimes intensively fed with legumes, and legumes and grain might on occasion be fed to
cattle, sheep, poultry, and even rabbits.[18] Collectively, grain consumed as
fodder accounted for perhaps a further 4 to 5 per cent of the harvest by value
(significantly more on some individual demesnes).[19]

Successful arable cultivation entailed far more, however, than merely
ploughing the land and then sowing and in due course harvesting the crop.
Yields were likely to be neither satisfactory nor sustained unless effective
measures were taken to conserve soil fertility and counter pests and diseases.
Better standards of ground preparation, for instance, via multiple ploughings,
facilitated germination and helped prevent weed growth. The latter was one of
the most obstinate problems facing medieval cultivators and a potentially
serious drain upon scarce supplies of soil nitrogen.[20] Apart from summer
ploughing of the fallow it was best dealt with by weeding, which was extravagant of labour and far from completely effective where seed was sown broadcast. The depredations of birds and animals were similarly dealt with by using
labour to scare off predators and by rotations which separated crops in space
and time. According to one school of thought the scattering of strips in open
fields was a response to the need to minimise this kind of risk.[21]

Heavy and poorly drained soils presented a different kind of problem.
Medieval cultivators generally obtained their worst results from these types of
soil, whose wetness compounded the dwindling store of phosphorus and
potassium within the soil to the detriment of plant growth.[22] Only the careful
digging and maintenance of ditches and construction of crude under-drains
could mitigate this situation.[23] Soil acidity was a bigger problem, especially in
the rainy north and west. It was best dealt with by applications of marl, lime,

---

[18] Biddick, 'Pig husbandry'; Brandon, 'Demesne arable farming', p. 123; Bailey, *A marginal economy?*, p. 135.

[19] This figure is based on the following six manors during the period 1375–1400: Wargrave, Berks.
– 1.9 per cent (Hants. RO, 11M59 B1/123–43); Birdbrook, Essex – 8.2 per cent (WAM,
25469–89); Sayesbury in Sawbridgeworth, Herts. – 5.0 per cent (WAM, 26305–29); Westerham,
Kent – 4.05 per cent (WAM, 26460–506); Hyde, Middlesex – 1.6 per cent (WAM, 27077–100);
Adderbury, Oxfordshire – 5.8 per cent (Hants. RO, 11M59 B1/123–43). On the estate of
Peterborough Abbey in 1300–1 livestock consumed 21 per cent of the total harvest: Biddick,
The other economy, p. 72. Straw, although widely sold, was of low intrinsic worth: Campbell,
'Measuring commercialisation', pp. 148–9. Its high market price reflects its bulk and hence high
transport costs: P. Glennie, 'Measuring crop yields in early modern England', in Campbell and
Overton (eds.), *Land, labour and livestock*, p. 266.

[20] Postles, 'Cleaning the arable'; Shiel, 'Improving soil fertility', p. 62.

[21] D. N. McCloskey, 'The open fields of England: rent, risk, and the rate of interest, 1300–1815',
in D. W. Galenson (ed.), *Markets in history* (Cambridge, 1989), pp. 5–51.

[22] On the low yields obtained from heavy soils see Chapter 7, pp. 336, 354; Chapter 9, pp. 418–19.

[23] In eastern Sweden extensive ditching using iron-shod spades appears to have made a significant
contribution to the agricultural expansion of the twelfth and thirteenth centuries: J. Myrdal,
Medieval arable farming in Sweden (Stockholm, 1986), pp. 268–9. Cf. J. L. Langdon,
'Agricultural equipment', in Astill and Grant (eds.), *Countryside*, pp. 98–9.

or sea sand; all labour intensive and therefore, for demesnes, expensive tasks unless undertaken by servile labour.[24] Conversely, in the drier south and east, especially on free-draining sandy and chalk soils, water was sometimes the limiting factor.

It was, in fact, whichever factor was in relative minimum – Liebig's Law – which ultimately set the limit to crop growth. Among these limiting factors nitrogen is usually identified as the single most common source of variations in plant growth, although this is often compounded by deficiencies in other nutrients, particularly phosphorus and potassium.[25] As R. S. Shiel observes, 'Crops use more nitrogen than any other nutrient, and unless nitrogen is replaced, a shortage in supply will limit growth after only a few years of arable cropping.'[26] Medieval farmers applied a variety of fertilising materials which contributed either directly (in the case of manure and nightsoil) or indirectly (in the case of marl, sand, and lime) to the supply of nitrogen in the soil. Crop rotations further served to manipulate nitrogen budgets, but also led to improvements in soil structure and helped reduce problems from weeds, pests, and diseases. Legumes – beans, peas, and vetches – here potentially performed a vital role for they helped to 'fix' atmospheric nitrogen in the soil and thereby partially replenished what other crops had removed.[27] Bare fallows served much the same purpose, albeit far less effectively. They were usually followed by a winter crop, sown before too much of the labile nitrogen was lost by leaching. More serious losses of nitrogen were incurred by sales and transfers of crops off the manor. The greater that loss the more active the measures it was necessary to take to counteract it.

Conserving and re-using existing nitrogen supplies provided one solution. Effective re-use of soil nitrogen depended upon returning the greatest possible amount of crop waste to the soil and ensuring that it was applied to maximum advantage. Accelerating the decomposition of crop wastes by processing them through animals into manure further facilitated the release of mineral nitrogen to the soil.[28] Thus, a given amount of nitrogen in the form of manure ploughed into the soil will give a larger growth effect in the year of application, and a more rapid decay of residues, than an equal amount of nitrogen added as undigested plant remains. Timing the application of manure to coincide with the onset of plant growth greatly aided the efficiency of the operation, as did the night housing and winter housing of livestock the initial collection and storage of manure.[29] Nocturnal folding of the arable with sheep

---

[24] Smith, *Canterbury Cathedral Priory*, pp. 133–8; Hallam (ed.), *AHEW*, vol. II, pp. 285–7, 323, 346–8, 388, 404, 435–42.

[25] Shiel, 'Improving soil fertility'; Newman and Harvey, 'Did soil fertility decline?'.

[26] Shiel, 'Improving soil fertility', p. 53.

[27] G. P. H. Chorley, 'The agricultural revolution in northern Europe, 1750–1880: nitrogen, legumes and crop productivity', *EcHR* 34 (1981), 71–93. At Rimpton in Somerset it is explicitly stated that vetches were sown on the fallow in order to 'compost' the land: Thornton, 'Determinants of productivity', p. 196.   [28] Chorley, 'Agricultural revolution', p. 64.

[29] Walter of Henley gave advice on these points: *Walter of Henley*, pp. 327–9, 337–9.

which had fed on permanent pastures and sheep walks during the day performed much the same function (and greatly improved the structure of light soils) but did so by transferring additional nutrients from elsewhere. Other transfers of external sources of manure, such as seaweed or nightsoil from towns, served the same purpose.[30] But for most demesnes the principal reserve of stored nitrogen was represented by permanent and temporary pastures. Converting them into arable, or bringing them into temporary cultivation via various forms of ley and outfield cultivation, released an immediate shot of nitrogen into the system.[31]

Maintaining the sustainability of arable production was therefore a demanding task and became increasingly so the more difficult the environmental conditions with which cultivators had to contend and the more that was demanded from the soil. These considerations exercised a direct influence upon the choice of crops grown and types of rotation that were operated. Intensive production of wheat, for instance, may have been contingent upon sowing a larger acreage with legumes and closer integration of arable and pastoral husbandry. The most effective systems were those in which the whole technological complex represented far more than the sum of the individual parts. Sustainability was not easily attained and involved a more or less continuous process of trial and error. In the thirteenth and fourteenth centuries it is plain that this was an on-going process.

5.12 Consumption

Once the requirements of sustainability had been satisfied lords used their demesnes either to satisfy immediate consumption needs or to generate a cash income through the sale of produce. The concept that lords should be able to live off their estates was of considerable antiquity. It found explicit expression in the system of annual food farms at one time operated on many ecclesiastical estates.[32] Individual manors were charged with the responsibility of producing and delivering to the central household a week's supply of food at a specified time of the year. Usually the provisions requested took account of the natural resources of the manor in question and the task of carrying them to the central household was imposed as a service upon the servile tenants of the manor. On certain estates this system persisted in modified form until well

---

[30] Campbell, 'Agricultural progress', pp. 34–5.
[31] E.g. T. A. M. Bishop, 'The rotation of crops at Westerham, 1297–1350', *Economic History Review* 9 (1938), 38–44; P. F. Brandon, 'Arable farming in a Sussex scarp-foot parish during the late Middle Ages', *Sussex Archaeological Collections* 100 (1962), 60–72; Thornton, 'Determinants of productivity', pp. 196–8; B. Harrison, 'Field systems and demesne farming on the Wiltshire estates of Saint Swithun's Priory, Winchester, 1248–1340', *AHR* 43 (1995), 7–9.
[32] R. V. Lennard, *Rural England 1086–1135* (Oxford, 1959), pp. 130–38; Raftis, *Estates of Ramsey Abbey*, pp. 61, 309–13; E. Miller, *The abbey and bishopric of Ely* (Cambridge, 1951), pp. 38–41; Harvey, *Westminster Abbey*, pp. 80–1; Biddick, *The other economy*, pp. 36–40.

Table 5.01. Aggregate disposal of crops (net of tithe) in monetary value by lord and ownership type within the FTC counties, 1288–1315

Lord (by ownership type)	No. of manors[a]	Aggregate crop receipt[b] (£)	Sown as % of aggregate receipt	Aggregate net crop receipt[c] (£)	Percentage of aggregate net crop receipt (adjusted for grain converted to malt)[d]				
					(Malted)	Retained	Transferred	Sold	Disposed of[e]
Bec Abbey	6	142	19	116	(11)	58	5	38	43
Bicester Priory	5	26	31	18	(1)	35	56	8	64
Boxley Abbey	1	5	40	3	(13)	75	7	16	23
Canterbury Cathedral Priory	33	282	35	183	(<1)	41	22	37	59
Crowland Abbey	3	36	25	27	(34)	45	40	15	55
Merton College, Oxford	8	87	28	63	(1)	60	1	38	39
Oseney Abbey	2	12	35	8	(13)	48	34	17	51
Peterborough Abbey	18	301	22	234	(25)	49	42	9	51
Ramsey Abbey	1	26	24	20	(22)	54	31	13	44
Titchfield Abbey	1	11	29	8	(1)	62	11	26	37
Waltham Abbey	1	16	28	11	(1)	58	33	8	41
Westminster Abbey (abbot)	7	61	35	40	(3)	70	13	17	30
Westminster Abbey (convent)	20	220	33	148	(3)	53	26	21	47
Winchester Cathedral Priory	1	11	22	9	(0)	45	3	52	55
Conventual and collegiate	**107**	**1,236**	**28**	**886**	**(10)**	**51**	**25**	**23**	**48**
Archbishopric of Canterbury	5	65	32	44	(0)	57	2	41	43
Bishopric of London	1	8	21	7	(0)	48	<1	51	52
Bishopric of Winchester	13	118	29	84	(0)	36	0	64	64
Episcopal	**19**	**192**	**30**	**135**	**(0)**	**43**	**1**	**56**	**57**

D'Argentine	1	20	27	14	(14)	72	12	17	29
De Barley	1	1	54	1	(0)	90	6	4	10
De Clare	1	16	37	10	(0)	61	6	33	39
De Cobham	1	36	29	26	(7)	84	5	9	14
De Fortibus	1	13	24	10	(12)	27	5	68	73
De Hamelton	1	15	17	12	(0)	37	0	63	63
Earldom of Cornwall	7	62	32	42	(0)	41	0	59	59
Earldom of Lincoln	9	68	27	50	(1)	33	1	65	66
Earldom of Norfolk	4	69	30	48	(5)	50	5	46	51
Le Ferrers	1	31	12	28	(27)	37	9	54	63
Lay	*27*	*332*	*27*	*241*	*(6)*	*48*	*4*	*48*	*52*
King	16	150	36	96	(0)	42	3	55	58
Late Earldom of Norfolk	1	24	23	18	(3)	34	5	61	66
Late Holy Trinity Abbey, Caen	1	19	30	13	(0)	50	0	50	50
Late Knights Templar	15	119	39	73	(0)	46	1	53	54
Late Walter de Langton	3	35	32	23	(0)	41	0	59	59
Queen	3	34	40	20	(0)	47	4	49	53
Royal	*39*	*381*	*36*	*244*	*(<1)*	*42*	*2*	*55*	*57*
All[a]	**190**	**12,747**	**28**	**9,136**	**(7)**	**44**	**17**	**38**	**55**

Notes:

[a] Weston, Herts., is counted both as a possession of the Earldom of Norfolk and of the king
Isleworth, Middx., is counted both as a possession of the Earldom of Cornwall and of the king
Upper Heyford, Oxon., is counted both as a possession of Isabella de Fortibus and of the king

[b] Net of tithe: conversion of volume to value carried out using the following per bushel sale prices: wheat 8.6d., rye 7.1d., winter mixtures 6.9d., barley 6.5d., dredge 5.1d., oats 3.7d., beans 5.9d., peas 5.7d., vetches 5.1d., legumes 5.6d., grain–legume mixtures 5.4d.

[c] Net of tithe and seed

[d] On the assumption that 33.3% of malted grain was retained, 33.3% was transferred, and 33.3% was sold

[e] Transferred plus sold

Source: FTC1 accounts database.

Table 5.02. *Aggregate disposal of crops (net of tithe) in monetary value by lord and ownership type within the FTC counties, 1375–1400*

Lord (by ownership type)	No. of manors	Aggregate crop receipt[a] (£)	Sown as % of aggregate receipt	Aggregate net crop receipt[b] (£)	Percentage of aggregate net crop receipt (adjusted for grain converted to malt)[c]				
					(Malted)	Retained	Transferred	Sold	Disposed of[d]
Boxley Abbey	3	66	40	40	(0)	83	17	<1	17
Canons of St Paul's, London	1	34	29	24	(0)	47	6	47	53
Canterbury Cathedral Priory	23	1,110	37	695	(4)	45	20	35	55
Coggeshall Abbey	1	34	30	24	(7)	80	2	18	20
Edington Priory	1	64	26	48	(8)	54	5	41	46
Eynsham Abbey	1	37	21	30	(38)	71	12	16	28
Merton College, Oxford	5	206	24	157	(2)	41	1	58	59
New College, Oxford	7	342	18	281	(8)	39	5	57	61
Peterborough Abbey	3	258	34	171	(33)	62	26	12	38
Ramsey Abbey	1	62	31	43	(45)	64	15	21	36
Robertsbridge Abbey	1	29	36	19	(0)	59	35	6	41
Rochester Cathedral Priory	1	71	24	54	(4)	31	2	67	69
St Catherine's Priory, Rouen	2	136	22	107	(5)	35	2	63	65
Titchfield Abbey	1	36	28	26	(0)	80	11	9	20
Westminster Abbey (abbot)	5	240	28	173	(9)	46	23	30	54
Westminster Abbey (convent)	18	713	30	498	(6)	48	19	33	52
Winchester Cathedral Priory	1	39	26	29	(1)	42	<1	58	58
Windsor College	2	94	30	66	(0)	48	0	52	52
Conventual and collegiate	*81*	*3,716*	*31*	*2,583*	*(8)*	*48*	*14*	*38*	*52*

Archbishopric of Canterbury	2	141	29	100	(0)	57	11	32	43
Bishopric of Winchester	12	533	24	403	(20)	47	11	42	53
Episcopal	**14**	**674**	**25**	**503**	**(16)**	**49**	**11**	**40**	**51**
Beauchamp	1	139	23	106	(28)	33	9	57	66
Berners	2	91	28	65	(8)	46	12	41	54
Butler	1	54	26	40	(48)	30	16	54	70
Carew	1	135	31	94	(31)	52	35	13	48
Colepeper	1	38	29	27	(43)	65	14	21	35
De Bohun	1	46	26	34	(0)	23	0	78	78
De Burnell	1	44	37	28	(0)	44	0	56	56
De Gildeburgh	1	33	39	20	(4)	53	34	13	47
De Grey of Wilton	1	54	40	32	(38)	54	23	22	45
De la Lee	1	42	24	32	(0)	37	0	63	63
De la Pole	1	28	29	20	(16)	34	60	6	66
De Missenden	1	117	28	84	(18)	79	12	9	21
De Morle	1	30	48	16	(0)	79	0	21	21
De Seyton	1	73	32	49	(31)	48	27	25	51
De Sutton	2	55	29	39	(2)	46	13	41	54
De Waterton	1	54	36	34	(0)	46	0	54	54
De Wykeham	1	71	28	51	(0)	50	0	50	50
Doget	1	78	29	56	(30)	49	36	14	50
Duchy of Gloucester	2	102	28	73	(0)	54	14	33	46
Earldom of Arundel	2	31	32	22	(0)	54	8	38	46
Earldom of Ormond	1	82	33	55	(0)	48	0	53	53
Le Strange	1	52	23	40	(12)	46	7	47	54
Oddyngsels	1	34	21	27	(35)	40	34	26	60
Wanton	1	24	60	10	(0)	83	0	17	17
Unknown (1)	1	52	29	37	(0)	58	30	17	42
Unknown (2)	1	20	34	13	(0)	59	15	26	41
Lay	**30**	**1,581**	**30**	**1,105**	**(16)**	**49**	**16**	**35**	**51**

Table 5.02. (cont.)

Lord (by ownership type)	No. of manors	Aggregate crop receipt[a] (£)	Sown as % of aggregate receipt	Aggregate net crop receipt[b] (£)	Percentage of aggregate net crop receipt (adjusted for grain converted to malt)[c]				
					(Malted)	Retained	Transferred	Sold	Disposed of[d]
King	1	39	27	29	(0)	61	0	40	40
Late De Stonor	1	37	23	28	(40)	27	35	38	73
Royal	*2*	*76*	*25*	*57*	*(20)*	*44*	*17*	*39*	*56*
Unknown	1	60	31	42	(0)	43	0	57	57
All	*128*	*6,107*	*30*	*4,289*	*(11)*	*48*	*14*	*37*	*52*

Notes:

[a] Net of tithe: conversion of volume to value carried out using the following per bushel sale prices: wheat 7.9d., rye 4.1d., winter mixtures 5.8d., barley 6.0d., dredge 4.6d., oats 3.4d., beans 5.8d., peas 5.3d., vetches 5.6d., legumes 5.5d., grain–legume mixtures 5.2d.

[b] Net of tithe and seed

[c] Adjusted on the basis that 29.1% of malted grain was retained, 32.6% was transferred, and 36.5% was sold

[d] Transferred plus sold

Source: FTC2 accounts database.

into the fourteenth century, although renders in kind were increasingly commuted for payments in cash and customary carrying services were diverted from delivering goods to the household to taking them to market.[33] From the mid-thirteenth century, following the shift from leasing to direct management and the creation of the first manorial accounts, the extent to which lords were using their estates to satisfy their own consumption requirements comes more clearly into view.

Manorial accounts and central household accounts confirm that throughout the era of direct management most lords continued to draw upon their estates for a proportion of their consumption requirements. A portion of that consumption took place immediately on the manor in the form of direct food liveries to estate and manorial officials, farm servants, servile tenants and others. For example, in 1300–1 manorial workers on the estates of Peterborough Abbey consumed 10 per cent of the total harvest.[34] Notwithstanding the increasing availability of cash, such food liveries long remained a preferred form of payment and it was a rare manor which did not use them in some measure. Indeed, in the era of rising wage rates and falling food prices which set in after 1375, paying workers in kind rather than cash made financial good sense.[35] In addition, some grain was often consumed *in situ* by seigniorial households either permanently or temporarily resident on their manors. For instance, the households of John de Cobham and Roger de Barley, respectively consumed most or all of the net crop receipts of their home manors of Cobham, Kent in 1290–1 and Wicken Bonhunt, Essex in 1314–15 (Table 5.01), as did William de Fiennes at Wendover, Buckinghamshire in 1296–7.[36] At Stebbing, Essex in 1377–8 and Walkern, Hertfordshire in 1390–1 the bulk of available produce seems similarly to have been consumed by the households of William Wanton and Thomas de Morle (Table 5.02).[37] In other instances – Edmund de Missenden at Quainton, Buckinghamshire in the period 1379–84 was possibly one – lords and their households took up temporary residence on a manor for as long as its provisions lasted.[38]

Among major lay lords the itinerant habit lingered long. The earl and countess of Norfolk provide a notable example from the close of the thirteenth century. Although they possessed several major seats, most notably the great castles at Framlingham, Suffolk and Chepstow, Monmouthshire, they

[33] Campbell and others, *Medieval capital*, p. 149; D. Postles, 'Customary carrying services', *Journal of Transport History*, 3rd series, 5 (1984), 1–15.

[34] Biddick, *The other economy*, p. 72.

[35] Poos, *Rural society*, pp. 218–28. Cf. A. Kussmaul, *Servants in husbandry in early modern England* (Cambridge, 1981), pp. 97–119.

[36] BL, Harl. Roll D1; Essex RO, D/DU 36/12; PRO, SC 6/763/16.

[37] BL, Add. Roll 66016; Herts. RO, 9357.

[38] Bucks. RO, BASM Quainton 24, 31; D/BASM/9/14.

and their retinue regularly descended upon individual manors for periods at a time. In spring 1273 they spent nine weeks on their manor of Forncett, Norfolk, and the very large numbers of retainers and of horses that they brought with them were a heavy charge upon the estate.[39] At other times huntsmen, lawyers journeying to Norwich on the earl's behalf, itinerant bailiffs en route to other manors, grooms with their horses, and knights and clerks travelling on the earl's business found the manor a convenient resting place.[40] Other great lords and their officials similarly used strategically situated manors either as staging posts when they travelled on business or as refuges to retire to when they wished to escape the cares of office. The bishops of Winchester used Downton, Wiltshire, and Witney, Oxfordshire in this way, as is reflected in the high on-the-manor consumption of oats as fodder for the bishop's riding horses.[41] Similarly, the abbots of Westminster paid regular visits of a month or even longer to La Neyte in Eye (Middlesex), Pyrford (Surrey), Denham (Buckinghamshire), Islip (Oxfordshire), and Sutton-under-Brailes (Warwickshire).[42]

The mobility of lay and episcopal lords was obviously denied the households of conventual institutions. When the latter chose to consume their own estate produce it was usually necessary to transfer the relevant provisions off the manor to a central monastic granary. In the case of major households such as Peterborough Abbey, Westminster Abbey, Canterbury Cathedral Priory, or Norwich Cathedral Priory the quantities involved could be considerable. Nevertheless surviving granary accounts indicate that the bulk of the grain received was intended for consumption not subsequent sale. At Norwich Cathedral Priory between 1263 and 1300 the granger annually accounted for up to 675 quarters of wheat and 2,020 quarters of malted barley. While even Sedgeford and Gnatingdon, a 40-mile cart journey from the priory, contributed at least some grain, it was the cathedral priory's most productive manor, Hemsby, where the prior owned the tithes as well as a substantial demesne, that contributed most.[43] Although it was more distant from the priory than several of its other properties, access to direct water transport using a sailing barge expressly maintained by the prior for this purpose greatly assisted the bulk transfer of grain.[44] Other manors along the River Yare were also regular suppliers to the granary, especially Martham, Plumstead, and Newton.

Analysis of the two FTC accounts databases confirms that although transfers of grain were a feature of all types of estate they were most characteristic of manors in the ownership of conventual and collegiate institutions (Tables 5.01 and 5.02). In the period 1288–1315 the latter transferred 25 per cent of net grain receipts on average compared with a mean of 5 per cent from manors

---

[39] Davenport, *Norfolk manor*, p. 24. Other Norfolk manors were similarly visited.
[40] Davenport, *Norfolk manor*, p. 23.   [41] Biddick, 'Agrarian productivity', pp. 109, 112.
[42] Harvey, *Westminster Abbey*, pp. 132–3.   [43] NRO, DCN 1/1/1–15.
[44] NRO, DCN 1/1/28, 67.

belonging to all other types of lord. By the end of the fourteenth century, however, the importance of such transfers appears to have diminished. In the period 1375–1400 the sampled conventual and collegiate manors transferred only 14 per cent of their grain compared with 7 per cent from all other manors. Moreover, it was much less usual for transfer to represent the dominant form of disposal: whereas 12 per cent of sampled manors transferred over half of net grain receipts in the period 1288–1315 the equivalent proportion in 1375–1400 was only 5 per cent. Nor, in either period, was even the most self-sufficient estate averse to selling a portion of its crops and purchasing at least part of its provisions on the market. Canterbury Cathedral Priory purchased just over a quarter of its grain needs at the beginning of the fourteenth century at an average outlay of £156 a year, and Westminster Abbey purchased an even larger share of its grain needs at an annual outlay of about £238. Both these major religious houses had access to well-provisioned markets on their door-steps. Significantly, Peterborough Abbey did not enjoy such an advantage. Its purchases of grain were far smaller and were mostly made in local markets, as required.[45]

This continued reliance upon transfer rather than purchase for many essential provisions provides an interesting commentary upon the perceived costs and reliability of grain markets. Conventual institutions with major consumption needs to meet may have found it cheaper and less risky to store and consume their own grain than purchase it on the market. They may have been encouraged in this by heavy investment in barns and other storage facilities, a steepening seasonal price gradient, and heightened annual price fluctuations.[46] Faced, in many cases, by cash-flow problems, they may have feared putting their faith in volatile markets, prone to fluctuations in both price and supply.[47] Possibly, too, the relative transaction costs of large-scale market purchase were as yet uneconomically high. Access to carrying services may also have served as a transport subsidy to the direct provisioning of households.[48] Significantly, when labour services began to decline in the aftermath of the Black Death, and especially following the Peasants' Revolt of 1381, this old

---

[45] Campbell and others, *Medieval capital*, p. 153.

[46] On seigniorial investment in storage see N. Brady, 'The gothic barn of England: icon of prestige and authority', in E. Smith and M. Wolfe (eds.), *Technology and resource use in medieval Europe* (Aldershot, 1997), pp. 76–105. For conventional economic views on storage costs see D. N. McCloskey and J. Nash, 'Corn at interest: the extent and cost of grain storage in medieval England', *American Economic Review* 74 (1984), 174–87. An alternative view is offered by N. Poynder, 'Grain provision and conventual economics in medieval England', paper presented at the Annual Conference of the Economic History Society, Leeds (1998). On the volatility of early-fourteenth-century prices see Fischer, *Great wave*, pp. 30–6; Bailey, 'Peasant welfare', pp. 234–52.

[47] Biddick, *The other economy*, pp. 50–77; Biddick, 'Agrarian productivity', pp. 98–106.

[48] Farmer, 'Marketing', pp. 347–50. In the twelfth century, for instance, tenants of Ramsey Abbey manors in Huntingdonshire had owed carrying services to Bury St Edmunds, Cambridge, Colchester, Ipswich and London: Postles, 'Customary carrying services', pp. 14–15.

reliance upon direct consumption proved harder to sustain and the disposal of grain by transfer became a diminishing feature of production on most demesnes. By the close of the fourteenth century the proportion of net grain receipts disposed of by transfer in the FTC counties had slipped from 17 per cent to 14 per cent (Table 5.05).

So long as consumption remained an important principle of production the nature of the household to be provisioned, size and composition of estate, and location of a manor within the estate network exercised a bearing upon what was produced and the manner of its production. To keep production as 'costless' as possible carrying services along with other labour services remained important components of such consumption-orientated systems. Often demesnes were expected to specialise according to their natural advantages (although this is likely to have been reinforced by the comparative advantages that stemmed from commercialised production). Some demesnes performed specific functions. On the estates of Peterborough Abbey, for instance, much of the rye required for the food liveries of those who laboured on the abbey's demesnes was produced at Kettering and thence regularly transferred to between seven and twelve other manors on the estate.[49] *Circa* 1300 the eighteen Peterborough demesnes in Northamptonshire transferred virtually all the grain that was not directly consumed on the manor either to other manors on the estate or to the central monastic household at Peterborough. Only on the abbey's outlying manors of Collingham (Nottinghamshire) and Fiskerton, Scotter, and Walcot (all Lincolnshire) was a different policy pursued, whereby surplus grain was disposed of by sale so that cash rather than provisions was transferred to Peterborough.[50] Within the FTC counties Bicester Priory, Crowland Abbey, and Oseney Abbey appear to have pursued similar consumption-orientated strategies, each selling on average less than 10 per cent of the net grain receipt on the sampled manors. Few other estates, however, chose to isolate themselves so completely from the market. In the period 1288–1315 conventual and collegiate manors in the FTC counties transferred 25 per cent and sold 23 per cent of their net grain receipt on average (Table 5.01). By 1375–1400, the balance had tipped firmly in favour of sale; the proportion transferred fell to 14 per cent (including grain transferred *ad hospicium domini*) whereas that sold rose to 38 per cent (Table 5.02).[51] Cheaper grain and less volatile prices were rendering direct provisioning less attractive as an option. Across all the sampled demesnes in these two periods twice as much grain was

---

[49] Campbell and others, *Medieval capital*, p. 151.   [50] Biddick, *The other economy*, p. 76.

[51] Campbell, 'Measuring commercialisation', pp. 141–2. In aggregate, 4.4 per cent of net grain receipts were transferred *ad hospicium domini* in the period 1375–1400. On individual estates, however, such as those of the abbot of Westminster and the archbishop of Canterbury, this proportion might be in excess of 10 per cent, and on particular manors, such as those belonging to the Carews and the Oddyngsels, it was sometimes in excess of 25 per cent: FTC2 accounts database.

sold as transferred at the beginning of the fourteenth century, and two-and-a-half times as much at the end (ratios which take no account of grain sold subsequent to transfer). Important as consumption strategies may have been, market exchange was plainly already the greater influence upon production by the close of the thirteenth century and it was to become even more so throughout the following century.

5.13 Exchange

If lords needed provisions in great quantities for themselves, their households, their officials, and their workers, they were also by the late thirteenth century, no matter what their station, in urgent and constant need of cash. The economy at large was becoming more monetised, a widening range of consumer goods was becoming available, and as warfare escalated the tax demands of the crown were becoming ever more onerous.[52] The great attraction of cash lay, of course, in its liquidity. By resuming direct management of their estates during the thirteenth century lords were able to strike that balance between production for consumption and production for exchange which best suited their current needs. As the amount of money in circulation increased, as local, regional, national, and international food markets expanded, and as prices rose, so the temptation to dispose of a growing proportion of arable production on the market became irresistible.[53]

Sales might either occur incidentally, as a means of disposing of crops surplus to consumption, or as a result of a deliberate process of commercial specialisation.[54] Where the latter was the case it was the market via its influence upon economic rent which largely determined the crops produced and intensity of their production. There was an incentive here for demesnes to capitalise upon whatever comparative advantage they may have enjoyed in producing particular crops. Distance from markets was also important, since some crops, such as wheat, were better able to withstand the costs of carriage than others, most notably oats (Table 5.04). Nevertheless, it was imprudent to take market specialisation further than the sustainability of production would allow, hence it was usually expedient for those demesnes which intensified in response to market opportunity to expand their cultivation of nitrogenous and fodder crops. Developing commercial considerations consequently had implications for the entire production system far beyond the specific crops intended for sale.

It follows that there are two key dimensions to commercialisation, neither

---

[52] Mayhew, 'Modelling medieval monetisation'; C. Dyer, 'The consumer and the market in the later Middle Ages', *EcHR* 42 (1989), 305–27; Ormrod, 'Crown and economy'.

[53] Snooks, 'Dynamic role', pp. 39–40, has hypothesised that the seigniorial sector's commercial involvement may have been at least as high in 1086 as in 1300.

[54] Cf. Farmer, 'Woodland and pasture sales'.

entirely straightforward of measurement. First, there is the proportion of net disposable produce that was sold; second, there is the income such sales yielded per unit area. At one extreme, extensive cropping systems in areas of low economic rent relatively remote from markets may have sold a high proportion of their produce but at a low rate of sale per unit area. At the other, the high rates of sale that arose from intensive methods of production closer to markets were only sustainable if significant proportions of gross output were recycled on the farm as fodder and food liveries. Whereas measuring *relative* specialisation in commercial production involves estimating the proportion of the total net disposable crop that was sold, measuring the *intensity* of that specialisation entails calculating the cash yield from crop sales per arable acre. Unfortunately, detailed as is the information contained in manorial accounts, neither aspect can be measured with complete precision.[55]

Costing the quantities of each crop sold and unsold poses immediate problems. Discontinuities in account survival mean that it is necessary to use relative rather than absolute prices. Although relative prices unique to each location are to be preferred, only average relative prices are generally available. Thus mean relative prices for the FTC counties are potentially misleading close to London where rye and oats – cheap and bulky crops demanded in quantity by that great city – commanded a higher relative price than at a distance.[56] Defining and measuring the 'net disposable crop' poses even greater problems. For instance, within the FTC counties crop sales represented 27 per cent of gross receipts, 38 per cent of gross receipts net of seed, and 50 per cent of gross receipts net of seed, fodder, and food liveries to farm workers. The last is the most precise of these three measures but makes the most exacting demands on the evidence. The second is more readily calculated but understates the significance of sales on those demesnes which made fullest use of fodder and farm labour. In all three cases allowance has to be made for that 7–11 per cent of grain consumed, transferred, or sold subsequent to malting.[57] No such allowance can however be made for that tenth of gross output 'topsliced' as tithe, whose commercial potential was considerable.[58] Consequently, the measure of relative commercial specialisation employed here is the value of crops sold as a proportion of total crop receipts, net of both tithes and seed (Tables 5.01, 5.02, and 5.03).[59]

---

[55] Campbell, 'Measuring commercialisation'.

[56] Campbell and others, *Medieval capital*, pp. 111–25. E.g. at Fulham, within a few miles of London, rye was valued above wheat in 1304: p. 124, n. 36.

[57] Campbell and others, *Medieval capital*, pp. 146–7.

[58] Tithe owners feature prominently among those known to have been active in provisioning fourteenth-century Exeter with grain, and rectors were active participants in the grain trade of the London region in much the same period: M. Kowaleski, 'The grain trade in fourteenth-century Devon', in DeWindt (ed.), *The salt of common life*, pp. 30–1, 35–8; Campbell and others, *Medieval capital*, p. 74, n. 114.

[59] For the prices used in converting volume to value see Table 5.07.

Calculating income per unit area from crop sales is equally problematic. Some accounts record internal transfers as proxy sales.[60] As these are not true sales they need to be discounted; nevertheless, it is only possible to do so when they are explicitly recorded as such. Customary acres are even more elusive of detection and introduce a mostly unquantifiable margin of error. A more systematic bias arises from the failure of all but a handful of accounts to record the total arable area. Without such information only the sales income per *cropped* acre can usually be calculated, even though this exaggerates rates of sale on those demesnes which made greatest use of fallows. Substantial sales of non-demesne grain are a further potential source of inflation; accounts which record 25 per cent or more of gross receipts from such sources have therefore been excluded from analysis.

Applying these two measures of commercialisation to the two FTC accounts databases reveals marginally higher levels of commercialisation at the beginning of the fourteenth century than at the end (Table 5.03). Whereas in the period 1288–1315 demesnes in these ten counties sold 40 per cent of their net crop receipt and received a gross mean income of £8.8 per 100 sown acres, in the period 1375–1400 they sold on average 36 per cent of their net crop receipt and received £8.1 from crop sales.[61] The comparison should not, however, be pressed too far, for neither sample is random and in both periods certain demesnes and particular estates were conspicuously more commercialised than others. Moreover, falling prices depressed cash revenues in the final quarter of the fourteenth century.

The most strongly commercialised demesnes were obviously those which sold a majority of their produce and thereby generated a substantial cash revenue per acre sown (Table 5.03). In the earlier period just over a third of demesnes may be classified as strongly commercialised, and of these a further third were distinguished by particularly high levels of commercialisation. Six belonged to the bishop of Winchester, four were in royal hands, three were possessions of Canterbury Cathedral Priory, two were held by the earl of Lincoln, and one each were properties of Westminster Abbey and Bec Abbey. In the later period less than a quarter of demesnes may be classified as strongly commercialised, and on only two of these twenty-three demesnes – one (Appledore in Kent) a possession of Canterbury Cathedral Priory, the other a property of Rochester Cathedral Priory – was commercialisation maintained at the very highest level. A common denominator of both periods, however, is that the most strongly commercialised demesnes invariably belonged to major lay or ecclesiastical magnates, more in need of cash than provisions. Bec, a relatively minor alien monastery, is the exception which

---

[60] Farmer, 'Marketing', pp. 359–61.
[61] Fifty-four manors common to the FTC1 and FTC2 accounts databases sold 38 per cent and 36 per cent respectively of their net crop receipts at rates of £10.2 and £9.3 per 100 sown acres.

Table 5.03. *Alternative measures of commercialisation in crop production on manors in the FTC counties, 1288–1315 and 1375–1400 (by ownership type)*

		No. of manors[a]								
		FTC1, 1288–1315 Ownership type					FTC2, 1375–1400 Ownership type			
Measure of commercialisation	All	Conv./ Coll.	Episc.	Lay	Royal	All	Conv./ Coll.	Episc.	Lay	Royal
Sales as % of aggregate net crop receipt (£):										
80% +	9	2	3	2	2	1	1	0	0	0
60% – <80%	20	6	4	1	9	11	8	0	3	0
40% – <60%	51	27	7	6	11	36	21	6	9	0
20% – <40%	23	14	1	2	6	33	17	6	9	1
0% – <20%	33	32	0	0	1	28	23	0	5	0
Mean %	*39.8*	*31.4*	*63.1*	*49.0*	*52.1*	*36.1*	*34.5*	*42.3*	*37.2*	*39.5*
Std.	*23.5*	*21.5*	*17.1*	*17.9*	*18.2*	*20.4*	*22.2*	*8.7*	*19.1*	*0.0*
Income from sale of field crops per 100 sown ac.:										
£15.0+	23	8	8	3	4	8	4	2	2	0
£12.5 – <£15.0	9	3	3	0	3	8	6	2	0	0
£10.0 – <£12.5	17	10	2	2	3	18	9	6	3	0
£7.5 – <£10.0	18	13	1	1	3	15	9	1	5	0
£5.0 – <£7.5	18	8	0	2	8	22	15	0	6	1
£2.5 – <£5.0	20	10	1	2	7	19	13	1	5	0
£0.0 – <£2.5	31	29	0	1	1	19	14	0	5	0
Mean £	*8.8*	*6.8*	*18.9*	*9.8*	*8.9*	*8.1*	*6.9*	*16.8*	*7.1*	*7.0*
Std.	*7.3*	*6.3*	*7.3*	*5.8*	*5.4*	*7.1*	*4.8*	*13.4*	*5.2*	*0.0*

Combined commercialisation index[b]:

Very strongly commercialised	17	5	6	2	4	2	2	0	0	0
Strongly commercialised	30	14	7	3	6	23	13	6	4	0
Intermediate	35	18	1	4	12	32	19	4	9	0
Weakly commercialised	22	12	1	2	7	26	15	2	8	1
Very weakly commercialised	32	32	0	0	0	26	21	0	5	0
Total	136	81	15	11	29	109	70	12	26	1

Notes:

[a] Excluding manors deriving less than 75 per cent of gross receipts (net of tithe) from the harvest

[b] Very strongly commercialised = selling at least 60%; receiving at least £15.0 per 100 sown ac.

Strongly commercialised = selling at least 40%; receiving at least £10.0 per 100 sown ac.

Intermediate = *either*, selling at least 40%; receiving less than £10.0 per 100 sown ac.

or, selling less than 40%; receiving at least £10.0 per 100 sown ac.

Weakly commercialised = Selling less than 40%; receiving less than £10.0 per 100 sown ac.

Very weakly commercialised = Selling less than 20%; receiving less than £5.0 per 100 sown ac.

Source: FTC1 and FTC2 accounts databases.

proves the rule, since cash rather than provisions was obviously a higher priority for an institution whose estate headquarters was in Normandy. The three other conventual landlords on this list all held extensive English estates and, as with Bec, it tended to be those demesnes too remote to provision the household that were geared most directly towards market production. On the Canterbury Cathedral Priory estate these were the Romney Marsh demesnes of Appledore and Fairfield and the north Kent demesne of Cliffe, all at least 25 miles distant from Canterbury and well placed to take advantage of major grain markets. The Westminster Abbey demesne of Birdbrook in Essex and Rochester Cathedral Priory demesne of Cuddington in Buckinghamshire were at even further removes from their respective estate headquarters.

The majority of conventual demesnes were, however, weakly rather than strongly commercialised. All thirty-two of the demesnes which in the period 1288–1315 sold less than 20 per cent of their net crop receipt and received less than £5.0 per 100 sown acres from crop sales belonged to conventual estates (Table 5.03). The list includes properties of the abbeys of Boxley, Crowland, Oseney, Peterborough (thirteen demesnes), Waltham, and Westminster (eight demesnes) and the priories of Bicester and Canterbury. In the period 1375–1400 twenty of the twenty-six demesnes which were similarly least commercialised were likewise in conventual hands. Again, properties of Boxley Abbey, Canterbury Cathedral Priory, Peterborough Abbey, and Westminster Abbey feature prominently, along with others belonging to Battle Abbey, Robertsbridge Abbey, St Catherine's Priory, Rouen, and Titchfield Abbey. They are joined by the demesne at Drayton in Berkshire belonging to New College, Oxford, and five demesnes in the ownership of such lay lords as John Doget, John de Gildeburgh, Edmund de la Pole, and William Wanton. The appearance of these lay demesnes on the later list is a reminder that many a lesser lay household remained as dependent upon its own estates, or at least its home farm, for its provisions as its conventual counterparts. On the other hand, for few lay estates is a full profile available of production and disposal strategies across the estate as a whole. The documentary prominence of demesnes managed as home farms may possibly exaggerate the importance of management strategies geared towards self-sufficiency.[62]

It will be noted that strongly and weakly commercialised demesnes co-existed on certain estates, notably those of Canterbury Cathedral Priory and Westminster Abbey. These were large and geographically extensive estates on which it made sense to manage some demesnes to yield provisions, some to yield cash, and others to yield a combination of the two. Much consequently hinged upon the place of the individual demesne within the overall estate production system. It is therefore hardly surprising that neighbouring demesnes

---

[62] Cf. M. R. Livingstone, 'Sir John Pulteney's landed estates: the acquisition and management of land by a London merchant', MA thesis, University of London (1991).

sharing effectively the same location and physical characteristics and employing much the same cropping system nevertheless often displayed fundamentally different commercialisation profiles depending on the estate to which they belonged and their role within it. This is most readily illustrated with reference to Essex and Hertfordshire, two counties close to London whose predominantly heavy soils supported a strikingly uniform pattern of cropping. In the period 1375–1400 no fewer than thirty documented demesnes in these two counties practised some version of three-course cropping with wheat and oats. Ten were possessions of Westminster Abbey; the other twenty were divided between fifteen different conventual, collegiate, and lay lords. Significantly, disposal strategies mirrored this diversity of ownership rather than the uniformity of production. Individual demesnes sold anything from 6 to 63 per cent of their net crop receipts, generating (with one exception) a sales income per 100 sown acres of £0.1 to £9.2. Even Westminster Abbey's ten demesnes sold 9 to 57 per cent of their net crop receipts at rates ranging from £1.0 to £8.8 per 100 sown acres. The four Hertfordshire demesnes of Aldenham, Kinsbourne, Stevenage, and Wheathampstead were the least commercialised of this group, no doubt because they were the best placed to send provisions to Westminster.[63] Feering, Kelvedon, and Moulsham in mid-Essex were further removed and consequently more actively involved in supplying the market. Most commercialised of all were, however, Bekeswell, a close neighbour of Moulsham in mid-Essex, Sawbridgeworth in Hertfordshire, convenient to the navigable River Lea, and Birdbrook in north Essex, remotest of all from Westminster.

Commercialised Birdbrook – 20 miles from Cambridge and Colchester, 30 miles from Ipswich, and almost 50 miles from London – demonstrates that proximity to urban markets was not a precondition for a disposal strategy based upon sale. Avington in south-west Berkshire, Broadwell in west Oxfordshire, and Adderbury and Middleton Stoney in north Oxfordshire likewise confirm that commercial opportunities were widely available when lords chose to exploit them.[64] Concentrated commercial opportunities did nevertheless encourage some demesnes to become more actively involved in supplying the market. In this context access to those riverine and coastal arteries which serviced either the metropolitan or the national and international grain trades was a crucial advantage.[65] For instance, Brightwell, Harwell, and Wantage in Berkshire – all highly commercialised demesnes – were convenient to

---

[63] For a case study of Kinsbourne see Stern, 'Hertfordshire manor'.

[64] Rates of sale on these demesnes remain impressive even when allowance is made for the possibility that between a third and a half of the arable is likely to have lain fallow.

[65] For an early analysis of those trades, see N. S. B. Gras, *The evolution of the English corn market from the twelfth to the eighteenth century* (Cambridge, Mass., 1915). For a recent analysis of the metropolitan grain trade, see Campbell and others, *Medieval capital*. Also Farmer, 'Marketing', pp. 358–77.

Abingdon whence they could supply either Oxford or London via the Thames. Lower down the river Billingbear, Waltham St Lawrence, and Wargrave also in Berkshire were even better placed to provision London owing to their proximity to the major grain entrepot of Henley, much frequented by London cornmongers.[66] Several of these Thames-side manors are known to have engaged directly in the metropolitan grain trade.[67] Downstream of the city, Eastwood and West Thurrock in south Essex and Cliffe and Ospringe in north Kent appear to have taken similar advantage of the lively grain trade – partly metropolitan and partly national and international – that focused upon the Thames estuary.[68] Other less strongly commercialised demesnes reinforce this pattern and define the Thames Valley both upstream and downstream of London as the major axis of commercialisation within the south-east (Figure 5.01). Elsewhere, the lure of the coastal and cross-Channel grain trades may possibly explain why the Romney Marsh demesnes of Appledore and Fairfield represent a similar focus of more commercialised arable production.[69]

Evidently there were certain localities where the commercial pulse beat faster and drew a greater proportion and volume of production to market. True, most of the demesnes which took greatest advantage of those opportunities belonged to those lords most interested in exploiting their estates as a source of cash rather than provisions, hence it was clearly not good enough merely to be in the right place. It was being in the right place *and* belonging to the right estate that made the difference. This makes it particularly difficult to gauge the general level of commercial activity in those localities such as eastern Kent and the Soke of Peterborough where the bulk of the available evidence comes from demesnes largely geared towards direct consumption by major conventual households. Perhaps other producers in these two localities were energetically engaged in producing for the market, but in the absence of surviving accounts from a sufficient cross-section of estates it is possible only to speculate. Sometimes, of course, a weakly commercialised arable sector was a concomitant of strong commercial specialisation elsewhere. Near to London, for instance, sales of wood and pastoral products tended to eclipse those of field crops, as was consistent with prevailing patterns of land-use and levels of economic rent.[70]

It would require a far more robust sample of demesnes to bring this picture of arable commercialisation into sharper focus, especially as it was patently a picture which changed over time. The kinds of institutional and locational distinction which were clearly such a feature at the beginning of the fourteenth century were far less sharply etched by the close of that century (Figures 5.01

---

[66] Campbell and others, *Medieval capital*, pp. 47–9, 51–5, 76–7, 92–3, 101–2.
[67] Farmer, 'Marketing', pp. 367–8, 371.
[68] Campbell and others, *Medieval capital*, pp. 68–9, 92–4, 169, 181–2.
[69] Campbell, 'Tradable surpluses'.
[70] Campbell, 'Measuring commercialisation', pp. 181–4; Galloway and others, 'Fuelling the city'.

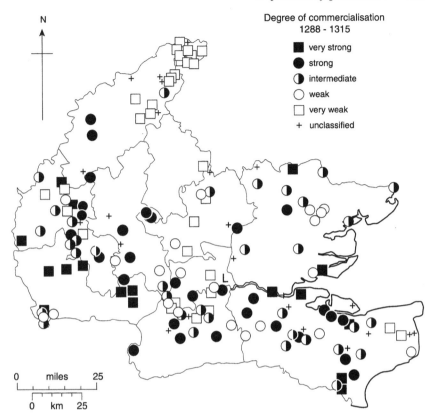

Fig. 5.01. Levels of arable commercialisation within the FTC counties, 1288–1315 (L = London) (source: FTC1 accounts database).

and 5.02 and Table 5.03). While differences between individual estates remained important, differences between broad ownership types were less so as direct provisioning of seigniorial and especially conventual households declined in significance (Table 5.02). Overall, the more extreme forms of commercial specialisation and intensification tended to disappear, as market participation quietened down (Table 5.03 and Figure 5.02). No longer does the Thames Valley upstream and downstream of London stand out as necessarily superior in commercial activity, at least so far as grain is concerned. Instead, it is the vale country north of the Chiltern scarp that emerges most strongly, together with parts of Kent. This mirrors the changing scale and structure of market demand and underscores the dynamic character of arable husbandry throughout this period.[71] Sale gained relative to transfer as a

---

[71] See below, pp. 231–48.

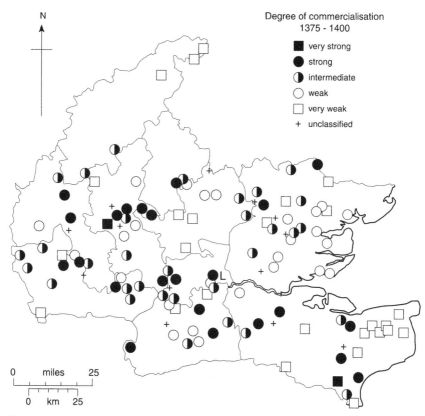

Fig. 5.02. Levels of arable commercialisation within the FTC counties, 1375–1400 (L = London) (source: FTC2 accounts database).

method of disposal but declined in absolute importance. Proportionately, there were as many weakly commercialised demesnes at the end of the four-teenth century as at the beginning, but fewer that were strongly commercial-ised. In the twilight of direct management economic autarky remained a cherished principle on at least one in four of demesnes that remained in hand. This is hardly surprising for by 1400 the grain trade was flowing down fewer, smaller arteries and provided a weakening incentive to commit a greater volume of seigniorial production to the market. Animals rather than crops now offered the best commercial opportunities and expanding fodder con-sumption may be one reason why a greater share of net crop production was now being retained and consumed on the manor.[72]

---

[72] See Chapter 4, pp. 183–6, and below, pp. 246–7.

5.2   The principal field crops and their attributes

The greater the rate at which lords wished to consume and/or sell, the more elaborate the precautions they had to take to ensure that production was sustainable. The most successful cropping systems were those which took maximum advantage of the different botanical, economic, and physical attributes of the principal crops. Medieval demesne managers sowed their arable with either grains (wheat, rye, barley, oats, and various mixtures of the same) or legumes (beans, peas, and vetches, either singly or in combination). Occasionally, they also sowed mixtures of the two. Field cultivation of root crops was unknown at that time in England and although many medieval farmers practised some form of convertible husbandry there is no evidence of the sowing of leys with artificial grasses.[73] Flax and hemp were extensively grown on peasant holdings but their appearance on demesnes is rare and then usually confined to gardens.[74] Other documented horticultural crops include teasels, madder, woad, vegetables, nuts, and fruit.[75] On individual demesnes these could be a lucrative source of income, especially when there was a major urban market on the doorstep.[76]

Although it was in gardens that several of the crops which subsequently were to have such an impact upon English agriculture were first tried out, between 1250 and 1450 no significant new crop escaped from the garden into the field.[77] The range of field crops available to cultivators at the end of the

[73] On the introduction of root crops see E. Kerridge, *The farmers of old England* (London, 1973), pp. 118–24; M. Overton, 'The diffusion of agricultural innovations in early modern England: turnips and clover in Norfolk and Suffolk 1580–1740', *TIBG*, new series, 10 (1985), 205–21. On medieval convertible husbandry see Bishop, 'Rotation of crops'; Finberg, *Tavistock Abbey*, pp. 103–7; Searle, *Lordship and community*, pp. 272–86; Dyer, *Warwickshire farming*, pp. 13–14; Hallam (ed.), *AHEW*, vol. II, pp. 387–8; Chapter 6, pp. 268, 293–7, 299.

[74] The *Nonae* Rolls of 1340–1 record the tithes paid on flax and hemp: e.g. N. Evans, *The East Anglian linen industry* (Aldershot, 1985), pp. 41–6. Examples of the demesne cultivation of hemp include Acle, Aldeby, Attlebridge, Catton, Costessey, Eaton, Halvergate, Hanworth, Hemsby, Martham, Monks Granges, Newton-by-Norwich, Plumstead, and Taverham, all in Norfolk, dated between 1260 and 1340. Halvergate also grew flax. The quantities were in every case small, and at Catton, Eaton, and Halvergate it is clear that these crops were being grown on the demesne curtilage. BLO, MS rolls Norfolk 47; NRO, DCN 60/2/1; 60/4/35–6; 60/8/1, 8–9, 23; 60/23/4; 60/26/7; 60/28/4; 60/29/13; 60/35/6; 61/12; PRO, SC 6/929/2–3; 6/933/14; 6/936/16, 18, 20; 6/937/3, 7, 9. Hemp's horticultural status is revealed by the fact that in 1329–30 and 1339–40 hemp seed was sold from the garden of Norwich Cathedral Priory: NRO, DCN 1/11/1, 1A. See also Sutton, 'Early linen industry'.

[75] Campbell, 'Agricultural progress', p. 41; J. Greig, 'Plant resources', in Astill and Grant (eds.), *Countryside*, pp. 113–18.

[76] For examples of commercial horticulture see Campbell, 'Agricultural progress', 41; J. A. Galloway and M. Murphy, 'Feeding the city: medieval London and its agrarian hinterland', *London Journal* 16 (1991), 7–8; 'Norwich Cathedral Priory gardeners' accounts, 1329–1530', ed. C. Nobel, in *Farming and gardening in late medieval Norfolk* (Norwich, 1997), pp. 1–93. According to Thünen, *Isolated state*, pp. 9–11, commercial horticulture was a characteristic land-use specialism of a narrow zone within a few miles of pre-industrial cities.

[77] Kerridge, *Farmers*, pp. 118–24.

period was essentially unchanged from that which had existed at the beginning. This does not mean to say that these crops had not themselves undergone modification through a process of seed selection, intentional or unintentional, but as yet there is insufficient archaeobotanical or other evidence to form a verdict.[78] Of the precise botanical character of the crops grown on medieval demesnes the litany of Latin names endlessly repeated in manorial accounts – *frumentum* (wheat), *siligo* (rye), *ordeum* (barley), *avena* (oats), *faba* (beans), *pisa* (peas), and *vicia* (vetches) – reveals nothing. Yet it would be incredible if this uniformity of vocabulary did not mask considerable botanical diversity. Certainly, by 1523 John Fitzherbert was able to name seven different kinds of wheat in his *Boke of Husbondrye*, of which bread wheat (*triticum aestivum*) and rivet wheat (*triticum turgidum*) were the most important.[79] Tall-growing varieties would have offered advantages in weed-infested fields and are both implied by the medieval practice of harvesting the grain high, near the ear, and borne out by preserved examples of medieval thatch.[80] The remaining trash of straw and weeds could then have been fed to livestock. Rye, the tallest grain, could grow to 5 or 6 feet. The fact that it was often sown mixed with wheat implies that the latter may sometimes have been almost as tall. Multi-tillering varieties, by providing a denser crop, would have stood the greatest chance of competing successfully against couch, thistles, and other weeds. On the other hand low yield-to-seed ratios imply small ears with few grain.[81]

5.21 Wheat

Wheat was the premier bread grain of medieval England, preferred by virtually all who could afford it. Its gluten content meant that it rose more than any other flour, making a lighter loaf. With rare exception it commanded a higher price than any other field crop, as was consistent with its exceptional baking qualities.[82] It was the densest and, therefore, the heaviest of the grains, yielded more kilocalories per bushel than any other, and shed fewest of those kilocalories when milled into flour (Table 5.04). Because of its weight, most medieval carts could carry only 3 quarters of wheat, compared with 3.5 quarters of barley and 4 quarters of oats. The higher unit carriage costs which this

[78] Greig, 'Plant resources', pp. 108–14. Analysis of preserved medieval thatch may help to provide some of the answers.   [79] Cited in Greig, 'Plant resources', p. 108.

[80] N. Hawkes, 'Secrets of medieval life entwined in country thatches', *The Times* (12 August 1995), p. 3; J. B. Letts, *Smoke blackened thatch* (Reading, 1999), pp. 24–7, 35–41.

[81] Personal communication, Philip Brooks (retired inter-war farmer who cultivated land formerly part of the bishop of Winchester's demesne at Churt, Farnham, Surrey); Harwood Long, 'Low yields of corn', p. 469.

[82] Price inversions between wheat and its cheaper alternatives were normal close to major urban centres due to differences in economic rent: Campbell and others, *Medieval capital*, p. 124, n. 36.

Table 5.04. *Absolute and relative weight, food value, extraction rate, price, and cartage costs of the principal field crops, c. 1300*

Crop	Weight c. 1300 lb. per bus.	kcal per lb.	kcal per bus. c. 1300	Extraction rate 1801 %	Price c. 1300 pence per stone	Price c. 1300 pence per bus.	Cartage cost per mile c. 1300[a] pence per qtr.	Cartage cost over 10 miles as % of price per bus. %
Absolute:								
Wheat	53	1,520	80,560	80	2.34	8.88	0.30	34
Rye	51	1,520	77,520		1.92	6.98	0.29	42
Barley	46	1,452	66,792	78	1.97	6.47	0.26	40
Oats	36	1,676	60,336	56	1.37	3.53	0.23	65
Peas (fresh/dried)		304–1,300				6.32		
Relative:								
Wheat	1.00	1.00	1.00	1.00	1.00	1.00	1.00	1.00
Rye	0.96	1.00	0.96		0.82	0.79	0.97	1.23
Barley	0.87	0.96	0.83	0.98	0.84	0.73	0.87	1.19
Oats	0.68	1.10	0.75	0.70	0.59	0.40	0.77	1.93
Peas (fresh/dried)		0.20–0.86				0.71		

Note:
[a] FTC counties

Source: B. M. S. Campbell, J. A. Galloway, D. J. Keene, and M. Murphy, *A medieval capital and its grain supply* (n.p., 1993), pp. 41, 191, 196; D. L. Farmer, 'Prices and wages', in Hallam (ed.), *AHEW*, vol. II, p. 734; A. A. Paul and D. A. T. Southgate, *McCance and Widdowson's 'The composition of foods'* (London, 1978).

Table 5.05. *Disposal of aggregate net crop receipts within the FTC counties, 1288–1315 and 1375–1400*

Crop	No. of manors	Aggregate net crop receipt[a]	% of aggregate net crop receipt malted	% of aggregate net crop receipt (adjusted for grain converted to malt)[b]		
				Retained	Transferred[c]	Sold
FTCI, 1288–1315:						
Individual crops (quarters):						
Wheat	188	13,328	1	31	21	48
Rye	99	2,640	0	65	6	28
Winter mixtures	55	1,596	0	64	1	35
Barley	156	7,971	21	46	19	34
Dredge	111	3,866	30	27	24	49
Oats	189	12,911	2	67	11	22
Miscellaneous grains	10	302	1	34	42	25
Legumes	177	2,506	0	59	8	34
Legume–grain mixtures	10	65	0	65	0	35
All crops (£):[d]	190	£9,136	7	44	17	39

FTC2, 1375–1400:

Individual crops (quarters):

Wheat	126	6,144	1	48	18	34
Rye	25	1,351	0	92	2	6
Winter mixtures	26	560	7	91	2	7
Barley	113	6,067	23	44	15	41
Dredge	65	2,261	45	23	18	58
Oats	122	4,212	1	53	10	37
Legumes	121	2,287	0	63	3	35
Legume–grain mixtures	21	172	0	61	0	40
Malt	60			29	33	37
All crops (£):[d]	128	£4,289	11	48	14	37

Notes:

[a] Net of tithe and seed

[b] FTC1 adjusted on the assumption that 33.3% of malted grain was retained, 33.3% was transferred, and 33.3% was sold; FTC2 adjusted on the basis that 23.1% of malted grain was transferred, 9.5% was transferred *ad hospicium domini*, 36.5% was sold, and 29.1% was retained

[c] Including 'sales' *ad hospicium domini*

[d] Conversion of volume to value carried out using the following per bushel sale prices:
1288–1315: wheat 8.6d., rye 7.1d., winter mixtures 6.9d., barley 6.5d., dredge 5.1d., oats 3.7d., beans 5.9d., peas 5.7d., vetches 5.1d., legumes 5.6d., grain–legume mixtures 5.4d.
1375–1400: wheat 7.9d., rye 4.1d., winter mixtures 5.8d., barley 6.0d., dredge 4.6d., oats 3.4d., beans 5.8d., peas 5.3d., vetches 5.6d., legumes 5.5d., grain–legume mixtures 5.2d.

Source: FTC1 and FTC2 accounts databases.

imposed were nevertheless more than offset by the superior price which wheat generally commanded, which rendered it the most transportable crop of all.[83] Cleaning the threshed grain further enhanced both its price and transportability.[84]

Wheat was the most demanding crop to grow. It required more nitrogen pound for pound than any other crop and for that reason was usually sown as the lead crop of rotations, immediately following the fallow and the replenishment of soil nitrogen which that allowed. It was invariably winter sown and unlike some other grains was virtually never sown on the same land in consecutive years. Sandy and acidic soils were unsuited to its cultivation and low temperatures created problems of both germination and ripening. Until growing conditions could be modified physically, and hardier strains of seed developed, wheat therefore remained environmentally more circumscribed in its cultivation than any other crop. Thus, it was grown on only a limited scale, if at all, in the north-western counties of Cumberland and Lancashire, parts of Devon and Cornwall, and on the poorest sandy soils of Nottinghamshire, Norfolk, and Suffolk, where rye superseded it as the principal winter grain and bread grain.[85] Spelt wheat (*triticum spelta*) was the hardiest available wheat, most tolerant of a cool, wet climate, but unambiguous evidence of its medieval cultivation is scarce.[86] Instead, rivet wheat was generally grown. It is a tall and productive grain and is practically immune to rust fungi, but it is sensitive to bad weather and poor soil, and grows slowly. According to Lord Ernle 'Red rivet, or a lost white variety, was then recommended for wheat-sowing on light land, red or white pollard for heavy soils, "gray" wheat for clays.'[87]

The accounts are more illuminating about the uses to which wheat was put than the species of wheat cultivated. Although it was not unknown for wheat to be used as animal fodder the amounts involved were always small and usually confined to the *curallum* or inferior wheat.[88] The malting of wheat for brewing was equally exceptional, for all that it is capable of yielding a distinctive and high-quality ale.[89] Instead, wheat more than any other grain was grown as a human foodstuff. This is apparent from the significant quantities included within the liveries paid to workers on some demesnes, especially as

---

[83] In 1300 a bushel of wheat commanded a price 28 per cent greater than that of its closest substitute, rye. In 1400 its price was 43 per cent higher. See Table 5.07.

[84] G. W. Grantham, 'Jean Meuvret and the subsistence problem in early modern France', *JEH* 49 (1989), 188.

[85] National accounts database; Hallam (ed.), *AHEW*, vol. II, pp. 390–3, 405–7; Miller (ed.), *AHEW*, vol. III, pp. 177–8, 186–8, 303–5; Chapter 6, pp. 267–9, 289–90.

[86] Greig, 'Plant resources', pp. 109–10.   [87] Ernle, *English farming*, p. 8.

[88] On the Abbey of Bury St Edmunds's manors of Thorpe Abbotts and Tivetshall in south Norfolk, for example, approximately 3 per cent of wheat receipts (net of tithe) were consumed as fodder in the fourteenth century: NRO, WAL 274×6/478, 288×1–3/1245.

[89] On the evidence of the FTC2 accounts database, 15 per cent of manors malted wheat, the quantities ranging from 2 quarters at Woolstone in Berkshire to 160 quarters at Beddington in Surrey: PRO, SC 6/757/8–9, 14; Surrey RO, 2163/1/11.

the dietary quality of liveries improved in the aftermath of the Black Death.[90] On five Norfolk demesnes and twelve Breckland demesnes just under a quarter of all wheat receipts after deduction of tithes and seed were retained and paid as food liveries in the first half of the fourteenth century.[91] By the final quarter of that century the equivalent proportion was over 40 per cent on six demesnes in the FTC counties.[92] Within these ten counties wheat, along with dredge, was also the crop most likely to be transferred to estate headquarters for consumption, since wheaten bread was a staple of monastic diets. Substantial proportions were also sold, the bulk of it presumably purchased – as in London – for bread making (Table 5.05).[93]

5.22 *Rye*

Rye, the nearest alternative bread grain, was clearly regarded as inferior to wheat. Although pound for pound it was as nutritious as wheat and per bushel was only marginally lighter, it commanded a price only 78 per cent that of wheat *c.* 1300, falling to 70 per cent that of wheat *c.* 1400 (Table 5.07). W. Ashley believed that rye was the staple bread grain of the medieval peasantry and that on the demesnes much of that produced was certainly destined for consumption by the manorial workforce.[94] In the period 1375–1400 respectively 48 per cent, 50 per cent, 88 per cent, and 93 per cent of net rye receipts on the manors of Adderbury (Oxfordshire), Westerham (Kent), Wargrave (Berkshire), and Hyde (Middlesex) were consumed directly as food liveries.[95] This compares with the 71 per cent similarly consumed at Thorpe Abbotts, Norfolk in the period 1336–79 and average of 52 per cent consumed on eleven Breckland demesnes in the period 1350–99.[96] By the final quarter of the fourteenth century over 90 per cent of the residual quantities of rye being sown by demesnes in the FTC counties was destined for consumption on the manor. Yet, even as a food livery it was declining. Workers were increasingly demanding and receiving liveries of wheat rather than rye, with the result that a growing share of net wheat receipts had to be retained for consumption on the manor (Table 5.05). A century earlier, when labour had been less able to dictate its dietary terms, not only was rye's domination of grain liveries far greater but off-the-manor demand was conspicuously stronger. *Circa* 1300, 65 per cent of a far larger volume of net receipts was retained by the FTC demesnes for consumption on the manor, the remaining 35 per cent being

[90] Dyer, 'English diet', pp. 213–14; Dyer, 'Changes in diet', p. 28.
[91] The Norfolk demesnes are Hemsby, Martham, Sedgeford, Thorpe Abbotts, and Tivetshall: NRO, DCN 60/15/1–16, 60/23/1–25, 60/33/1–30; WAL 274×6/478, 288×1/1245. For the Breckland demesnes see Bailey, *A marginal economy?*, pp. 241–4.
[92] See above, n. 19.   [93] Campbell and others, *Medieval capital*, pp. 24–36.
[94] W. Ashley, *The bread of our forefathers* (Oxford, 1928), pp. 86–100.
[95] See above, n. 19.   [96] See above, n. 91.

disposed of by transfer and sale (Table 5.05). The range over which rye could be effectively marketed was, however, less than that of wheat, since its comparable weight but lower price meant that it was less able to bear the costs of carriage. Thus, demesnes sowing substantial acreages of rye and rye mixtures for the lucrative London market were concentrated immediately upstream of the city within a 10-mile cart journey of either the city or the navigable River Thames, whereas commercial wheat producers were located at a greater riverine and overland distance from the city.[97]

Rye, unlike wheat, seems almost never to have been malted but it was on occasion fed as fodder to animals. At Adderbury in Oxfordshire, for instance, 3 per cent of net rye receipts were fed to animals in the period 1375–1400 and this proportion rose significantly higher in the principal rye-producing districts.[98] Thus, animals consumed an eighth of net rye receipts on manors in the East Anglian Breckland in the second half of the fourteenth century, a seventh of those on the manor of Thorpe Abbotts in south Norfolk, and a third of those at Sedgeford in north-west Norfolk.[99] The length of rye straw meant that it was especially valued for bedding and fodder and ensured that it was also in demand for thatching and the making of harness and halters. In fact, in more recent times it has not been unknown for the straw to be worth more than the grain.[100]

Since demesnes usually grew rye as a subsistence rather than commercial crop it rarely occupied more than a modest share of the sown acreage. The exceptions were either in the immediate vicinity of major urban markets, where there was strong demand for a cheap alternative to wheat, or where environmental conditions were unsuited to wheat.[101] Rye is much hardier than wheat and can be grown with success in colder and more exposed places, where it will germinate quicker and at lower temperatures. In much of Devon and Cornwall, for instance, it long remained the standard bread grain.[102] It requires less nitrogen, can be produced with less fertiliser, and does not exhaust the supply of nitrogen as much as wheat. Sandy and acidic soils therefore pose no great obstacle to its cultivation and it is with these that its cultivation in the Middle Ages is particularly associated, as in the arid Breckland of East Anglia.[103] Rye also has the virtue of being less susceptible to attack by insects and diseases than wheat, especially as the crop usually matures before

---

[97] Campbell and others, *Medieval capital*, pp. 121–3.

[98] Hants. RO, 11M59 B1/123–43.

[99] Bailey, *A marginal economy?*, pp. 242–3; NRO, DCN 60/33/30; L'Estrange IB 1/4, 3/4; WAL 1245/288 × 1.

[100] Ashley, *Bread of our forefathers*, p. 200. Westminster Abbey's manor of Knightsbridge in Middlesex sold rye straw worth 3 shillings in 1306–7: WAM, 16396.

[101] Campbell and others, *Medieval capital*, pp. 121–3; Chapter 6, pp. 267–8, 289–90.

[102] Miller (ed.), *AHEW*, vol. III, pp. 303–4.

[103] Bailey, *A marginal economy?*, pp. 209–13, 237–40.

rust becomes severe.[104] Its greatest drawback was its susceptibility to ergot, but that was probably unappreciated at the time.

5.23 Winter mixtures

It was commonplace in the Middle Ages to bake bread from a combination of wheat and rye and sometimes other grains as well.[105] Towards that end wheat and rye were often sown together as a mixture. There were problems in so doing of synchronising the ripening of the two grains, hence in Devon wheat and rye seem always to have been sown separately rather than together. The same was true of much of western and south-western England. In East Anglia and the midlands, on the other hand, the wheat–rye mixture known as maslin appears to have been grown in modest quantities wherever rye was cultivated as a crop. In fact, in the midlands its cultivation was something of a specialism. In 1309–10 at Chilvers Coton, Cubbington, Fletchamstead, Sherbourne, and Warwick maslin occupied 30–50 per cent of the grain acreage.[106] The accounts for these manors, like many others, refer to this wheat–rye mixture as *mixtilio*, which became corrupted into the Old French *miscelin* and thence into the English 'maslin'.[107] Some accounts, however, describe it as 'mancorn', a derivation from the Teutonic *man, mun,* or *meng,* meaning mingled or mixed. This, for instance, was the practice on many of the Berkshire, Buckinghamshire, Hampshire, and Wiltshire manors of the bishopric of Winchester.[108] Indeed, use of the term 'mancorn' on the FTC sample of demesnes is confined to these Winchester-owned manors plus Eynsham Abbey's manor of South Stoke, Oxfordshire.[109] Far more unusual is the term *sprigitum* used on William de Fiennes's manor of Wendover and Merton College's manor of Ibstone (both in Buckinghamshire) at the end of the thirteenth century, where it plainly describes the wheat–rye mixture normally known as maslin.[110]

Maslin/mancorn was not the only potential winter mixture, for barley was also sown as a winter crop, either on its own or as a mixture. Demesnes in Kent and to a lesser extent in Surrey commonly grew both the winter and spring varieties of barley, and they are distinguished as such in the accounts. Elsewhere, on the evidence of the FTC samples of accounts, the same clear distinction was drawn on only a handful of manors in Essex, Buckinghamshire, and Berkshire belonging to estates with their headquarters in Kent and Sussex, notably Canterbury Cathedral Priory and Battle Abbey.[111]

---

[104] Ashley, *Bread of our forefathers,* pp. 199–200.
[105] Ashley, *Bread of our forefathers,* pp. 95–100; Campbell and others, *Medieval capital,* p. 26.
[106] PRO, SC 6/1039/11, 6/1040/18.   [107] Ashley, *Bread of our forefathers,* p. 16.
[108] Titow, 'Land and population', pp. 33–5.   [109] BLO, MSS DD CH CH M93.
[110] PRO, SC 6/763/5; Merton College, Oxford, MM 5070.
[111] CCA, DCc/Borley 8, Lalling 17, Risborough 5, Southchurch 5; PRO, SC 6/742/26–9.

By implication, other manors in these counties may also have been sowing both varieties of barley but without distinguishing between them in the accounts. As at Westerham in Kent in 1307–8, winter barley may occasionally be described as 'berecorn', a term which occurs at Coombe, Hampstead Norreys, Inkpen, Templeton, and Woolstone (all Berkshire), Ivinghoe (Buckinghamshire), and Temple Dinsley (Hertfordshire).[112] The distribution of these demesnes certainly complements those that are known to have grown winter barley. On the other hand, at Coombe in Berkshire and Farnham in Surrey the term 'beremancorn' is used as a variant upon 'berecorn', plainly implying that this was a mixed rather than pure crop based on 'bere' or winter barley (presumably four-rowed barley).[113] In fact, on John le Ferrers's demesne of Hampstead Norreys (Berkshire) in 1300–1 the latter was almost certainly the case, for wheat, rye, maslin, winter barley, and spring barley are all additionally mentioned as separate crops.[114] By a process of elimination 'berecorn' must here have been a winter-barley-based mixture. The term is particularly characteristic of properties belonging to estates based in Hampshire and Berkshire and was variously used to describe either winter barley on its own, or winter barley mixed with rye and/or, less probably, wheat. It provided an alternative to rye and maslin as a coarse, cheap, bread grain and in its pure form could also be malted for brewing. If the example of Westerham is at all representative, by the close of the fourteenth century winter barley and its mixtures only survived (like rye and maslin) as a crop largely intended for the food liveries of manorial workers.[115]

5.24 Barley

Where the accounts refer solely to barley and make no distinction as to variety it is most likely to have been the two-rowed spring-sown variety, which was commonly sown in succession to a winter course, and, in the most intensive cropping systems of all, was sometimes sown on the same land in consecutive years (Figures 6.25–6.27). It was not dependent upon heavy manuring and too much nitrogen could diminish its suitability for brewing.[116] The latter quality meant that it tended to be favoured by farmers on medium to light land. Since it contains fractionally fewer kilocalories per pound than wheat or rye and in the Middle Ages was significantly less dense as a grain, barley yielded approximately a sixth fewer kilocalories per bushel (Table 5.04). While husked barley can be consumed whole as pottage, at little if any loss in food value, milling it into flour results in a loss of approximately 22 per cent of available kilocalo-

---

[112] Berks. RO, D/EC M66; Herts. RO, 11M59/B1/54, 65; Kings College, Cambridge, COM/59; PRO, SC 6/756/3, 6/865/13, 6/1122/26; WAM, 4535, 26405.

[113] King's College, Cambridge, COM/56; Hants. RO, 11M59/B1/54; Pretty, 'Sustainable agriculture', p. 5.   [114] PRO, SC 6/748/28.   [115] WAM, 26460–501.

[116] Bailey, *A marginal economy?*, p. 140.

ries. Malting and then brewing it – the most common use – is even more waste-ful, since about 70 per cent of the kilocalories present in the raw grain will probably be lost in its conversion to ale.[117] Nevertheless, the ale that it pro-duced was highly prized and during the course of the Middle Ages barley established itself as the premier brewing grain.[118]

Just over a fifth of net barley receipts on the FTC demesnes were malted on the manor and much of the barley transferred elsewhere or sold must subse-quently have been malted (Table 5.05). Norfolk was the country's greatest barley-producing county. Its demesnes devoted on average just under half of their grain acreage to barley before 1350 (Figure 6.01) and over half after 1350 (Figure 6.13) and typically malted much of it on the manor.[119] All the barley received by Norwich Cathedral Priory came in ready-malted form, its manors of Sedgeford, Martham, and Hemsby malting respectively 33, 57, and 70 per cent of their net barley receipts.[120] Some Breckland demesnes, which produced a barley particularly well suited to brewing owing to its low nitrogen content, malted similarly large proportions of their barley: 43 per cent at Brandon and 61 per cent at Fornham All Saints in the first half of the fourteenth century.[121] Here, malting reduced barley's weight and added value, thereby raising its capacity to bear the costs of transport (a bushel of malt weighed on average 25 per cent less than a bushel of barley but commanded a price as much as 20 per cent higher). Even in its unmalted state it could be marketed at a greater distance than rye (Table 5.04) and the best-quality Norfolk barley malt rivalled wheat in the range at which it could be sold.[122] As well as brewing, barley was an important ingredient of the coarser and cheaper breads, was consumed as pottage, and very occasionally was used as fodder for livestock (although its days as an important foodstuff for fattening bullocks lay in the future).[123]

---

[117] A portion of the kilocalories contained in the by-products of milling and brewing could, of course, be reclaimed either by recycling them as livestock feed or applying them as a soil dress-ing: Campbell and others, *Medieval capital*, p. 205.
[118] In an age of untreated water alcoholic beverages of varying strengths were a much healthier source of liquid.   [119] Campbell and Overton, 'New perspective', pp. 55–7.
[120] NRO, DCN 1/1/1–40, 60/15/1–16, 60/23/1–22, 60/33/2–27, 62/2.
[121] Bailey, *A marginal economy?*, pp. 243–4.
[122] During the first half of the fourteenth century a quarter of Norfolk malt was worth 94 per cent of a quarter of Norfolk wheat, and was substantially cheaper to transport: calculated from London School of Economics Library, unpublished Beveridge price data, Box G9.
[123] On four demesnes of Westminster Abbey and two demesnes of the bishopric of Winchester in the period 1375–1400 animals consumed only 0.05 per cent of net barley receipts (see above, n. 19). On Norwich Cathedral Priory's Norfolk manors of Hemsby, Martham, and Sedgeford the equivalent proportion was a little over 2 per cent, but on the Abbey of Bury St Edmunds's Norfolk demesnes of Tivetshall and Thorpe Abbotts it rose to 6 per cent and 8 per cent: NRO, DCN 60/15/1–16, 60/23/1–23, 60/33/1–31; NNAS 5890–903 20 D1, 5904–15 20 D2, 5916–17 20 D3; L'Estrange IB 1–3/4; Raynham Hall, Norfolk, Townshend MSS.

5.25 Oats

Only oats were more versatile than barley, serving both as a pottage grain and a drink grain, a source of human food and of livestock fodder. Oats also contained about 15 per cent more kilocalories per pound than barley, but had a lower density than any other grain and bushel for bushel weighed only four-fifths the weight of barley and two-thirds the weight of wheat (Table 5.04). Their food value per bushel was therefore the lowest of any grain and this was compounded by an extraction rate of only 56 per cent when oats were milled into flour and probably only 30 per cent when they were malted and brewed into ale.[124] Contemporaries clearly rated oats as a third-rate bread grain and second-rate brewing grain, as witnessed by the verdict of a sixteenth-century visitor that Cornish ale brewed from oats was 'lyke wash as pygges had wrestled dryn'.[125] Nor did their widespread use as a fodder crop enhance their esteem as an ingredient of human diets.[126] Consequently, both per bushel and per pound, they were the cheapest of the grains, a bushel of oats selling for less than half the price of a bushel of wheat throughout the fourteenth century, declining to only a third the price by the mid-fifteenth century.[127] This greatly restricted the range at which they could be marketed since they were low in value relative to their bulk. Commercial production therefore had to be geared in the main towards the provisioning of local markets. Where, as in the case of major urban centres, oats were required in bulk that meant devoting a substantial proportion of the immediate hinterland to their cultivation.[128] At Ruislip, Edgware, and Hampstead (Middlesex), in the early fourteenth century respectively half to two-thirds of the grain acreage was devoted to oats, most of it, no doubt, either intended for London or the heavy traffic generated by that great city.[129] In the same period Sandford-on-Thames, 3 miles outside Oxford, and Blean, 2½ miles outside Canterbury, also grew more oats than any other grain.[130]

The range of uses to which oats could be put, their value as a cheap foodstuff for the poor, and their vital importance as a fodder crop for horses, nevertheless ensured them a ready market. In the FTC counties demesnes sold

---

[124] Only modest amounts of oats were malted on demesnes in the FTC counties (Table 5.05) even though ale brewed from oats continued to be widely drunk in the vicinity of London: Campbell and others, *Medieval capital*, p. 25. The amounts malted became insignificant in the aftermath of the Black Death as consumers were able to afford a higher-quality product.
[125] Cited in Miller (ed.), *AHEW*, vol. III, p. 304.
[126] A twelfth-century chronicler noted contemptuously that Exeter men and beasts fed on the same grain: M. Kowaleski, *Local markets and regional trade in medieval Exeter* (Cambridge, 1995), p. 14.   [127] Farmer, 'Prices and wages, 1350–1500', p. 447.
[128] Campbell and others, *Medieval capital*, pp. 116–18, 160–61.
[129] *Select documents of the English lands of the Abbey of Bec*, ed. M. Chibnall (London, 1951); PRO, DL29/1/1–2; WAM, 32404, 32373, 32401; Campbell and others, *Medieval capital*, pp. 31, 34–5.   [130] PRO, E358/19–20; CCA, DCc/Blean 1(2).

over a fifth of net oats receipts in the period 1288–1315 and over a third in the period 1375–1400. In both periods a further tenth of net receipts was transferred off the manor (Table 5.05). Of that portion of the crop retained on the manor some was paid as a livery to the workforce, probably to be consumed as pottage or brewed into ale, and substantially more was consumed as fodder, primarily by the draught horses and the riding horses of visiting officials. Wargrave in Berkshire in the final quarter of the fourteenth century fed a fifth of its net oats receipts to its workers and a further quarter to its livestock.[131] According to the scale of the oat crop the proportion consumed as fodder could be substantially higher. On the five Norfolk demesnes of Hemsby, Martham, Sedgeford, Thorpe Abbotts, and Tivetshall, for instance, an average of 15 per cent of net oats receipts were used as liveries and 75 per cent as fodder.[132] These demesnes all made considerable use of horses but nevertheless sowed only a relatively small share of their sown acreages with oats. Conversely, in the ox-dominated north and west of the country the opposite conditions prevailed and it tended to be for human rather than animal consumption that oats were principally sown.[133]

Oats were more tolerant of difficult growing conditions than any other crop. They were the standard spring-sown crop on lowland demesnes wherever soils were cold, stiff, and heavy, as was conspicuously the case on the heavy boulder-clay soils of central Essex and Hertfordshire.[134] They were also the staple grain in much of the north and west of England where low temperatures and high rainfall hindered other grains from germinating and ripening. They were grown in quantity on most documented demesnes in the Marcher counties of Cheshire, Shropshire, Herefordshire, and Monmouthshire, and the south-western counties of Cornwall and Devon, and were the dominant grain crop on many demesnes in the West Riding of Yorkshire as well as on the Berwickshire lands of Coldingham Priory. At Birkby, Cockermouth, and Bolton in Cumberland and West Derby in Lancashire they occupied over two-thirds of the demesne grain acreage.[135] They fared well even on water-logged soils and hence were grown on a large scale in many areas of reclaimed marshland, only being superseded by other crops as the land dried out and became desalinated. This explains the importance of oats on manors with land in Romney Marsh and the coastal marshes of Essex.[136] Elsewhere, because they required lower levels of soil nitrogen than almost any other crop, they were often sown as the final course in the most intensive rotations, before the land was fallowed, manured, and stirred (Figure 6.28). Often, where they succeeded

[131] Hants. RO, 11M59 B1/123–43.   [132] See above, n. 123.
[133] P. F. Brandon, 'New settlement: south-eastern England', in Miller (ed.), *AHEW*, vol. III, pp. 177, 186–8.   [134] Campbell and others, *Medieval capital*, pp. 116–18.
[135] National accounts database; PRO, SC 6/824/2, 8; 6/1094/11.
[136] Smith, *Canterbury Cathedral Priory*, pp. 137, 178; Campbell and others, *Medieval capital*, pp. 116–18.

other crops, it was necessary to sow them thickly to choke out weed growth. Instances occur of seed being sown as thickly as 8 bushels per acre.[137]

Archaeobotanical evidence indicates that both the common oat (*avena sativa*) and bristle oat (*avena strigosa*) were widely cultivated, although insofar as accounts make any distinction it is between great oats and small oats (the naked oats or 'pillcorn' of later centuries).[138] The latter were very poor oats that had nearly deteriorated back to black hairy oats.[139] Both types are recorded at Billingbear in Berkshire at the beginning of the fourteenth century and at Ashford in Middlesex at the end but otherwise the distinction is rarely encountered in the FTC counties.[140] On the poor, sandy soils of the East Anglian Breckland, however, it was standard practice to distinguish between these two types from which it appears to have been small oats rather than great oats that were most commonly sown.[141] The same was evidently true of demesnes on the upland margins of Devon and Cornwall.[142] These may have resembled the 'grey-awned, thin, and poor' oats described by Ernle.[143]

5.26 Spring mixtures

Oats and barley were often sown together as a mixture known as dredge (or drage). About 30 per cent of demesnes grew dredge, many of them concentrated in the midland counties. Bedfordshire above all, along with neighbouring portions of Cambridgeshire and Northamptonshire, seems to have specialised in the cultivation of this crop: at Houghton, Sundon, and Eaton Bray in the second half of the thirteenth century over 40 per cent of the grain acreage was sown with dredge.[144] In later centuries crushed dredge was prized as an animal feed but in the Middle Ages its use as a fodder crop was decidedly limited. At Adderbury (Oxfordshire), Hyde (Middlesex), and Wargrave (Berkshire), for instance, 1 per cent or less of net dredge receipts were fed to livestock.[145] Instead, dredge appears to have been most highly prized as a brewing grain. In fact, demesnes in the FTC counties malted larger proportions of their dredge – 30 per cent in the period 1288–1315, 45 per cent in the period 1375–1400 – than their barley (Table 5.05). Larger proportions of dredge were also sold than any other crop, presumably to satisfy the demands of commercial maltsters and brewers. For this reason it rarely shows up in the liveries paid to workers. On this criterion it was the most commercialised of all crops. As dietary standards and expectations rose in the aftermath of the

---

[137] Campbell, 'Arable productivity', pp. 387–8.
[138] Greig, 'Plant resources', p. 111; Hatcher, 'Farming: south-western England', pp. 392–3.
[139] I. Adams, *Agrarian landscape terms* (London, 1976), p. 137.
[140] Hants. RO, 11M59/B1/54, 62, 65; WAM, 26804–5, 7, 13.
[141] Bailey, *A marginal economy?*, pp. 238–40.
[142] Finberg, *Tavistock Abbey*, pp. 95–7.
[143] Miller (ed.), *AHEW*, vol. III, p. 303; Ernle, *English farming*, p. 9.
[144] *Ministers' accounts*, vol. I, pp. 6–12; PRO, SC 6/1094/11.   [145] See above, n. 19.

Black Death it also gained as the nearest substitute for oats, which bore the brunt of the waning demand for the cheapest grains.

5.27 Grain–legume mixtures

Oats were also sometimes sown mixed with legumes. The acreages involved were never large and the crop was intended exclusively as fodder, to be fed either threshed or unthreshed. The oats served as physical support for the legumes while the legumes helped fix atmospheric nitrogen in the soil. Later centuries were to know these kinds of mixture as 'horsemeat': a name which clearly denotes their function. Threshed and coarsely ground they could be processed into a particularly high-valued livestock feed known as horse-bread, which was generally reserved for cart-horses and riding horses. The earliest medieval references to horse-bread date from the very beginning of the fourteenth century.[146] Later in the century William Langland's *Piers Plowman* advised beggars capable of work to stave off starvation with 'hounds' bread and horse-bread'.[147]

Manorial accounts refer to oat–legume mixtures by a variety of names. Within the FTC counties they occur as *harascum* at Bulmer and Messing Hall (Essex), Harmondsworth (Middlesex), and Barton, Copton, Eastry, Ham, Ickham, Newnham Court, and Sharpness (Kent); as 'bulmong/bullimong' at Berwick Berners, Birchanger, Hornchurch, and Writtle (Essex); as 'pesemong' at High Easter (Essex); as 'benmong' at Sawbridgeworth (Hertfordshire); as 'drogman' at Kinsbourne, Stevenage, and Wheathampstead (Hertfordshire) (all manors of Westminster Abbey); and as *pulmentum* at Maidwell (Northamptonshire), where it is defined as a mixture of vetches and 'the lighter grain of oats'.[148] As these names imply, the precise composition of the mixture varied. Sometimes it was pea based, sometimes bean based, and sometimes vetches were included, but the grain component was almost invariably oats. The strong association of this brand of mixed crop with East Anglia and the south-east is striking. All but one of the FTC examples occur in Essex (seven), Hertfordshire (four), Middlesex (one), and Kent (seven). Additional East Anglian examples occur at Hinderclay, Redgrave, and Rickinghall

---

[146] Campbell and others, *Medieval capital*, p. 27.

[147] *Visions from Piers Plowman taken from the poem of William Langland*, trans. N. Coghill (London, 1949), p. 53.

[148] CCA, DCc/Barton Carucate 14–15; Copton 33–4, 36; Ickham 56–7; Essex RO, D/DH X 19, 21; D/DHf M45; Guildhall Library, London, 25404/44; Herts. RO, D/ELW/M183, 185, 192–3; Lambeth Palace Library ED386; New College, Oxford, 5785; 6386–8; 7312–14, 23; N'hants. RO, Finch Hatton 475, 482; PRO, SC 6/1245/9; 6/892/1–2; 6/893/27–8; 6/897/7, 9, 11, 12; DL 29/42/817; WAM, 8842, 45, 48, 55; 26309, 13, 17, 24, 55, 58; Winchester College, 11501–2; Miller (ed.), *AHEW*, vol. III, p. 218. From 1361 small acreages of *pulmentum* – a mixture of rye and peas destined to be fed unthreshed to the stots – were sown at Sedgeford, Norfolk: NRO, L'Estrange IB 1/4.

(Suffolk), and Felbrigg, Little Fransham, Syderstone, Thorpe Abbotts, and Tivetshall (Norfolk), but none pre-dates 1347, when 'bulmong' appears first to have been sown at Rickinghall.[149] Nor does the FTC1 sample of accounts furnish any earlier examples. Possibly the earliest references are those at Clavering and Thurrocks in Essex which both date from the 1330s.[150] It would seem that grain–legume mixtures were either a mid-fourteenth-century innovation on demesnes, or perhaps it was only from that time – as the employment of horses rose in these counties – that devoting land specifically to the cultivation of 'horse-meat' became worthwhile

5.28 Legumes

Legumes – peas, vetches, and beans – were grown by 69 per cent of demesnes in the period 1250–1349 and 81 per cent of demesnes in the period 1350–1449. Peas were far more generally grown than either vetches or beans and when distinctions were drawn between their type it was mainly on the basis of colour.[151] Within the FTC counties grey peas were rarest, recorded on only one demesne (Elverton, Kent), followed by green peas (six demesnes), black peas (ten demesnes), and white peas (thirteen demesnes).[152] Mention of type is, however, rare outside the metropolitan counties of Essex, Hertfordshire, Middlesex, and Kent. It is likely that the black and white peas were small and hard, with a low water content which aided drying, threshing, and storage. Without fuller information it is difficult to hazard an estimate of their food value, which may have been as low as 304 kilocalories per pound or as high as 1,300 kilocalories if dried (Table 5.04). Nevertheless, it is plain that compared with the grains they were less important as a source of energy than of protein, since they contain amino acids which combine with the elements in grain to produce the kind of protein that grain alone cannot supply.[153] No doubt this was one reason why peas and beans were an important component of the diet of the poor, consumed either whole usually as an ingredient of pottage, or dried, coarsely ground and used as an ingredient in the cheapest breads.[154]

Demesnes grew legumes for consumption by seigniorial households, for sale, as a food livery for manorial workers, and as a fodder crop. Peas were

---

[149] BL, Add. Roll 63525–43, 52–62; CUL, Cholmondley (Houghton) MSS, reeves' and bailiffs' accounts 29; Chicago UL, Bacon Roll 337–9, 344–65, 496–7, 499, 500–508; NRO, WAL 274 ×3, 274×6, 288×2; WKC 2/130/398×6; MS 13127 40 A5.

[150] Miller (ed.), *AHEW*, vol. III, pp. 280, 303.

[151] In the period 1250–1349 42 per cent of demesnes grew peas, 22 per cent vetches, and 17 per cent beans. The equivalent proportions in the period 1350–1449 were 52 per cent, 37 per cent and 14 per cent, respectively. The marked increase in the proportion of demesnes growing vetches is a function of the active diffusion of this crop during the fourteenth century: Campbell, 'Diffusion of vetches'.   [152] CCA, DCc/Elverton 38–40.

[153] R. Tannahill, *Food in history*, 2nd edition (London, 1988), p. 157.

[154] Dyer, 'English diet', pp. 200–1; Dyer, *Standards of living*, pp. 153, 240.

indispensable to the intensive sty management of swine and with vetches provided an important alternative feed for horses where meadows and consequently hay were scarce.[155] Whether legumes were grown primarily for human
or animal consumption varied considerably, according to local circumstance.
For example, in the final quarter of the fourteenth century 5 per cent of the
net legume receipts at Westerham in Kent were fed to livestock and 50 per cent
to the manorial workforce, whereas at Adderbury in Oxfordshire 45 per cent
of legumes were fed to the livestock and none to the workforce.[156] Much
depended upon the intensity of the husbandry regime and extent to which the
pastoral sector was dependent upon the arable for fodder crops.[157] Above all,
the more that was demanded from the soil the greater became the necessity to
expand the relative share of the cropped acreage sown with legumes.

Unlike the grains, whose cultivation diminished the store of nitrogen available within the soil, legumes actually contributed to the replenishment of
nitrogen supplies by fixing it from the atmosphere. They were therefore a
crucial component of the all-important nitrogen cycle. Many centuries later
they would be superseded in this function by clover and sainfoin, which were
more efficient nitrogen fixers mainly because they stayed in the ground
longer.[158] Sometimes, when the demand for livestock and their products was
stronger than that for crops legumes were merely incorporated into existing
rotations as an alternative to grain.[159] In that context their cultivation should
be regarded primarily as an adjunct of pastoral husbandry. Where grain production was the objective, however, they tended to be substituted for bare
fallows, sometimes simply sown on an *inhok* from the arable, at others integrated into more intensive rotational schemes in which fallowing occurred
much less frequently.[160] Thus, legumes might be sown to provide a nitrogen

---

[155] Trow-Smith, *Livestock husbandry*, pp. 117–18; Biddick, 'Pig husbandry'; Biddick, *The other
economy*, pp. 122–3, 132; Campbell, 'Diffusion of vetches'.   [156] See above, n. 19.
[157] On the exceptionally intensively cultivated demesnes of Hemsby and Martham in east
Norfolk, where the arable and pastoral sectors were closely integrated, legumes occupied
between a fifth and a quarter of the sown acreage in 1300–49 and were either fed to livestock
or sold: NRO, DCN 60/15, 23.
[158] Although clover was known and cultivated from the 1660s it was not until well into the eighteenth century that its share of the arable began to exceed that of the traditional medieval
legumes: Campbell and Overton, 'New perspective', pp. 54–5, 58–60. Clover and sainfoin were
less easily assimilated into rotations than the medieval triumvirate of peas, beans, and vetches
and were contingent upon the development of a specialised international trade in seed:
M. Ambrosioli, *The wild and the sown*, trans. M. M. Salvatorelli (Cambridge, 1997), pp.
337–98, 431–3.
[159] At Tingewick in Buckinghamshire, for example, legumes expanded from 4 per cent to 47 per
cent of the sown area between 1312 and 1380, mostly at the expense of oats, which declined
from 52 per cent to 10 per cent: New College, Oxford, 7086–8.
[160] Ravensdale, *Liable to floods*, pp. 116–18; Campbell, 'Agricultural progress', pp. 31–3. At
Hemsby in east Norfolk, where cropping became virtually continuous by the beginning of the
fourteenth century, legumes increased their share of the sown acreage from 16 per cent in the
1270s to 22 per cent in the 1320s: NRO, DCN 60/15.

boost between successive grain crops and they might be sown again at the end of the rotation in order to promote a better temporary sward on the ensuing fallow (Figures 6.26 and 6.27).[161]

Their benefits extended further than this, for the fodder they provided was more nutritious than that available on the bare fallows they replaced. The result was higher stocking levels and improved manure supplies, especially when the animals were stall fed and the resultant farmyard manure was systematically spread upon the fields.[162] According to the cost and availability of labour, legumes could be cut and threshed, cut and fed in the sheaf (*in siliquis*), grazed on the stem in the field, or ploughed in as a green manure. They therefore suited farming systems of varying degrees of intensity, except the most extensive where labour was at a premium and livestock had to make shift as far as possible with natural grazing. Beans fared particularly well on heavy land, peas and vetches on medium to heavy land.[163] Only on the lightest soils, most deficient in nitrogen, did they do badly. Here, therefore, alternative strategies of maintaining the nitrogen cycle – such as sheep-corn husbandry – had to be adopted.[164] Elsewhere, they were instrumental in allowing medieval cultivators to evolve intensive, productive, and sustainable systems of cropping.

5.29 Winter- versus spring-sown crops

Since the principal field crops were suited to different functions, varied in their food value, and differed in their growth requirements they offered cultivators a remarkably wide range of options. Because the winter-sown crops – wheat, rye, winter barley, and their various mixtures – were the most demanding they were generally placed first in rotations, immediately following the fallow. They thus received maximum benefit from the nitrogen fixation, casual and systematic manuring, and multiple ploughings which had occurred during the fallow year or years. These were also the crops which generally commanded the highest prices and were best suited to the production of bread, the staple foodstuff of the period. In a simple two-course rotation it was notionally possible to indulge in monoculture and grow nothing other than these winter-sown bread grains. But in practice that rarely happened because of the need to maintain a seasonally balanced distribution of work and produce a variety of crops suited to a range of functions. Most rotations therefore combined the cultivation of both winter- and spring-sown crops.

The spring crops included both the principal brewing grains – barley and dredge – and the main pottage and fodder crops – oats and legumes. From the

---

[161] Farmer, 'Grain yields on Westminster Abbey manors', pp. 346–7, was puzzled why several Westminster Abbey demesnes sowed their legumes immediately before the fallow.

[162] *Walter of Henley*, pp. 327–9; Shiel, 'Improving soil fertility', p. 64.

[163] For instance, Glastonbury Abbey's Sedgemoor demesnes of Brent and Zoy devoted substantial acreages to beans: Postan, *Economy and society*, pp. 51–2.

[164] Bailey, *A marginal economy?*, pp. 56–85, 237–40.

former were manufactured the ale that was the universal drink of almost all classes while the latter comprised a vital food source for both humans and animals. Dietary preferences rated pottage low, for all that it involved least wastage of available kilocalories, and ale high, notwithstanding its extravagantly low extraction rate. Because these brewing, pottage, and fodder crops were all mostly spring sown and matured at different rates they helped spread the peak labour demands of ploughing and harvesting. Since they made lighter demands on the soil and allowed at least a half-yearly fallow during the winter they could be sown either in consecutive years or as a spring course following the winter course. This was the rationale of three-course rotations, where the sequence was winter-course, spring-course, fallow. Such rotations offered cultivators less flexibility in their choice of crops than two-course rotations but sustained a significantly higher intensity of cropping, offering real output gains provided that the soil was equal to the increased demands made upon it. At Podimore in Somerset, for instance, where a two-field was converted to a three-field system immediately following the harvest of 1333, Fox reckons that the annual gain in output was worth £9 to Glastonbury Abbey.[165]

When medieval cultivators endeavoured to raise output further they generally chose to do so by cropping land more frequently rather than raising yields *per se*.[166] In all the most successful systems this was generally achieved by introducing additional spring courses and offsetting them with more systematic manuring, larger sowings of legumes (themselves usually spring sown), and better preparation of the seed bed.[167] A marked bias in favour of spring cropping in conjunction with large-scale legume cultivation is therefore a good indication of intensive cropping with a low incidence of fallowing.[168] Such systems should not however be mistaken for those in which spring-cropping predominated merely by force of circumstance. Oats, for instance, were sometimes grown to the exclusion of virtually all else under environmental conditions where no other grain crop would succeed. Here, their versatility of function came into its own and oats became the all-purpose crop, a substitute for most of the others.[169]

5.3   Trends in cropping, 1250–1449

Agricultural producers have rarely experienced such marked changes in the scale and structure of demand as those that occurred between 1250 and 1450. Until the second decade of the fourteenth century there were more mouths to

[165] Fox, 'Alleged transformation to three-field systems', pp. 533–8.

[166] B. M. S. Campbell, 'Land, labour, livestock, and productivity trends in English seignorial agriculture, 1208–1450,' in Campbell and Overton (eds.), *Land, labour and livestock*, pp. 159–74.

[167] Brandon, 'Demesne arable farming'; Campbell, 'Agricultural progress'; Mate, 'Agrarian practices'.   [168] Campbell and others, *Medieval capital*, pp. 128–38.

[169] E.g. Kershaw, *Bolton Priory*, pp. 38–9, 72.

be fed but at a deteriorating dietary standard as mean living standards fell.[170] Commencing with the Great European Famine, a succession of mortality crises then dramatically reduced the population to be supported and eventually transformed land scarcity into land abundance. As mean living standards rose, standards of nutrition recovered and consumers were better able to indulge their dietary preferences (Figure 1.01). Demesne managers responded to these changes in the scale and composition of demand by adjusting the scale and composition of their arable production.

5.31 *The scale of cultivation*

Arable producers benefited from rising demand until at least the second decade of the fourteenth century.[171] As the population pressed ever harder on the land not only was more food in aggregate required but grain consumption in various forms increasingly dominated diets. High and rising land values also offered more favourable returns to investment and encouraged demesne lords to make the most of their arable resources. Whether that was best achieved by continuing to keep their demesnes in hand and managing them directly or by leasing portions or the whole of them to others was another matter. Men were eager to take land on lease at the end of the thirteenth century and it was usually possible for landlords to negotiate very advantageous terms. Indeed, it is far from clear why so many lords persevered in managing their estates directly.[172] From the final quarter of the thirteenth century the bishops of Winchester were curtailing the scale of their arable operations and leasing portions of their demesnes piecemeal to tenants.[173] Other landlords were also rationalising the scale of their arable operations. The priors of Norwich, for instance, curtailed production on their Norfolk manors of Catton, Gateley, Hindringham, and Hindolveston, while either maintaining or expanding it on their eleven other Norfolk properties. Between *c.* 1265 and *c.* 1305 they thereby raised the total acreage under crop by an eighth from an estimated 2,583 acres to 2,928 acres (Figure 5.03).[174] Over the same period, in

---

[170] See Chapter 1, pp. 4–6.   [171] Kershaw, 'The Great Famine'.
[172] Fenoaltea, 'Authority'.   [173] Titow, 'Land and population', pp. 15–29.
[174] The following account of the estate of Norwich Cathedral Priory is based upon all extant accounts for the manors of Catton, Eaton, Gateley, Gnatingdon, Hemsby, 'Heythe', Hindolveston, Hindringham, Martham, Monks Granges, Newton-by-Norwich, North Elmham, Plumstead, Sedgeford, Taverham and Thornham, plus the *Proficuum maneriorum*, which records total sown acreages on many of the manors between 1292–3 and 1306–7 and between 1324–5 and 1339–40, and the central accounts of the prior's own office (that of the 'master of the cellar'): BLO, MS rolls Norfolk 20–47; NRO, DCN 1/1; 40/13; 60/4, 8, 10, 13, 14, 15, 18, 20, 23, 26, 28, 29, 33, 35, 37; 61/35–6; 62/1–2; DCN R233B 4626; L'Estrange IB 1/4, 3/4, 4/4; NNAS 5890–918 20 D1–3; Raynham Hall, Norfolk, Townshend MSS. The occasional use of long hundreds presents a problem, although the discontinuous nature of most of the manorial records poses greater difficulties. For no single year are accounts available for each manor belonging to the estate. After 1340 the records for several manors are especially patchy.

contrast, Roger Bigod, earl of Norfolk, maintained cultivation on his fifteen Norfolk manors at more or less its existing level: some individual demesnes temporarily ploughed up additional land when grain prices made it profitable to do so but this was then withdrawn from cultivation when prices fell.[175]

The price inflation of the long thirteenth century climaxed during the famine years 1315–22 (Figure 1.01).[176] Subsequently, prices for arable crops registered their first sustained fall since the onset of inflation in the 1180s. At much the same time, heavy livestock losses from murrain and rinderpest depleted flocks and herds thereby diminishing the capacity to fertilise and till the land.[177] Nor did on-going taxation and purveyancing help producers cope with the situation. Low prices can sometimes stimulate farmers to produce more not less in an endeavour to maintain income levels, but this strategy is only feasible where opportunities for expansion exist, there is no constraint upon the supply of inputs, and the return from the land is sufficient to cover the costs of production.[178]

The initial reaction of Norwich Cathedral Priory to these adverse economic conditions was to reduce the overall area under crop by some 3 per cent from its pre-Famine peak. The brunt of that reduction was borne by the monks' north-western and central-Norfolk manors and several of their properties near Norwich. Meanwhile, production was maintained at just below existing levels on their two highly productive eastern manors of Hemsby and Martham and was actually expanded at Newton near Norwich through a vigorous policy of land acquisition.[179] Then, in 1333, as economic conditions worsened, they leased their two central Norfolk demesnes of Hindringham and Hindolveston, followed in 1334 by their smaller north-western demesne of Thornham.[180] At the same time, however, they continued to acquire land at Newton-by-Norwich and also created a small new demesne at 'Heythe', a few

The total sown acreage shown in Figure 5.03 is therefore an estimate, based on such information as is available. For case studies of the estate see Saunders, *Norwich Cathedral Priory*;
Stone, 'Estates of Norwich Cathedral Priory'; Campbell, 'Field systems', pp. 24–144 (manor
of Martham); Virgoe, 'Estates of Norwich Cathedral Priory'.

[175] The manors are Acle, Attleborough, Bressingham, Caister-cum-Markshall, Ditchingham,
Earsham, Forncett, Framingham, Halvergate, Hanworth, Loddon, Seething, South Lopham,
South Walsham and Suffield: PRO, SC 6/929/1–7, 14–21; 6/931/21–3; 6/932/11–26;
6/933/20–29; 6/934/1–39; 6/935/2–37; 6/936/2–32; 6/937/1–10, 22–33; 6/938/1–11; 6/943/10–11;
6/944/1–10, 21–31. For a case study of Forncett see Davenport, *Norfolk manor*. At Hanworth
between 142 acres and 192½ acres were sown, depending upon the price of grain: the 50 acres
in question alternated between arable and pasture. The management of the Bigod estate is discussed in Denholm-Young, *Seignorial administration*, pp. 45, 123, 138–41.

[176] Kershaw, 'The Great Famine'; Smith, 'Demographic developments'; Jordan, *The Great
Famine*.

[177] Baker, 'Contracting arable lands'; Campbell (ed.), *Before the Black Death*; Campbell, 'Ecology
versus economics', pp. 77, 95.

[178] E.g. Campbell and Overton, 'New perspective', pp. 87–8, 91–2.

[179] Many small purchases of land at Newton-by-Norwich are recorded in the prior's central
accounts: NRO, DCN 1/1.   [180] NRO, DCN R232C 5176; DCN 1/1/32–4, 62/2, 60/18/29.

miles east of Norwich.[181] The net effect was to concentrate the share of production taking place on their demesnes near or to the east of Norwich from 50 per cent to 58 per cent while reducing the overall scale of their operations by 10 per cent. Nevertheless, this initial flirtation with leasing was evidently not an unqualified success for both Hindringham and Hindolveston were taken back in hand in 1339, followed by Thornham a few years later, with the result that by the eve of the Black Death the cropped acreage on the priory's estate may have been restored to its pre-Famine level (Figure 5.03).[182]

The Black Death shattered whatever fragile equilibrium landlords may have succeeded in re-establishing in the aftermath of the Great Famine. Suddenly there were at least a third fewer people to feed and correspondingly fewer hands to work the land. Between 1315 and 1375 the cumulative reduction in population may have been approximately 40 per cent.[183] Wage rates rose inexorably and compounded the impact of population decline upon the arable sector by facilitating a shift away from heavily grain-based diets.[184] For a generation after the Black Death price inflation and wage restraint combined to shelter arable producers from the worst effects of the contraction in demand, but in the mid-1370s grain prices collapsed, never fully to recover, while wages climbed ever upward (Figure 1.01).[185] By the century's close wage rates had risen by approximately one-third relative to grain prices. Profit margins for arable producers were thus seriously eroded. At the same time, reduced land values depressed the returns on investment. Poor land and expensive land became unprofitable to work. Meanwhile, lower production costs in pastoral husbandry coupled with a more buoyant market for pastoral products encouraged the conversion of tillage to grass. Lords responded by leasing many of their demesnes wholesale to tenants and curtailing the scale of their arable operations on those of their demesnes which they still retained in hand.

---

[181] Between 1273–4 and 1332–3 the sown acreage at Newton-by-Norwich expanded by 64 per cent, from approximately 194 acres to 318 acres: NRO, DCN 60/28/1, 40/13. 'Heythe' appears for the first time in 1333–4 with a sown acreage of 44 acres: NRO, DCN 62/2.

[182] The record of the 1340s is tantalisingly incomplete. 1339–40 is the last year for which virtually complete information is available, thanks in the main to the enrolment of sown acreages in the *Proficuum maneriorum*. With Thornham still at farm, a total of 2,900¾ acres were sown, just 35¼ acres short of the maximum previously sown on the estate in 1306–7. By the mid-1340s Thornham was once more back in hand, while the eight demesnes for which accounts are available had increased their aggregate sown acreage by 4.5 per cent. The estate's sown area may therefore briefly have peaked at approximately 3,100 acres. NRO, DCN 1/1/36–42, 40/13–14, 60/4/37–42, 60/10/24, 60/13/26, 60/18/30, 60/20/25, 60/23/11, 60/29/25, 60/35/28–9, 61/36; L'Estrange IB 1/4; NNAS 5890 20 D1. In the Norfolk and Suffolk Breckland 'the area under cultivation was often higher in the 1340s than in any other documented decade': Bailey, *A marginal economy?*, p. 203.

[183] Chapter 8, p. 402. R. M. Smith, 'Human resources in rural England', in Astill and Grant (eds.), *Countryside*, pp. 191–3, suggests a far greater decline of 60%. The decline is likely to have been exceptionally pronounced in Norfolk, where plague mortality was above average: Shrewsbury, *Bubonic plague*, pp. 94–9; Campbell, 'Population pressure', pp. 96–100.

[184] Dyer, 'Changes in diet', pp. 25–32; Dyer, *Standards of living*, pp. 157–60.

[185] Farmer, 'Prices and wages, 1350–1500', pp. 441–2, 434–6, 444, 471.

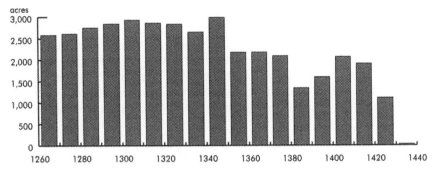

Fig. 5.03. Total sown acreage on the estate of Norwich Cathedral Priory, 1260–1440 (source: BLO, MS Rolls, Norfolk 20–47; NRO, DCN 1/1, 40/13, 60/4, 8, 10, 13, 14, 15, 18, 20, 23, 26, 28, 29, 33, 35, 37; DCN 61/35–6; DCN 62/1–2; DCN R233 B 4626; L'Estrange IB 1/4, 3/4, 4/4; NNAS 5890–918 20 D1–3; Raynham Hall, Norfolk, Townshend MSS).

On the estate of Norwich Cathedral Priory the 1350s brought an immediate and lasting reduction in the acreage in hand and under crop (Figure 5.03). The demesne at Hindolveston was again leased out and by the end of the decade two of the priory's smallest demesnes, Thornham and North Elmham, were also set at farm (a fate probably shared by the two small demesnes at Catton and Gateley).[186] On all other demesnes the acreage under crop was curtailed. By the mid-1370s the area under direct management on the estate had contracted by 70 per cent, while the eleven demesnes probably still in hand were sowing on average only three-quarters of the area they had sown on the eve of the Black Death.[187] This contraction was achieved partly by expanding the proportion of fallow, partly by converting arable to pasture, and partly through the piecemeal leasing of demesne arable to tenants.

The abrupt deterioration in economic circumstances which occurred in the mid-1370s precipitated a further spate of leasing. In the early 1380s both Hindringham and Hindolveston were again set at farm and Taverham, Monks Granges, Plumstead, and probably Eaton were leased for the first time.[188] By the final decade of the fourteenth century the monks were cropping barely half the area they had cropped on the eve of the Black Death, although those demesnes which they retained in hand had suffered no further erosion of their cropped acreage.[189] With the opening of the fifteenth century that policy was reversed. Most of the demesnes which had been farmed were taken back in hand as their leases fell in. At Martham, which had always remained in hand,

[186] NRO, DCN 1/1/45–6.
[187] In contrast, the bishops of Winchester were still cultivating approximately 85 per cent of their pre-Black Death demesne acreage: Farmer, 'Grain yields on Winchester manors', pp. 560–1.
[188] NRO, DCN 1/1/59–66; 60/18/44, 49; 60/20/33, 35.
[189] Those demesnes still in hand on the Winchester estate *c.* 1410 were still cultivating 74 per cent of their pre-1348 sown acreage: Farmer, 'Grain yields on Winchester manors', p. 562.

piecemeal leasing was also phased out. Briefly, cultivation was restored to perhaps 64 per cent of its pre-Black Death level. Within a decade, however, wholesale leasing had been resumed and by 1420 the decision had clearly been taken to abandon direct management entirely (Figure 5.03). By 1424 Martham (the priory's second most productive demesne) had been farmed and by 1427 only Sedgeford (its largest demesne) remained in hand.[190] In 1431 it, too, was leased, thereby terminating direct management on this estate.[191]

The changing scale of cultivation on the estates of Norwich Cathedral Priory exemplifies the vicissitudes experienced by the seigniorial sector at large throughout much of the fourteenth and early fifteenth centuries. It also illustrates the divergent paths that might be followed even on demesnes belonging to the same estate. Plainly, aggregate trends were the net outcome of a diversity of individual responses. On the estate of Norwich Cathedral Priory, as presumably on many others, the second quarter of the fourteenth century brought the first real setback to arable production, which until then had been expanding. Permanent decline, however, only set in following the Black Death. During the second half of the fourteenth century the scale of cultivation on those of the priory's demesnes that remained in hand was reduced, on average, by 25 per cent. This is consistent with the general scale of decline on demesnes within the county. Thus, in the opening quarter of the fourteenth century the mean cropped acreage of fifty-eight Norfolk demesnes was 162 acres whereas by the final quarter of the century the mean cropped acreage of fifty-six Norfolk demesnes was 124 acres, a decline of 29 per cent.[192] As on the estate of Norwich Cathedral Priory, the first real reduction in the mean sown area on demesnes kept in hand appears to have taken place between the first and second quarters of the century, decline subsequently becoming more or less continuous (Figure 5.03 and Table 5.06).

Analysis of the National accounts database and FTC accounts databases reveals much the same story (Table 5.06). Nationally, the price collapse of the second quarter of the fourteenth century again appears to have elicited a particularly sharp cut-back in production, the downward trend thereby initiated continuing for the remainder of the century. Overall, the mean cropped acreage fell by 25 per cent between *c.* 1300 and *c.* 1400, from 193 acres to 145 acres. Comparison of the two FTC samples of demesnes yields a 21 per cent decline in mean cropped acreage from 224 acres in 1288–1315 to 178 acres in

---

[190] NRO, DCN 1/1/75–80, 60/18/62, 60/20/39; L'Estrange IB 3/4; NNAS 5918 20 D3.

[191] In fact, the prior continued to keep his flocks in hand. These had been reorganised on a centralised basis in 1392–3 and thereafter were accounted for centrally: NRO, DCN 1/1/68–80; L'Estrange IB 3/4. Seventeen sheep accounts are extant for the period 1485–1524: NRO, DCN 62/2, 17, 19, 22–3, 25, 28–9; 64/1–5. See Allison, 'Flock management'.

[192] To control for the changing spatial distribution of extant accounts, these means are weighted means based upon four sub-regional means.

Table 5.06. *Trends in the mean sown acreage of demesnes in hand: England, Norfolk, and the FTC counties, 1250–1449*

	Mean sown acreage		
Years	England	Norfolk	FTC counties
1250–1299	189.2	172.9	
1275–1324	193.4	171.1	223.7[a]
1300–1349	172.1	146.0	
1325–1374	156.4	132.8	
1350–1399	147.1	126.8	178.4[b]
1375–1424	144.7	136.9	
1400–1449	142.8	158.6	

Notes:
[a]   1288–1315
[b]   1375–1400
Method:
All Norfolk means are the product of four regional sub-means weighted equally. All national means are the weighted product of six regional means, comprising the weighted mean for Norfolk ×0.081; the eastern counties (Cambs., Essex, Herts., Hunts., Lincs., Middx., Suffolk) ×0.214; the south-east (Hants., Kent, Surrey, Sussex) ×0.12; the midlands (Beds., Berks., Bucks., Leics., N'hants., Oxon., Rut., Warks.) ×0.164; the south-west (Devon, Dorset, Cornwall, Gloucs., Heref., Mon., Somerset, Wilts., Worcs.) ×0.209; and the north (Berwick., Ches., Cumb., Derbs., Durham, Lancs., Northumb., Notts., Salop., Staffs., Westmor., Yorks.) ×0.213. The weightings are based on each region's share of assessed lay wealth in 1334 and poll tax population in 1377.
Sources: National accounts database; Norfolk accounts database; FTC1 and FTC2 accounts databases.

1375–1400. The equivalent figure for sixty-one demesnes common to both FTC samples (most of them in conventual or collegiate ownership) is 29 per cent.[193] From the National accounts database it would appear that contraction was most marked on demesnes in southern England, about average in the midlands, and below average in both the eastern counties and the north of England.[194] Evidently the declining demand for arable products was itself regionally selective as was the readiness with which land could be withdrawn from arable production and put to profitable alternative use.[195] In much of

[193] On the estates of the bishopric of Winchester the equivalent contraction was 26 per cent: Farmer, 'Grain yields on Winchester manors', p. 562.
[194] For the midlands cf. C. C. Dyer, 'The occupation of the land: the west midlands', in Miller (ed.), *AHEW*, vol. III, pp. 77–81.
[195] For an instructive case study see Fox, 'Occupation of land: Devon and Cornwall', pp. 152–63.

eastern England, for example, arable did not convert well to grass, while in the north, where arable was in limited supply, there may have been a reluctance to take it out of production, especially as grassland was abundant.

These estimates vary in their precision but all point to a contraction in mean demesne cropped acreage between the beginning and end of the fourteenth century of 20–30 per cent (and an even more dramatic reduction in the area directly in hand). Why, at a time when the population may have fallen by as much as 60 per cent and diets became less rather than more dependent upon grain, was the contraction in mean demesne cropped acreage not greater, especially given that it was achieved in part by leasing portions of demesne arable piecemeal to tenants? The selective nature of arable contraction provides part of the explanation: the retreat from arable production may have been most pronounced on soils and in terrains where demesne agriculture had never been well developed. A reluctance to forgo customary labour services at a time when labour was appreciating in value may also have endowed seigniorial arable production with a degree of inertia. For all producers it made sense to substitute land which was falling in price for labour which was rising. Many demesnes were being cultivated less intensively at the end of the century than they had been at the beginning.[196] Crop yields consequently tended to decline, with the result that arable output fell by more than the reduction in cultivated area.[197] But the crop mix also changed as land-extensive drink and leguminous fodder crops expanded their respective shares of the cultivated area (Table 5.08). These kinds of development endowed the arable sector with a degree of resilience in the face of demographic decline and ensured that there was no simple direct relationship between the population to be supported and the area under crop. Just as more people could be accommodated on the land by modifying diets and shifting to more food-productive crops – food crops rather than fodder crops, bread grains rather than brewing grains – so when the population fell it was possible to revert once more to crops and methods of production that were more extravagant of land. The prime constraint now became less the supply of land than the labour to work it.

5.32 Bread grains

Bread, the staff of life, formed the foundation of medieval diets. Rich and poor alike ate it, combining it with other foodstuffs according to their means.[198] It is a measure of its central dietary importance that the principal bread grains

---

[196] Campbell, 'Agricultural progress', pp. 38–9; Thornton, 'Determinants of productivity', pp. 205–7; Chapter 6, pp. 301–2.

[197] Campbell, 'Land, labour, livestock', pp. 160–5, 171; M. Overton and B. M. S. Campbell, 'Production et productivité dans l'agriculture anglaise, 1086–1871', *Histoire et Mesure*, 11, 3/4 (1996), 290–7; Chapter 7, pp. 370–85.   [198] Dyer, *Standards of living*, pp. 55–7, 153, 157–9.

Table 5.07. *Trends in crop prices relative to wheat: England, Norfolk, and the FTC counties, 1250–1449*

| Years | | Price relative to wheat: | | | |
	Wheat	Rye	Barley	Oats	Peas
England:					
1250–1299	1.00	0.80	0.70	0.40	0.70
1275–1324	1.00	0.78	0.71	0.40	0.68
1300–1349	1.00	0.77	0.72	0.42	0.67
1325–1374	1.00	0.72	0.72	0.42	0.64
1350–1399	1.00	0.69	0.72	0.42	0.62
1375–1424	1.00	0.70	0.68	0.40	0.61
1400–1449	1.00	0.71	0.60	0.36	0.58
Norfolk:					
1250–1299	1.00	0.65	0.72	0.40	0.65
1275–1324	1.00	0.72	0.77	0.44	0.71
1300–1349	1.00	0.67	0.74	0.49	0.72
1325–1374	1.00	0.61	0.65	0.46	0.67
1350–1399	1.00	0.68	0.55	0.36	0.53
1375–1424	1.00	0.71	0.56	0.43	0.60
1400–1449	1.00	0.58	0.53	0.38	0.63
FTC counties:					
1288–1315	1.00	0.83	0.76	0.43	0.66
1375–1400	1.00	0.52	0.76	0.43	0.67

Source: D. L. Farmer, 'Prices and wages', in Hallam (ed.), *AHEW*, vol. II, pp. 793–5; D. L. Farmer, 'Prices and wages, 1350–1500', in Miller (ed.), *AHEW*, vol. III, pp. 502–4; London School of Economics, unpublished Beveridge price data, Box G9; FTC1 and FTC2 accounts databases.

– wheat, rye, and maslin/mancorn – consistently occupied at least 40 per cent of the national demesne grain acreage and, until 1400, at least a third of the national demesne cropped acreage (Table 5.08). Within the FTC counties bread grains assumed an even greater importance, for these counties had to feed London as well as themselves, and London's appetite for bread was voracious.[199] At the opening of the fourteenth century, when the city attained its medieval peak in commercial activity, demesnes in the FTC counties devoted 47 per cent of their grain acreage and 43 per cent of their cropped acreage to

[199] Campbell and others, *Medieval capital*, pp. 24–36.

Table 5.08. Trends in seigniorial crop production: England, Norfolk, and the FTC counties, 1250–1449 (demesne means)

Years	Bread grains[a] as % of total cropped area			Brewing grains[b] as % of total cropped area			Pottage/fodder grains[c] as % of total cropped area		
	England	Norfolk	FTC counties	England	Norfolk	FTC counties	England	Norfolk	FTC counties
1250–1299	38.6	24.3		16.5	44.6		44.4	31.1	
1275–1324	40.3	24.9	43.1[d]	18.2	45.4	17.8[d]	40.6	29.7	39.2[d]
1300–1349	42.9	25.9		19.3	46.9		37.2	27.2	
1325–1374	39.4	25.6		22.0	47.3		37.6	27.1	
1350–1399	35.6	21.5	36.6[e]	25.5	51.3	26.5[e]	37.9	27.2	37.0[e]
1375–1424	33.6	18.5		27.3	54.0		36.6	27.4	
1400–1449	32.5	20.3		27.8	52.3		38.3	27.3	

Years	Rye and winter mixtures as % of bread-grain area			Barley as % of brewing-grain area			Legumes as % of pottage/fodder-grain area		
	England	Norfolk	FTC counties	England	Norfolk	FTC counties	England	Norfolk	FTC counties
1250–1299	16.8	40.4		85.3	99.2		14.1	44.1	
1275–1324	17.2	44.6	23.9[d]	82.1	99.5	62.9[d]	19.1	46.1	23.5[d]
1300–1349	17.7	50.2		77.9	99.4		26.6	49.0	
1325–1374	15.0	41.4		72.5	98.8		35.8	48.8	
1350–1399	12.4	30.4	12.3[e]	74.4	99.1	69.1[e]	35.7	47.0	38.9[e]
1375–1424	7.3	29.9		75.7	99.9		41.8	44.8	
1400–1449	5.0	26.6		79.9	100.0		48.5	41.4	

Notes:

a Wheat + winter mixtures + rye

b Barley + dredge

c Oats + legumes + legume–grain mixtures

d 1288–1315

e 1375–1400

Method:

All Norfolk means are the product of four regional sub-means weighted equally. All national means are the weighted product of six regional means, comprising the weighted mean for Norfolk ×0.081; the eastern counties (Cambs., Essex, Herts., Hunts., Lincs., Middx., Suffolk) ×0.214; the south-east (Hants., Kent, Surrey, Sussex) ×0.12; the midlands (Beds., Berks., Bucks., Leics., N'hants., Oxon., Rut., Warks.) ×0.164; the south-west (Devon, Dorset, Cornwall, Gloucs., Heref., Mon., Somerset, Wilts., Worcs.) ×0.209; and the north (Berwick., Ches., Cumb., Derbs., Durham, Lancs., Northumb., Notts., Salop., Staffs., Westmor., Yorks.) ×0.213. The weightings are based on each region's share of assessed lay wealth in 1334 and poll tax population in 1377.

Sources: National accounts database; Norfolk accounts database; FTC1 and FTC2 accounts databases.

bread grains. A century later these grains still occupied 43 per cent of the grain acreage and 37 per cent of the cropped acreage in these counties. Evidently, the area devoted to the conventional bread grains waxed and waned as the demand for them rose and fell. Until the aftermath of the Great Famine, *c.* 1325, they occupied an expanding share of the cropped acreage; thereafter, and especially during the half century following the Black Death, that share contracted by just over a fifth. The same was true of Norfolk, although here, because barley – grown in abundance in the county – was widely used as a bread grain, the conventional bread grains never accounted for more than 30 per cent of the grain acreage and 26 per cent of the cropped acreage.[200]

Notwithstanding the universal preference for refined wheaten bread, only the wealthy and inmates of religious houses could afford to indulge that preference on a regular basis. Apart from anything else, the extra refining it required was too wasteful of available kilocalories. Wholemeal bread was therefore much more widely eaten and many people contented themselves with far cheaper and coarser alternatives in which rye, maslin, barley, and peas featured as ingredients.[201] The London assise recognised three standards of bread: white wheaten bread, wheaten bread made of whole meal and of mixtures of brown and white meal, and bread made from other grains. Although in fixing the price and weight of bread according to the assise the city authorities were primarily concerned with the price of wheat, they are known to have purchased maslin on occasion and a maslin dealer is recorded as early as the late twelfth century. At the beginning of the fourteenth century the city's brown bakers almost certainly outnumbered its white.[202]

Maslin bread, rye bread, and breads which incorporated barley, oats, and legumes were all much cheaper than pure wheaten bread. People tended to trade down to them whenever they could no longer afford the superior wheaten variety. Significantly, both in Norfolk, where rye and maslin were well established as demesne crops, and in the country as a whole, where on average they were less than half as important, these cheaper bread grains occupied an increasing share of the expanding bread-grain acreage between 1275 and *c.* 1325 (Table 5.08). They fared better than wheat on the inferior soils being brought into production during these years and were a staple component of the food liveries paid to manorial workers, whose wage rates and dietary standards were being progressively eroded. By growing rye and maslin on their poorer land and feeding this to their workers lords were able to concentrate wheat production on their better soils and maximise the quantities available either for their own consumption or sale on the market. Occasionally one demesne specialised in producing most of the rye and maslin required for food liveries by manors elsewhere on the estate.[203] Others, especially near major

---

[200] Dyer, 'Changes in diet', p. 28.   [201] Ashley, *Bread of our forefathers*, pp. 53–4.
[202] Campbell and others, *Medieval capital*, p. 26   [203] See above, p. 202.

cities or in localities unsuited to wheat, took advantage of the expanding urban and rural markets for these grains.[204]

There was less need to cultivate rye and maslin on such a scale once the population fell, dietary standards began to improve, and cultivation was withdrawn from the poorer soils. From *c.* 1325, and especially from 1350, fewer demesnes cultivated these crops, which occupied dwindling shares of both the cropped acreage and that proportion of it devoted to bread grains.[205] In the FTC counties rye and maslin's share of the bread-grain acreage was halved from 24 per cent in the period 1288–1315 to 12 per cent in the period 1375–1400. In the country as a whole cultivation of these two crops contracted even more dramatically, from 17 per cent of the bread-grain area *c.* 1300 to 7 per cent *c.* 1400 (Table 5.08). Notwithstanding this pronounced cutback in production, rye suffered a 10 per cent fall in price relative to wheat (Table 5.07). No clearer proof is required of the evaporation of demand for these cheaper bread grains in the second half of the fourteenth century.

Compared with rye and maslin, wheat was much more successful in maintaining both its price and its share of the cropped acreage. Before the Black Death it benefited from the expanding role of bread within diets. Thereafter it remained in demand as consumers substituted wheaten bread for that made from rye, maslin and other grains. As the modest but expanding proportions of wheat fed as liveries to manorial workers indicate, wheat increasingly was being consumed by socio-economic groups which had hitherto been unable to afford it. The steadily diminishing importance of the cheaper bread grains is all the more pronounced when set against the buoyancy of demand for wheat. Collectively, the principal bread grains' share of the cropped acreage was reduced by 15 per cent in the FTC counties between the periods 1288–1315 and 1375–1400, and by 24 per cent in the country as a whole between the periods 1300–49 and 1400–49 (Table 5.08).

5.33 Brewing grains

The story of the principal brewing grains – barley and dredge – is somewhat different. Before 1350 they too expanded their share of the cropped area in tandem with the cheaper bread grains, as consumers traded down to the cheapest forms of bread and from bread to pottage. Barley long remained the bread grain of the poor and barley and dredge were standard ingredients of pottage, the great staple of the poor. Pottage had the merit that the grain could be used whole with merely sufficient milling to remove the husk, hence food extraction rates were maximised. The opposite was true when the same grains

---

[204] Campbell and others, *Medieval capital*, pp. 121–3.
[205] The proportion of demesnes growing rye shrank from a third in the period 1250–1349 to a fifth in 1350–1449.

were malted and brewed into ale. It took between a fifth and a quarter of the cropped area to produce sufficient grain to supply 10 per cent of grain-derived kilocalories when these were consumed in the liquid form of ale. The scarcer that land became, therefore, the more that ale became a luxury. Mounting population pressure before 1315 was consequently associated with a trading down both to cheaper and weaker ales (brewed increasingly from dredge and oats rather than the more expensive barley), and to unpasteurised milk and unpurified water, with all the potential health problems associated with the latter.[206]

Sheer shortage of land prevented high *per capita* rates of ale consumption before 1350.[207] Once that constraint was relaxed, however, and especially as living standards rose, more land could be devoted to barley and dredge, more of their harvest could be malted and brewed (Table 5.05), and *per caput* consumption of ale could rise.[208] Increasingly, too, it was quality as well as quantity that was wanted. Hence the progressive elimination of oats as a brewing grain in many parts of the country and its replacement with either dredge or barley. Yet whereas before 1350 it was dredge, the cheaper of these two brewing grains, that gained most, it was now barley whose cultivation expanded most vigorously since it brewed the better ale. Production grew both in Norfolk and the south-east, where barley had long been a prominent crop, and in several new areas – the valleys of the Thames and Severn and the east coast from the Humber to the Tweed – where it had formerly been of limited significance. By the close of the fourteenth century it had expanded to occupy over two-thirds of the brewing-grain acreage in the FTC counties, three-quarters of that acreage in the country as a whole, and, effectively, the entire brewing-grain acreage in Norfolk, the country's greatest malt-producing county now producing well in excess of its own needs (Table 5.08).[209] Simultaneously, the brewing grains increased their share of the cropped acreage by approximately a fifth in Norfolk and a half in both the FTC counties and the country as a whole. So great a relative increase suggests that, against the prevailing arable trend, the absolute acreage sown with barley and dredge may have been greater at the close of the fourteenth century than at the beginning.[210] Seigniorial producers were obviously playing to the most

---

[206] Dyer, 'English diet', pp. 203–4; Campbell and others, *Medieval capital*, p. 33.

[207] A *per caput* ale consumption of a quart a day has been posited by J. Bennett, *Ale, beer, and brewsters in England* (Oxford, 1996), p. 17. She further extrapolates a total annual ale output of 17m. barrels *c.* 1300 on the assumption of a population of 6m. Such an estimate is irreconcilable with the acreages and yields of brewing grains recorded on seigniorial demesnes. These suggest a maximum annual ale output in the range 7.75–10.5m. barrels and a maximum *per caput* consumption for a population of 4.35m. of 1.3–1.7 pints per day.

[208] Galloway, 'London's grain supply'; J. A. Galloway, 'Driven by drink? Ale consumption and the agrarian economy of the London region, *c.* 1300–1400', in M. Carlin and J. T. Rosenthal (eds.), *Food and eating in medieval Europe* (London, 1998), pp. 92–100.

[209] Saul, 'Great Yarmouth', pp. 226, 368–71.

[210] E.g. Bailey, *A marginal economy?*, p. 237.

buoyant component of arable demand. It is a measure of the strength of that demand that until almost the very end of the fourteenth century the relative price of barley proved remarkably resilient to the massive growth in supply (Table 5.07). Norfolk is the exception. Formerly unique in its concentration upon malting barley, its barley producers now had to face much stiffer competition in regional, national, and international markets and consequently had to accept relative barley prices in the second half of the fourteenth century which were substantially down on those to which they had hitherto been accustomed.[211]

5.34 Pottage and fodder crops

Between 37 per cent and 44 per cent of the cropped area was consistently devoted to crops which might be used interchangeably for human or animal consumption. These included oats, legumes (peas, beans, and vetches), and various oats–legume mixtures.[212] As human food these were staple ingredients of pottage, which, as already noted, had higher kilocalorie extraction rates than either bread or ale. Small amounts were also included within the cheaper and coarser forms of bread. In addition, oats were regularly malted and brewed and, in non-wheat- or rye-producing areas, especially the extreme north and west of the country, oat-cakes usually served as the closest substitute for bread among the poorer classes. Oats had the merit of being both nutritious and cheap; within the demesne sector probably half to two-thirds of those produced were destined for human rather than animal consumption. Nevertheless, over the period 1250–1349 (and despite increased use of the oats-fed horse) their share of the demesne cropped acreage contracted by 30 per cent (Table 5.08).

Oats declined because they were substituted with other crops – primarily dredge – in brewing and pottage. Moreover, legumes increasingly took their place in rotations. Before 1350 this was mainly because of the nitrifying properties of legumes, which, like oats, might be fed to humans or animals, although in this period they appear to have been more important as a foodstuff than a fodder crop.[213] In effect, there was a change in the choice of pottage crop from one extractive to one restorative of soil nitrogen. This made sound agricultural sense and was crucial to the maintenance and improvement of arable productivity. Significantly, legumes occupied a greater share of the

---

[211] Between the 1350s and the 1400s barley prices in Norfolk fell by 31 per cent compared with 18 per cent in the country as a whole: London School of Economics Library, unpublished Beveridge price data, Box G9; Farmer, 'Prices and wages, 1350–1500', p. 444.

[212] See above, pp. 227–8.

[213] In nine east-Norfolk townships legumes increased their share of the demesne cropped acreage from approximately 12 per cent in the period 1238–46 to approximately 19 per cent in the mid-fourteenth century when arable husbandry was at its fullest stretch in this most intensively cultivated of localities: Campbell, 'Agricultural progress'; NRO, Diocesan Est/1–2, 9–13; Church Commissioners 101426 2/13, 3/13, 5/13, 7/13; PRO, SC 6/944/21–31.

pottage/fodder-crop acreage in the FTC counties than the country as a whole, and an even greater share in Norfolk, arguably the most intensively cropped county of all.[214] Moreover, it was on the most intensively cultivated of these Norfolk demesnes that oats occupied their smallest and legumes their greatest share of the cropped acreage (Table 6.01 and Figure 6.01).[215] For instance, at Hemsby in the 1320s a mere 3 per cent of the cropped area was sown with oats compared with 22 per cent sown with legumes. Of the oats, 36 per cent of the net receipt was consumed as food liveries, 54 per cent as fodder, and the remaining 10 per cent was sold: of the legumes, 1 per cent was consumed as food liveries, 8 per cent as fodder, and 91 per cent was sold.[216] Plainly, legumes were here being grown more for their nitrifying properties and commercial value than as a fodder crop.

After 1350, although arable farming in general became less intensive, legumes continued their inexorable advance (Table 5.08). The brunt of that advance now occurred not in areas of intensive arable production (where, on the contrary, the legume acreage tended to contract) but in localities and regions traditionally associated with more extensive methods of production. Thus, they became a major crop in many parts of the midlands which had long been ecologically strait-jacketed by a shortage of permanent grassland.[217] Here in the heartland of the commonfield system the opportunity of sowing more land with legumes solved many inter-related problems. Legumes also began to be grown in substantial quantities in both the north-west and north-east. In fact, on a few demesnes they became the single largest crop in terms of acreage.[218] Their appeal lay as a source of fodder to support the growing numbers of animals now being stocked, for whose meat and dairy produce there was a steadily strengthening demand (Table 4.03). Converting crops into pastoral foodstuffs has the lowest kilocalorie extraction rate of all and is therefore even more extravagant in its land-use requirements than ale production.[219] The doubling in legumes' share of the pottage/fodder-grains acreage thus bears testimony to the falling unit value of land. So great was this expansion that by the end of the fourteenth century the relative price of peas was beginning to sag. Oats, too, suffered a lowering in relative price, notwithstanding a massive cutback in supply. When times had been hard they had been

[214] Chapter 6, pp. 269–72.   [215] Campbell, 'Arable productivity', pp. 392–5.
[216] NRO, DCN 60/15/12–15.
[217] R. H. Hilton, *The economic development of some Leicestershire estates in the fourteenth and fifteenth centuries* (Oxford, 1947), pp. 65–6; C. Howell, *Land, family and inheritance in transition* (Cambridge, 1983), pp. 96–9; M. P. Hogan, 'Clays, *culturae* and the cultivator's wisdom: management efficiency at fourteenth-century Wistow', *AHR* 36 (1988), 117–31; Miller (ed.), *AHEW*, vol. III, pp. 213–14, 215, 229–30.
[218] Miller (ed.), *AHEW*, vol. III, pp. 178, 187; N. Morimoto, 'Arable farming of Durham Cathedral Priory in the fourteenth century', *Nagoya Gakuin University Review* 11 (1975), 137–331; Durham, Dean and Chapter, Cell accounts; PRO, SC 6/1083/4, 6/1144/10.
[219] Simmons, *Ecology*, 2nd edition (London, 1981), pp. 170–3.

much in demand as a cheap, nutritious, and versatile foodstuff but as living standards improved demand for oats progressively narrowed until it was as fodder for horses that they were chiefly cultivated on most lowland demesnes.

5.35 Net change

At the beginning of the fourteenth century bread grains, brewing grains, and pottage/fodder crops were grown in the ratio 41:18:41. A century later the equivalent ratio was 35:28:38. Less land was being devoted to bread grains, significantly more to brewing grains, and the slight contraction in pottage/fodder crops was accompanied by a marked increase in the proportions fed to animals rather than humans. The fortunes of individual crops changed even more markedly, with the higher-quality bread and brewing grains faring conspicuously better than their lower-quality alternatives. Amidst such changes the comparative stability of relative prices over the course of the century is remarkable (Table 5.07). It implies that demesne managers were sufficiently market sensitive to keep what they produced more or less in line with what was in demand. With wheat that meant holding its share of total output reasonably steady, with rye and oats it entailed cutting relative output back hard, and with barley and peas it meant expanding production. These, however, are aggregate trends and as such the outcome of many separate production decisions, each taken in the context of the husbandry system peculiar to each demesne. Collectively, these decisions transformed the configuration of production. They also transformed the rate of food output per cropped acre.

Before 1350 arable land-use was increasingly geared towards crops with high food extraction rates; thereafter there was a progressive shift towards crops with lower kilocalorie extraction rates. On the assumption that all bread grains were processed and consumed as bread, all brewing grains as ale, and two-thirds of the pottage/fodder crops as pottage *c.* 1300 and one-third as pottage *c.* 1400; and on the further assumption that the bread grains, brewing grains, and pottage crops had respective mean kilocalorie extraction rates of 80 per cent, 30 per cent, and 100 per cent; then at constant relative yields the mean kilocalorie extraction rate per cropped acre would have been approximately 25 per cent lower at the end of the fourteenth century than the beginning. This goes a long way to explaining why the contraction in mean cropped acreage was not greater after 1350. Together, a 25 per cent reduction in the cropped area and 25 per cent reduction in mean extraction rates would have been sufficient to reduce available food kilocalories by over 40 per cent. Any fall in mean crop yields would have reduced it further. Whether crops were consumed as pottage, bread, ale, or milk/meat therefore made a material difference to the size of population that a given land area could support and the character and composition of that population's diet. One of the reasons why early-fourteenth-century England was able to support a relatively high population density was because it maximised the share of the arable devoted to bread grains and pottage crops and minimised that sown with brewing grains and fodder crops.[220] Unfortunately, that meant relatively poor and monotonous diets for the majority. Those diets were only able to improve once scarcity of land ceased to be a problem. Then, as living standards rose, so, too, did the cropped area *per caput*.

[220] Overton and Campbell, 'Production et productivité', p. 292.

6

Crop specialisation and cropping systems

The limited choice of field crops available to demesne managers was capable of combination into a variety of different cropping types: key sources of difference were the relative scale upon which individual crops were grown, the duration and sequencing of rotations, and the intensity of capital and labour inputs. Crops were not grown in isolation but as components of cropping systems whose character reflected the influence of environmental, institutional, and economic factors. Since the latter were never constant, cropping systems varied across space and changed over time. In fact, the capacity for variation was almost infinite with the result that no two farms were ever exactly identical in the character of their arable husbandry. Hence the need for some system of classification in order to reduce this diversity to its broad essentials.

The classification employed here is that outlined in Chapter 2. It is a national scheme, insofar as it is based upon two national samples of demesnes for the periods 1250–1349 and 1350–1449, and has been derived by the application of cluster analysis (Relocation method) to data on the percentage share of the sown acreage occupied by each of the six principal crops (wheat, rye, barley, oats, grain mixtures, and legumes).[1] Ideally, the share of the arable left uncultivated as fallow or ley ought to be included as a seventh variable but too few accounts record this on a systematic basis. Seven basic cropping systems are distinguished in the earlier period and six in the later (Tables 6.01 and 6.02). These are not the only classifications possible; a different method, or alternative choice and specification of variables, would undoubtedly yield different results.[2] They should not therefore be regarded as hard and fast. Rather, they serve as one indicator of the principal ways in which the various crops were combined into cropping systems and provide some guide to the main spatial variations in arable farming systems. The spatial focus of this

[1] Campbell and Power, 'Mapping'.
[2] Power and Campbell, 'Cluster analysis', pp. 232–42; Chapter 2, pp. 44–6.

Table 6.01. *National classification of cropping types, 1250–1349 (national, Norfolk, and FTC1 samples of demesnes combined)*

Variable	National cropping type							Overall
	1	2	3	4	5	6	7	
Means:								
Mean % of total sown area:								
Wheat	22.9	8.5	29.2	35.0	44.0	70.3	12.0	32.1
Rye	3.5	27.8	2.1	3.2	1.5	0.0	11.8	5.7
Barley	39.7	35.6	6.2	17.7	3.7	8.4	3.9	16.6
Oats	11.8	19.5	18.4	27.7	44.3	12.6	66.6	29.6
Grain mixtures	1.4	1.9	35.5	4.4	2.7	4.2	3.2	6.8
Legumes	20.8	6.8	8.6	12.0	3.8	4.6	2.6	9.3
Mean total sown acres	156.2	181.6	186.9	219.0	214.3	145.6	158.3	191.4
No. of demesnes	109	70	81	157	173	37	52	679
Standard deviations:								
Mean % of total sown area:								
Wheat	10.6	7.9	14.0	7.6	8.0	13.5	9.6	17.8
Rye	4.8	12.3	5.7	5.5	3.0	0.0	13.2	10.4
Barley	13.0	16.8	8.1	8.9	4.3	8.7	5.5	17.0
Oats	7.1	10.6	13.6	7.8	7.2	8.8	12.7	18.4
Grain mixtures	3.2	4.2	12.5	6.0	4.7	7.4	8.8	12.5
Legumes	9.2	4.9	8.2	8.8	3.9	5.1	4.9	9.2

Mean total sown acres	88.2	137.7	91.1	137.5	139.3	87.9	110.8	124.9
No. of demesnes	109	70	81	157	173	37	52	679
% of total classified:								
Demesnes	*16.1*	*10.3*	*11.9*	*23.1*	*25.5*	*5.5*	*7.7*	*100.0*
Area	*13.1*	*9.8*	*11.7*	*26.5*	*28.5*	*4.2*	*6.3*	*100.0*

Note:
The demesnes are listed in Appendix 1.
Method:
B. M. S. Campbell and J. P. Power, 'Mapping the agricultural geography of medieval England', *JHG* 15 (1989), 24–39.
Source: National accounts database; FTC1 accounts database; Norfolk accounts database.

exercise is further enhanced by the incorporation of the FTC and Norfolk samples of demesnes into the same classifications using discriminant functions calculated on each national cluster grouping (Figures 6.01–6.07 and 6.13–6.18).[3] Chronologically, however, both classifications are essentially static analyses of a dynamic situation (Table 5.08).

Reconstructions of rotations on individual well-documented demesnes provide further detailed insights into the character of some of these cropping systems and the similarities and differences between them (Figures 6.08–6.12, 6.19 and 6.23–6.27). These reconstructions are based upon consecutive runs of accounts which record the names and acreages of the fields and furlongs in which the individual crops were sown. By linking the evidence of consecutive accounts it becomes possible to infer which portions of the demesne were left unsown and for how long. Each of the rotations illustrated here has been drawn to scale using the key shown in Figure 6.08. It is thus possible to follow the sequence of cropping both on the demesne as a whole and on its component parcels.[4]

6.1   Seigniorial cropping systems, 1250–1349

6.11   Individual cropping systems

Extensive cultivation of oats (type 7)
Probably the simplest and most extensive cropping systems were those in which the greater part of the sown area was devoted to oats (cropping type 7). Nationally, perhaps one in ten of all arable demesnes were of this type. They are immediately recognisable from the fact that oats occupied well over half and, usually, nearer two-thirds of the entire cropped area, with either rye or wheat occupying most of the remainder. Other crops were rarely grown on any significant scale and legumes in particular were of small importance (Table 6.01). This was a cropping regime admirably suited to the hard winters, cool summers, high rainfall, and acid soils of much of the north and west of the country, where environmental circumstances exercised a powerful influence upon the scale and character of arable husbandry. Unfortunately, such localities have furnished few extant manorial accounts. Those that survive almost invariably document some version of this type of cropping system. Examples occur on the Scottish border, in the lowlands of the Lake District, throughout the flanks and dales of the Pennines and Peak District, and scattered through the south-western counties of Devon and Cornwall (Figure 6.07). This broadly corresponds with those areas where oats are known to have comprised the staple foodstuff. Oats were grown by a variety of methods depending upon

---

[3] Bartley, 'Classifying the past'.
[4] Cf. P. F. Brandon, 'Agriculture and the effects of floods and weather at Barnhorne, Sussex, during the late Middle Ages', *Sussex Archaeological Collections* 109 (1971), 72–3.

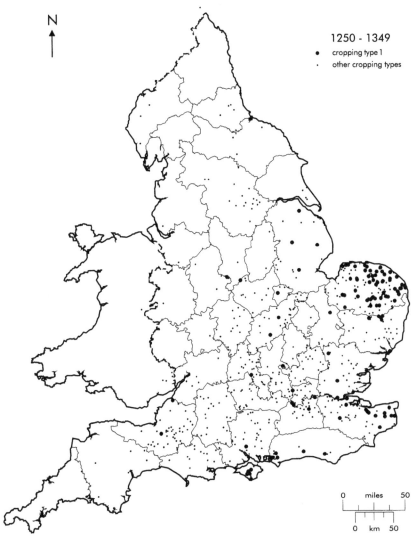

N

1250 - 1349

• cropping type 1

· other cropping types

0   miles   50

0   km   50

Fig. 6.01. Demesnes practising cropping type 1, 1250–1349 (see Table 6.01 and pp. 269–72 for explanation of cropping type) (source: National, Norfolk and FTC1 accounts databases).

Fig. 6.02. Demesnes practising cropping type 2, 1250–1349 (see Table 6.01 and pp. 267–9 for explanation of cropping type) (source: National, Norfolk, and FTC1 accounts databases).

Fig. 6.03. Demesnes practising cropping type 3, 1250–1349 (see Table 6.01 and pp. 266–7 for explanation of cropping type) (source: National, Norfolk, and FTC1 accounts databases).

Fig. 6.04. Demesnes practising cropping type 4, 1250–1349 (see Table 6.01 and pp. 263–4 for explanation of cropping type) (source: National, Norfolk, and FTC1 accounts databases).

N

1250 - 1349
• cropping type 5
· other cropping types

0   miles   50

0   km   50

Fig. 6.05. Demesnes practising cropping type 5, 1250–1349 (see Table 6.01 and pp. 262–3 for explanation of cropping type) (source: National, Norfolk, and FTC1 accounts databases).

Fig. 6.06. Demesnes practising cropping type 6, 1250–1349 (see Table 6.01 and pp. 261–2 for explanation of cropping type) (source: National, Norfolk, and FTC1 accounts databases).

Fig. 6.07. Demesnes practising cropping type 7, 1250–1349 (see Table 6.01 and pp. 250–61 for explanation of cropping type) (source: National, Norfolk, and FTC1 accounts databases).

what the land might stand. For instance, they might be grown intensively on heavily manured infields and/or extensively on occasionally cultivated outfields.[5] The general scarcity of arable land in these regions (few arable demesnes were of great size) ensured that the bulk of the crops that were produced were destined for consumption locally.[6] Indeed, except at coastal locations, oats could not have borne the cost of transport over any distance (Table 5.04).

Environmental circumstances similarly account for the occurrence of oats-dominated cropping systems on several coastal demesnes in the east and south-east of the country. In Kent, Leysdown on the Isle of Sheppey and Appledore and Ebony on the Isle of Oxney all operated arable regimes in which the majority of the cropped area was devoted to oats.[7] The common denominator here is reclaimed marshland. Oats were generally the first crop sown as the land dried out and was upgraded from pasture to arable. This has been documented in some detail on the estate of Canterbury Cathedral Priory, to which the manors of Leysdown, Appledore, and Ebony all belonged.[8] Once reclaimed, high water tables ensured that oats remained the dominant spring-sown crop on these heavy alluvial soils.

An association between heavy, water-retentive soils and large-scale oats cultivation shows up in several other lowland contexts. On Battle Abbey's home farm of Marley in Sussex, for instance, an emphasis upon oats was consistent with the application of an extensive convertible rotational scheme to stiff clay soils. The combined fodder demands of the nearby monastery and the specialist cattle rearing undertaken on the manor further reinforced this specialism.[9] Such individual circumstances possibly explain the occurrence of this cropping type at Doulting (Somerset) and several other scattered locations.[10] Elsewhere, as at Beauworth and East and West Meon in Hampshire, it can be merely a manifestation of two-course cropping on relatively poor soils.[11] On a group of demesnes near London, however, it is clearly an expression of market specialisation.[12]

High relative transport costs limited commercial production of oats – the bulkiest of crops – to locations close to the market being provisioned. There was no greater single market for oats (or any other grain) than London. They were required in quantity in this period both as pottage for the capital's teeming population of urban poor and as fodder for the thousands of horses

---

[5] E.g. Finberg, *Tavistock Abbey*, pp. 104–8; Kershaw, *Bolton Priory*, pp. 31–2, 39–40; Miller, 'Farming in northern England', pp. 9–10; Hatcher, 'Farming: south-western England', pp. 387–8, 392; Miller, 'Farming: northern England', pp. 404–6.

[6] Miller, 'Farming: northern England', p. 407.

[7] CCA, DCc/Appledore 6, 9, 15; DCc/Ebony 8, 18, 27; DCc/Leysdown 8, 14, 16.

[8] Smith, *Canterbury Cathedral Priory*, pp. 138–9; Hallam, *Settlement and society*, pp. 179, 196.

[9] Searle, *Lordship and community*, pp. 272–91.

[10] Keil, 'The estates of Glastonbury Abbey', pp. 91–4.

[11] Hants. RO, Eccles. 2 159302, 8, 15, 29; Titow, 'Land and population', pp. 17–18.

[12] Campbell and others, *Medieval capital*, pp. 116–18.

which daily carried goods and travellers to and from the city.[13] Such was the traffic generated by the metropolis that fodder was almost as much in demand from the many small towns in London's immediate hinterland as from the city itself: hence the large acreages sown with oats on a concentration of demesnes in Middlesex and northern Surrey where there were no over-riding environmental reasons for this specialism. What determined the crop's pre-eminence here was its superior economic rent. The same circumstance should also have promoted relatively intensive methods of production, although whether that was in fact the case remains to be established.[14] Yields of oats, when they can be calculated, were certainly unremarkable on most of this metropolitan group of demesnes.[15] Significantly, Blean outside Canterbury and Nuneham Courtenay outside Oxford display the same specialism (Figure 6.07).[16] On the other hand, there is not a single example of this cropping type in the well-documented environs of Norwich, England's second city, nor, for that matter, within Norfolk as a whole. The presence of concentrated urban demand was therefore no guarantee of large-scale oats production.

Extensive cultivation of wheat (type 6)
Large-scale specialisation in wheat, the most demanding of the grains and the one best able to bear the costs of carriage, was altogether more unusual. Such specialisation was only possible in lowland arable contexts where growing conditions were well suited to wheat and arable regimes remained relatively extensive with frequent fallowing. In practice, that meant areas of strong soil and traditional two-course cropping. Demesnes of this type – cropping type 6 – commonly sowed well over half and often more than two-thirds of their cropped area with wheat, devoting the remainder to an assortment of spring-sown crops. It is this overwhelming emphasis upon wheat that is their distinguishing feature (Table 6.01). Such demesnes were largely confined to areas of regular commonfield systems, particularly those dominated by two-field systems (Figure 6.06). They show up most strongly in Somerset and Dorset, plus the clay-vale country of northern Wiltshire extending as far east as Wantage and Harwell in Berkshire's Vale of the White Horse.[17]

---

[13] Approximately 6,000–12,000 grain-carrying carts entered London through Newgate each year at the beginning of the fourteenth century, when the city was at the height of its medieval prosperity: Campbell and others, *Medieval capital*, p. 31; Keene, *Cheapside*. Some of these carts would of course have been ox-drawn; those that were horse-drawn would have required from one to four animals: Langdon, *Horses*, pp. 223–5.

[14] At Teddington, Middlesex (adjoining the Thames 20 miles upstream of London), 'much of the demesne arable was under continual cultivation between 1345 and 1370': Brandon, 'Farming: south-eastern England', p. 320.   [15] Campbell and others, *Medieval capital*, pp. 125–8.

[16] CCA, DCc/Blean 1(2); BL, Harl. Roll K29.

[17] Gray, *English field systems*, pp. 452–4, 461–2, 494–6, 501–2; I. Keil, 'Farming on the Dorset estates of Glastonbury Abbey in the early fourteenth century', *Proceedings of the Dorset Natural History and Archaeology Society* 87 (1966), 234–50; H. E. Hallam, 'Farming techniques: southern England', in Hallam (ed.), *AHEW*, vol. II, pp. 341–5, 354, 358, 364–5; *Select documents*; Hants. RO, 11M59/B1/54, 62, 65.

Elsewhere, examples of this cropping type are few and far between and mostly represent local deviations from prevailing husbandry norms where the latter were either two-course rotations in which the spring-sown crops usually predominated (cropping type 4) or three-course rotations which struck a more even balance between the areas devoted to wheat and the various spring-sown crops (cropping type 5). In the midlands, too, demesnes commonly sowed a combination of wheat and maslin (a wheat–rye mixture) rather than wheat alone. These demesnes show up as cropping type 3 (Figure 6.03). Wheat, of course, had the greatest marketing potential of any grain and it is possible that some of that produced on such an extensive scale in Somerset was destined for 'export' via the port of Bridgewater, whence purveyance accounts record the purchase and transhipment of wheat to the English forces in Scotland at the close of the thirteenth century.[18]

Three-course cropping of wheat and oats (type 5)

Demesnes which devoted the majority of their cropped acreage to either wheat or oats were outnumbered by those which grew both in roughly equal proportions within a classic three-course rotation (cropping type 5: Figure 6.05). In effect, wheat comprised the winter course and oats the spring, with all other crops relegated to a comparatively minor role. Demesnes in commonfield and non-commonfield areas alike operated some version of this cropping system, from Bamburgh (Northumberland) and Birkby (Cumberland), in the far north-east and north-west, to Plympton (Devon) and Otford (Kent), in the extreme south-west and south-east.[19] Since three-course cropping of wheat and oats was especially well suited to heavy land it was particularly well represented in lowland areas of clay soil. On the heavy boulder-clay soils of Hertfordshire, Essex, and parts of Suffolk, for instance, demesnes rarely departed from some version of this course of husbandry.[20] It was also commonplace in the Vale of York, the midlands, and many parts of the southern counties from Hampshire in the east to east Devon in the west.[21] In fact, in Somerset, Dorset, Wiltshire, and Monmouthshire the distribution of these three-course demesnes complemented the distribution of those two-course demesnes which devoted such a large share of their cropped acreage to wheat

---

[18] In 1333, for example, the sheriff of Somerset assembled 131½ quarters of wheat and 138 quarters of beans at Bridgewater for shipment up the Irish Sea to the military depot of Skinburness on the English shore of the Solway Firth (a journey which, in this instance, took 12 weeks): PRO, E358/2(2).

[19] BL, Add. MS 29794; Lambeth Palace Library ED 381; PRO, SC 6/824/8, 6/827/39, 6/1089/19.

[20] Roden, 'Demesne farming'; R. H. Britnell, 'Agriculture in a region of ancient enclosure, 1185–1500', *Nottingham Medieval Studies* 27 (1983), 37–55; Campbell and others, *Medieval capital*, pp. 116–18, 123–4.

[21] Hare, 'Wiltshire agriculture', pp. 3–4; Hallam (ed.), 'Farming techniques', p. 356; Miller, 'Farming: northern England', p. 399; Jack, 'Farming: Wales and the Marches', pp. 422–4, 469–70.

(cropping type 6: Figure 6.06). Geographically, therefore, it was one of the most widespread of all cropping systems. Both nationally and within the FTC counties over a quarter of sampled demesnes were of this type (Table 6.01). Nevertheless, widespread as was this arable system, examples of it are rare in both Norfolk (where Hingham, on heavy soils in mid-Norfolk is the sole example) and neighbouring Cambridgeshire.[22] With the exception of a group of demesnes in eastern Surrey and western Kent, examples are also scarce in the counties of the extreme south-east.

Spring-sown crops predominant (type 4)
Better represented in these eastern and south-eastern counties were cropping systems that might be two-course, three-course, or some more intensive and irregular alternative, in which wheat and oats were less overwhelmingly important and rather more space was devoted to other crops, particularly barley and legumes (cropping type 4: Figure 6.04). These, therefore, were arable regimes in which spring- exceeded winter-sown crops in area. Unlike national cropping types 7, 6, and 5, this cropping type was operated on a significant number of Norfolk demesnes (Figure 6.04). A group of ten demesnes on fairly heavy soils in the south-east of the county (including two just over the county boundary into Suffolk), four more also on heavyish soils in mid-Norfolk, and four demesnes on the silt soils of the Norfolk Fens were all of this type. Reconstruction of the rotations followed on the abbey of Bury St Edmunds's demesnes of Redgrave and Rickinghall during the years 1338–51 and 1334–46 demonstrates that their distinctive overall mix of crops arose from the concurrent application of different rotations to different portions of the demesne (Figures 6.09 and 6.10).[23] On both demesnes, conventional three-course rotations of wheat, followed by oats (or occasionally barley or legumes), and then fallow were applied to a substantial proportion of the arable area. On a further significant portion rye and barley were sown in alternation, usually without the relief of either annual fallows or the occasional nitrifying course of legumes (a relentless rotational regime of dubious agronomic wisdom). Finally, on a third substantial portion, more intensive and flexible rotations were followed, usually featuring some variant of the sequence wheat, barley, legumes. Fallowing occurred only when absolutely necessary and usually only after half a dozen or so years of continuous cropping. Not surprisingly, neither demesne was notable for its yields.[24]

Presumably other Suffolk and Norfolk demesnes of this type likewise

[22] BL, Campb. IX 8; NRO, Kimberley MAC/B/1.
[23] BL, Add. Roll 63372–5, 63513–24; Chicago UL, Bacon Roll 329–36.
[24] In the period 1322–50 Redgrave obtained mean yields per acre (net of tithes and seed) as follows: wheat 6.4 bus., rye 5.3 bus., barley 8.4 bus., oats 5.7 bus. Corresponding yields at Rickinghall in the period 1331–50 were wheat 3.1 bus., rye 5.3 bus., barley 5.3 bus., oats 4.9 bus. Chicago UL, Bacon Roll 325–35; BL, Add. Roll 63372–5, 63513–27, 63445.

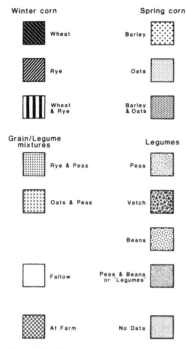

Fig. 6.08. Key to crop rotation diagrams

employed different rotational sequences on different parts of the arable to achieve the desired overall mix of crops. Apart from anything else, this was a means of taking advantage of soils of differing qualities. Many Kentish demesnes straddled a range of different soils suited to rotations of varying intensity and contrasting combinations of crops. It is therefore no coincidence that here, too, and especially in the centre and north of the county, demesnes of this cropping type were present in some numbers (Figure 6.04). Like their East Anglian counterparts, they did not have to conform to a regular, communal rotation and were therefore free to develop more intensive and irregular systems of cropping.[25]

It was also the case that many demesnes within the zone of regular commonfield systems grew this basic combination of crops. Examples occur in a broad band extending diagonally across the country from Jarrow (Co. Durham) in the north-east to Ashbury and Plympton (Devon) in the south-west (Figure 6.04).[26] Within this zone they are particularly well represented in

[25] Campbell, 'Regional uniqueness'; Baker, 'Field systems', pp. 393–418.
[26] *The inventories and account rolls of the Benedictine houses or cells of Jarrow and Monk-Wearmouth*, ed. J. Raine (Newcastle upon Tyne, 1854), pp. 1, 2, 12; Keil, 'The estates of Glastonbury Abbey', pp. 91–4.

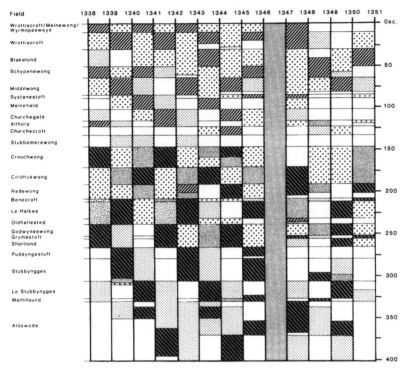

Fig. 6.09. Rotation of crops on 405 acres at Redgrave, Suffolk, 1338–51 (source: BL, Add. Rolls 63372–5; Chicago UL, Bacon Rolls 329–36). For key see Figure 6.08.

the east midlands, especially on the Northamptonshire manors of Peterborough Abbey, and in the southern counties of Sussex, Hampshire, Wiltshire, and Dorset. Virtually all demesnes in the downland country of south Wiltshire and eastern Dorset were of this type, providing a striking contrast with the vale country of northern Wiltshire dominated by cropping types 5 and 6 (compare Figures 6.05 and 6.06).[27] Many occur in two-course townships, whose flexibility of cropping allowed a greater area to be spring than winter sown. Nor was this cropping type incompatible with three-course cropping, especially when some of the barley was winter sown and the legumes were sown as an *inhok* from the fallow. A number of the east midland examples certainly appear to be associated with a greater differentiation of rotational practice as furlongs rather than fields became the basic units of cropping.[28] In the Soke of Peterborough there are also indications that arable

---

[27] Harrison, 'Field systems', pp. 7–13.

[28] J. Thirsk, 'Field systems of the east midlands', in Baker and Butlin (eds.), *Field systems*, pp. 255–62; D. N. Hall, *Medieval fields* (Princes Risborough, 1982); E. King, 'Farming practice and techniques: the east midlands', in Miller (ed.), *AHEW*, vol. III, pp. 212–14.

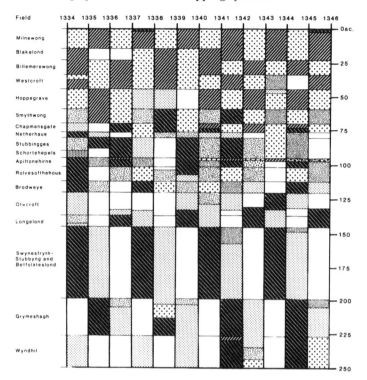

Fig. 6.10. Rotation of crops on 250 acres at Rickinghall, Suffolk, 1334–46 (source: BL, Add. Roll 63513–24). For key see Figure 6.08.

husbandry was becoming more intensive, partly in response to mounting demand within the region but also owing to the stimulus of the commercial grain trade focusing on the ports of the Wash whose hinterland penetrated quite deeply into the east midlands via the navigable rivers Nene, Welland, and Ouse.[29]

Cultivation of mixed grains (type 3)

The midlands are also the main focus of cropping type 3, whose principal distinguishing feature is the high proportion of the cropped area devoted to grain mixtures often in conjunction with a general bias towards spring cropping (Table 6.01). Spring-sown dredge was especially popular on a group of demesnes in Bedfordshire and its immediately adjacent counties in the east midlands where, on the example of the FTC sample of accounts, it was of con-

---

[29] Raftis, *Estates of Ramsey Abbey*, pp. 184–6; Biddick, *The other economy*, pp. 67–72; Campbell and others, *Medieval capital*, pp. 132–3, 139, 140, 143. On the grain trade of King's Lynn see Gras, *The corn market*, pp. 62–4, 110–11. For navigable rivers see Langdon, 'Inland water transport', pp. 3–5.

siderable commercial significance.[30] This was one of the most closely settled arable regions in medieval England and was also linked to external markets via the ports of the Wash. Further west, on a notable group of Warwickshire demesnes, dredge and maslin were both important, often grown in a three-course rotation. As this was the most land-locked part of England production of these two crops was here almost certainly intended for local markets.[31] In the south midlands, however, a specialism in mixed grains on demesnes focusing upon the middle Thames Valley in general and the busy grain entrepot of Henley in particular was almost certainly a response to the substantial demand for cheap bread grain from London (Figure 6.03).[32] Here, therefore, it was mainly maslin or mancorn that was grown, usually within more flexible and irregular rotations than those encountered further west.

Rye with barley (type 2)
The group of mid-Thames Valley demesnes which specialised in growing maslin/mancorn were associated with a smaller and more tightly focused group of demesnes slightly downstream and nearer London upon which rye was the dominant crop, grown in conjunction with barley and oats (Figure 6.02). The cultivation of rye rather than wheat, and barley rather than oats, plus usually only limited quantities of legumes were the hallmarks of cropping type 2 (Table 6.01). Geographically, this was one of the most regionally specific of all arable systems, being well represented in some parts of the country but almost entirely absent from most others (Figure 6.02). It shows up, as already noted, on a group of demesnes well within provisioning range of London, to which should be added West Thurrock (Essex) at the mouth of the Thames estuary, Great Amwell (Hertfordshire) at the head of the navigable River Lea, and Byfleet (Surrey) a dozen miles south of the capital by road.[33] A number of demesnes in, or accessible to, the Trent Valley – Walcot and Scotter in Lincolnshire and Collingham in Nottinghamshire – may have developed a similar specialism in response to the commercial opportunities offered by cheap water transport in combination with relatively poor soils.[34]

Such a conjunction of locational and environmental factors undoubtedly explains why so many Norfolk and Suffolk demesnes were of this type. In Norfolk, in fact, this was the second most common cropping type, showing up as the dominant arable specialism of several sub-regions within the county (Table 6.01 and Figure 6.02). The latter include the sandy Breckland and its margins in the south-west of the county, linked to the port of King's Lynn by the navigable Little Ouse and Great Ouse rivers; a scatter of demesnes on the

---

[30] Campbell and others, *Medieval capital*, pp. 119–21, 161–2.
[31] Langdon, 'Inland water transport', p. 4.
[32] Chapter 9, pp. 425–6; Campbell and others, *Medieval capital*, pp. 47–9, 114–15, 121–3, 164–7.
   [33] BL, Add. Roll 66722; PRO, E358/20; SC 6/1011/3–4; WAM, 26145.
[34] N'hants. RO, Fitzwilliam charters 2389; PRO, SC 6/1077/13.

Fen edge in the west of the county similarly accessible to Lynn via the Great Ouse and its lesser tributaries; a tight cluster of demesnes on the 'good sands' of the north-west of the county within reach of several of the lesser creeks of the Norfolk coast; a light scatter of demesnes across the centre of the county; and, finally (echoing the close association of the same cropping type with metropolitan demand), a major concentration of demesnes in the immediate hinterland of Norwich on the light sandy soils which occur to the south and especially to the north of the city.[35] All these demesnes display a strong commercial orientation and appear to have turned their predominantly light sandy soils to material advantage. The same appears to have been true of their Suffolk counterparts. Concentrations of demesnes of this type show up in the Suffolk Breckland in the north-west of the county, and in the Sandlings region of similarly poor sandy soils in the south-east, where they were within striking range of Woodbridge and several other of the smaller ports of this part of the East Anglian coast (Figure 6.02).

Norfolk demesnes of this cropping type generally grew much more barley than the national norm, and less oats, as was consistent with the county's status as England's premier barley-producing county. Rotational practice varied a good deal. Brandon, just over the county boundary in Suffolk, in the heart of the Breckland, cropped its land for just one or two years and then rested it for several years in succession, when it reverted to temporary pasture for sheep.[36] Bircham, in north-west Norfolk, operated a related but more intensive convertible regime. In the period 1341–9 barley was normally sown first, followed by rye and oats, after which the land was left uncultivated for the next three or four years in succession (Figure 6.11).[37] At Taverham, a few miles west of Norwich on soils derived from glacial sands and gravels, rotations appear to have been yet more intensive. If the rotations in operation at the beginning of the fifteenth century are representative of earlier practice, the best and most intensively manured and marled land was subject to almost continuous cropping following no regular predetermined scheme, except that there were com-

---

[35] Bailey, *A marginal economy?*, pp. 153–7; Langdon, 'Inland water transport', pp. 3–5. Purveyance accounts of the early fourteenth century indicate that in west Norfolk provisions from Mileham, Southmere, and Walsingham were carted to King's Lynn, grain from Gooderstone was carted to Oxborough and then boated down the Wissey and Great Ouse to Lynn, and grain from Culford was carted to Santon Downham and from Dalham was carted to Lakenheath and then boated down the Little Ouse and Great Ouse to Lynn. In north Norfolk grain from Thornage was carted to Snitterley (Blakeney) and then shipped to Newcastle, while grain from Stiffkey was carted to Holkham and then shipped around the coast to Lynn. In east Norfolk grain from a variety of vills was assembled at Norwich and then boated down the Yare to Yarmouth, where it was joined for onward shipment by grain boated down the Bure from Wroxham. In north Suffolk, just over the county boundary from Norfolk, grain from Hoxne was carted to Beccles and then boated down the Waveney to Yarmouth for onward shipment. PRO, E101/574/4, 7–8, 11, 25–6, 32–3; E101/575/13.
[36] Bailey, *A marginal economy?*, pp. 60–2; Figure 6.23; below, pp. 293–4, 296.
[37] PRO, SC 6/930/17–23.

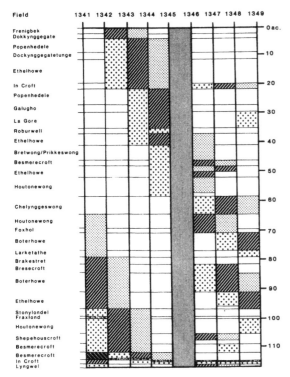

Fig. 6.11. Rotation of crops on 117¼ acres at Bircham, Norfolk, 1341–9 (source: PRO, SC 6/930/17–23). For key see Figure 6.08.

monly two or three consecutive courses (and sometimes more) of barley. Legumes were often, but not always, sown prior to wheat and rye, and from time to time the land was bare fallowed to help cleanse it of accumulated weed growth. Other land might be cropped for three, four, or five years and then left unsown for the next two or three in an irregular and rather debased convertible regime.[38] So intensive a regime on such poor soils was only justifiable given the powerful commercial incentive provided by proximity to Norwich and its thirst for malted barley with a low nitrogen content for brewing. It demonstrates the role of concentrated demand in stimulating the development of cropping regimes that were both more specialised and intensive.

Intensive cultivation with legumes (type 1)

The most specialised and intensive cropping systems of all were those which featured wheat and barley as their main winter and spring grains (respectively, the most highly prized bread and brewing grains), devoted a significantly

[38] NRO, DCN 60/35/43–50.

greater area to spring- than winter-sown crops, and used substantial sowings of legumes as a substitute for bare fallows. Half of all documented Norfolk demesnes employed some version of this cropping system (cropping type 1), compared with perhaps an eighth of those nationally. In Norfolk they occur throughout the county, except in the Fens and on the lightest and the heaviest soils (Figure 6.01). They were particularly well represented in a coastal arc of country extending from Hunstanton in the north-west to the exceptional fertile 'island' of Flegg in the east, and were therefore well placed to take advantage of the trading opportunities afforded by such ports as Brancaster, Wells, Blakeney, and Yarmouth.[39] The last conducted a lively trade in the late thirteenth and fourteenth centuries both up and down the English coast and across the North Sea to Flanders and Norway.[40] Moreover, demesnes in its immediate hinterland enjoyed the double advantage of cheap and convenient access to Norwich via the navigable rivers Bure, Thurne, and Yare.[41] This cropping type was already in existence on the east Norfolk demesnes of St Benet's Abbey by 1238–46.[42] Another group of demesnes in south-central Norfolk, 10–15 miles west and south-west of Norwich, developed the same specialism, but nearer the city it was confined to the better soils, as at Arminghall, Heigham, and Plumstead.[43] Sometimes the accounts record the amount of land left fallow on these demesnes, from which it is plain that the majority of the arable was under crop each year.[44]

The best arable on the most intensively cropped demesnes was often subject to virtually continuous cropping while elsewhere land was typically cropped

[39] See above, n. 35.

[40] Campbell and others, *Medieval capital*, pp. 69–71, 85–6, 175; Campbell, 'Tradable surpluses'.

[41] Langdon, 'Inland water transport', pp. 4–5. The cost of carrying grain the 30 leagues (45 miles) from Norwich to Yarmouth via the River Yare ranged from ½d. to 2d. per quarter in the first half of the fourteenth century. The equivalent cost for the 24 leagues (36 miles) from Wroxham to Yarmouth via the River Bure was 1d. per quarter in 1345. Transhipment costs at each end – loading from the granary into the boats and from the boats into the granary – added 1¾–4d. per 10 quarters. In 1319–20 the total shipping cost from Yarmouth to Newcastle upon Tyne was 4d. per quarter. In 1340 a granary was hired at Yarmouth to accommodate grain assembled from a 400-square-mile hinterland embracing the city of Norwich and the hundreds of East and West Flegg, Happing, Tunstead, Loddon, and Clavering: PRO, E101/574/7, 25, 33; E101/575/13. In 1298–9 the prior of Norwich spent 11s. on wages for a boatman, 13s. on a new boat, 16s. on a new 'galeyam' and repairing the 'little galeye', and 4s. 6d. on repairs to the great boat. In 1314–15 a great boat, great 'galeye', little 'galey', boat, marsh boat, and boat of Plumstead were all accounted for, together with the wages of two boatmen. Replacing the great boat cost £12 18s. 6d. in 1320–1. In 1412–13 the river traffic between Norwich and Yarmouth included a prefabricated cart-house for the demesne at Martham and window tracery for the chancel of Hemsby church sent downstream, and three shipments of wheat and malt brought upstream. In 1419–20 the upstream cargo included coal and Spanish iron: NRO, DCN 1/1/14, 24, 28, 72, 74.   [42] NRO, Diocesan Est/1, Est/2/1.

[43] NRO, DCN 60/29/1–26, 61/7; Diocesan Est/1, Est/2/1.

[44] South Walsham cropped 93 per cent of its arable, Acle 95 per cent, Heigham-by-Norwich 97 per cent, Guton Hall in Brandiston 98 per cent, and Halvergate and Flegg 100 per cent in the period 1250–1349: Campbell, 'Arable productivity', p. 393.

for at least four years out of every five.[45] Cropping sequences followed no prescribed course but instead displayed great variety and flexibility. Wheat generally received priority insofar as it was usually only sown following bare fallows or a nitrifying course of legumes. It was commonly succeeded by one, two, or even three courses of barley, sown either consecutively or with a nitrifying course of legumes. Oats when sown were invariably placed at the end of the rotation, immediately before the land was bare fallowed and repeatedly summer ploughed. Fallows were chiefly important as a means of cleansing the land of weed growth, which could be a considerable problem whenever cropping was so continuous.[46] Competition with weeds was one reason why oats, often the last course in the rotation, were sown so thickly (Table 7.02). A brief glimpse of this type of rotation in operation may be obtained at Little Ellingham in south-central Norfolk over the three-year period 1342–5 (Figure 6.12).[47] Although situated on relatively light soils of limited fertility, approximately seven-eighths of this demesne's arable area was cropped each year. Such a demanding regime was obviously in danger of exhausting the land unless appropriate compensatory measures were adopted to maintain the nitrogen cycle: hence the emphasis placed by many demesnes of this type upon regular manuring and marling, the folding of sheep upon the arable, and systematic weeding of the growing crop.[48]

Outside of Norfolk a similar system of cultivation prevailed on many demesnes in northern and eastern Kent, an area which enjoyed locational advantages at least equal to those of eastern Norfolk and which also shared easily worked loam soils and an abundant labour supply (Figure 6.01). Substantial sowings of legumes – peas and especially vetches – were here an integral part of rotations whose aim seems to have been to keep the maximum possible area under crop each year. On the most intensively cultivated demesnes this resulted in virtually continuous cropping.[49] Wheat and barley were again the leading grain crops and much effort was expended upon methods intended to maintain the fertility of the soil, including manuring, folding, marling, and liming. A string of demesnes in coastal Sussex, extending as far west as southern Hampshire and the Isle of Wight, operated a similar system in which fallows were either partially or wholly abolished (Figure 6.01).[50] Again, the demesnes in question mostly enjoyed ready access to the various small trading ports of the Channel coast.

[45] Campbell, 'Agricultural progress', pp. 28–9.
[46] In the late fourteenth century bare fallows were still normally subject to between three and six ploughings: Campbell, 'Agricultural progress', p. 29, n. 15. Walter of Henley recommended two fallow ploughings: *Walter of Henley*, p. 315.
[47] Nottingham UL, Manvers Collection 24–6.
[48] Campbell, 'Agricultural progress', pp. 32–9.
[49] Gray, *English field systems*, p. 302; Smith, *Canterbury Cathedral Priory*, pp. 133–41; Campbell, 'Agricultural progress', pp. 41–2, n. 59; Campbell and others, *Medieval capital*, pp. 128–44.
[50] Brandon, 'Farming: south-eastern England', pp. 318–20; Mate, 'Profit and productivity', pp. 331–4.

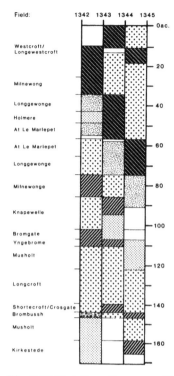

Fig. 6.12. Rotation of crops on 171 acres at Little Ellingham, Norfolk, 1342–5 (source: Nottingham UL, Manvers Collection 24–6). For key see Figure 6.08.

Examples of this cropping type outside of Norfolk, eastern Kent, and coastal Sussex were mostly few and far between. A few demesnes located on the more fertile soils of the lower Thames Valley probably owed their adoption of this cropping system to the stimulus of the London market (Figure 6.01).[51] Elsewhere occurrences of this cropping type mostly reflect a greater than average emphasis upon legumes. Beans, for instance, were grown in quantity on the alluvial soils of Zoy in the Somerset Levels, which, as a home demesne of Glastonbury Abbey, was also a major producer of malting barley for the convent.[52] A light scatter of demesnes in the east midlands were similarly notable for the scale on which they grew legumes. They represent isolated outliers of a cropping regime otherwise normally confined to certain privileged localities in eastern and south-eastern England and anticipate the extensive fodder cropping with legumes which was to become a feature of so many of these midland demesnes in the post-Black Death period (Figure 6.18).

[51] Campbell and others, *Medieval capital*, pp. 135, 137, 140.
[52] Keil, 'The estates of Glastonbury Abbey', pp. 77–8, 91–4.

6.12 The overall configuration of cropping systems pre-1350

Expanding rural and urban food markets and slowly changing agricultural technology together account for the marked differentiation of cropping systems apparent in the period 1250–1349. Deteriorating mean living standards nevertheless ensured that the production of crops for human consumption took precedence over that for animals, and placed a mounting premium upon the bread and pottage grains over those grown primarily for brewing. Demesne cropping systems in this period consequently range from the simple to the complex, from those extensive in their use of land to those that exploited it intensively. The upshot was a widening divergence between systems in the volume and value of their output per unit area (Table 7.08).[53]

Overall, relatively simple and extensive systems of cropping clearly predominated in this period. On the evidence of the national accounts database, over 40 per cent of demesnes sowed some combination of winter corn and oats, with few legumes, only minor quantities of other crops, and, usually, biennial or triennial fallows (cropping types 7, 6, and 5) (Table 6.01). Such demesnes were widely distributed throughout the country and were especially characteristic of two particular types of location. First, they formed the predominant arable type in much of the north, west and south-west of the country, where there were either environmental or economic constraints upon the development of more intensive and diversified cropping regimes. Second, they were also the typical cropping type in many lowland areas of heavy soil and moderate population density in the south and south-east of the country, showing up particularly strongly in the immediate hinterland of London where they may have benefited from the substantial metropolitan demand for wheat and oats. Elsewhere in the south and east, notably in Kent, Sussex, and the Isle of Wight, examples of these cropping types were rare, and generally owed their existence to locally specific environmental and economic circumstances. These were areas of higher than average population density and more advanced economic development as reflected in high levels of assessed wealth per unit area.[54] The same applied to Norfolk, north Suffolk, Cambridgeshire, northeast Northamptonshire (including the Soke of Peterborough), and much of Lincolnshire, where more intensive cropping systems were also the order of the day.

At the opposite extreme, the two most complex and intensive cropping types – cropping types 1 and 2 – were practised by just over 20 per cent of all demesnes. They were characterised by a strong emphasis upon spring-sown crops, especially barley, and, on the most intensively cropped demesnes of all, substantial sowings of legumes. Examples were rare outside the extreme

[53] Chapter 7, pp. 334–45.
[54] R. E. Glasscock, 'England circa 1334', in Darby (ed.), *New historical geography*, pp. 137–45.

south-east of England, Lincolnshire, Norfolk, and Suffolk, and can usually be attributed to specific economic, environmental, and institutional circumstances. Although these types of cropping system were successfully operated within areas of regular commonfields, they occurred in greatest number in areas of irregular field systems, to the east of the main zone of two- and three-field systems (Figures 6.01 and 6.02).[55] They attained their fullest development where agrarian institutions, in the form of field systems and manorial structures, allowed individuals greatest freedom and enterprise in the management of their land, and especially where cultivators enjoyed the triple advantage of naturally fertile and easily cultivated soils, an abundant labour force, and cheap and convenient access to concentrated centres of demand both at home and overseas. Eastern Norfolk and eastern Kent stand out in this regard and on the evidence of their cropping systems were the localities where levels of economic rent attained their pre-Black Death maximum.[56] Significantly, the areas which they faced across the North Sea appear to have shared many of the same agricultural characteristics.[57]

The remaining 35 per cent of demesnes operated cropping regimes of intermediate complexity and intensity and, appropriately, mostly occupied intermediate locations between the more complex and intensive regimes of East Anglia and the south-east and the simpler and more extensive regimes of the north, west, and south-west. In the southern counties these cropping types represented light-land alternatives to the wheat and oats cultivation of the clay vales. Nearer to London, in the Thames Valley and south-east midlands, they are also a manifestation of market specialisation, especially in the cheaper bread and brewing grains. The same may also be true of the many demesnes of this type in Kent, East Anglia, and north-eastern Northamptonshire. But at the same time these also appear to have been the demesnes most actively engaged in modifying conventional cropping regimes and adopting new methods. Their distribution therefore helps to pinpoint likely areas of technological innovation and intensification.[58] The latter were all specific in location and limited in extent: early-fourteenth-century England remained a country more extensive than intensive and more conservative than innovative in its demesne cropping systems.[59]

[55] This bears out the observation initially made by Hilton, *Economic development*, p. 152, and re-emphasised by Fox, 'Alleged transformation to three-field systems', pp. 527–9. For a classification of field systems see B. M. S. Campbell, 'Commonfield origins – the regional dimension', in T. Rowley (ed.), *The origins of open-field agriculture* (London, 1981), pp. 113–17.
[56] B. M. S. Campbell, 'Economic rent and the intensification of English agriculture, 1086–1350', in Astill and Langdon (eds.), *Medieval farming*, pp. 238–9.
[57] A. Verhulst, 'L'Intensification et la commercialisation de l'agriculture dans les Pays-Bas méridionaux au XIIIe siècle', in Verhulst, *La Belgique rurale* (Brussels, 1985), pp. 89–100; E. Thoen, 'The birth of "the Flemish husbandry": agricultural technology in medieval Flanders', in Astill and Langdon (eds.), *Medieval farming*, pp. 69–88.
[58] Raftis, *Estates of Ramsey Abbey*, pp. 184–6; Hogan, 'Clays, *culturae*'.
[59] Campbell and others, *Medieval capital*, pp. 138–42; Campbell, 'Economic rent'.

Superimposed upon these broad regional variations in cropping systems was much local variation which requires a much denser coverage of data to be brought properly into focus. The juxtaposition of wold and vale, of light land and heavy land, and of upland and reclaimed marshland almost everywhere found expression in a differentiation of cropping regime.[60] The home demesnes of seigniorial households also sometimes developed particular specialisms, as did demesnes close to major urban centres. In the latter regard, demesnes with convenient access to riverine or coastal transport enjoyed a cost advantage over those solely reliant on overland carriage. Counties such as Kent, Somerset, and Lincolnshire, which embody most of these contrasts, consequently contained an exceptionally wide range of cropping types. In contrast, cropping in the neighbouring counties of Essex, Devon, and Norfolk was altogether more uniform. Some of the greatest variation over the shortest distances arose as a result of the differential impact of commercial forces; hence the wide array of cropping systems to be found within the middle and lower Thames Valley, upstream of London. The wide hinterland of the busy grain entrepot of King's Lynn similarly contained many different cropping types. Nevertheless, neither hinterland was yet wide enough to generate spatial differentiation on a grander scale, with the result that the bulk of the country remained outside their respective Thünen 'fields of force'.[61] It took the massive metropolitan growth of the seventeenth century to transform the situation and generate a nationally more integrated geography of cropping systems.[62]

6.2 Seigniorial cropping systems, 1350–1449

The demographic haemorrhage precipitated by the Black Death and its subsequent manifestations caused a reduction in demand for arable products that was both massive and selective. The trend towards a selective intensification of cropping systems was thereby reversed. A general lowering of Ricardian and Thünen economic rent – as subsistence pressures receded and urban populations contracted – in conjunction with altered factor costs (as labour costs rose and land and, eventually, capital costs fell) encouraged a return to more extensive forms of land-use and a search for more efficient and cost-effective forms of production. At the same time, changes in the character of demand induced a reorientation of production from lower-quality to higher-quality foodstuffs, from bread grains to brewing grains, and from human food to live-

---

[60] Many of these contrasts are to be seen in Norfolk: Figures 6.20–6.22.
[61] Cf. J. Bieleman, 'Dutch agriculture in the Golden Age, 1570–1660', in K. Davids and L. Noordegraaf (eds.), *The Dutch economy in the Golden Age* (Amsterdam, 1993), pp. 159–85.
[62] Langton and Hoppe, *Town and country*, pp. 30–41; E. A. Wrigley, 'Urban growth and agricultural change: England and the Continent in the early modern period', *Journal of Interdisciplinary History* 15 (1985).

stock fodder (Table 5.08).[63] The upshot was a reconfiguration of cropping systems as demesne managers altered both the balance and range of crops grown and the intensity of their cultivation.

With certain notable exceptions, cropping became more diversified and it was increasingly unusual for any one crop to dominate production. There was also an escalating and selective retreat from direct management with the result that available samples of documented demesnes are smaller and less representative, home farms featuring more prominently than ideally they ought. Any reconstruction of the geography of cropping systems therefore suffers from a loss of focus. This may partially account for the fact that six rather than seven main cropping types may be distinguished in the period 1350–1449. Former cropping types 7 and 6, specialising respectively in the extensive production of oats and wheat, effectively disappear after 1349. Cropping types 5, 4, 3, 2, and 1 all persist, albeit in altered form and distribution, while a new cropping type distinguished by the large-scale cultivation of legumes – cropping type 8 – appears for the first time (Table 6.02).

6.21 Individual cropping systems

Extensive cultivation of legumes (type 8)
Although peas, beans, and vetches, singly and in combination, were widely grown before the Black Death it was exceptional for any demesne to devote more than a quarter of its sown acreage to these crops. Moreover, the demesnes which grew legumes on the largest scale (cropping type 1: Figure 6.01) were almost invariably those that were most intensively cultivated, since the smaller the area fallowed the greater the need to sow legumes as a nitrifying substitute. After 1349, however, large-scale legume cultivation ceased to be the almost exclusive preserve of these intensive cropping systems (whose intensity in any case proved difficult to sustain).[64] Instead, legumes were increasingly taken up by demesnes operating comparatively extensive rotational regimes, some of which began to devote well over a third of their sown acreage to legumes (cropping type 8: Figure 6.18).[65] These demesnes were using legumes to underpin an expansion in pastoral husbandry. In effect, land was diverted from feeding humans to feeding animals.[66] Typically, those

---

[63] Chapter 5, pp. 238–48.

[64] Campbell, 'Fair field', p. 63; R. H. Britnell, 'The occupation of the land: eastern England', in Miller (ed.), *AHEW*, vol. III, pp. 60, 63; M. Mate, 'The occupation of the land: Kent and Sussex', in Miller (ed.), *AHEW*, vol. III, p. 121; Harvey, 'Farming: Home Counties', p. 260; Mate, 'Farming: Kent and Sussex', pp. 269, 271.

[65] King, 'Farming: east midlands', pp. 213–14, 215; C. C. Dyer, 'Farming practice and technology: the west midlands', in Miller (ed.), *AHEW*, vol. III, pp. 229–30; Harvey, 'Farming: Home Counties', p. 260.

[66] King, 'Farming: east midlands', p. 216, and Dyer, 'Farming: west midlands', p. 229, confirm that in both the east and west midlands demesnes were primarily growing legumes as a fodder crop.

Table 6.02. *National classification of cropping types, 1250–1* (combined)

Variable	National cropping type						
	1	2	3	4	5	8	Overall
Means:							
Mean % of total sown area:							
Wheat	*12.9*	*12.4*	*23.1*	*37.4*	*37.8*	*29.7*	*28.7*
Rye	*5.6*	*21.4*	*1.9*	*1.4*	*1.1*	*0.6*	*3.4*
Barley	*53.8*	*21.6*	*10.3*	*28.9*	*7.4*	*19.2*	*25.2*
Oats	*13.0*	*29.5*	*10.7*	*16.3*	*43.1*	*10.1*	*22.1*
Grain mixtures	*1.5*	*2.2*	*39.8*	*2.8*	*2.6*	*4.5*	*6.4*
Legumes	*13.1*	*5.7*	*14.2*	*13.2*	*8.0*	*35.8*	*13.7*
Mean total sown acres	130.8	136.7	157.0	162.0	150.5	165.5	151.0
No. of demesnes	88	28	44	121	111	43	435
Standard deviations:							
Mean % of total sown area:							
Wheat	*8.0*	*9.2*	*10.3*	*11.7*	*11.5*	*10.0*	*14.9*
Rye	*8.0*	*11.7*	*6.4*	*4.4*	*3.3*	*2.8*	*7.8*
Barley	*8.4*	*14.0*	*10.7*	*9.4*	*6.7*	*10.9*	*19.1*
Oats	*8.2*	*17.3*	*10.9*	*9.4*	*10.9*	*8.6*	*16.7*
Grain mixtures	*4.1*	*5.2*	*13.7*	*5.2*	*4.7*	*7.8*	*13.0*
Legumes	*6.9*	*4.7*	*11.2*	*6.6*	*6.9*	*9.9*	*10.9*
Mean total sown acres	74.8	82.0	89.8	93.8	81.8	78.1	85.4
No. of demesnes	88	28	44	121	111	43	435
% of total classified:							
Demesnes	*20.2*	*6.4*	*10.0*	*27.8*	*25.5*	*9.9*	*100.0*
Area	*17.5*	*5.8*	*10.5*	*29.8*	*25.4*	*10.8*	*100.0*

Note:
The demesnes are listed in Appendix 1.
Method:
B. M. S. Campbell and J. P. Power, 'Mapping the agricultural geography of medieval England', *JHG* 15 (1989), 24–39.
Source: National accounts database; FTC2 accounts database; Norfolk accounts database.

Fig. 6.13. Demesnes practising cropping type 1, 1350–1449 (see Table 6.02 and pp. 290–301 for explanation of cropping type) (source: National, Norfolk, and FTC2 accounts databases).

Fig. 6.14. Demesnes practising cropping type 2, 1350–1449 (see Table 6.02 and pp. 289–90 for explanation of cropping type) (source: National, Norfolk, and FTC2 accounts databases).

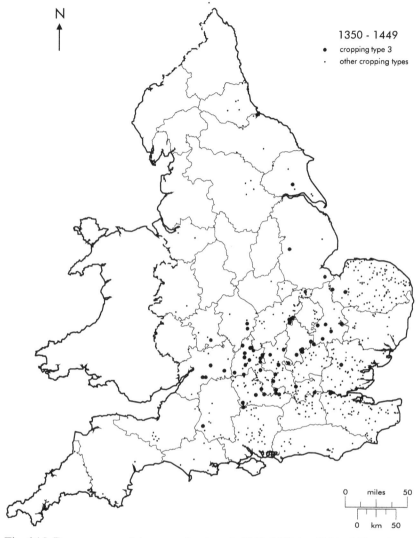

Fig. 6.15. Demesnes practising cropping type 3, 1350–1449 (see Table 6.03 and pp. 288–9 for explanation of cropping type) (source: National, Norfolk, and FTC2 accounts databases).

Fig. 6.16. Demesnes practising cropping type 4, 1350–1449 (see Table 6.04 and pp. 286–8 for explanation of cropping type) (source: National, Norfolk, and FTC2 accounts databases).

Fig. 6.17. Demesnes practising cropping type 5, 1350–1449 (see Table 6.05 and p. 285 for explanation of cropping type) (source: National, Norfolk, and FTC2 accounts databases).

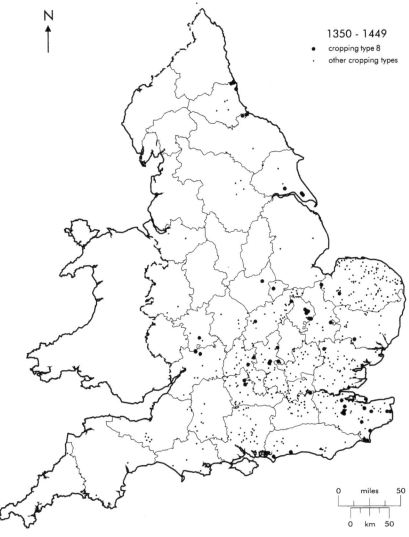

N

1350 - 1449
• cropping type 8
· other cropping types

0   miles   50

0   km   50

Fig. 6.18. Demesnes practising cropping type 8, 1350–1449 (see Table 6.02 and pp. 276–85 for explanation of cropping type) (source: National, Norfolk, and FTC2 accounts databases).

demesnes which made this switch were those either unwilling or unable to take the ultimate step and convert land directly from tillage to grass. Many were commonfield demesnes locked into an arable regime from which land could not easily be withdrawn. For them large-scale fodder cropping may have been a compromise measure. They combined it with the cultivation of wheat for bread and barley for ale, together with lesser quantities of oats and the various grain mixtures.

In Kent and coastal Sussex legumes had long featured prominently in rotations, hence it is no great surprise to find several examples of cropping type 8 in these two counties (Table 6.02 and Figure 6.18). Notwithstanding a general diminution in the intensity of cultivation, there were some demesnes which maintained or even expanded the area devoted to legumes, although this rarely exceeded a third of their sown acreage. The marshland demesnes of Dengemarsh and Orgarswick stand out as the two most conspicuous exceptions, explicable in terms of the suitability of their alluvial soils to legumes, the desirability of supplementing extensive summer marshland grazings with adequate supplies of produced fodder for winter feed, and the commercial potential of marketing legumes via the Channel ports.[67] Many other demesnes with reclaimed alluvial soils likewise became large-scale producers of legumes.[68] They show up in and around the East Anglian fens in Norfolk, Cambridgeshire, and Huntingdonshire, adjoining the Humber and lower Tees marshes in the north-east, on the Fylde coast of Lancashire, and adjacent to the alluvial levels of the lower Severn Valley.

It was, however, in non-pasture-rich locations that this 'new' cropping type tended to assume its most extreme form, with at least 40 per cent of the sown acreage devoted to legumes. This was conspicuously the case on a number of demesnes scattered in a broad swathe of country stretching through the midlands (Figure 6.18). These demesnes were well placed to draw in young stock from breeders further north and west, rear them up, and then sell them on, either as mature working animals or for fattening, to farmers further south and east. Along the south coast some of the Kentish and Sussex demesnes operating this cropping type may have plied a similar trade, breeding and rearing livestock on the coastal marshes and then selling them inland.[69] Certainly, the emergence of this cropping type in several widely removed locations bears testimony to growing specialisation within the pastoral sector and, on some demesnes, the subordination of the arable sector to the pastoral.

[67] PRO, SC 6/889–90/27, 4, 5; CCA, DCc/Agney 56. In the same area the marshland demesnes of Appledore and Scotney-in-Lydd made substantial sales of beans: CCA, DCc/Appledore 51, 54, 56; Lambeth Palace Library ED 194; BLO, MS DD All Souls C183 SC.

[68] On the association between legumes (especially beans) and reclaimed marshland, see J. Thirsk, *Fenland farming in the sixteenth century* (Leicester, 1965), p. 38; Brandon, 'Agriculture at Barnhorne', pp. 71–8; Postan, *Economy and society*, p. 51. For a survey of medieval marshland drainage see R. A. Donkin, 'Changes in the early Middle Ages', in Darby (ed.), *New historical geography*, pp. 104–6.   [69] Farmer, 'Marketing', pp. 384–5.

The preference on these demesnes was for legumes rather than oats as the chief fodder crop, a trend symptomatic of the final eclipse of oats as a leading crop in this period. No longer do oats-growing demesnes show up as an independent cropping type. Nor, notwithstanding the greater affluence and, presumably, heightened *per capita* reliance upon horsepower by those frequenting London, did oats remain such an overwhelming specialism of demesnes in the immediate hinterland of that city. Indeed, for any crop to occupy more than half the sown acreage was increasingly unusual. Thus, demesnes specialising in the extensive production of wheat effectively disappear as a type (Tables 6.01 and 6.02). Demesnes with a dual emphasis upon wheat and oats sown predominantly in a three-course rotation (cropping type 5) did, however, survive and remain widespread, possibly gaining from the conversion of two- to three-field systems which appears to have gathered momentum during the fourteenth century as population decline created opportunities for field and holding rationalisation.[70]

Three-course cropping of wheat and oats (type 5)

Over a quarter of sampled demesnes operated some form of three-course cropping with wheat and oats, its precise character varying according to environmental, economic, and institutional circumstances (Table 6.02).[71] As in the period 1250–1349, examples show up in much of England north of the Trent, in the west and north-west midlands, the south-western counties of Cornwall, Devon, and Somerset, parts of Hampshire, Surrey, Kent, and Sussex, and most of Hertfordshire, Essex, and southern Suffolk (Figure 6.17). They occur in many areas which practised regular commonfield agriculture and others which did not, and display a particular association with areas of medium to heavy soil. Exceptions include Holderness, where this cropping type had formerly been well represented; most of the south and east midlands, where a wider range of crops was now generally cultivated; and Norfolk, where Mileham and Newton on heavy soils in the west of the county are the sole examples (Figure 6.17).[72]

---

[70] Fox, 'Alleged transformation to three-field systems'; Campbell, 'Fair field', p. 65.

[71] A few demesnes which grew rye rather than wheat in combination with oats are also subsumed into this cropping type.

[72] Mileham was the home farm of the earls of Arundel whose household at Mileham Castle probably required substantial quantities of oats for fodder: Holkham Hall, Norfolk, Estate Records, Tittleshall bundles 16, 17. Newton, to the west, bears every semblance of having been cultivated according to a three-course rotation comprising a winter course, spring course, then fallow: CUL, Cholmondley (Houghton) MSS 32; PRO, C135 File 51(10). Several other demesnes in this western part of Norfolk seem to have been similarly cropped, e.g. Mundford, Larling, Rushworth, Raynham, Gayton, Helhoughton, and Syderstone: PRO, C135 Files 1(12), 35(25), 45(18), 46(3), 51(10).

Spring-sown crops predominant (type 4)

Equally numerous after 1349 were demesnes which combined a marked spring bias to their cropping with the cultivation of a wider array of crops (cropping type 4: Figure 6.16). Some operated basic two-course rotations, others employed less regular and more intensive systems. Typically, they grew wheat and barley in above average, legumes in average, and oats in below average quantities; a combination best suited to medium rather than heavy soils. The equivalent system before the Black Death was identical in almost every respect, except that oats was more prominent than barley (Table 6.01). These demesnes were therefore those most closely associated with the wider diffusion of barley cultivation which was such a feature of the era of rising living standards which followed the Black Death (Table 5.08).[73] Since such demesnes are most strongly represented in southern and eastern England (Figure 6.16) it was presumably here that barley cultivation was making its greatest advances.

Along the south coast, from Kent, through coastal Sussex and southern Hampshire as far as the Isle of Wight, the spread of this cropping type also arose from the general shift towards more extensive forms of arable cultivation, as demesnes reduced their sowings of legumes and thereby shed much of their formerly distinctive character. By this combination of developments, except on the very heaviest soils, cropping type 4 spread to become the characteristic cropping type throughout the counties south of the Thames. North of the Thames Valley, apart from a couple of cases on the northern edge of the Cotswolds and a scattering of examples stretching up the eastern flank of England as far as Holy Island off the Northumberland coast, the only significant concentration of demesnes employing this cropping type was in East Anglia.[74] Fifteen of the seventeen Norfolk examples were concentrated on the heavier soils of the south-east of the county, whence this cropping type stretched southwards across the boulder-clay plateau of High Suffolk into south-east Cambridgeshire, northern Hertfordshire, and northern Essex (Figure 6.16). In these areas irregular and often quite intensive rotations were the norm, with different cropping sequences often being followed on different parts of the arable area.[75] Thorpe Abbotts, on clay soils in south Norfolk, was one such demesne. Between 1356 and 1363 it sowed wheat, rye, barley, oats, and legumes in an almost bewildering variety of combinations (Figure 6.19).[76] Fallowing occurred with the utmost irregularity and often only after land had been sown for four, five, or six years in succession. Whatever underlying logic the system possessed defies detection.

In according pride of place to wheat and barley these demesnes were concentrating production upon those bread and brewing grains which were most

---

[73] Chapter 5, pp. 244–5.   [74] Durham, Dean and Chapter, Cell accounts.
[75] See above, pp. 263–4; Figures 6.09, 6.10.
[76] NRO, WAL 274×6/479, 274×6/488–91; Elveden Hall, Suffolk, Iveagh Collection, Cornwallis (Bateman) MS, box 60 no. 4.

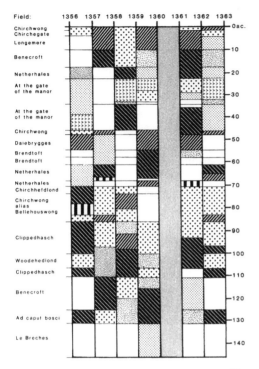

Fig. 6.19. Rotation of crops on 152 acres at Thorpe Abbotts, Norfolk, 1356–63 (source: NRO, WAL 274 × 6/479, 488–91; Elveden Hall, Suffolk, Iveagh Collection, Cornwallis (Bateman) MS, box 60 no. 4). For key see Figure 6.08.

esteemed, commanded the highest price, and were in greatest demand (Tables 5.04 and 5.07). On a number of manors these crops were plainly grown with a view to satisfying the subsistence requirements of seigniorial households, but on many more they were grown with an eye to commercial opportunity. In the FTC counties many of the demesnes operating this cropping type followed the commercial artery of the Thames whence access was obtained to the lucrative metropolitan market (Figure 6.16). A notable concentration shows up in the vicinity of Henley, the most important grain entrepot upstream of London (albeit of dwindling importance in this period).[77] This cropping type was also widely practised in eastern Kent, a region exposed to powerful external commercial influences and a major market area in its own right.[78] The same was true of East Anglia where population densities were well above the national average and a high proportion of the rural population engaged in non-agricultural activities.[79] Under such circumstances access to Norwich and

[77] Farmer, 'Marketing', pp. 372–3.   [78] Campbell and others, *Medieval capital*, pp. 179–82.
[79] Baker, 'Changes', pp. 190–2; R. H. Hilton, *Bond men made free* (London, 1973), pp. 171–2.

to riverine and coastal ports was a bonus. Demesnes in south-east Norfolk, for instance, were well within provisioning range of Norwich and some enjoyed the additional option of sending grain to Yarmouth via the rivers Yare and Waveney.[80] Significantly, Yarmouth drove a more active grain trade in the second half of the fourteenth century than it had done in the first, mainly because with a smaller population to support Norfolk now had a larger grain surplus for disposal.[81]

Cultivation of mixed grains (type 3)
The southern bias to the national distribution of these wheat- and barley-pro-ducing demesnes was complemented and extended to the north and west by demesnes devoting a substantial proportion of their sown acreage to mixed grain crops, particularly dredge, which was benefiting from the expanding demand for brewing grains in this period (cropping type 3: Figure 6.15).[82] This cropping type shows up strongly in a diagonal band of country extending from the Gloucestershire Cotswolds in the west, through the commonfield country of Warwickshire, Oxfordshire, Northamptonshire, north Bucking-hamshire, Bedfordshire, and south Cambridgeshire, as far east as Exning and Acton in south-west Suffolk, Stradsett in west Norfolk, and Popenhoe and Gedney in the East Anglian Fens.[83] As with cropping type 4, several additional examples extend the distribution northwards up the eastern side of England as far as Fulwell and Monkwearmouth in Co. Durham.[84] The affinities between cropping types 3 and 4 were functional as well as geographical: both grew wheat in some quantity, the latter especially so, and where the latter spe-cialised in barley the former specialised in dredge. In addition, both grew legumes in average, and oats in below average quantities (Table 6.02). Although each was characterised by a bias towards spring cropping, that bias was on average more pronounced in the case of cropping type 3 than cropping type 4 (especially since some demesnes practising the latter grew both winter and spring varieties of barley). Since most examples of cropping type 3 fall within the midland zone of regular commonfield systems this emphasis upon spring cropping implies an association in many cases with two-course regimes.

Within the east midlands many of these demesnes may have become engaged in supplying malted dredge to King's Lynn whence it could have been sent up or down the east coast or across the North Sea. Certainly, in the south midlands, the growth of both dredge and barley production on a string of demesnes just north of the Chiltern scarp can be linked to the expanding met-ropolitan demand for brewing grain generated by rising living standards. These demesnes lay just within range of the London market, especially when

[80] See above, n. 41.   [81] Saul, 'Great Yarmouth', pp. 226, 368–71, 374.
[82] Chapter 5, p. 244.
[83] PRO, SC 6/989/1, 6/996/9; DL 29/242/3888; SC 6/942/17, 6/943/16.
[84] Durham, Dean and Chapter, Bursar's accounts; *Inventories and account rolls*, pp. 152–93.

the transportability of the grain had been improved by on-the-manor malting, as seems to have been increasingly the case in this period (Table 5.05). The strengthening of this particular spatial specialism is paralleled by the greater prominence within the urban records of specialist dealers in malt. Many of these 'maltmen' operated out of southern Hertfordshire and hence were located between these specialist dredge producers and the metropolitan market.[85] A further group of mixed-grain-producing demesnes in the middle Thames Valley also owed their specialism to the London market (Figure 6.15). In their case the preferred mixed grain was maslin/mancorn rather than dredge, intended for baking into inferior qualities of bread rather than brewing into inferior qualities of ale. Yet whereas the market for brewing grain was an expanding one, that for bread grain, and especially the cheaper bread grains, was contracting.[86] This specialism is therefore less marked than before the Black Death and the demesnes in question were even more dependent upon the low-cost bulk carriage afforded by the Thames for getting their cheap grain to market.[87]

Rye with barley and/or oats (type 2)
Low transport costs were most essential to those demesnes which specialised in producing the cheapest grains of all – rye and oats – for the metropolitan market (cropping type 2) (Table 6.02 and Figure 6.14). Within the FTC counties examples of this distinctive cropping type are concentrated almost exclusively along the Thames axis within 25 miles of London (Figure 6.14). Such demesnes characteristically grew more rye than wheat, substantial quantities of oats and often quite significant quantities of barley, but few legumes. As such they shared certain features in common with cropping types 7 and 2 of pre-1350 (Table 6.01) and occupied much the same inner metropolitan zone. Elsewhere, within the arable heartland of central and southern England, a number of light-land demesnes displayed much the same cropping specialism, especially where rye continued to form the principal ingredient of food liveries.

As in the earlier period, nowhere is this association between light land and cropping type 2 more apparent than in East Anglia. Norfolk furnishes no fewer than eleven examples, all on light soils and mostly concentrated in the centre, west, and south-west of the county where winter-sown crops in general were most prominent at that time (Figure 6.14). Two more show up in Suffolk – one in Breckland and the other in the Sandlings – and two in north-east Essex (Wix and Wrabness) in very similar agricultural contexts.[88] Finally, in

[85] Galloway, 'London's grain supply', pp. 32–3; Galloway, 'Driven by drink?'.
[86] Chapter 5, pp. 242–3.
[87] Cf. Campbell and others, *Medieval capital*, pp. 121–3. Transport costs of firewood – an even bulkier commodity than grain – doubled in the vicinity of London between the second and fourth quarters of the fourteenth century: Galloway and others, 'Fuelling the city', p. 458.
[88] PRO, SC 6/849/15, 19; WAM, 3229.

the north-east of the country, whether for reasons of climate, soils, or conservatism, the same basic cropping type was traditionally employed as an alternative to the more demanding combination of wheat and oats.

Intensive cultivation with legumes (type 1)
Cropping type 1 was formerly the most intensive cropping type and in both periods the only one in which barley was almost invariably the single most important crop (Tables 6.01 and 6.02). Before the Black Death versions of this system were well established in eastern Norfolk and eastern Kent and were also to be found at a wide scatter of other locations in eastern England (Figure 6.01). After the Black Death a modified version of this system became almost exclusively confined to Norfolk, within which county its relative dominance was enhanced (Figure 6.13). Norfolk had long been England's premier arable county, rivalled in the intensity and productivity of its husbandry only by Kent. Eastern Norfolk and eastern Kent shared similar advantages of deep and easily cultivated loam soils, a lack of institutional constraints, an abundance of cheap labour, an enterprising peasantry, and ready access to major urban markets both within and beyond the region.[89] In response, both had developed very similar husbandry systems in the period 1250–1349 which achieved almost continuous cropping of the arable by substituting legumes for fallows as one element within a sophisticated technological complex which successfully sustained soil fertility.[90] After 1349, and especially after 1375, as the prices of most crops fell, land declined in value, and labour climbed in cost, the economics of this intensive regime were undermined. As the intensity of arable husbandry was necessarily reduced so the character of husbandry in Kent and Norfolk increasingly diverged. Kent, always a mixed county agriculturally, remained so. A few demesnes boosted the acreage devoted to legumes and became exemplars of the new fodder-orientated cropping type 8. Most, however, especially in the eastern half of the county, reduced their sowings of legumes and thereby became virtually indistinguishable in their cropping from the majority of demesnes in the south and east of England, operating some version of either cropping type 5 (dominated by wheat and oats) or cropping type 4 (dominated by wheat and barley). A few only retained a strong affinity with the kind of cropping regime now almost exclusively confined to Norfolk, in which spring-sown crops occupied by far the greater part of the sown area with barley pre-eminent among them.[91]

   In the period 1250–1349 the two most common cropping systems in Norfolk had been an intensive system and a light-land version of the same (cropping types 1 and 2) (Table 6.01 and Figures 6.01 and 6.02). Neither system was unique to the county, versions of both occurring in quite widely

[89] Campbell, 'Economic rent'.   [90] Campbell, 'Agricultural progress', pp. 41–3.
[91] Examples include Agney-and-Orgarswick, Barton, Bekesbourne, Ham, Monkton, and Sharpness: BL, Harl. Roll Z3–4; CCA, Bedels Rolls; DCc/Agney 56, Barton Carucate 14–15; PRO, SC 6/892/1–2; 6/897/7, 9, 11–12.

separate locations elsewhere in eastern and south-eastern England. After 1349 these two systems coalesced into one, which became almost exclusively confined to Norfolk where it was taken up in some form or other by 70 per cent of demesnes (Figure 6.13). The handful of examples outside the county include Soham in the Cambridgeshire fens, half a dozen residual demesnes in northern and eastern Kent, Hambledon in southern Hampshire, a loose scatter of demesnes in Surrey and the lower Thames Valley near London, and Wittenham and Speen in Berkshire. For the most part, these were all localities exposed to strong commercial impulses where there was an established tradition of either large-scale barley cultivation or relatively intensive methods of production.[92] Elsewhere, the few stray examples of this cropping type at Dorking and West Horsley in Surrey and Speen and Long Wittenham in Berkshire are explicable in terms of an unusually heavy emphasis upon barley.[93]

Most of these non-Norfolk examples of this cropping type embodied its main characteristics in diluted form. Only in Norfolk were its distinctive traits taken to an extreme. For instance, many Norfolk demesnes displayed a stronger than average bias towards spring crops, reinforced by a corresponding concentration upon barley (Figure 6.13).[94] Thus, on the most specialised demesnes, mostly concentrated in the north-east of the county, upwards of 80 per cent of the cropped area was spring sown and upwards of 80 per cent of that was devoted to barley (Figure 6.21).[95] A concomitant of barley's pre-eminence was that oats were of less importance than in almost any other cropping system. Many demesnes evidently grew only enough to satisfy the essential fodder requirements of their work-horses. On light land where plough-teams were small these requirements were often quite modest. Consequently, over a quarter of Norfolk demesnes sowed less than a tenth of their cropped area with oats (Figure 6.21). Many demesnes used legumes as a supplementary source of fodder and, except on the lightest land where legumes fared badly, grew them on a comparable or greater scale (Figure 6.22). Since fallowing was

[92] Above, n. 89; PRO, DL 29/288/4721; N'hants. RO, PDC AR/1/4; WAM, 26933, 36–7, 47, 49; 27054, 34, 38, 40, 44; Hants. RO, Eccles. 2 159388.

[93] Arundel Castle, Sussex, A1778–80, 82; New College, Oxford, 9145–7; PRO, SC 6/750/26; 6/1013/12, 15–16.

[94] A group of demesnes in the north-east of the county commonly devoted at least 90 per cent of their sown acreage to spring crops, notably Calthorpe, Costessey, Gimingham, Horsham, Hoveton, Little Hautbois, North Walsham, Saxthorpe, Scottow, Thurning, Thwaite, and Tunstead: NRO, Case 24, Shelf C; MS 6001 16 A6; NRS 11331–2 26 B6, 11058 60 25 E2, 11069 25 E3, 19517 42 C6, 19650 54–8 42 D7, 19690 42 E4, 19677 42 E3, 2797–9 12 E2; Church Commissioners 101426 2/13, 7–8/13, 11/13; Diocesan Est/2 2/15–17, 21; Est/11; Est/12; Lambeth Palace Library ED 479; PRO, DL 29/288/4719–20, 22, 34, 44, 52.

[95] As at Calthorpe, Costessey, Heigham-by-Norwich, Horsham, North Elmham, and Thwaite: above, n. 94; NRO, Diocesan Est/2 2/20; DCN 60/10/25–6. On the other hand, no documented Norfolk demesne went as far as Holywell, just outside Oxford, which in 1391–2 sowed nothing but barley. It was, presumably, destined for ale to slake the thirsts of the fellows and undergraduates of Merton College: Merton College, Oxford, MM 4533.

often relatively infrequent legumes also performed a vital role in the replenishment of soil nitrogen, the most intensive regimes sowing them in some quantity, although rarely on a par with that which had prevailed during the climax of the high-farming era before the Black Death.[96] Only a handful of demesnes sowed more than a quarter of their cropped area with legumes (twice the norm for this cropping type), almost all of them on the intensively cultivated loam soils of the east.[97] By concentrating on spring crops, which allowed half-year fallows and could withstand being sown as consecutive courses, demesnes often succeeded in keeping land under crop for at least five or six years on end. Consequently wheat, the most demanding of crops, was relegated to a smaller share of the sown acreage than in any other cropping system, although as a winter crop it was sometimes augmented by modest sowings of rye (Figure 6.20).

Rotationally, the hallmarks of this Norfolk system of cropping were variety and irregularity. Indeed, it could only flourish where there were few if any institutional constraints upon the sequencing of cropping and fallowing. Norfolk field systems with their minimal regulations were consequently peculiarly sympathetic to the operation of this cropping type.[98] In its pronounced seasonal asymmetry of ploughing and sowing it was also dependent upon adequate supplies of labour (as was that labour of alternative employment opportunities). It is therefore no coincidence that it was so closely associated with what the 1377 poll tax returns reveal to have been England's most densely populated county together with what the 1332 lay subsidy reveals to have been the most closely settled districts of that county.[99] When precise rotational sequences can be reconstructed, as in about a dozen cases when runs of consecutive accounts record the names of the plots and fields sown, the lack of any regular and consistent plan becomes immediately apparent.[100]

---

[96] Campbell, 'Agricultural progress', p. 33. The St Benet's Abbey demesne of Flegg, for instance, had sown 31 per cent of its cropped area with legumes in 1341, whereas in 1351–1428 the equivalent area was generally less than 25 per cent and often less than 20 per cent: NRO, Diocesan Est/9–10, Est/58/8.

[97] Namely Burgh-in-Flegg, Potter Heigham, Scottow, and Tunstead, plus Stradsett and West Walton in west Norfolk: PRO, SC 6/931/28, 6/943/16; DL 29/288/4719–20, 22, 24; NRO, Diocesan Est/11; Hare 210×2–3/4018–30.

[98] M. R. Postgate, 'Field systems of East Anglia', in Baker and Butlin (eds.), *Field systems*, pp. 293–305; Campbell, 'Regional uniqueness'.

[99] Baker, 'Changes', pp. 190–2. I am grateful to Dr R. E. Glasscock of St John's College, Cambridge for supplying vill-by-vill data of the number of taxpayers in 1332 and to K. C. Bartley for undertaking their analysis.

[100] Rotations have been reconstructed as follows: Ashill (1357–62); Brandon, Suffolk (1366–75), Figure 6.23; Felbrigg (1400–8), Figure 6.24; Hinderclay, Suffolk (1379–88); Keswick (1370–7), Figure 6.26; Langham (1364–9), Figure 6.25; Martham (1412–23); Ormesby (1423–31); Reedham (1377–85), Figure 6.27; Taverham (1413–21); Thornage (1370–81): BL, Add. Charter 26852–8; Chicago UL, Bacon Roll 491–8, 529–34, 643; NRO, DCN 60/35/43–50; NNAS 5909–16 20 D2–3; NRS 21162 45 A5, 23358 Z 98; WKC 2/130/398×6; PRO, SC 6/939/1–8A, 6/1304/30–36; Raynham Hall, Norfolk, Townshend MSS.

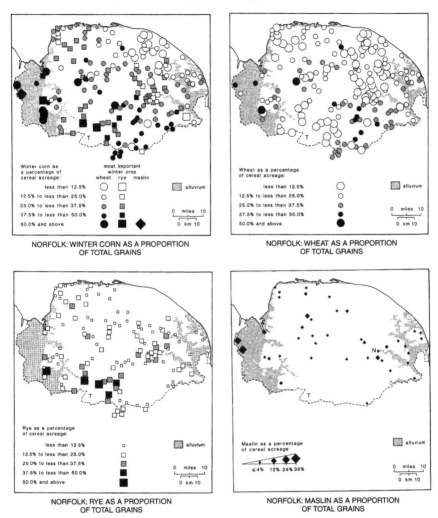

Fig. 6.20. Cultivation of winter-sown grains on Norfolk demesnes, 1250–1449
(L = Lynn; N = Norfolk; T = Thetford; Y = Yarmouth) (source: Norfolk accounts
database).

The common denominators of Norfolk rotations were several. First, the
duration of individual cropping sequences and frequency and duration of
fallows were matters of almost infinite variation, both within and between
demesnes. Cropping sequences were shortest and the duration of fallows
longest on the lightest and poorest soils, as on former 'light-land' demesnes
such as Brandon (Suffolk) in the heart of the East Anglian Breckland. Here,
between 1366 and 1375, land in some named fields was never or rarely

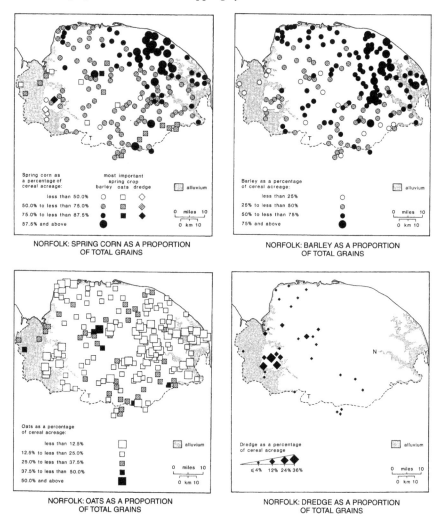

Fig. 6.21. Cultivation of spring-sown grains on Norfolk demesnes, 1250–1449
(L = Lynn; N = Norfolk; T = Thetford; Y = Yarmouth) (source: Norfolk accounts
database).

cultivated, while other fields bore crops only once or twice. On the other hand,
the better land carried crops in five out of seven recorded years. In no recorded
instance, however, was land sown for more than three years in succession
(Figure 6.23). Soils were almost equally light and poor on the Holt–Cromer
ridge, a glacial end-moraine in the extreme north-east of Norfolk. Here, too,
fallows were more likely to be of several years' duration rather than just one,
although occasional instances of the latter do occur. At Felbrigg a fully

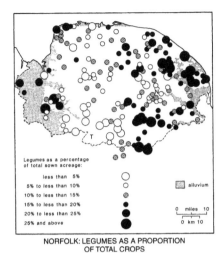

Legumes as a percentage of total sown acreage:

less than 5%	○
5% to less than 10%	○
10% to less than 15%	◐
15% to less than 20%	●
20% to less than 25%	●
25% and above	●

alluvium

0   miles   10
0   km   10

NORFOLK: LEGUMES AS A PROPORTION
OF TOTAL CROPS

Fig. 6.22. Cultivation of legumes on Norfolk demesnes, 1250–1449 (L = Lynn;
N = Norfolk; T = Thetford; Y = Yarmouth) (source: Norfolk accounts database).

fledged convertible regime can be observed in operation between 1400 and
1408 (Figure 6.24). Land might be sown for anything from one to five years,
but with a norm of three, and then left unsown for an equivalent period during
which it served as temporary pasture. Leys of one, two, or three years' duration can be similarly observed at Taverham between 1413 and 1421 (another
former 'light-land' demesne). Its soils, derived from glacial outwash sands and
gravels, were also poor but cropping sequences were generally longer than at
Felbrigg, possibly owing to the commercial stimulus of the nearby Norwich
market. Land was usually sown for at least three consecutive years, and some
was sown for as many as eight and possibly more. Ashill, Martham, and
Thornage also furnish examples of leys of two or three years' duration
between longer bouts of cultivation.

The aim of most demesne managers, at least on the better soils, nevertheless seems to have been to minimise the frequency and duration of fallows,
employing bare fallows in the main as a means of controlling weed growth. At
Langham, on relatively heavy soils in north Norfolk, land was being cropped
for four consecutive years and then fallowed in the fifth between 1364 and
1369 (Figure 6.25). At Thornage, not so very far away, the cropping cycle
between 1370 and 1381 was much less regular, although at least part of the
arable seems to have been subject to a similar 'five-course' cycle. At Keswick,
just south of Norwich, and Reedham, on the western edge of the Broadland
marshes, rotations were a degree more intensive (Figures 6.26 and 6.27). On
the latter between 1377 and 1385 land was usually sown for five out of every

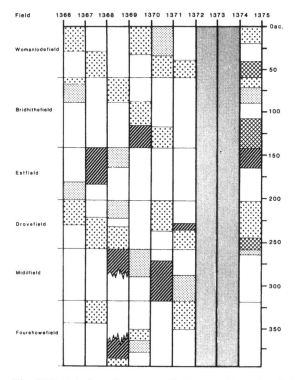

Fig. 6.23. Rotation of crops on 395 acres at Brandon, Suffolk, 1366–75 (source: PRO, SC 6/1304/30–6; Chicago UL, Bacon Roll 643). For key see Figure 6.08.

six years. On the former, on in fact poorer soils but, like Taverham, within the commercial penumbra of Norwich, land was sown for at least six out of every seven years between 1370 and 1377. Similar, if not more intensive, rotations seem to have been in operation on portions of the demesnes at Martham and Ormesby in the periods 1412–23 and 1423–31 (and it can safely be assumed that they were formerly even more intensive), although both demesnes still retained periodic annual fallows as an integral element of these rotations.[101]

It was immediately after the fallow that wheat and rye were most usually sown. At Felbrigg, Langham, Reedham, and Thornage wheat was rarely sown other than as the first crop in the rotation. At Ashill, Hinderclay (Suffolk), Keswick, Martham, Ormesby, and Taverham, where rotations often contained more courses, wheat and rye were also sown immediately after relieving courses of legumes, and occasionally even – as at Taverham – following barley. At Ashill, Keswick, and Taverham, in fact, barley regularly took precedence

---

[101] R. H. Britnell, 'Farming practice and techniques: eastern England', in Miller (ed.), *AHEW*, vol. III, pp. 203–4.

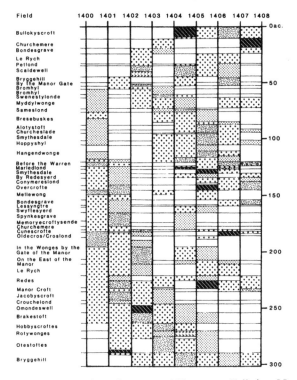

Fig. 6.24. Rotation of crops on 303 acres at Felbrigg, Norfolk, 1400–8 (source: NRO, WKC 2/130/398×6). For key see Figure 6.08.

over rye. Sometimes, especially where the acreage sown with winter crops was small, barley was also sown as an alternative first course in the rotation, as at Brandon, Felbrigg, Keswick, Taverham, and Thornage. Whether sown first or second in rotations, that initial course of barley was typically succeeded by another. Double-cropping of barley was one of the most universal features of this Norfolk system, showing up on every demesne for which rotations can be reconstructed. Indeed, it remained a distinctive feature of Norfolk husbandry until well into the eighteenth century.[102] In the period 1350–1449 there were even instances when land was subjected to a third consecutive course of barley, as at Ashill, Keswick, and Taverham. But if the first three courses of the rotation comprised some combination of wheat (or rye) and barley the norm was to devote the fourth course to legumes or oats. That was certainly the case at Langham where there usually were only five courses, comprising wheat, barley, barley, oats or legumes, and finally fallow. With rotations that

---

[102] At Hunstanton in north-west Norfolk double-cropping of barley is recorded in a detailed crop book of 1705–11: NRO, L'Estrange BH/4. I am grateful to Professor Mark Overton for this reference.

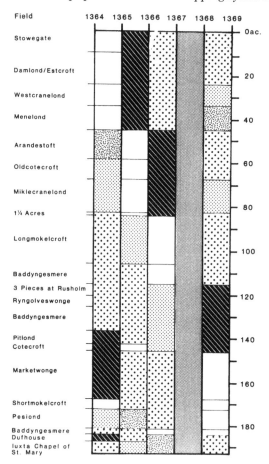

Fig. 6.25. Rotation of crops on 193 acres at Langham, Norfolk, 1364–9 (source: Raynham Hall, Townshend MSS). For key see Figure 6.08.

contained more than five courses there was an obvious advantage in using legumes as a nitrifying crop between successive grain crops. They were regularly sown in this position at Reedham, but were sometimes also sown as the final course in the rotation. How much they could be used to provide a respite from grain crops nevertheless depended upon their relative share of the cropped acreage. At Reedham this was well above average for this cropping type, as it was at Martham where legumes were likewise a prominent component of prevailing rotations. Here they were often used in preparation for a further course of wheat. The same practice was employed at Ashill, Hinderclay (Suffolk), Keswick, and, occasionally, Taverham.

The benefits of sowing legumes between successive courses of grain are

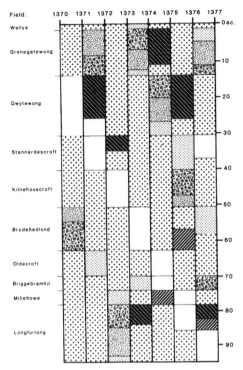

Fig. 6.26. Rotation of crops on 94½ acres at Keswick, Norfolk, 1370–7 (source: NRO, NRS 23358 Z 98). For key see Figure 6.08.

fairly self-evident: the benefits of sowing them at the very end of the rotation, before the fallow, are much less so. Indeed, Farmer was perplexed by the latter practice on the East Anglian demesnes of Westminster Abbey.[103] Several otherwise seemingly advanced Norfolk manors also sowed some of their legumes last. This was most obviously the case at Felbrigg where they were often the final crop before a three- or four-year ley and as such may have been sown with the object of providing a nitrogen boost to land which immediately afterwards was to be allowed to tumble down to grass. Perhaps a quicker and better sward was thereby established. On other demesnes it was an occasional rather than regular practice and as such more probably reflects the difficulty of matching output needs to such an infinitely varied rotational regime. Most commonly, however, it was oats that were sown last. This was the case at Ashill, Brandon, Felbrigg, Hinderclay (Suffolk), Martham, Langham, Reedham, and Thornage. Only at Keswick and Taverham – both within a few miles of Norwich – were oats sometimes accorded a higher priority. On these, as on

---

[103] Farmer, 'Grain yields on Westminster Abbey manors', pp. 346–7.

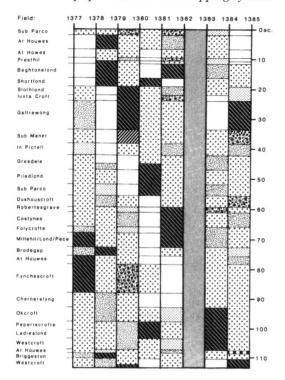

Fig. 6.27. Rotation of crops on 114 acres at Reedham, Norfolk, 1377–85 (source: BL, Add. Charter 26852–8). For key see Figure 6.08.

several other major barley-producing demesnes, the final course before the fallow was as likely to be devoted to barley.

Demesnes operating this Norfolk system of cropping plainly observed no hard and fast rules. Cropping sequences varied almost as much within demesnes as between demesnes. The one consistent denominator of crop rotations was flexibility. It was because of this that rotations were capable of being adapted to such a variety of environmental and economic circumstances within the county, thereby endowing this cropping type with the almost unique quality that it existed in both extensive and intensive versions.[104] That it was not more widely represented geographically was a function of the fact that this flexibility was harnessed to a comparatively unusual specialism, namely the mass production of barley. This reflected the unique dietary dominance of barley within Norfolk as a food as well as a drink grain, together with the prominence of Norfolk as an exporter of malted barley to other regions. Paradoxically, far from curtailing their barley production after 1350

[104] Campbell, 'Land, labour, livestock', pp. 174–8.

as other parts of the country became more self-sufficient in this crop, Norfolk demesne producers actually expanded the relative scale on which they grew it. From 45 per cent of the county's demesne sown acreage at the opening of the fourteenth century it expanded to 54 per cent at the close (Table 5.08). This pronounced specialism was long to remain one of the most immutable features of the county's husbandry.[105]

6.22 The overall configuration of cropping systems post-1349

The crystallisation, after 1349, of cropping type 1 as more a regional than a national farming type, often, but not invariably, more intensive than other cropping types, is symptomatic of the general reconfiguration of cropping types taking place within the country as a whole. Whereas before 1350 it is possible to distinguish a loose hierarchy of cropping types, differentiated by the range of crops grown and intensity of their cultivation, after 1349 cropping types represent more of a continuum, differentiated in the main by the extent and character of crop specialisation. The gap between the most and the least intensive systems was a narrowing one. As grain prices fell and unit labour costs rose so there was a general crying down in the intensity of the most intensive systems. Nowhere is this more apparent than in eastern Norfolk, eastern Kent, and coastal Sussex. The upshot was that cropping in all three areas became less distinctive. Thus, cropping in eastern Norfolk ceased to stand out from that to be found in much of the rest of the county, while cropping in Kent and Sussex took on many of the characteristics of that prevailing elsewhere in southern England. At the same time, the simplest and most extensive cropping types ceased to be so well represented among those demesnes that remained under direct seigniorial management. Demesnes growing mainly oats – the cheapest and least demanding of crops – disappeared as a distinctive type, as poor land was increasingly withdrawn from cultivation and converted to pasture, and estates pursued a selective policy of leasing. Demesnes operating basic two-course rotations in which wheat was the principal crop (cropping type 7) also seem to have become comparative rarities (Table 6.01; cf. Table 6.02, from which cropping type 7 has disappeared altogether).

In the period 1350–1449 the vast majority of demesnes therefore operated cropping systems which were neither intensive nor extensive but of an intermediate character, and it is this proliferation of intermediate systems that most distinguishes it. In a few localities this represented a downgrading of existing systems whose intensity it was no longer viable to sustain, but in many more instances it constituted an improvement upon (or at the very least a diversification of) previous practice. Often it was associated with the cultivation of a wider range of crops and the allocation of a larger share of the

---

[105] Campbell and Overton, 'New perspective', pp. 54–7.

cropped acreage to barley and legumes, both of which tended to expand at the expense of oats. The upshot was a range of cropping types, variants of each of which might be found in quite contrasting geographical contexts irrespective of differences in economic rent and types of field system. With the conspicuous exception of the barley-dominated demesnes of cropping type 1, which were almost exclusively confined to areas with flexible and irregular commonfield regimes, field systems appear to have exercised little influence upon the nature of prevailing cropping systems.[106] No doubt this was in part because individual commonfield cultivators enjoyed greater scope for flexibility and experimentation now that the acute land hunger of the pre-plague period had abated. This would appear to be borne out by the fact that there is more evidence of the reorganisation of commonfield systems after 1349 than before.[107]

Concentrated market demand was also of lesser influence after 1349 than before. London, for instance, reduced in size but *per caput* thirstier and better fed, appears to have exercised a less pronounced impact upon the configuration of cropping types within the Thames basin than formerly. Similarly, there is less evidence that overseas demand was structuring agricultural production along England's North Sea littoral. By national standards cropping systems were less distinctive than formerly in both the south-east and those counties of the east midlands that focused on the Wash and its extensive network of navigable rivers. Norfolk alone retained and enhanced its already distinctive arable identity. All of this points to a flattening and lowering of the contours of economic rent as the main centres of urban demand contracted. The more that Thünen economic rent receded, the more that Ricardian economic rent – reflecting differences in land quality and population density – came to the fore.[108] Demesne managers seem increasingly to have attuned their cultivation to those crops best suited to prevailing environmental circumstances, subject to the estate's own consumption needs and local and regional (rather than national and international) market opportunities. Production was pitched at a level consistent with local land values, themselves a function of land quality and the demand for land. The overall pattern may consequently be interpreted as a reversion to one less structured by wider and more powerful market forces. This is consistent with the narrowing commercial differential already observed within the FTC counties, as the extreme levels of commercial and non-commercial activity encountered before the Black Death effectively disappeared (Table 5.03, Figures 5.01 and 5.02).

---

[106] But see R. L. Hopcroft, 'The social origins of agrarian change in late medieval England', *American Journal of Sociology* 99 (1994), 1,559–95.

[107] Britnell, 'Farming: eastern England', p. 198; Dyer, 'Farming: west midlands', pp. 223–4; Harvey, 'Farming: Home Counties', pp. 254–5; Fox, 'Alleged transformation to three-field systems'.   [108] Grigg, *Dynamics*, pp. 50–1, 134–40.

6.3   Crop specialisation and commercialisation

Were certain of these cropping types intrinsically more commercialised than others? On the evidence of the demesnes within the FTC counties it would appear not. Producing what was consistent with prevailing levels of economic rent was one thing, deciding whether or not to sell what was produced was another. The theories of Ricardo and Thünen seek to explain production, not disposal.[109] As already seen, decisions about the latter were strongly influenced by institutional factors, especially the type of estate, overall estate management policy, and place of the individual demesne within the overall estate structure (Tables 5.01, 5.02, and 5.03). They were also influenced by transaction and storage costs and the seasonal and annual volatility of prices, for these largely determined the extent to which production for exchange was a more profitable, efficient, and reliable alternative to production for consumption where the estates in question had high internal consumption needs to satisfy. For many lords it always made good sense to provision their workers and their households, in part at least, from the produce of their own estates. The market had yet to attain a level of development where it was advantageous to do otherwise.

The dichotomy which existed between the rationales of production and disposal is brought out by Table 6.03. This analyses whether choice of cropping type was a significant influence upon levels of commercialisation within the FTC counties. Unfortunately, not all cropping types are equally well represented, hence direct comparison is difficult. In particular, mean percentages sold and mean rates of sale tend to be misleading, as the mostly high standard deviations indicate. In the period 1288–1315 very strongly commercialised demesnes selling at least 60 per cent of their net receipts and receiving at least £15 per 100 sown acres from crop sales were characteristic of all cropping types. The same is equally true (with the exception of cropping type 6 of which there are only three examples) of very weakly commercialised demesnes selling less than 20 per cent of their net receipts and receiving less than £5 per 100 sown acres from crop sales, and this remains the case in the period 1375–1400. By that time very strongly commercialised demesnes had become relatively rare (partly because revenues per acre were depressed by falling prices), although cropping types 3, 4, 5, and 6 all furnish examples of demesnes which sold at least 40 per cent of their net receipts and received at least £10 per 100 sown acres from crop sales. In neither period do the intensively cultivated demesnes of cropping type 1 appear to have been intrinsically more commercialised than the more extensively cultivated demesnes of cropping type 5, which grew mostly wheat and oats. Nor do the specialised

---

[109] M. Chisholm, *Rural settlement and land-use* (London, 1962), pp. 20–32; Grigg, *Dynamics*, pp. 50–1, 134–40.

Table 6.03. *Proportions sold and rates of sale by cropping type within the FTC counties, 1288–1315 and 1375–1400*

National cropping type	No. of demesnes	% aggregate net crop receipt[a] (£ sold)				Income from sale of field crops (£ per 100 sown acres)			
		Mean	*Std.*	*Min.*	*Max.*	*Mean*	*Std.*	*Min.*	*Max.*
FTC1, 1288–1315:									
Cropping type 1	12	43.1	24.5	3.7	75.8	11.3	7.5	1.2	24.9
Cropping type 2	8	41.2	30.7	2.9	92.7	7.0	6.3	0.8	19.5
Cropping type 3	26	50.7	26.5	0.0	89.9	11.0	8.5	0.4	28.2
Cropping type 4	41	37.1	22.3	0.2	93.0	8.0	6.8	0.0	28.3
Cropping type 5	37	36.6	19.5	0.0	82.4	7.1	5.1	0.0	21.0
Cropping type 6	3	67.2	15.4	46.0	82.1	24.7	8.5	13.8	34.5
Cropping type 7	8	42.6	18.8	12.0	71.7	8.2	6.6	1.2	22.8
FTC2, 1375–1400:									
Cropping type 1	7	23.6	12.9	3.5	41.0	5.3	2.9	1.2	10.8
Cropping type 2	4	35.7	10.0	19.8	47.2	5.8	2.5	1.9	8.6
Cropping type 3	13	34.3	14.9	6.1	53.9	11.6	6.3	0.5	24.3
Cropping type 4	29	38.7	22.7	0.0	77.5	9.2	10.8	0.2	54.9
Cropping type 5	36	35.4	18.3	5.9	70.5	6.1	3.7	0.2	16.2
Cropping type 8	12	40.1	29.1	0.5	100.0	8.9	6.2	0.0	20.4

Note:
[a] Conversion of volume to value carried out using the following per bushel sale prices:
1288–1315: wheat 8.6d., rye 7.1d., winter-mixtures 6.9d., barley 6.5d., dredge 5.1d., oats 3.7d., beans 5.9d., peas 5.7d., vetches 5.1d., legumes 5.6d., grain–legume mixtures 5.4d.
1375–1400: wheat 7.9d., rye 4.1d., winter-mixtures 5.8d., barley 6.0d., dredge 4.6d., oats 3.4d., beans 5.8d., peas 5.3d., vetches 5.6d., legumes 5.5d., grain–legume mixtures 5.2d.

Source: FTC1 and FTC2 accounts databases.

pottage- and fodder-producing demesnes of cropping types 7 (1288–1315) and 8 (1375–1400), or the cheap bread-grain producing demesnes of cropping type 2, appear to have been necessarily more or less highly commercialised than others. Perhaps a fuller and more consistent sample of demesnes from each cropping type might yield a different result, although analysis of the combined proportion of net receipts transferred and sold suggests otherwise. The inescapable conclusion would seem to be that as far as arable production was concerned a demesne's choice of cropping type was of less moment than the type of estate to which it belonged in determining the extent of its involvement in the market. Although the character of that involvement certainly changed over the course of the fourteenth century, the net effect was to render all cropping systems generally less commercialised at the end of the century than they had been at the beginning (Table 6.03).

7
Arable productivity

7.1 Productivity as an issue

The productivity of arable husbandry has been one of the most debated aspects of medieval agriculture. In part this reflects the abundance of detailed yield information provided by manorial accounts, which is unmatched in quantity and quality until the nineteenth century, but it is also a function of the importance attached to the issue by those explanatory models of medieval socio-economic development which emphasise the adverse consequences of over-expansion when undertaken in conjunction with under-investment. Nor, once productivity decline had set in, was it easily reversed: there was no 'quick fix' to reduced nitrogen balances and depleted phosphorus levels. A more optimistic interpretation is offered by those who stress the growing commercialisation of the medieval economy and the productivity gains which thereby accrued from involution, innovation, and market specialisation. That is not to say that ecological limits were never transgressed, but rather that such transgressions were more the exception than the norm with the result that there was a net overall gain in land productivity. On this line of reasoning the crucial limits to sustained productivity growth lay less within the agricultural sector *per se* than within the wider commercial economy. Once recession replaced expansion, markets contracted, and trade subsided, so land productivity, with certain notable exceptions, also declined, not to recover until the next wave of commercial and demographic growth in the sixteenth century.[1]

Resolving these alternative interpretations is ultimately more a matter of evidence than ideology, although the lack of direct productivity evidence for the majority peasant sector will always be a serious lacuna. Even with good evidence measuring productivity poses problems.[2] Productivity may be defined as the ratio of outputs to inputs. Where the ratio is that between total

---

[1] See Chapter 1, pp. 16–22, for a review of this debate.
[2] Overton and Campbell, 'Productivity change'; Overton, *Agricultural revolution*, pp. 70–88; Overton and Campbell, 'Production et productivité', pp. 256–60, 288–9.

Table 7.01. *Definitions of agricultural area, land productivity, and the rate of yield*

Variables

A	Agricultural land	H	Non-agricultural consumption
Ar	Arable	i	Individual crop
C	Crops in arable rotations	N	Next harvest
Cf	Leguminous crops	O	Off the farm
Cg	Grain crops	U	Unsown arable
O	Agricultural products	Q	Total quantity
P	Pasturage	R	Rate of yield
S	Seed for current harvest	T	Total area
S^N	Seed for next harvest	Y	Yield
W	Workers in agriculture		

Area

1	Total agricultural area	TA	$= T(Ar + P)$
2	Total arable area	TAr	$= T(Cf + Cg + Ar^U)$
3	Total crop area	TC	$= T(Cf + Cg)$
4	Total grain area	TCg	$= T(Cg^i)$

Land productivity

5	Gross agricultural land productivity	YA	$= QO \div TA$
6	Net agricultural land productivity	YA^H	$= QO^H \div TA$
7	Gross aggregate crop yield per unit arable area	YAr	$= QC \div TAr$
8	Net aggregate crop yield per unit arable area	YAr^H	$= QC^H \div TAr$
9	Gross aggregate grain yield per unit grain area	YCg	$= QCg \div TCg$
10	Net aggregate grain yield per unit grain area	YCg^H	$= QCg^H \div TCg$
11	Gross individual crop yield per unit cropped area	YC^i	$= QC^i \div TC^i$
12	Net individual crop yield per unit cropped area	YC^{Hi}	$= QC^{Hi} \div TC^i$

Rate of yield

13	Gross individual crop yield per seed	RC^i	$= QC^i \div QS^i$
14	Net individual crop yield per seed	RC^{Hi}	$= QC^{Hi} \div QS^i$

outputs and total inputs it is total factor productivity that is being measured. In many respects this is the key productivity, since it measures the gains that accrued not from increases in land, labour, and capital *per se* but from the improved combination of those factors of production via better technology, new knowledge, and superior organisation and efficiency in production. Historically, this is the hardest aspect of productivity to measure owing to the demands that are placed upon the quality and quantity of data. Recently, Persson has attempted to derive indirect estimates of total factor productivity growth in English agriculture between 1250 and 1450 using the wage and price series constructed by Farmer. These estimates are methodologically experimental and historically controversial, not least because contrary to the reasoning of Postan and R. Brenner they indicate that total factor productivity increased throughout this period and at a higher rate before 1350 than after (0.25–0.33 per cent compared with 0.07–0.15 per cent). Persson suggests that while institutional changes associated with the modification and decline of feudal socio-property relations may have been a factor throughout these 200 years, growing scarcity of resources may have been a spur to greater efficiency in their use before the Black Death.[3]

The fact that total factor productivity may have been steadily improving does not necessarily mean that the separate productivities of land, labour, and capital displayed the same trend. Indeed, each was capable of behaving very differently from the others; for example, higher land productivity may have been bought at the expense of lower labour productivity and vice versa. Thus, on the bishop of Winchester's manor of Rimpton in Somerset C. Thornton has shown how a 40 per cent rise in the number of days worked per arable acre between *c.* 1230 and *c.* 1300 was accompanied by a 37 per cent decline in crop yield per labour unit. Conversely, between *c.* 1300 and *c.* 1375 a 22 per cent reduction in labour inputs was matched by a 19 per cent recovery in crop yield per labour unit, notwithstanding that yields themselves registered a 37 per cent decline (Table 7.15).[4] Unfortunately, Thornton's painstaking micro-scale calculations depend upon the availability of long runs of exceptionally detailed accounts and currently stand alone, although the process of factor substitution which they document is likely to have been widespread. It is also important to recognise that labour productivity as measured here is both farm and product specific. Nevertheless, the trends that Thornton identifies are consistent with national trends in real wage rates, which fell significantly in the 1270s, sank to their nadir during the famine years from 1315 to 1322, but thereafter recovered by fits and starts as demographic decline rendered labour increasingly scarce (Figure 1.01).[5]

An alternative approach and interpretation are offered by Persson, an exponent of the productivity benefits of commercial specialisation. Arguing from macro-scale changes in occupational structure as implied by the growing proportion of the total population resident in towns, he claims that 'slow growth of labour productivity and income occurred in the most advanced regions in Europe (particularly Tuscany and the "Low Countries" but also parts of northern France and southern England) at least up to the early fourteenth century'.[6] He is the first to acknowledge the empirical flimsiness of his data

---

[3] K. G. Persson, *Total factor productivity growth in English Agriculture, 1250–1450* (Copenhagen, 1994).   [4] Thornton, 'Determinants of productivity', pp. 205–7.

[5] Phelps Brown and Hopkins, 'Seven centuries of prices'; May, 'Index of impoverishment'.

[6] K. G. Persson, 'Labour productivity in medieval agriculture: Tuscany and the "Low Countries"', in Campbell and Overton (eds.), *Land, labour and livestock*, p. 140; Persson, *Pre-industrial growth*, pp. 114–18.

and hence offers a range of possible labour-productivity estimates for the thir-teenth century, notably annual growth rates of 0.15–0.35 per cent for Tuscany and 0.10–0.25 per cent for the 'Low Countries'. The source of this slow but steady growth in labour productivity lay, he suggests, in 'an increase in the hours worked, as market opportunities promoted the adoption of more labour-intensive methods of husbandry and the cultivation of industrial and horticultural plants with their greater labour demands'.[7] In England corre-sponding levels of urbanisation, although much debated, were undoubtedly lower, implying a greater occupational bias towards agriculture and hence lower levels of labour productivity than in Tuscany and the 'Low Countries'.[8] Clark's courageous macro-estimates of agricultural labour productivity suggest that output per worker in English agriculture was only a quarter in 1300 what it was to be in 1851.[9] This, however, is consistent with a four-fold increase over the same period in the proportion of the population resident in towns and therefore lends some support to Persson's reliance upon levels of urbanisation as a surrogate measure of aggregate labour productivity. For Clark the root cause of low medieval labour productivity lay in 'low levels of work intensity owing to inefficiency and underemployment', for which struc-tural, institutional, and demographic factors all played their part.[10] On peasant farms work intensity was handicapped by a sub-optimal ratio of labour to land, with much consequent under-employment of labour. In this context, post-medieval evidence points to a positive correlation between labour efficiency and farm size.[11] Whether such factors were responsible for raising or reducing agricultural labour productivity during the century or so before 1300 must be for further research to resolve.

Even less can currently be said about trends in capital productivity, which are yet more elusive of direct measurement. The one conspicuous exception is the return upon the seed sown, commonly known as the yield ratio.

7.2 Crop yields

7.21 Seeding rates

Seedcorn represented capital and was also the most indispensable of all forms of investment since upon it depended in large measure the next year's harvest. Seed was invested in the land at different rates depending upon the crop being sown, its place within the prevailing rotation, and the quality of the land. As a

---

[7] Persson, 'Labour productivity', p. 139.

[8] K. G. Persson, 'Was there a productivity gap between fourteenth-century Italy and England?', *EcHR* 46 (1993), 105–14; Dyer, 'How urbanized was medieval England?'.

[9] Clark, 'Labour productivity', p. 221.   [10] Clark, 'Labour productivity', p. 235.

[11] R. C. Allen, 'The two English agricultural revolutions, 1459–1850', in Campbell and Overton (eds.), *Land, labour and livestock*, pp. 249–52; Allen, *Enclosure and the yeoman*, pp. 211–31.

general rule the winter-sown grains were less thickly sown than the spring-sown grains. Peas and vetches usually conformed to the seeding rates of the winter grains and beans to the seeding rates of the spring. Nationally, moderate to low seeding rates of 2.0–2.9 bushels per acre for wheat, rye, maslin, peas, and vetches and 2.0–3.9 bushels per acre for barley, oats, dredge, and beans predominated. Thinner seeding rates of less than 2 bushels an acre were sometimes employed but, at least in Norfolk and the FTC counties, were very much the exception, confined either to manors with abnormally small customary acres or those which cultivated the poorest and most unrewarding soils (Table 7.02). On several Berkshire demesnes, for instance, small customary acres are betrayed by seeding rates of less than 1.5 bushels for wheat, rye, maslin, and peas and 3.0 bushels for barley and oats.[12] In Norfolk and Suffolk such minimal rates were virtually unknown except on the light and sterile soils of Breckland and the Breck edge where the winter crops were commonly sown at rates of 2 bushels an acre or less and the spring at 3.5 bushels or less.[13] Such soils did not repay the heavier capital investment represented by thicker seeding rates.

All moderate to low seeding rates ran the risk that the growing crops would be choked and smothered by rampant weed growth, especially given their botanical predisposition to yield poorly.[14] The fact that this risk so rarely materialised was probably owing to the superior height to which medieval cereal plants grew, thereby enabling them to overtop most weeds – hence the practice of reaping grain close to the ears and leaving the stubble and trash as fodder for livestock and thatch for roofing (preserved medieval thatch bears testimony to high levels of weed infestation).[15] It was partly as a check to weed growth that some demesnes regularly sowed their seed more thickly, at rates of 3 bushels an acre or more for wheat, rye, maslin, peas, and vetches, and 4 bushels an acre for barley, oats, dredge, and beans. Maximum rates of 4 bushels an acre for wheat, rye, maslin, and peas, 6 bushels an acre for barley, and 8 bushels an acre for oats are recorded on those demesnes which seeded their crops most thickly. Such high rates – far higher than those noted by the agricultural commentators of the early nineteenth century when seed was increasingly drilled rather than sown broadcast – were characteristic of certain specific localities. Indeed, seeding rates in general seem to have conformed to local rather than estate practice with the result that geographically they are one of the most differentiated of all farming attributes (see Figures 7.01, 7.02, 7.03, 7.06 and 7.08).[16]

---

[12] Examples include Bec Abbey's manor of Coombe, Winchester Cathedral Priory's manor of Woolstone, Battle Abbey's manor of Brightwalton, and New College, Oxford's manor of Long Wittenham.   [13] Bailey, *A marginal economy?*, pp. 105–8.   [14] Postles, 'Cleaning the arable'.

[15] Hawkes, 'Country thatches'; Letts, *Smoke blackened thatch*, pp. 24–7, 35–41.

[16] Campbell, 'Arable productivity', pp. 386–8; Mate, 'Agrarian practices', pp. 25–7; M. Mate, 'Agricultural technology in south-east England, 1348–1530', in Astill and Langdon (eds.), *Medieval farming*, pp. 251–74.

In Norfolk high seeding rates were one of the most conspicuous features of the intensive mixed-husbandry practised in one form or another by demesnes in the east and north-east of the county on soils ranging from deep and fertile loams to light and intrinsically less rewarding sands. In every case, however, the high seeding rates accompanied long and flexible rotations, with the heaviest seeding rates of all reserved for those crops – typically barley and oats – which occupied the final courses of the rotation at the point just before levels of weed infestation required a cleansing fallow course. In the late fourteenth century these self-same demesnes commonly subjected bare fallows to at least three, and, on occasion, as many as eight, summer ploughings, with the object of eradicating weed growth.[17] Northern and eastern Kent – where demesnes employed correspondingly long and intensive rotations – were similarly characterised by high seeding rates.[18] In both Norfolk and Kent the long association of these high rates with particular localities rather than specific institutions bears testimony to the enduring importance both of local husbandry traditions and those economic opportunities which made such heavy seeding rates rational and worthwhile.

In Norfolk seeding rates (and, presumably, fallow ploughings) attained their maxima when arable husbandry was at its most intensive (Table 7.13), with high inputs of capital and labour being lavished on the land in order to secure correspondingly high outputs. The climax of such 'high farming' was attained in the decades immediately prior to the agrarian crisis of 1315–20. Thereafter seeding rates were lowered. They recovered somewhat with the return of better harvests in the 1330s but were curtailed again, this time more severely, in the immediate aftermath of the Black Death, when the economics of high farming suffered a serious and lasting setback. Thereafter, when seeding rates were adjusted, it was almost invariably downwards. Across Norfolk as a whole the reduction in seeding rates between the opening and close of the fourteenth century amounted to 3 per cent for wheat, 6 per cent for barley, 9 per cent for oats, and 14 per cent for rye. This physical reduction in seeding rates was naturally most pronounced on those demesnes where the potential for reduction was greatest. At Martham, for instance, one of the most intensively cultivated demesnes in the country where the arable was under virtually continuous cultivation by the beginning of the fourteenth century, wheat, maslin, and peas were rarely sown at less than 4 bushels an acre between 1303 and 1339, while seeding rates of barley never fell below 6 bushels an acre and those of oats rarely below a hefty 8 bushels. Yet a century later mean seeding rates per acre of wheat, barley, oats, and peas had been reduced, respectively, by 20, 15, 33, and 30 per cent.[19] A similar downward adjustment of seeding rates occurred in the FTC counties. Here it was the seeding rates of

[17] Campbell, 'Agricultural progress', p. 29.
[18] Campbell and others, *Medieval capital*, pp. 131, 136–8.
[19] NRO, DCN 60/23; NNAS 20 D1–3.

Table 7.02 Individual and composite seeding rates: Norfolk and the FTC counties, pre- and post-1350

	Wheat		Rye		Barley		Oats		Peas		Overall	
Seeding rate[a]	pre-1350[b]	post-1350[c]	pre-1350[b]	post-1350[c]	pre-1350[b]	post-1350[c]	pre-1350[b]	post-1350[c]	pre-1350[b]	post-1350[c]	pre-1350[b]	post-1350[c]
Norfolk:												
(% of demesnes):												
Very low	0.9	1.3	9.2	6.6	0.0	0.0	0.9	0.0	7.2	7.6	16.0	11.9
Low	41.1	43.0	51.7	68.9	2.7	1.1	5.4	9.9	43.3	44.3	22.1	24.8
Moderate	19.6	22.8	17.2	13.3	10.7	20.4	13.4	23.1	14.4	22.8	25.2	30.3
High	18.7	19.0	11.5	11.1	53.6	50.5	46.4	37.4	14.4	19.8	16.0	11.0
Very high	19.6	18.9	10.4	0.0	33.0	28.0	33.9	29.7	20.6	6.3	20.6	22.0
n	107	79	87	45	112	93	112	91	97	79	131	109
Bushels per acre:												
Min.	1.8	1.6	1.5	1.4	2.0	2.6	1.9	2.1	1.5	1.6		
Mean	2.7	2.7	2.5	2.3	4.5	4.4	4.7	4.3	2.7	2.5		
Max.	4.0	4.1	4.0	3.2	7.0	6.2	7.8	6.8	4.6	4.0		

FTC counties:
(% of demesnes):

Very low	4.5	3.7	7.2	3.8	3.1	4.2	0.0	3.0	9.7	2.1	4.9	3.7
Low	33.1	36.6	37.7	50.0	4.6	71.7	5.9	73.5	38.1	35.1	19.8	44.1
Moderate	34.4	36.6	23.2	30.8	18.5	14.2	22.9	14.4	17.7	36.2	50.6	35.3
High	5.8	8.2	8.7	3.8	51.5	9.2	32.0	8.3	9.7	10.6	5.6	8.1
Very high	22.1	14.9	23.2	11.5	22.3	0.8	39.2	0.8	24.8	16.0	19.1	8.8
n	154	134	69	26	130	120	153	132	113	94	162	136

Bushels per acre:

Min.	1.3	1.3	1.7	1.3	1.6	1.3	2.1	1.3	1.1	1.3
Mean	2.8	2.7	2.8	2.6	4.2	2.8	4.8	2.8	2.8	2.8
Max.	4.8	8.0	4.5	4.0	6.6	8.0	8.0	8.0	5.0	6.0

Notes:

a Seeding rate:

	wheat, rye, and peas	barley and oats
Very low	<2.0 bus./ac.	<2.0 bus./ac.
Low	2.0–<2.5 bus./ac.	2.0–<3.0 bus./ac.
Moderate	2.5–<3.0 bus./ac.	3.0–<4.0 bus./ac.
High	3.0–<3.5 bus./ac.	4.0–<5.0 bus./ac.
Very high	3.5+ bus./ac.	5.0+ bus./ac.

b Norfolk 1250–1349; FTC counties 1288–1315

c Norfolk 1350–1449; FTC counties 1375–1400

Sources: Norfolk accounts database; FTC1 and FTC2 accounts databases.

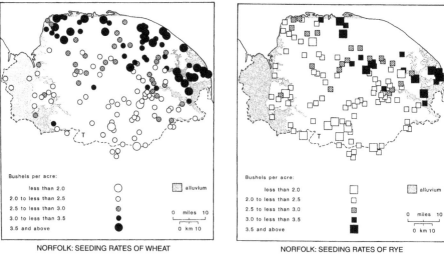

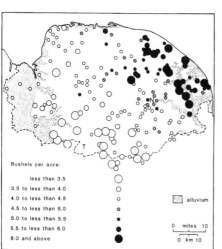

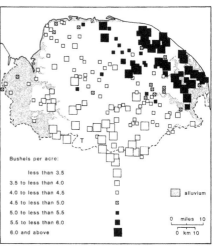

Fig. 7.01. Seeding rates of individual grains: Norfolk, 1250–1449 (N = Norwich; T = Thetford; Y = Yarmouth) (source: Norfolk accounts database).

the spring-sown grains that were curtailed most drastically, those of barley being reduced, on average, by 34 per cent and those of oats by 42 per cent. The net result was a doubling of the proportion of demesnes employing low or very low seeding rates relative to the proportion employing moderate to high ones. This is consistent with the poorer return to be obtained from capital investment in arable husbandry by the final quarter of the fourteenth century,

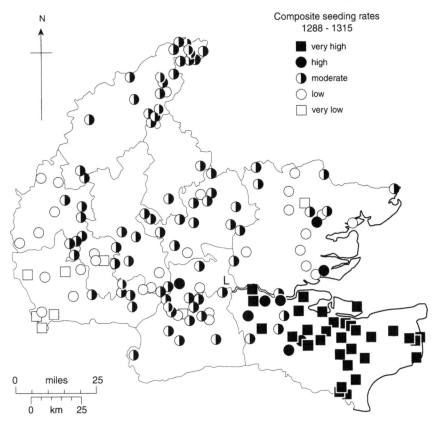

Fig. 7.02. Composite seeding rates: the FTC counties, 1288–1315 (seeding rates are quantified in Table 7.02; L = London) (source: FTC1 accounts database).

especially from production of the cheaper grains whose prices tended to fall relative to those for wheat (Table 5.07).[20]

7.22 Yields per seed

Methods of calculating yield ratios
What physical return could landlords expect from the seed sown on their demesnes? Between 1295 and 1308 the prior of Norwich was sufficiently interested to compile a central record of the annual yield ratio of each crop on each of his demesnes.[21] Many a medieval auditor, equally aware of the significance of this ratio, added a calculation of the yield ratio to the relevant section of the annual manorial account. Often the yield actually obtained was measured

[20] Clark, 'Cost of capital', pp. 268–73.   [21] NRO, DCN R236A; DCN 66.

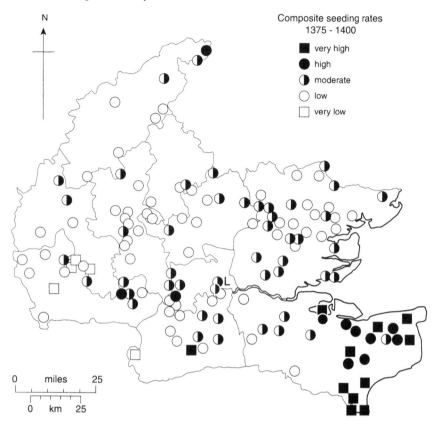

Fig. 7.03. Composite seeding rates: the FTC counties, 1375–1400 (seeding rates are quantified in Table 7.02; L = London) (source: FTC2 accounts database).

against that which could normally be expected and the reeve held personally responsible for any shortfall.[22] Among contemporary agricultural treatises the *Husbandry* (probably written by or for the use of an auditor at the end of the thirteenth century) famously stated the yield ratios that landlords might expect.[23] These were wheat 5.0, rye 7.0, maslin 6.0, barley 8.0, oats 4.0, and dredge 6.0 (no ratios are given for peas and beans). At the beginning of the fourteenth century such yield ratios were regularly exceeded on several well-documented seigniorial estates in Artois, which obtained yield ratios for wheat and oats in the range 7.3–16.0 and 2.6–8.2 respectively.[24] Similarly high yield

[22] J. S. Drew, 'Manorial accounts of St Swithun's Priory, Winchester', in E. M. Carus-Wilson (ed.), *Essays in economic history* (London, 1962), vol. II, pp. 12–30.
[23] *Walter of Henley*, pp. 200–1, 419.
[24] J. M. Richard, 'Thierry d'Hireçon, agriculteur artésien', *Bibliothèque de l'école des chartes* 53 (1892), 383–416, 571–604; B. H. Slicher Van Bath, *The agrarian history of western Europe A.D. 500–1850*, trans. O. Ordish (London, 1963), pp. 175–7.

ratios have also been reported from the Ile-de-Paris and Beauce.[25] Probably nowhere else in northern Europe at this date were the economic incentives to raise the productivity of arable husbandry so great, nor were the results obtained so impressive. Certainly no English landlord could boast of yield ratios to rival those obtained in Artois by Thierry d'Hireçon, possibly because the strains of seed they sowed were intrinsically lower yielding. In fact, few if any English demesnes measured up to the standard of productivity laid down by the *Husbandry*.[26]

Manorial accounts provide all the information necessary for the accurate calculation of yield ratios and many examples have been published.[27] Such ratios hold the great attraction to historians of immunity to the distorting effects of customary measures of area and volume. Normally two consecutive accounts are required, the earlier recording the amount of seed sown, the later recording the amount of grain harvested. Sometimes, however, the medieval auditors have already made the calculation and entered the yield ratio as a marginal note, in which case a single account will suffice. This is of great value when, as is most often the case, surviving series of accounts are discontinuous. Both these methods provide information of the precise harvest of a particular crop in a specific year. When neither consecutive accounts nor auditors' yield calculations are available, the *mean* yield ratio can nevertheless be estimated from discontinuous accounts using the internal evidence of grain harvested one year and seed sown the next, which all accounts record as a matter of course. The results are obviously less reliable since they rely on the assumption that *on average* the amounts sown of each crop varied relatively little from one year to the next.[28] The accuracy of such 'internal' yields improves, as the number of accounts upon which they are based increases. Although an imperfect method it does offer the prospect of estimating yield levels for parts of the country where runs of consecutive accounts are either sparse or non-existent. Here it has been applied to those FTC manors with at least three sampled accounts.

[25] G. Fourquin, *Les Campagnes de la région Parisienne à la fin du Moyen Age* (Paris, 1964); A. Chédeville, *Chartres et ses campagnes* (Paris, 1973), pp. 211–13. I am grateful to Gérard Béaur for these references.

[26] Among the very few that did were several of the Devon demesnes of Tavistock Abbey which, during the fifteenth century, reaped the productivity benefits of a flexible system of convertible husbandry: Finberg, *Tavistock Abbey*, 2nd edition (Cambridge, 1969), pp. 114–15.

[27] R. V. Lennard, 'Statistics of corn yields in medieval England: some critical questions', *Economic History* 3, 11 (1936), 173–92. For a compilation of yield statistics see B. H. Slicher Van Bath, 'The yields of different crops, mainly cereals in relation to the seed c. 810–1820', *Acta Historiae Neerlandica* 2 (Leiden, 1967), 78–97. For the greatest single body of published yield data see Titow, *Winchester yields*.

[28] R. V. Lennard, 'The alleged exhaustion of the soil in medieval England', *Economic Journal* 32 (1922), 12–27; Campbell and others, *Medieval capital*, pp. 39–40. Estimating yields from probate inventories rests on even greater assumptions and for the farms in question is always restricted to the evidence of a single harvest year: Glennie, 'Measuring crop yields'.

Table 7.03 *Frequency distribution of mean gross yields per seed (net and gross of tithe) by crop: Norfolk and the FTC counties, pre- and post-1350*

Gross yield per seed Norfolk:	Wheat		Rye		Barley		Oats		Peas	
	1250–1349	1350–1449	1250–1349	1350–1449	1250–1349	1350–1449	1250–1349	1350–1449	1250–1349	1350–1449
	%	%	%	%	%	%	%	%	%	%
Consecutive and auditors' yields:[a]										
<1	0.0	0.0	0.0	0.0	0.0	0.0	0.0	0.0	0.0	0.0
1–<2	0.0	0.0	0.0	8.7	0.0	0.0	4.0	7.9	15.6	11.1
2–<3	4.3	8.8	11.1	13.0	44.2	31.0	68.0	71.1	57.8	33.3
3–<4	28.3	38.2	61.1	43.5	40.4	66.7	24.0	18.4	22.2	52.8
4–<5	34.8	44.1	22.2	26.1	13.5	2.4	4.0	2.6	4.4	2.8
5–<6	23.9	8.8	5.6	8.7	1.9	0.0	0.0	0.0	0.0	0.0
6–<7	6.5	0.0	0.0	0.0	0.0	0.0	0.0	0.0	0.0	0.0
7+	2.2	0.0	0.0	0.0	0.0	0.0	0.0	0.0	0.0	0.0
No. of demesnes[b]	46	34	36	23	52	42	50	38	45	36
Min.	2.3	2.0	2.4	1.6	2.3	2.3	1.7	1.4	1.3	1.6
Mean	4.6	4.0	3.6	3.4	3.3	3.2	2.6	2.6	2.6	2.9
Max.	7.5	5.2	5.6	5.6	5.9	4.3	4.2	4.0	4.7	4.1
Mean[c]	5.1	4.5	3.9	3.8	3.6	3.6	2.9	2.8	2.9	3.3

FTC counties:	1288–1315	1375–1400	1288–1315	1375–1400	1288–1315	1375–1400	1288–1315	1375–1400	1288–1315	1375–1400
	%	%	%	%	%	%	%	%	%	%
Internal yields:[a]										
<1	0.0	0.0	0.0	0.0	0.0	0.0	0.0	0.0	0.0	0.0
1–<2	6.2	8.7	13.3	0.0	10.5	0.0	33.9	7.0	29.2	32.3
2–<3	44.6	37.0	20.0	40.0	36.8	20.0	52.3	58.1	35.4	22.6
3–<4	33.9	39.1	23.3	40.0	24.6	45.0	10.8	25.6	14.6	25.8
4–<5	7.7	13.0	13.3	20.0	15.8	15.0	1.5	7.0	6.3	9.7
5–<6	7.7	4.4	10.0	0.0	8.8	10.0	1.5	2.3	8.3	0.0
6–<7	0.0	0.0	6.7	0.0	3.5	5.0	0.0	0.0	2.1	0.0
7+	0.0	0.0	13.3	0.0	0.0	5.0	0.0	0.0	4.2	3.2
No. of demesnes[b]	65	46	30	5	57	40	65	43	48	31
Min.	1.4	1.1	1.2	2.1	1.2	2.4	1.1	1.4	1.1	0.7
Mean	3.2	3.2	4.2	3.3	3.3	4.1	2.4	2.9	3.0	2.8
Max.	5.5	5.7	8.5	4.2	6.2	10.0	5.0	5.8	8.3	10.0
Mean[c]	3.5	3.5	4.6	3.7	3.7	4.6	2.6	3.2	3.3	3.1
Auditors' yields:										
Mean[c]	4.1	3.6	4.0	4.0	4.1	4.2	2.6	3.0	4.6	3.1
No. of demesnes[d]	55	72	25	13	46	75	55	79	40	57

Notes:
For definition and calculation of 'consecutive', 'internal yields', and 'auditors' yields' see p. 317.
[a] Net of tithe
[b] Demesnes with a minimum of three recorded harvests
[c] Gross of tithe
[d] All demesnes irrespective of number of recorded harvests
Sources: Norfolk accounts database; FTC1 and FTC2 accounts databases.

Table 7.04 *Frequency distribution of gross yields per seed and per acre (net of tithe) on the estate of the bishopric of Winchester, 1209–1349*

Yield	Wheat		Barley		Oats	
	% of demesnes	% of harvests	% of demesnes	% of harvests	% of demesnes	% of harvests
Gross yield per seed (net of tithe):						
<1	0.0	0.1	0.0	0.3	0.0	1.8
1–<2	0.0	5.4	0.0	5.5	12.2	33.6
2–<3	7.5	21.8	7.3	24.4	80.5	49.7
3–<4	57.5	31.7	63.4	33.4	7.3	12.1
4–<5	27.5	22.9	19.5	21.4	0.0	2.1
5–<6	7.5	11.0	9.8	9.1	0.0	0.5
6–<7	0.0	4.5	0.0	4.0	0.0	0.2
7+	0.0	2.6	0.0	1.8	0.0	0.1
n	40	2,855	41	2,697	41	2,751
Min.	2.61		2.79		1.79	1.79
Mean	3.85		3.77		2.35	2.35
Max.	5.34		5.55		3.40	3.40
Gross yield per acre (net of tithe):						
<8 bus.	12.2	35.0	0.0	7.3	2.4	25.9
8–<12 bus.	78.0	44.6	15.0	25.1	80.5	44.1
12–<16 bus.	9.8	16.1	60.0	31.9	12.2	21.6
16–<20 bus.	0.0	3.6	17.5	20.4	4.9	6.7
20–<24 bus.	0.0	0.4	2.5	8.4	0.0	1.2
24 bus.+	0.0	0.3	5.0	6.8	0.0	0.6
n	41	2,199	40	2,072	41	2,156
Min. (bus.)	5.8		11.0		7.5	
Mean (bus.)	9.6		15.3		10.7	
Max. (bus.)	13.8		27.6		16.0	
Mean net yield per acre (bus. net of tithe):	7.00		11.2		5.7	

Source: J. Z. Titow, *Winchester yields* (Cambridge, 1972), pp. 13–14.

Yields of individual crops

Table 7.03 summarises the mean gross yield ratios of the principal crops obtained on demesnes in Norfolk and the ten FTC counties. The former – mapped in Figures 7.04 and 7.06 – are based exclusively on the consecutive and auditors' methods, while the latter employ the auditors' and internal methods. Table 7.04 gives corresponding yield ratios for the forty-one demesnes of the bishop of Winchester as calculated by Titow from consecutive accounts.[29] This is the most robust of these three independent sets of yield statistics which together span a wide geographical area and diversity of farming systems. Notwithstanding the obvious differences between them there is a remarkable consensus in the overall picture which they convey. In the period before 1350 mean gross yield ratios in the range 2 to less than 5 were characteristic of at least half of all demesnes cultivating rye, two-thirds of all demesnes cultivating wheat and oats, and three-quarters of all demesnes cultivating barley. The overall mean ratios for Norfolk, the FTC counties, and the Winchester estate range between 3.2 and 4.6 for wheat and rye, 3.3 and 3.8 for barley, and 2.4 and 2.6 for oats. These results leave no doubt that most lowland demesnes could normally expect a two-and-a-half- to four-and-a-half-fold return on the seed sown, the return for wheat, rye, and barley usually being significantly better than that for oats. Yields for peas fall into the lower half of this same general range but as such understate the true return since not all accounts record an estimation of the peas fed unthreshed to livestock. This is even more the case with vetches, which were grown almost exclusively as livestock fodder, only enough being threshed to provide seed for sale and for the following year.

On this evidence the ratios recommended by the author of the *Husbandry* were almost wholly unrealistic, at least as far as demesne managers were concerned. While there was some prospect of matching the five-fold target yield for wheat – a feat achieved by a quarter of all Norfolk demesnes – only a tiny handful of demesnes attained the four-fold target yield for oats, while the seven-fold and eight-fold yields for rye and barley remained wholly out of range, except in the most bountiful of years (Figure 7.04). Those landlords who took the *Husbandry* as their guide were therefore doomed to be disappointed. High returns on the quantities of seed sown were not to be expected. Moreover, a grand slam of high yields across all crops was more or less out of the question. Good returns on one crop were often bought at the expense of inferior returns on another, as in the case of wheat versus barley in Norfolk. The more privileged position a crop occupied in the rotation the better it was likely to yield; hence the fact that the winter crops generally out-performed the spring. It was because oats were frequently placed last in rotations that they were the crop which usually yielded worst, a small but significant number of demesnes regularly securing yields of less than two-fold. Even wheat and barley returned yields of less than two-fold on occasion, indicating that partial failure was a periodic hazard. Complete crop failure, in contrast, was an extremely rare event.

After 1350 the general level of yields in Norfolk and the FTC counties remained much the same, except that the winter-sown crops tended to fare a little worse and the spring-sown crops to perform somewhat better. Rye was

---

[29] Titow, *Winchester yields.*

going out of favour as a crop; fewer demesnes cultivated it and it was increasingly confined to the poorest soils. Improved yields are not therefore to be expected. Wheat, on the other hand, remained in demand and was consumed at greater quantities per head by a wider cross-section of society. As the lead crop within most rotations, however, it bore the brunt of reductions in labour inputs and their effect upon standards of soil preparation. Marling and manuring in preparation for the wheat crop became increasingly expensive and in Norfolk, where the retreat from labour-intensive methods was especially pronounced, wheat yields registered a particularly marked decline. Wheat yields held up rather better in the FTC counties, although here too there was a tendency for them to sag. Yields of barley and oats, in contrast, were altogether more resilient, the former crop tending to expand its share of the cropped acreage at the expense of the latter but very much maintaining its yields in the process. Oats, on the other hand (unlike rye), gained from being grown on a smaller scale. As spring-sown crops, barley and oats also benefited from half-year fallowing and the opportunities this afforded for folding sheep upon the arable (a far less laborious method of fertilising the land than the manual spreading of muck). In much of Norfolk and Kent both crops also gained from shorter rotations with fewer courses and more frequent fallowing.

Weighted aggregate grain yields per seed

The divergence in the productivity performance of individual crops highlights the inadequacy of individual crop yields as an effective measure of productivity. Some more comprehensive productivity indicator is required which measures the overall return upon the total quantity of seed sown. Weighted aggregate grain yields (WAGY) provide one such measure, calculated by weighting the yield ratio of each grain crop according to its price relative to wheat and the proportion of the total grain acreage it occupied.[30] Legumes are excluded from the calculation for two reasons. First, because they were often fed unthreshed to livestock their true yield tends to be under-recorded. Second, many demesnes sowed them as a partial substitute for fallows; hence if legumes are included in the equation so too ought fallows. Aggregate crop yields weighted according to each crop's respective share of the total *arable* area (WACY yields) would certainly provide a superior measure of productivity but, unfortunately, the silence of most accounts on the areas fallowed and (in convertible systems) uncultivated means that it can only be calculated for a handful of demesnes (Table 7.08). The great advantage of WAGY, and especially WACY, yields is that they allow direct productivity comparison between demesnes operating very different cropping systems and rotations.

WAGY yield ratios calculated for Norfolk, the FTC counties, and the

[30] Campbell, 'Land, labour, livestock', pp. 165–74; Overton and Campbell, 'Production et productivité', pp. 256–60, 295.

Table 7.05 *Weighted aggregate grain yield (WAGY) per seed (gross of seed, net of tithe) in wheat equivalents: Norfolk, the FTC counties, and the estate of the bishopric of Winchester*

Gross WAGY per seed (net of tithe) in wheat equivalents	Norfolk		FTC counties		Winchester estate
	1250–1349 %	1350–1449 %	1288–1315 %	1375–1400 %	1209–1349 %
<1.50	0.0	7.9	0.0	4.7	2.5
1.50–<1.75	0.0	5.3	0.0	9.3	2.5
1.75–<2.00	11.9	23.7	17.6	7.0	12.5
2.00–<2.25	23.8	36.8	17.6	20.9	27.5
2.25–<2.50	23.8	15.8	21.6	23.3	25.0
2.50–<2.75	16.7	7.9	9.8	9.3	12.5
2.75–<3.00	9.5	2.6	15.7	9.3	7.5
3.00–<3.25	9.5	0.0	5.9	4.7	2.5
3.25+	4.8	0.0	11.8	11.6	7.5
No. of demesnes[a]	42	38	51	43	40
Min.	1.75	1.34	1.76	0.78	1.40
Mean	2.46	2.05	2.54	2.38	2.38
Max.	3.41	2.82	3.27	4.26	3.65
Mean gross of tithe	2.73	2.28	2.82	2.64	2.64

Note:
[a] Demesnes with a minimum of three recorded harvests.
Gross WAGY ratio $(RCg) = \sum(RC^i.P^i/Pw.TCg^i/TCg)$ where RC^i is the yield ratio of grain i, P^i is the price of grain i per bushel, Pw is the price of wheat per bushel, TCg^i is the acreage under grain i, and Tcg is the total area under grains. See Table 7.01.
Sources: Norfolk accounts database; FTC1 and FTC2 accounts databases; J. Z. Titow, *Winchester yields* (Cambridge, 1972); J. Z. Titow, 'Land and population on the bishop of Winchester's estates 1209–1350', PhD thesis, University of Cambridge (1962); Hants. RO, 11M59 B1/38, 43, 45, 53, 58, 76, 97.

Winchester estate are summarised in Table 7.05. Again, the general consistency of productivity levels in all three areas and across a range of different estates and cropping types is confirmed. All three sets of demesnes conform to the same relatively narrow yield range, the highest-yielding demesnes rarely being more than two to two-and-a-half times more productive than the lowest yielding. Before 1350 the mean WAGY yield ratio (gross of seed but net of tithe) of all 121 demesnes (twelve demesnes are common to the FTC and Winchester samples) is 2.4–2.5. Just under a quarter of these demesnes obtained high WAGY yield ratios of 3.0 and above, the maximum ratio being the 3.65 scored by the Winchester demesne of Rimpton in Somerset. A slightly larger proportion obtained low WAGY yield ratios of 2.0 or less, the lowest being the 1.40 scored by the Winchester demesne of Esher in Surrey.

The coexistence of both productivity extremes on the same estate reflects the realities of medieval estate management. Rimpton was one of four demesnes (including the great multi-demesne complex at Taunton) with high WAGY ratios, and Esher one of seven (the other six all being in central Hampshire) with low WAGY ratios. Likewise, in Norfolk the earl of Norfolk obtained a low WAGY ratio from his demesne at Caistor-cum-Markshall just south of Norwich but a high WAGY ratio from his demesne at Forncett, only 8 miles further up the Tas Valley (Figure 7.09).[31] Within the FTC counties the abbey of Westminster and priory of Canterbury similarly owned both high- and low-yielding demesnes. This does not necessarily mean that landlord policy and estate management had little effect upon productivity. Peterborough Abbey had conspicuous success and no obvious failures with its home demesnes in Northamptonshire and of Norwich Cathedral Priory's significant holding of demesnes in Norfolk none yielded poorly and Hemsby, Hindolveston, and North Elmham yielded well. The bishop of Norwich, in contrast, obtained but poor yields from his demesnes of Langham and Eccles, possibly because, unlike the monks of the cathedral priory, he still relied heavily upon customary rather than hired labour to work his demesnes.[32] Nor did the abbey of Bury St Edmunds fare any better with its clutch of demesnes in south Norfolk and north Suffolk: poor yields at Thorpe Abbotts, Tivetshall, and Rickinghall in the first half of the fourteenth century became bad yields in the second half of that century (Figure 7.15). To judge from these examples, standards of estate and demesne management were more likely to mar than to make productivity.

The same was true of the type of cropping system practised.[33] The bishop of Winchester cropped his Somerset demesnes of Rimpton and Taunton in much the same way as his Hampshire demesnes of Cheriton and Sutton, yet the former returned high WAGY ratios and the latter low. A similar productivity gulf separated his demesnes at Brightwell (Berkshire) and Crawley (Hampshire), even though both operated two-course rotations within which they practised cropping type 4.[34] In East Anglia and Kent similar crop combinations occurred in conjunction with far more flexible, irregular, and intensive systems of rotation. Some demesnes cropped in this way yielded impressively, notably Castle Acre Priory's demesne of Kempstone and Lewes Priory's demesne of West Walton in the Norfolk fens. Others – most conspicuously Bury St Edmunds's demesne of Rickinghall in Suffolk and Canterbury

[31] For a case study of Forncett, see Davenport, *Norfolk manor*.
[32] Stone, 'Hired and customary labour'.   [33] For types of cropping system see Chapter 6.
[34] See Chapter 6, pp. 263–4, 286–8, for a description of this cropping type.

Cathedral Priory's demesnes of Little Chart, Loose, and West Farleigh in Kent – did badly.

If cropping type 4 was as likely to deliver high as low WAGY ratios, cropping type 2 seems to have been predisposed towards the latter. This farming type was associated with the cultivation of substantial quantities of the cheaper grains, sometimes but not always on inferior soils. Because grains with low relative values loomed so large, high WAGY ratios were virtually unattainable (Norwich Cathedral Priory's small demesne at North Elmham in Norfolk is the sole high-yielding exception). Sometimes, however, these low WAGY ratios may be the product of using a single set of relative prices across a wide geographical area. In reality, relative prices will have varied with economic rent. Close to major markets, for instance, normal relations between relative prices were often inverted, thereby providing the necessary economic incentive for producers to specialise in these inferior grains.[35] The low WAGY ratios of Ashford (Middlesex) and Battersea (Surrey) are probably exaggerated by this price effect.

Only cropping type 1 was more likely to deliver high than low WAGY ratios. This was the most intensive of all cropping systems and was associated with substantial sowings of legumes as a partial or total substitute for fallows. It was a cropping type closely associated with Norfolk where six of the demesnes that practised it (Brandiston, Forncett, Hemsby, Hindolveston, Hunstanton, and Titchwell), belonging to an assortment of owners, scored high WAGY ratios. Langham – conservatively managed on behalf of the bishop of Norwich – is the sole Norfolk instance of this cropping type delivering a low rather than high WAGY ratio.

It follows that there was no *sine qua non* of high or low yield ratios. Nor can analysis be taken as far as might be wished, since for too few demesnes is a truly representative record of harvests extant. Only a close examination of the evidence can stand any chance of revealing the role played by the level and quality of labour inputs, the physical layout and ownership status of the land (whether consolidated or scattered, enclosed or common), and the character and quality of the soils. Certainly, geographical factors – human and physical – must have exercised some influence for there is a strong spatial dimension to the pattern of WAGY ratios. In Norfolk, for instance, with the conspicuous exception of Forncett, all the highest ratios occurred in the centre, north, and east of the county, whereas the lowest ratios were concentrated in the south and south-west on a variety of light and heavy soils (Figure 7.09). Similarly, there is a pronounced concentration of low-yielding Winchester demesnes in central Hampshire in marked contrast to the high yields obtained by the same estate in the Vale of Taunton in Somerset and the Vale of the White Horse in Berkshire. Within the FTC counties, contrary to locational theory, the

---

[35] Chapter 5, p. 204; Campbell and others, *Medieval capital*, pp. 111–25.

immediate hinterland of London was mostly characterised by moderate to low yield ratios, especially on the heavier soils of west Kent, Middlesex, and Hertfordshire. High-yielding demesnes were mostly located at some remove from the metropolis, the single most notable concentration occurring in and near the Soke of Peterborough in Northamptonshire. It is tempting to attribute these kinds of pattern to the influence of environmental factors, such as soils, but reality was more complex than this and soils alone cannot provide a complete explanation. Rather, it was the response to economic and environmental opportunities by those responsible for organising and undertaking production, within the context provided by prevailing estate structures, that determined the productivity performance of any given demesne.

WAGY ratios in Norfolk and the FTC counties after 1350 endorse these general observations (Table 7.05). Within both areas these ratios confirm that productivity levels tended if anything to sag. In Norfolk no demesne registered a WAGY ratio in excess of 3.0, whereas several continued to register ratios of less than 2.0, three of them possessions of the abbey of Bury St Edmunds on the Norfolk/Suffolk border (Figure 7.15). Within the FTC counties such low ratios similarly remained a feature of a number of demesnes in Hertfordshire, Essex, Middlesex, Surrey, and western Kent. But a loose scatter of demesnes continued to register high ratios in excess of 3.0, with Berkshire again well represented (Brightwalton, Harwell, and Woolstone). Differences in the spatial and estate coverage of the two Norfolk and FTC samples of demesnes pre- and post-1350 make straight comparison between them fraught with difficulties. Nevertheless, in both cases mean WAGY ratios are lower after 1350 than before. The decline is most marked if calculated on the net WAGY ratio (i.e. including tithe but excluding seed). This suggests that in the FTC counties and Norfolk yield ratios were respectively 10 and 26 per cent lower at the end of the fourteenth century than at the beginning. Such a finding is consistent with evidence from tax assessments which indicates that Norfolk was more adversely affected by post-plague economic contraction than the counties within the immediate vicinity of London.[36] Evidently the return on seed was greatest when the demand for grain was likewise at its maximum. As, after 1350, that demand fell, landlords not only sowed less seed but obtained a lower return on it. Output per unit of land suffered accordingly.

7.23 *Yields per unit area*

There are several different land productivities (Table 7.01).[37] The most difficult to measure from the historical record, especially for relatively early periods, is total agricultural output per unit area of farmland. One reason is that com-

[36] Schofield, 'Geographical distribution of wealth'; Darby and others, 'Distribution of wealth'.
[37] Overton, *Agricultural revolution*, pp. 70–4; Overton and Campbell, 'Production et productivité', pp. 256–60, 288–9.

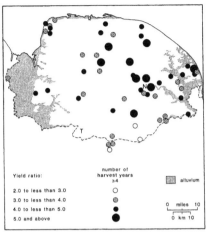

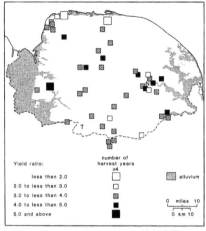

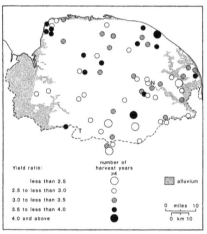

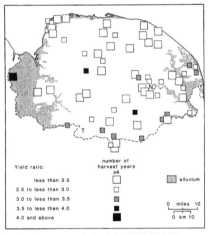

Fig. 7.04. Gross yields per seed of individual grains: Norfolk, 1250–1449 (L = Lynn; N = Norwich; T = Thetford; Y = Yarmouth) (source: Norfolk accounts database).

plete information on the full range of agricultural outputs – field and horticultural crops; hay, straw, grass, and other sources of fodder; timber and wood products; and animals and animal products – is in practice never available. Another is that the total area of all farmland is seldom recorded and often included grazing and other hard-to-quantify rights in common pastures and wastes. A third is that problems attach to the conversion of very different commodities into standard units of measurement. This is most obviously and

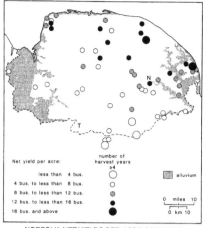

NORFOLK: NET YIELDS PER ACRE OF WHEAT

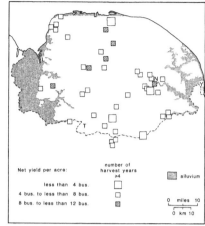

NORFOLK: NET YIELDS PER ACRE OF RYE

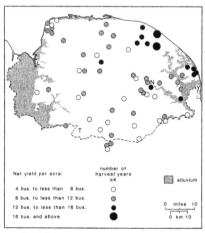

NORFOLK: NET YIELDS PER ACRE OF BARLEY

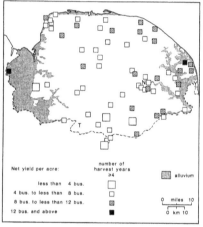

NORFOLK: NET YIELDS PER ACRE OF OATS

Fig. 7.05. Net yields per acre of individual grains: Norfolk, 1250–1449 (L = Lynn; N = Norwich; T = Thetford; Y = Yarmouth) (source: Norfolk accounts database).

easily done using prices, but rarely is a sufficient range of accurate price information available. For these reasons historians generally rely upon more limited measures of land productivity. The most specific, readily available, and widely used are crop yields per unit area devoted to that crop.

Most manorial accounts provide all the information necessary for the calculation of yields per unit area, although some state only the seed and not the area sown (which can, however, be estimated when it is known at what rate seed was sown) and it is rarely clear whether the units used are customary or

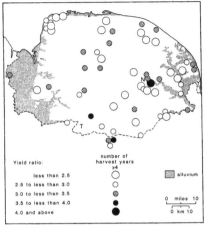

NORFOLK: GROSS YIELD RATIOS OF LEGUMES

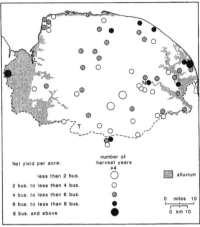

NORFOLK: NET YIELDS PER ACRE OF LEGUMES

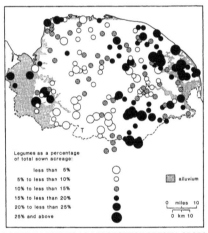

NORFOLK: LEGUMES AS A PROPORTION OF TOTAL CROPS

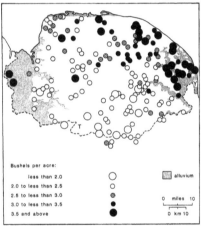

NORFOLK: SEEDING RATES OF LEGUMES

Fig. 7.06. Cultivation, seeding rates, and yields of legumes: Norfolk, 1250–1449
(L = Lynn; N = Norwich; T = Thetford; Y = Yarmouth) (source: Norfolk accounts
database).

statute. Yields per unit area can be measured gross or net of seed. For reasons
demonstrated by E. A. Wrigley, net yields are the truer indicator of land productivity and hence are much to be preferred.[38] Unless otherwise stated, it is
yields net of seed (and also of tithe – which is how most accounts record them)
that are presented and discussed here.

[38] E. A. Wrigley, 'Some reflections on corn yields and prices in pre-industrial economies', in
People, cities and wealth, pp. 94–7.

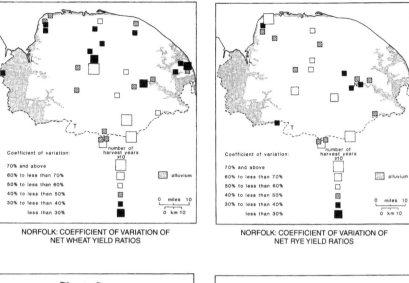

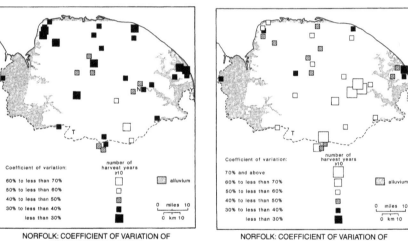

Fig. 7.07. Yield variability of individual grains: Norfolk, 1250–1449 (L = Lynn; N = Norwich; T = Thetford; Y = Yarmouth) (source: Norfolk accounts database).

Yields of individual crops

Tables 7.06 and 7.04 summarise mean net yields per acre for the principal crops in Norfolk, the FTC counties, and on the estate of the bishops of Winchester (the Norfolk yields are mapped in Figures 7.05 and 7.06). It should be noted that they reveal a wider productivity range than that observed for yields per seed, with a four- to five-fold differential between the highest and lowest yields. This is because yields per acre are the product of the yield per

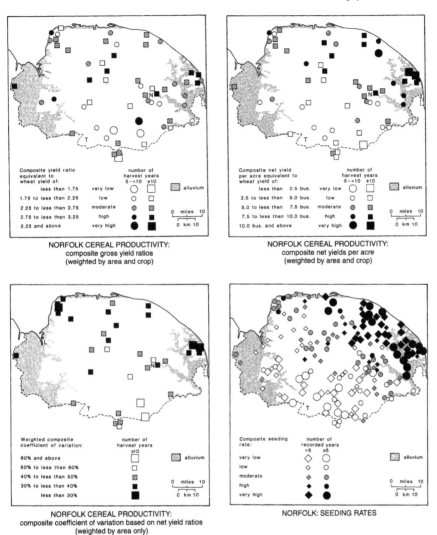

Fig. 7.08. Composite measures of grain productivity: Norfolk, 1250–1449 (L = Lynn; N = Norwich; T = Thetford; Y = Yarmouth) (source: Norfolk accounts database).

seed *and* the seed sown per acre, both of which varied in their own right. On the evidence of these three independent samples of demesnes the highest mean net yields per acre that could realistically be expected (including tithe but excluding seed) were 20 bushels for wheat, and 24 bushels for barley and oats. Such yields were attained by only a handful of demesnes and often depended upon heavy seeding rates. Including rather than excluding seed reveals that the

Table 7.06 *Frequency distribution of mean yields per acre (net of seed and net and gross of tithe): Norfolk and the FTC counties, pre- and post-1350*

Net yield per acre (bushels) Norfolk:	Wheat		Rye		Barley		Oats		Peas	
	1250–1349 %	1350–1449 %	1250–1349 %	1350–1449 %	1250–1349 %	1350–1449 %	1250–1349 %	1350–1449 %	1250–1349 %	1350–1449 %
Consecutive and auditors' yields:[a]										
<4	2.3	9.4	5.9	21.7	0.0	0.0	4.2	13.2	40.0	38.9
4–<8	36.4	50.0	70.6	69.6	30.6	34.1	60.4	52.6	57.8	58.3
8–<12	22.7	34.4	23.5	8.7	42.9	41.5	27.1	31.6	2.2	2.8
12–<16	31.8	6.3	0.0	0.0	22.4	17.1	6.3	2.6	0.0	0.0
16–<20	6.8	0.0	0.0	0.0	4.1	7.3	2.1	0.0	0.0	0.0
20–<24	0.0	0.0	0.0	0.0	0.0	0.0	0.0	0.0	0.0	0.0
24+	0.0	0.0	0.0	0.0	0.0	0.0	0.0	0.0	0.0	0.0
No. of demesnes[b]	44	32	34	23	49	41	48	38	45	36
Min.	3.1	2.1	3.1	1.1	5.2	4.2	3.8	1.6	0.6	1.3
Mean	10.1	8.5	6.5	5.4	10.1	10.2	7.6	6.9	4.2	4.6
Max.	18.6	14.3	9.5	8.9	19.7	18.6	16.4	13.4	9.2	8.7
Mean[c]	11.4	9.9	7.5	6.3	11.8	11.9	8.9	8.6	5.0	5.4

FTC counties:	1288–1315 %	1375–1400 %	1288–1315 %	1375–1400 %	1288–1315 %	1375–1400 %	1288–1315 %	1375–1400 %	1288–1315 %	1375–1400 %
Internal yields:[a]										
<4	26.2	26.1	20.0	40.0	7.0	0.0	23.1	14.0	47.9	51.6
4–<8	50.8	58.7	33.3	40.0	33.3	22.5	43.1	53.5	33.3	38.7
8–<12	20.0	15.2	13.3	20.0	28.1	42.5	18.5	27.9	10.4	6.5
12–<16	1.5	0.0	20.0	0.0	19.3	15.0	12.3	2.3	4.2	0.0
16–<20	1.5	0.0	13.3	0.0	8.8	7.5	0.0	2.3	4.2	3.2
20–<24	0.0	0.0	0.0	0.0	3.5	5.0	3.1	0.0	0.0	0.0
24+	0.0	0.0	0.0	0.0	0.0	7.5	0.0	0.0	0.0	0.0
No. of demesnes[b]	65	46	30	5	57	40	65	43	48	31
Min.	0.9	0.4	0.3	2.4	1.0	5.6	0.8	0.9	0.3	0.0
Mean	6.2	5.5	9.0	5.4	9.8	12.5	7.3	6.8	5.3	4.4
Max.	17.8	11.9	18.9	9.1	20.8	36.0	22.8	20.4	19.8	18.0
Mean[c]	7.2	6.1	10.4	6.0	11.3	13.9	8.6	7.6	6.2	4.9
Auditors' yields:										
Mean[c]	8.8	7.0	8.4	7.5	13.4	13.4	7.4	8.1	10.0	5.6
No. of demesnes[d]	55	72	25	13	46	75	55	79	40	57

Notes:
For definition and calculation of 'consecutive', 'internal yields', and 'auditors' yields' see p. 317.
[a] Net of tithe
[b] Demesnes with a minimum of three recorded harvests
[c] Gross of tithe
[d] All demesnes irrespective of number of recorded harvests
Sources: Norfolk accounts database; FTC1 and FTC2 accounts databases.

most productive medieval English demesnes rivalled the gross yield per unit area of their most productive continental counterparts, with average gross yields of up to 22 hectolitres per hectare for wheat, and 26 hectolitres per hectare for barley and oats.[39] In good harvest years, of course, yields were higher still. This demonstrates the maximum that could be achieved by this class of producer at that time. It anticipated the average performance of English agriculture some five centuries later, following the so-called 'agricultural revolution', when manure-intensive husbandry was taken to its ecological limits. In the fourteenth century it was a standard attained by only a tiny handful of highly distinctive demesnes in eastern Norfolk and northern and eastern Kent, both areas distinguished by exceptionally high levels of economic rent on account of their naturally fertile and easily cultivated soils, high population densities, and ready access to major concentrations of market demand.[40] These were the self-same circumstances that underpinned the high-yielding husbandry of Artois and adjacent parts of Flanders.

Against this maximum, the normal productivity performance of medieval English demesnes was far less impressive. Across the combined sample of over 150 demesnes in the period before 1350 net yields per acre averaged 9.1 bushels for wheat, 12.3 bushels for barley, and 8.0 bushels for oats. After deduction of both seed and tithes, over a third of Norfolk demesnes and three-quarters of FTC demesnes normally obtained less than 8 bushels an acre from wheat, while in both cases two-thirds of demesnes obtained less than 8 bushels an acre from oats and 12 bushels an acre from barley. Gross yields per acre imply a very similar situation on the Winchester estate. On the least productive demesnes lords sometimes had to be satisfied with a net yield of less than 4 bushels an acre from one or more of the crops sown. Poor medieval yields were indeed low and historians should never be surprised by how little the land sometimes yielded. Nevertheless, low yields did not end with the Middle Ages. Estimations of yields from crop valuations contained in probate inventories imply that many a seventeenth-century husbandman cultivated his land to no better effect and it was not until the eighteenth century that the mean and maximum yields of the Middle Ages were decisively exceeded.[41]

Weighted aggregate grain yields per unit area
Useful as such individual crop yields are, they provide only a partial measure of productivity, since the performance of any one crop was contingent upon

---

[39] Cf. A. Derville, 'Le Rendement du blé dans la région lilloise (1285–1541)', *Bulletin de la commission historique du département du nord* 40 (1975–6), 34–6; E. Thoen, *Landbouwekonomie en bevolking in Vlaanderen gedurende de late Middeleeuwen en het begin van de Moderne Tijden* (Ghent, 1988), pp. 818, 1,240–2.

[40] Campbell, 'Agricultural progress'; Campbell, 'Arable productivity'; Campbell, 'Economic rent'.

[41] Glennie, 'Measuring crop yields', pp. 261–6; Campbell and Overton, 'New perspective', pp. 74–5.

Table 7.07 *Weighted aggregate grain yield (WAGY) per acre (net of seed and tithe) in wheat equivalents: Norfolk, the FTC counties, and the estate of the bishopric of Winchester*

Net WAGY per acre (net of tithe) in wheat equivalents (bushels)	Norfolk		FTC counties		Winchester estate
	1250–1349 %	1350–1449 %	1288–1315 %	1375–1400 %	1209–1349 %
<2.5	0.0	8.1	5.9	2.3	2.5
2.5– <5.0	36.6	40.5	45.1	55.8	50.0
5.0– <7.5	31.7	37.8	33.3	30.2	42.5
7.5–<10.0	24.4	13.5	13.7	7.0	2.5
10.0+	7.3	0.0	2.0	4.7	2.5
No. of demesnes[a]	41	37	51	43	40
Min.	2.9	2.2	2.2	0.9	2.0
Mean	6.3	5.3	5.3	5.2	5.1
Max.	12.1	8.9	10.1	12.4	10.0
Mean gross of tithe	7.3	6.2	6.3	6.0	

Note:

[a] Demesnes with a minimum of three recorded harvests.

Net WAGY yield per acre $(YCg^H) = \sum(YCg^{Hi}.P^i/Pw.TCg^i/TCg)$ where YCg^{Hi} is the net yield in bushels per acre of grain i, P^i is the price of grain i per bushel, Pw is the price of wheat per bushel, TCg^i is the acreage under grain i, and TCg is the total area under grains. See Table 7.01.

Sources: Norfolk accounts database; FTC1 and FTC2 accounts databases; J. Z. Titow, *Winchester yields* (Cambridge, 1972); J. Z. Titow, 'Land and population on the bishop of Winchester's estates 1209–1350', PhD thesis, University of Cambridge (1962); Hants. RO, 11M59 B1/38, 43, 45, 53, 58, 76, 97.

its place within the overall cropping system. Weighted aggregate grain yields (WAGY) per unit area of grain – calculated on the same basis as WAGY ratios – are therefore to be preferred as a summary measure of grain productivity. Such yields, calculated for those Norfolk, FTC, and Winchester demesnes with at least three recorded harvests, are summarised in Table 7.07. This shows that the most productive demesnes obtained four to five times more per grain acre than the least productive demesnes.

Top-of-the-range demesnes obtained a WAGY yield per grain acre (net of both tithe and seed) of at least 10 bushels (8.7 hectolitres per hectare). Examples are very few, but in the period before 1350 include a duo of conventual demesnes – Hemsby and Martham belonging to the prior of Norwich Cathedral – in the populous and richly fertile Flegg district of east Norfolk, plus Peterborough Abbey's home demesne of Boroughbury, within a mile of the mother house in the similarly fertile Soke of Peterborough (Figures 7.15 and 7.10).[42] Only rarely did demesnes elsewhere match this exceptional standard of productivity. After 1350 the only sampled demesnes to do so were Canterbury Cathedral Priory's demesne of Elverton in north-eastern Kent (an area sharing many affinities with east Norfolk) and the bishop of Winchester's demesne of Harwell in Berkshire's fertile Vale of the White Horse. At the opposite extreme, Westminster Abbey's demesnes at Kinsbourne and Stevenage on the unrewarding clay-with-flints soils of the Chiltern Hills in Hertfordshire, the royal demesne at Wootton on shallow soils in mid-Oxfordshire, and the bishop of Winchester's demesne at Esher on the light sandy soils of the lower Thames Valley in Surrey, all obtained a WAGY yield per grain acre of less than 2.5 bushels (2.17 hectolitres per hectare) in the period before 1350.[43] Subsequently, equally low levels of productivity prevailed on the abbey of Bury St Edmunds's demesnes at Thorpe Abbotts and Tivetshall on the heavy boulder-clay soils of south Norfolk, on the bishop of Norwich's light-land demesne at Eccles in south-west Norfolk, and on Westminster Abbey's demesne of Pyrford in the infertile Wey Valley of Surrey.

In Norfolk high-yielding demesnes reaped a double productivity dividend and low-yielding demesnes a double productivity penalty. The weighted aggregate coefficient of variation calculated on the gross yield per seed for those demesnes with at least ten recorded harvests and correlated against their WAGY net yield per acre indicates a strong inverse correlation between the two of -0.79.[44] In other words, the higher the WAGY yield per acre, the more reliable the harvest (and vice versa). Indeed, yields to be high had to be sustained. In Norfolk harvests on the highest-yielding demesnes were twice as reliable as those on the lowest-yielding demesnes (Figures 7.07 and 7.08). It is small wonder, therefore, that Norwich Cathedral Priory chose to keep its premier demesnes of Martham and Hemsby in hand until the bitter end, whereas a demesne such as Monks Granges, closer to Norwich but with lower and less reliable yields, was set at farm almost immediately after the Black Death.

The vast majority of demesnes in the lowland east and south of England performed neither as well nor as badly. In fact, in the period before 1350 over 70 per cent of demesnes in Norfolk, 75 per cent in the FTC counties, and 90 per cent of those belonging to the bishop of Winchester obtained WAGY yields (net of seed and tithe) in the range 2.5 to 7.5 bushels per grain acre. Subsequently, these proportions rose to over 75 per cent in Norfolk and 85 per cent in the FTC counties. On average, WAGY yields per grain acre in Norfolk were 15 per cent lower in the period 1350–1449 than they had been during the

---

[42] Campbell, 'Agricultural progress'; Campbell, 'Arable productivity'; Biddick, *The other economy*, pp. 67–72.   [43] For a case study of Kinsbourne see Stern, 'Hertfordshire manor'.
[44] The weighted aggregate coefficient of variation is weighted by relative area alone.

period 1250–1349. They fell from above the national average, at just over 6 bushels, to what appears to have been the average for the east and south, of just over 5 bushels. Corresponding mean yields in the FTC counties held relatively stable (Table 7.07).

High WAGY yields per grain acre were not the preserve of any one cropping type, nor were low. The three spectacularly high-yielding Norfolk demesnes all practised a version of cropping type 1, the most intensive cropping system of all. Some other Norfolk demesnes which operated this cropping type also yielded impressively. The earl of Norfolk's demesnes at Halvergate and Hanworth, Norwich Cathedral Priory's at Hindringham, and the L'Estrange family's demesne at Hunstanton all obtained WAGY yields per grain acre in excess of 8 bushels before 1350. So, too, did Norwich Cathedral Priory's demesne of Hindolveston, which operated a slightly different version of the same cropping type, and its demesnes at Newton-by-Norwich and North Elmham, which operated a version of cropping type 2. The lowest mean yields obtained by any Norfolk demesne employing cropping type 1 were the 5.2 bushels which the bishop of Norwich obtained at Langham, a demesne already identified as performing well below the productivity norm for its locality.[45]

If cropping type 1 does seem to have delivered WAGY yields that were average or better, cropping type 2 provided no such guarantee. At Caister-cum-Markshall and East Wretham in Norfolk and at Brandon just over the county boundary into Suffolk, demesnes practising this cropping type obtained WAGY yields of less than 4 bushels. After 1350 Eccles (Norfolk) and Pyrford (Surrey) got even worse results from the equivalent cropping type, the latter obtaining a pitiful 0.9 bushels per acre (a figure possibly artificially depressed by the years represented). Altogether more securely documented is the 2.0 bushels obtained at Esher in Surrey before 1350 from oats-dominated cropping type 7. It provides probably the most reliable yardstick against which poor WAGY yields ought to be judged. Kinsbourne and Stevenage in Hertfordshire obtained only marginally better results from cropping type 5 (in which oats also featured prominently), as did Wootton in Oxfordshire – a major cultivator of dredge – from cropping type 3.

Of all the cropping types, it was cropping type 4 which appears to have given the most mixed results. This is probably accounted for by the fact that before 1350 it was one of the most widespread of all cropping types and after 1350 no other cropping type was practised by so large a number of sampled demesnes. Low-yielding Thorpe Abbotts and Tivetshall in Norfolk both practised variants of this system, but so, too, at the opposite extreme, did high-yielding Boroughbury (Northamptonshire) before 1350 and Harwell (Berkshire) after 1350. Chartham (Kent) and Upper Heyford (Oxfordshire)

[45] See above p. 324.

also got good results from versions of this system. At this later date the new cropping type 8, with its substantial sowings of nitrifying legumes, also supported high grain yields at Elverton (Kent).

Any correlation between cropping types and land productivity, as measured by WAGY yields, is therefore a fairly loose one. Some cropping systems were more likely to deliver good or bad yields than others, but there were invariably many exceptions. Certainly, it may have been with the aim of attaining a similar productivity result from dissimilar factor endowments that demesne managers varied the crops which they grew and the rotations they employed. The latter, after all, were a means and not an end. The above-average yields commonly obtained from a variety of cropping types by Peterborough Abbey's Soke of Peterborough demesnes are a case in point (Figure 7.10). In fact, the common denominator of demesnes in some localities and regions was often less their cropping system than their land productivity, a characteristic reinforced by the tendency for neighbouring demesnes to employ very similar seeding rates.

In Norfolk, with, on the one hand, its striking contrasts in soil types, market access, and cropping types and, on the other, its relatively high density of well-documented demesnes, spatial trends in land productivity are particularly apparent (Figure 7.09). They reveal themselves in a productivity gradient falling from maxima WAGY net yields per grain acre of 9–12 bushels in north-east and east Norfolk to minima WAGY yields of 2–4 bushels in the south and south-west of the county, a gradient almost as steep as that to be found within lowland England as a whole. Reinforcing this dichotomy in land productivity was an even more pronounced spatial differentiation of seeding rates. Evidently, Norfolk demesnes found it easier to manipulate their seeding rates than their yield ratios in order to achieve high net yields per acre. The earl of Norfolk's demesnes of Earsham and Hanworth, for instance, obtained almost identical and decidedly average WAGY yield ratios of 2.3, which low seeding rates at Earsham and high seeding rates at Hanworth then translated into contrasting WAGY net yields per acre of 4.9 bushels and 8.9 bushels respectively.

The spatial differentiation of arable productivity to be observed within Norfolk (Figure 7.09) mirrors that which existed within lowland England as a whole. In northern Northamptonshire the combination of above-average seeding rates and above-average yields per seed generated WAGY net yields that were consistently moderate or better. The same was true of a clear majority of demesnes in northern and eastern Kent, where seeding rates were mostly exceptionally heavy (Figure 7.10). Some demesnes in the vale country north-east and south-west of Oxford also secured yields that were well above average, as did several on the better soils of the mid-Thames Valley. Away from the commercial artery of the Thames, however, it was low rather than high yields that were very much the norm throughout London's hinterland. None of the sampled Essex demesnes, either before or after 1350, obtained WAGY net yields in excess of 5 bushels and, with one exception, the same was true of the sampled Hertfordshire

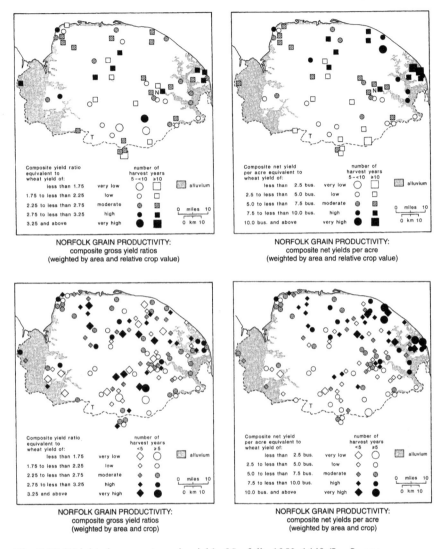

Fig. 7.09. Weighted aggregate grain yields: Norfolk, 1250–1449 (L = Lynn; N = Norwich; T = Thetford; Y = Yarmouth) (source: Norfolk accounts database).

demesnes. WAGY yields of less than 5 bushels also predominated in Middlesex and Surrey and extended eastwards into Kent. In fact, the steep north-east to south-west yield gradient within Kent echoed that of Norfolk.

Outside of these two counties such pronounced productivity contrasts were comparatively unusual. Not one of the twenty-four firmly documented Hampshire demesnes of the bishopric of Winchester secured WAGY yields of less than 2.5 or more than 7.5 bushels, and the great majority secured yields in

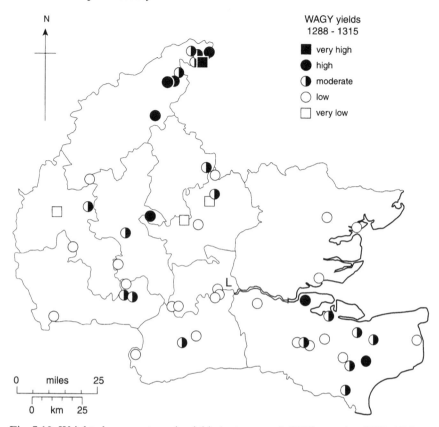

Fig. 7.10. Weighted aggregate grain yields (net per acre): FTC counties, 1288–1315 (L = London) (source: FTC1 accounts database).

the range 4 to 6 bushels (the exceptions are Ecchinswell, on the northern edge of the Hampshire Downs, with 6.8 bushels per grain acre, and Beauworth, Cheriton, and Sutton, on the south-eastern edge of the Downs, with 2.5–3.5 bushels per grain acre). Yields on the bishopric's Wiltshire and Somerset demesnes fall into the same narrow range, notwithstanding the impressive yield ratios secured at Taunton and Rimpton. In their cases, low seeding rates ensured that net yields per acre were no more than moderate. Plainly, yields of this modest order were the norm throughout the greater part of the commonfield arable country of southern England, superior yields being confined to those few rare localities where naturally fertile soils coincided with exceptionally good commercial opportunities.

Weighted aggregate crop yields
Effective though WAGY yields may be as a means of comparing grain yields per grain acre on farms employing very different cropping systems, they nev-

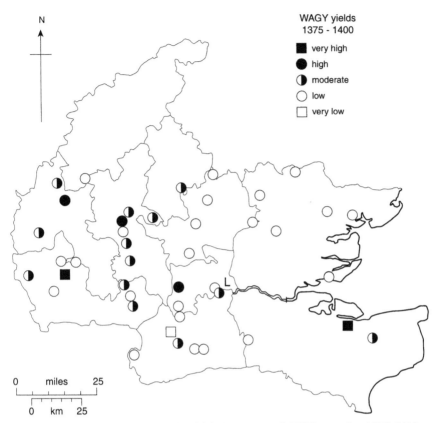

Fig. 7.11. Weighted aggregate grain yields (net per acre): FTC counties, 1375–1400 (L = London) (source: FTC2 accounts database).

ertheless fail to take account of one of the most important determinants of arable productivity, namely the frequency with which land was cropped. At constant yields, the more frequently that land was cropped, the higher its productivity. It might even be worth accepting lower yields per *grain* or *cropped* acre in order to secure higher yields per *arable* acre. Usually that meant replacing fallows with courses of legumes and expending more labour on manuring, ploughing, and weeding. Weighted aggregate crop yields (WACY) per unit area of arable therefore represent a superior measure of the true productivity of the arable sector. Calculating this measure naturally makes more exacting demands upon the evidence since it requires accounts which either record the area fallowed and/or left uncultivated or allow that area to be estimated via the reconstruction of rotations.

Table 7.08 summarises the seeding rates, crop yields, WAGY yields per seed and per grain acre, and WACY yields per seed and per arable acre, calculated for six well-documented demesnes operating contrasting cropping and

Table 7.08 *Alternative measures of arable productivity on six demesnes, c. 1300–49*

				Gross yield per seed			Grain area		Arable area	
Manor	Cropping type	Grain area as % arable	Composite seeding rate	Wheat/(Rye)	Barley	Oats	Relative value	WAGY per seed (gross)	Relative value	WACY per seed (gross)
Index: 100=		50.1%		3.4	2.8	1.9	64.1	2.1	34.8	1.1
Martham	1	**148**	**312**	168	114	147	**123**	152	**200**	241
Cuxham	5	122	175	**188**	**207**	163	119	**214**	144	**261**
Brightwell	4	*100*	200	132	196	142	**123**	177	118	173
Rimpton	5	122	*100*	165	175	**184**	117	187	140	227
Bircham	2	108	190	(135)	129	168	*100*	121	*100*	124
Cheriton	5	*100*	163	*100*	*100*	*100*	110	*100*	108	*100*

				Net yield per acre			Grain area		Arable area	
Manor	Cropping type	Cropped area as % arable	Composite seeding rate	Wheat/(Rye)	Barley	Oats	Relative value	WAGY per acre (net)	Relative value	WACY per acre (net)
Index: 100=		52.7%		5.1 bus.	7.3 bus.	3.8 bus.	64.1	3.7 bus.	34.8	2.0 bus.
Martham	1	**174**	**312**	**353**	211	**437**	**123**	**344**	**200**	**521**
Cuxham	5	127	175	282	271	242	119	288	144	349
Brightwell	4	*100*	200	135	**299**	255	**123**	236	118	229
Rimpton	5	126	*100*	137	162	192	117	144	140	177
Bircham	2	103	190	*(100)*	162	261	*100*	158	*100*	159
Cheriton	5	102	163	104	*100*	*100*	110	*100*	108	*100*

Notes:

Gross and net yields are both net of tithe

Gross WAGY ratio $(RCg) = \sum(RC^i.P^i/P_W.TCg^i/TCg)$ where RC^i is the yield ratio of grain i, P^i is the price of grain i per bushel, P_W is the price of wheat per bushel, TCg^i is the acreage under grain i, and TCg is the total area under grains. Gross WACY ratio $(RC) = \sum(RC^i.P^i/P_W.TC^i/TAr)$ where RC^i is the yield ratio of crop i, P^i is the price of crop i per bushel, P_W is the price of wheat per bushel, TC^i is the acreage under crop i, and TAr is the total arable area. See Table 7.01.

Net WAGY yield per acre $(YCg^H) = \sum(YC^{Hi}.P^i/P_W.TCg^i/TCg)$ where YC^{Hi} is the net yield in bushels per acre of grain i, P^i is the price of grain i per bushel, P_W is the price of wheat per bushel, TCg^i is the acreage under grain i, and TCg is the total area under grains. Net WACY yield per acre $(YC^H) = \sum(YC^{Hi}.P^i/P_W.TC^i/TAr)$ where YC^{Hi} is the net yield in bushels per acre of crop i, P^i is the price of crop i per bushel, P_W is the price of wheat per bushel, TC^i is the acreage under crop i, and TAr is the total arable area. See Table 7.01.

Bold = maximum; *italics* = minimum

Sources: B. M. S. Campbell, 'Arable productivity in medieval England: some evidence from Norfolk', *JEH* 43 (1983), 390–4; C. Thornton, 'The determinants of land productivity on the bishop of Winchester's demesne of Rimpton, 1208 to 1403', in Campbell and Overton (eds.), *Land, labour and livestock*, pp. 191–3; J. Z. Titow, 'Land and population on the bishop of Winchester's estates 1209–1350', PhD thesis, University of Cambridge (1962), pp. 17–18, 67; J. Z. Titow, *Winchester yields* (Cambridge, 1972). National accounts database; FTC1 accounts database; Table 5.07.

rotational systems over the period *c.* 1280 to *c.* 1360.[46] Martham (Norfolk), the only one of the six to have employed virtually continuous cultivation, is an example of cropping type 1 and was one of the most valued possessions of Norwich Cathedral Priory. Bircham, also in Norfolk, is on intrinsically poorer light land and hence was cultivated using a convertible regime whereby the arable was sown for three or four years and then fallowed for an equivalent period (Figure 6.11). It belonged to the de Clares, earls of Gloucester and Hertford, and is an example of cropping type 2. Cuxham in Oxfordshire, a property of Merton College, Oxford, and Rimpton in Somerset, part of the estate of the bishops of Winchester, both operated regular three-course rotations and are examples of cropping type 5.[47] Cheriton on the southern edge of the Hampshire Downs and Brightwell in Berkshire's Vale of the White Horse also belonged to the bishops of Winchester. The former is a third example of cropping type 5 whereas the latter is an example of cropping type 4. Both operated two-course rotations.

How great was the differential between the most and least productive of these six demesnes? As Table 7.08 reveals, the verdict depends to a great extent upon the choice of productivity measure. For instance, on the evidence of the gross yield per seed of individual crops the maximum productivity differential was barely two-fold. WAGY yields calculated on those yield ratios widen the gap slightly to a little more than two-fold and WACY yields, which take account of the frequency of cropping, widen it further to over two-and-a-half-fold. Gross yields per seed nevertheless understate the contrast. Thus, net yields per acre reveal a three- to four-fold difference in the output per unit area of the individual crops, the differential being greatest for oats and least for barley. Collectively, WAGY yields confirm a three-and-a-half-fold difference in the net output of grains. Yet even WAGY yields understate the true productivity gap, which WACY yields per arable acre show to have been in excess of five-fold: twice the maximum differential indicated by WACY gross yields per seed. Underpinning this contrast lay differences in the crops being cultivated, differences in the rates at which seed was sown, and differences in the proportions of the arable that were cropped. Plainly, individual crop yields alone are a very partial and inadequate measure of land productivity.

Nor is this the whole story. How productivity is measured also affects the verdicts passed on the respective performances of individual demesnes. Yield ratios suggest that Cuxham was the most productive, by a narrow margin, of these six demesnes. WAGY yields put it 14 per cent ahead of Rimpton and

---

[46] Christopher Thornton (personal communication) has calculated WACY net yields per acre for the five demesnes of the bishop of Winchester at Taunton in Somerset plus the demesne at Kinsbourne in Hertfordshire belonging to Westminster Abbey. All six of these demesnes operated three-course rotations and employed variants of cropping type 5. Their indexed WACY yields are as follows: Taunton Hull 203; Taunton Holway 202; Taunton Staplegrove 159; Kinsbourne 141; Taunton Nailsbourne 133; Taunton Poundisford 117.

[47] For detailed analysis and discussion of these two demesnes see Harvey, *Oxfordshire village*; *Manorial records of Cuxham*; Thornton, 'Determinants of productivity'.

WACY yields just 8 per cent ahead of Martham. Net yields per acre, on the other hand, leave no doubt of Martham's superiority. Although barley, its main crop, yielded 22 per cent lower than at Cuxham and 29 per cent lower than at Brightwell, its wheat and oats yields were respectively 25 and 67 per cent better than its nearest rivals'. WAGY yields per acre put it 19 per cent ahead of Cuxham and WACY yields widen this advantage to an impressive 49 per cent. Acre for acre, the monks of Norwich obtained half as much again from their demesne arable at Martham as did the fellows of Merton College, Oxford from theirs at Cuxham. Notwithstanding consistently inferior yield ratios, Martham emerges as by far the more productive demesne.

Similarly, whereas yield ratios suggest that three-course Rimpton out-performed two-course Brightwell by a substantial margin, net yields per acre reverse that advantage and put Brightwell 30 per cent ahead of Rimpton. The latter may have sown more of its arable and secured better returns on the seed sown, but Brightwell sowed its arable twice as thickly and thereby obtained significantly higher net yields per acre. Both productivity measures do, however, concur in placing Bircham consistently ahead of Cheriton. Neither had the benefit of good soils, with the result that both regularly sowed only half of their arable. Yet whereas Cheriton employed biennial fallowing, Bircham practised a form of convertible husbandry. Probably it was this, a function of its more flexible field system, which gave Bircham its productivity edge. Its WACY yields per seed and per acre were respectively 24 and 59 per cent higher than those at Cheriton.[48]

On the evidence of these six demesnes, provided that land quality and economic opportunities were propitious, increasing the proportion of the arable under crops, adopting more flexible rotations, and raising seeding rates provided medieval demesne managers with their best prospects of improving land productivity. Improving yield ratios *per se* was harder to achieve and did not necessarily guarantee higher net returns per acre of arable. Any demesne that matched the standard of arable productivity attained and sustained at Martham during the first half of the fourteenth century was doing outstandingly well. The neighbouring demesne of Hemsby, also in the possession of the prior of Norwich, may just have had the edge on Martham, but no more. Both cropped practically their entire arable area and on both the arable was expected to yield an annual net income or 'profit' of at least 28d. an acre.[49]

[48] Cheriton was also out-performed by all five of the Taunton demesnes of the bishop of Winchester and the Westminster Abbey demesne at Kinsbourne in Hertfordshire. These demesnes, like Cheriton, grew mainly wheat and oats but gained part of their productivity edge by doing so in three-course rather than two-course rotations. See above, n. 46.

[49] The arable was actually valued at 36d. per customary acre measured according to a perch of 18½ feet; i.e. each customary acre was equivalent to 1¼ statute acres: BL, Stowe MS 936 fol. 37. At about this time the actual purchase price of small parcels of arable in this fertile and densely populated district often exceeded £2 per acre. In 1284–5 the prior of Norwich paid £1 10s. 0d. per acre at Plumstead, £2 7s. 1d. at Hemsby, and £3 3s. 4d. at Scratby, but only 7s. 6d. at Hindolveston: NRO, DCN 1/1/8.

Valuations as high or higher do occur elsewhere but, on the evidence of *IPM* valuations, are exceptionally rare (Figure 7.12).[50] They are confined to a few equally distinctive localities – the silt fens of Lincolnshire and Norfolk, and the Isle of Thanet and Romney Marsh in Kent. Common denominators in each case include deep, naturally fertile and easily cultivated soils, freedom from communal rotations, an abundant labour force, and convenient access to major markets both locally and at a distance. While it is conceivable that Martham's standard of land productivity may have been exceeded by demesnes in these rival localities, it is inconceivable that it can have been by a significant margin. Certainly West Walton, the one relatively well-documented demesne in the Norfolk Fenland, secured WAGY net yields per acre that were a third lower than Martham's. Nor, on the limited FTC sample of accounts, does the Romney Marsh demesne of Appledore appear to have yielded any better. On the available evidence Martham and Hemsby win the laurels for the most productive arable demesnes currently known in pre-Black Death England.

Outside these few economically and environmentally privileged areas land productivity and arable valuations were both lower. Here there can be little doubt that poor as was Cheriton's productivity performance, there were many demesnes whose land productivity was even lower. Conspicuous among them were those light-land demesnes whose soils repaid only the lightest seeding rates and admitted little more than infrequent cultivation of the hardier grains, particularly rye and oats. Under these environmental constraints low yields per arable acre were practically unavoidable. The bishop of Ely's demesne at Brandon in the Suffolk Breckland and the bishop of Winchester's demesne at Esher in Surrey are good examples of this type of low-yielding demesne.[51] Many upland areas in the north and west imposed equivalent constraints, limiting the range of crops that could be grown and the intensity of their cultivation, as, for example, in the case of the predominantly oats-growing demesnes of Bolton Priory in the Pennine valleys of the Wharfe and Aire.[52] The maximum differential in arable land productivity between the most and least productive demesnes is therefore likely to have been considerably greater than the five-fold differential between Martham and Cheriton. On the poorest soils of west Norfolk, for instance, arable valuations fell to 2d. an acre

[50] In 1915 Gray, *English field systems*, p. 302, drew attention to 'the somewhat numerous Kentish manors on which in the middle of the fourteenth century all the acres of the demesne were sown yearly . . . under these conditions the value of an acre often became 12d.'. Notable examples include the lands of John Crul (1271–2) and Thomas de Morton (1293), and the demesnes at Ivychurch (1308–9), Westgate in Thanet, and Elmstone (both 1309–10) – all valued at 24d. to 29d. an acre: PRO, C132 File 41(16); C133 File 64(14); C134 Files 10(14), 17(7); the demesnes at Flete (1263–4) and Wickham (1330) – valued at 30d. to 35d. an acre: PRO, C132 File 31(1); C135 File 23; and the demesnes at Preston and Overland (1309–10) – both valued at 36d. an acre: PRO, C134 File 17(7).
[51] See Bailey, *A marginal economy?*, pp. 56–65, for an account of Breckland cropping systems.
[52] Kershaw, *Bolton Priory*, pp. 71–8; Miller, 'Farming in northern England'.

or less in contrast to the princely 28d. commanded by Martham.[53] Valuations as low or lower occur in many other parts of the country, especially in the poorer areas of the Welsh borderland and the south-west, where low levels of land productivity were presumably the norm (Figure 7.12).[54]

7.3   Unit land values

Most lords were as aware of the value of their arable as they were interested in the yield of their crop. Yet, whereas calculations of yield were usually entered by auditors in the margins of manorial accounts, estimations of the annual per acre income or 'rent' that could be expected from the arable were recorded in manorial extents.[55] The prior of Norwich was one of the very few landlords to maintain for a period a central record of both the yields and the 'profits' of his demesnes. This interest is first apparent under prior William de Kyrkeby in the 1280s, but it was his successor, the great high-farming landlord Henry de Lakenham, who between 1295 and 1309 took it furthest, initiating the remarkable register known as the *Proficuum maneriorum* as a central record of yields per seed and per acre and the profits obtained on each demesne.[56] Prior William de Claxton discontinued the practice of recording yields but maintained his predecessors' interest in profits or 'wainage', which he continued to have calculated and recorded for each of his demesnes until 1341. For almost fifty years, therefore, the *Proficuum maneriorum* provides an annual record of the anticipated profit per acre of arable of each of the prior's demesnes and the extent to which that was matched.

---

[53] Breckland: 2d. at Weeting in 1302–3, 1.9d. at Mundford in 1327–8, 1.5d. at Santon Downham in 1294–5, 1.3d. at Eriswell in 1308–9, 1d. at Euston in 1271–2, 0.8d. at Icklingham in 1296–7: PRO, C133 File 106 (20); C135 File 1(12); C133 File 69 (11); C134 File 8(11); C132 File 41(20); C133 File 79 (10). Sandlings: 2d. at North Glemham in 1286–7, Hollesley in 1306–7, Holbrook and Nacton in 1309–10, and Benhall in 1322–3; 1d. at Erwarton in 1286–7: PRO, C133 Files 48 (2), 127 (99); C134 Files 17 (2), 76 (11); C132 File 41 (20).

[54] In Shropshire, for instance, per acre arable valuations of ½d. are recorded at Cleobury and Leintwardine in 1304; 1d. at Alcaston, Clun, Meadowley, Shrawardine, and Stapleton in 1301–6; 1½d. at Worfield in 1306; and 2d. at Acton Round, Alveley, Boreton, Brompton, Buildwas, Claverley, Donnington, Ercall Magna, Fitz, Hatton, Isombridge, Kingsnordley, Ness, Rhiston, Stoke upon Tern, Upton, Wattlesborough, and Withington in 1300–9: PRO, C133 Files 114 (8), 104 (21), 103 (14); C134 File 4 (2); C133 Files 121 (14); 104 (21), (19); 119 (8); 98 (26), (30); 106 (14); 121 (15); 116 (15); 111 (11), (13); 93 (13); C134 Files 5 (1), 16 (6), 14 (19); C133 File 103 (14).

[55] These values pertain to an average type of grain (and price) on an average piece of land in a typical year. Allowance was sometimes made for different qualities of land and possibly even for the particular crops sown in the rotation course in the year of the extent. Examples are particularly common in Kent, for instance, Keston and Dachehirst 1296, Kingsdown 1306, Crateforde and Rombergh 1310–11, Graveshend 1314–15 and Egerton 1324–5: PRO, C133 Files 74 (23), 77 (3), 123 (8); C134 Files 21 (8), 37 (7), 83.

[56] NRO, DCN 66/1; DCN R236A. See E. Stone, 'Profit and loss accountancy at Norwich Cathedral Priory', *TRHS*, 5th series, 12 (1962), 25–48, for a discussion of the profit calculations and their significance. Also D. Postles, 'The perception of profit before the leasing of demesnes', *AHR* 34 (1986), 12–28.

Table 7.09 *Frequency distribution of mean value per (sown) acre of arable: England south of the Trent, 1300–49*

Mean value per acre of (sown) arable (d.)	No. of places
<1	5
1– <2	76
2– <3	495
3– <4	829
4– <5	1,148
5– <6	133
6– <7	548
7– <8	28
8– <9	111
9–<10	24
10–<11	24
11–<12	4
12–<15	126
15–<18	10
18–<21	19
21–<24	2
24–<30	22
30–<36	3
36+	6
n	3,613
Mean = 4.64d.	

Source: National *IPM* database.

These arable valuations were not fixed and immutable but were adjusted upwards or downwards as appropriate, according to prevailing price trends. In the 1290s many of these valuations were plainly unrealistically high, with the result that agricultural profits did not always match the level that was anticipated. Significantly, this was much less of a problem at Martham and Hemsby, where the land was valued at a staggering 36d. an acre (28d. per statute acre), than it was on demesnes such as Monks Granges and Taverham (both much closer to Norwich), where valuations were cut from 12d. to 9d. and 8d. respectively.[57] A profit of at least 12d. per sown acre was originally

[57] Taverham and Monks Granges were both valued at 12d. an acre in 1295–6, but this was reduced to 8d. at Taverham in 1300–1 and 9d. at Monks Granges in 1306–7. By 1324–5 both valuations had been raised – to 10d. at Taverham and back to 12d. at Monks Granges – but from 1330–1 at Taverham and 1335–6 at Monks Granges both were lowered once more to a common 8d. an acre: NRO, DCN R236A.

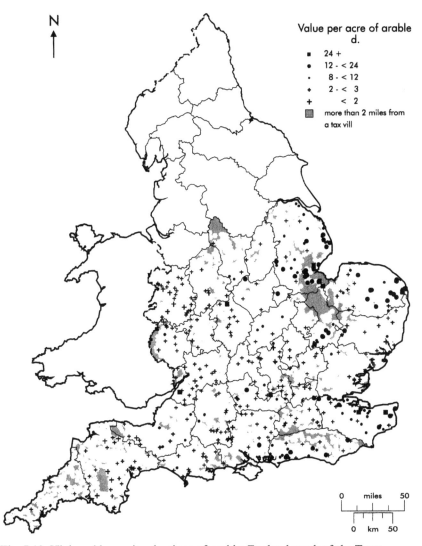

Fig. 7.12. High and low unit valuations of arable: England south of the Trent, 1300–49 (source: National *IPM* database).

expected from twelve of the seventeen demesnes on the estate. Only on the three western demesnes of Sedgeford, Gnatingdon, and Great Cressingham – where yields, population densities, and market opportunities were all much lower – were the values more or less consistently fixed at 8d. an acre.

The profit to be obtained from any given acre of arable was a function of the crop that it yielded times the price of that crop less production costs. It follows that a close association must therefore have existed between the rate at which arable yielded and the value placed upon it. Systematic statistical comparison of yields and valuations confirms such an association. For instance, correlating the value per sown acre against the WAGY net yield per acre on those fourteen of the prior's demesnes with contemporary evidence of both yields a clear positive correlation of + 0.79.[58] The strength of this correlation is all the more remarkable given the approximate nature of the valuations, the evidential shortcomings of the WAGY yields, and the differing unit production costs of these fourteen demesnes, with their contrasting soils, varying ploughing technologies, and differing labour processes. Perhaps that is why sown arable at Martham commanded a value over three times that at Sedgeford even though Martham's WAGY yields were barely 50 per cent better. A positive correlation between the productivity and value of arable was by no means confined to the Norfolk demesnes of Norwich Cathedral Priory. Within the county as a whole a correlation of WAGY net yields per acre on thirty-six demesnes against the mean value per acre of arable at those self-same locations, as estimated from *IPMs*, yields a positive correlation coefficient of + 0.56 (compare Figures 7.09 and 7.13). Extending this exercise to include the fifty demesnes belonging to the bishop of Winchester produces a marginally higher positive correlation coefficient of + 0.58. On this evidence values per acre as recorded by *IPM* extents can be used as a surrogate guide to variations in land productivity.[59]

7.31   *Unit land values as recorded in the* IPMs

Few valuations made for the royal escheators were as scrupulously estimated as those calculated on behalf of Norwich Cathedral Priory; some *IPM* valuations bear all the hallmarks of careful estimation, others are patently much more impressionistic. The strength of the *IPMs* therefore lies less in their precision than in their quantity. They are representative of a far wider cross-section of estates than accounts (albeit all of them in lay ownership) and are incomparably fuller in their geographical coverage, with over 3,750 separate locations represented in the counties south of the Trent. For all their inadequacies, they are the best available *indirect* guide to spatial variations in the productivity of arable in the half-century or so before the Black Death.[60]

Figure 7.13 depicts the mean value per acre of demesne arable in England south of the Trent during the period 1300–49. It is based upon all those *IPM* extents for which the per-acre value of arable either is recorded or can be calculated. When separate values are given for both the sown and the unsown arable only the former has been used. Where several values are available for

---

[58] The demesnes are Eaton, Gateley, Gnatingdon, Hemsby, Hindolveston, Hindringham, Martham, Monks Granges, Newton-by-Norwich, North Elmham, Plumstead, Sedgeford, Taverham, and Thornham. The WAGY yields are calculated from the yields recorded in the *Proficuum maneriorum* (NRO, DCN R236A) augmented by relevant manorial accounts.

[59] On the same reasoning Allen, *Enclosure and the yeoman*, pp. 176–9, has used rents as an index of productivity differences.   [60] Chapter 2, pp. 37–40.

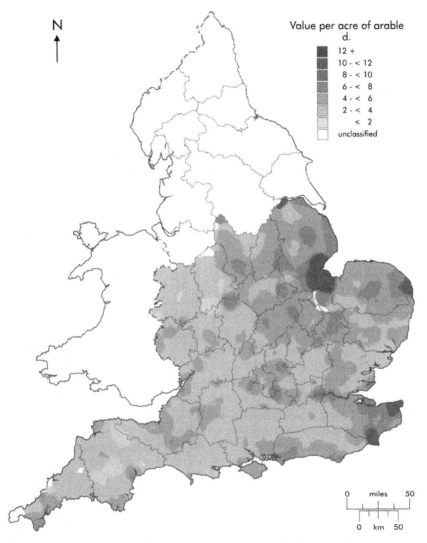

Fig. 7.13. Mean unit value of arable: England south of the Trent, 1300–49 (source: National *IPM* database).

the same location they have been averaged to provide a mean valuation for that location. Each location has been separately grid-referenced and a surface map of mean values has then been generated from these point data using SPANS Map. This allows genuine spatial variations within the data to emerge, irrespective of administrative boundaries.[61]

As will be noted, areas with exceptionally high mean arable valuations of at

---

[61] K. C. Bartley, 'Mapping medieval England', *Mapping Awareness* 10, 6 (1996), 34–6; Campbell and Bartley, *Lay lordship, land and wealth*.

least 12d. an acre were very circumscribed in distribution, occupying less than 2 per cent of the total area south of the Trent.[62] From north to south, they included the rich but low-lying alluvial land at the mouth of the Trent, the silt fens of Lincolnshire and Norfolk (by far the most extensive area of high-valued arable in the country), the Flegg area of east Norfolk (home to high-yielding Martham and Hemsby), and the Isle of Thanet in east, and Romney Marsh in south, Kent.[63] All were in coastal locations on the eastern side of England, several of them served by navigable rivers, and were well placed to exploit commercial opportunities both locally and further afield. All, also enjoyed the blessing of deep, easily worked and intrinsically fertile soils whose manner of cultivation was effectively free from institutional constraint. Undoubtedly, these were the areas where arable land-productivity attained its medieval maximum, being matched by concomitant arable valuations which peaked on a handful of individual manors at 36d. an acre (Figure 7.12).

A further 10 per cent of the country south of the Trent had arable worth at least 6d. but less than 12d. an acre. In a less spectacular way, this too was valuable and productive land. It was characteristic of several localities celebrated for the progressiveness and productivity of their husbandry: the upland margins of the Wash, much of eastern Norfolk, the greater part of north-eastern and eastern Kent (plus a small portion of Essex on the opposite shore of the Thames), and the flat coastal plain of south-west Sussex around Chichester. Again, these were intrinsically fertile localities, populous, and well placed to participate in the coastal and overseas grain trades. Portions of the Thames Valley – probably fourteenth-century England's greatest single trading artery – also commanded correspondingly high valuations: notably, the areas immediately upstream and downstream of London most completely under that city's commercial influence, the hinterland of Henley (probably the foremost inland grain entrepot serving the metropolis), and the Vale of the White Horse in Berkshire, opportunely placed where it could take advantage of both the London and the Oxford markets. The Vale of Aylesbury was similarly valued and also lay on the north-western edge of London's grain-provisioning zone.

All other areas of above-average arable valuations lay beyond regular grain-provisioning range of that city and therefore owed their superior valuations to other stimuli, both local and distant. The most land-locked included the better soils of mid-Norfolk (where manorial structures were also particularly weak

---

[62] In both Norfolk and Kent arable valued at 12d. an acre was described as being capable of being sown every year: PRO, C134 File 83; C135 File 56(1). Raftis, *Assart data*, p. 20, thought that valuations of 10d. or higher resulted from a different system of evaluation; while this was sometimes the case it was clearly not so in these particular instances.

[63] Also in east Norfolk, but outside Flegg, non-*IPM* extents record valuations in the range 24–30d. an acre at Alby, Thwaite, and Guton Hall in Brandiston: NRO, Diocesan Est/2; Magdalen College, Oxford, Estate Papers 130/16.

and fragmented), within easy provisioning range of Norwich, and the country immediately north of Lincoln. In contrast, the navigable rivers Ouse, Nene, Welland, and Trent provided some of the more populous and fertile parts of the commonfield country of the east midlands with a vital commercial lifeline to a wider commercial world. These included south-central Cambridgeshire, the Nene valley of Northamptonshire (the Soke of Peterborough does not show up because it was in almost exclusively ecclesiastical ownership and hence is poorly represented by *IPM*s), eastern Rutland, and the environs of Newark in the mid-Trent Valley. All are distinguished by above-average arable valuations, good land, and real commercial opportunities.[64] Such above-average valuations were not unknown elsewhere, but they tended in the main to show up as minor islands within a sea of lower values and sometimes, as in Derbyshire, probably owed as much to large customary acres as to high levels of productivity.

According to the *IPM*s, arable land south of the Trent had an average value of 4.9d. an acre. Over a third of the country had arable valued within a penny's range of this figure (4d. to less than 6d. an acre).[65] In a majority of cases this was associated with at least some land of above-average value. Most of the south coast from the Isle of Wight east, all but the south-western quadrant of Kent, both shores of the lower and much of the mid-Thames Valley, the vale country east and west of Oxford, western Essex, and the greater parts of Suffolk, Norfolk, Cambridgeshire, Huntingdonshire, north Bedfordshire, Northamptonshire, south Leicestershire and Rutland, Lincolnshire, and eastern Nottinghamshire had arable worth at least 4d. an acre (Figure 7.13). The concentration of such valuations in much of East Anglia and the east midlands is striking and reinforces the yield evidence from Norfolk (Figure 7.09) and the Soke of Peterborough (Figure 7.10) in identifying this as an area of above-average arable productivity, a finding consistent with what is known about the nature and distribution of cropping systems in the region. Significantly, this is the very area where the earliest reliable national yield statistics – the agricultural statistics of the late nineteenth century – show that yields of wheat and barley were highest.[66] In the west and south-west of England mean arable valuations of 4d. an acre or better were more the exception than the rule. They show up, as might be expected, in the more favoured areas: the Avon Valley around Trowbridge in Wiltshire, central Somerset, portions of the mid- and lower Severn Valley, and the lower Exe Valley in Devon. Collectively, just under half of all demesne arable was valued at 4d. an acre

---

[64] Raftis, *Assart data*, p. 21.

[65] This chimes with the request made by the peasants in 1381 that henceforth all rents should be pegged at 4d. an acre: *Peasants' Revolt*, p. 161.

[66] M. Overton, 'Agriculture', in J. Langton and R. J. Morris (eds.), *Atlas of industrialising Britain 1780–1914* (London, 1986), pp. 51–2. This remained the case a century later: Coppock, *Agricultural atlas*, pp. 70–80.

or more; the remainder – over half of the total – was valued at less. Eye-catching as the high values and high yields may be, there is no escaping the fact that low land values and by implication low arable productivity were the norm throughout the greater part of lowland England in the first half of the fourteenth century. This was as true of areas immediately adjacent to London – the kingdom's single greatest concentration of demand for arable produce – as it was of areas far distant from metropolitan influence. Thus, below-average valuations prevailed in much of Middlesex (beyond 10 miles from London), most of Hertfordshire and the neighbouring Chiltern portion of Buckinghamshire, and practically the whole of Surrey except the extreme north-eastern tip of the county on London's immediate doorstep. Low valuations were characteristic of the whole of Wealden Surrey and Sussex, an area notorious for its poor soils and roads, together with the heavy boulder-clay soils of eastern and northern Essex extending into neighbouring Suffolk. Downland Berkshire, Cotswold Oxfordshire, and north Buckinghamshire were similarly valued, which helps to explain why the mean value of arable land in the ten FTC counties was 16 per cent below the national (south of Trent) average.

Heavy clay soils and thin limestone soils alike depressed the productivity and the value of arable, especially when they lacked the bonus of cheap water access to major markets. It was good river communications which helped offset the inherent infertility of the sandy soils of Breckland in south-west Norfolk and north-west Suffolk. Arable here was valued at well below the regional average but on a par with that in many intrinsically more fertile areas. Throughout the greater part of the west midlands and south-central and southern England, however, mean arable values varied between much narrower limits, almost regardless of whether the land in question was wold or vale, coastal or inland. The monotonously low arable valuations which dominated the greater part of southern and western England are both striking and remarkable (Figure 7.13). Significantly, such valuations are characteristic of the core area of the Winchester estate in central and northern Hampshire and southern Wiltshire, which helps to put that estate's generally indifferent productivity performance in clearer perspective.[67]

Just as arable valuations rose to their highest in the extreme east of England, so they fell to their lowest in parts of the far west. Here was to be found that 2 per cent of the country with arable worth less than a paltry 2d. an acre. The localities concerned were almost invariably both land-locked and environmentally constrained. Dartmoor and Exmoor in Devon, neither ever kindly to the cultivator, are conspicuous; so, too, are Cannock Chase in southern Staffordshire and the Long Mynd in central Shropshire. Arable producers in these portions of the north-west midlands clearly laboured under a major

---

[67] Biddick, 'Agrarian productivity', pp. 95–8.

commercial disadvantage, so remote were they from major markets. Low land values are symptomatic of low prices, and low prices, as Farmer has pointed out, of limited commercial opportunities. It is small wonder that in the later Middle Ages this area diversified into pastoral husbandry and turned its low cost of subsistence to economic advantage by using cheap labour to produce manufactured goods for wider markets.[68]

The case of the west midlands serves as a reminder that arable valuations reflect more than land productivity alone. Prices and production costs both exercised a direct influence on the net profit or 'rent' that land was capable of yielding.

7.32 Grain prices and land values

High prices could inflate and low prices deflate the value of land independently of any real difference in physical productivity. Moreover, the spatial range in prices could be considerable. In a low-price region such as the east midlands mean wheat prices were on average 26 per cent below the national average c. 1300, whereas in the high-price region of south Hampshire they were 18 per cent above that average. The range in barley and oats prices was as wide, with the lowest prices again prevailing in the east midlands and the highest in Suffolk and Essex, where barley and oats prices were respectively 19 and 24 per cent above the national average.[69] On Farmer's figures prices were lowest in inland areas at some remove from major markets and, often, with no cheap means of getting their grain to them, as in the case of the east midland counties of Leicestershire and Northamptonshire and the Cotswolds of Oxfordshire and Gloucestershire. More surprisingly, low prices were a feature of Cambridgeshire and the lower Severn Valley, both of which, and especially the latter, enjoyed good riverine communications. Cambridgeshire was certainly involved in the wider grain trade, sending wheat and other grains to King's Lynn, which was probably the single most important grain entrepot on the east coast with well-established trading links northwards to Scotland and eastwards across the North Sea to Flanders and Norway.[70] In the first half of the fourteenth century King's Lynn was regularly used by the royal purveyors as a depot for assembling the provisions required by the king's army in Scotland.[71] Low prices were probably a precondition of participation in that trade since they were necessary to offset high transport costs.

The influence of transport costs upon local prices is clearly to be seen within the FTC counties where, within the wheat-supply zone of London, the

---

[68] Dyer, *Warwickshire farming*; W. H. B. Court, *The rise of the midland industries, 1600–1838* (London, 1938), pp. 33–44.   [69] Farmer, 'Prices and wages', p. 744.

[70] Gras, *The corn market*, pp. 171–6; R. A. Pelham, 'Medieval foreign trade: eastern ports', in Darby (ed.), *Historical geography*, p. 301; Farmer, 'Marketing', pp. 356, 372.

[71] Campbell, 'Ecology versus economics', pp. 81, 95–6.

difference between the local and the London wheat price was commonly the cost of transporting the grain to that city.[72] Nevertheless, there were also areas within these counties where price was generally too high to admit of profitable sale in London and others where prices were so low that they appear to have been beyond the capital's influence. Thus, at one extreme, wheat prices in central and eastern Kent were over 30 per cent higher than the October price prevailing in London, while at the other, prices in northern Oxfordshire and central Northamptonshire were 30 per cent below the London price.[73] This is a considerable differential, of a magnitude sufficient to have had a real impact on relative land values.

Much work remains to be done before it will be possible to reconstruct price surfaces in this period in any degree of detail. The patchiness of the available data will always pose problems but the task is further complicated by the fact that different regions grew different combinations of grains, while individual grains were characterised by different orbits of exchange. What is already clear from the work of N. S. B. Gras and Farmer is that several areas with above-average land values did owe those values as much to superior prices as to superior physical productivity. Such areas included coastal Hampshire and West Sussex, eastern and northern Kent, much of the Thames Valley, Suffolk, and parts of Norfolk. Conversely, low land values in many areas were as much a function of low prices as of low land productivity, as most conspicuously in the case of the Cotswolds and Severn Valley. There were, however, some areas – Cambridgeshire and the east midlands are prime examples – where below-average prices coincided with above-average land values; in these cases there is a strong implication that land productivity was well above average. It can be no coincidence that it was precisely these areas which most attracted the king's purveyors during the first half of the fourteenth century in their constant quest for military provisions.[74] Their attraction apparently lay in an abundance of relatively cheap grain.

7.33 *Production costs and the annual net value of land*

Variations in production costs further complicate this picture. The three principal costs to arable producers were labour, draught power, and capital inputs, notably seed and fertiliser. Their sum total per acre could be considerable. Walter of Henley reckoned that it required at least a three-fold yield to defray the costs of cultivation, on which estimate many arable demesnes can barely have broken even.[75] Whether Walter's estimate is high or low there is no doubt that cultivation costs were several times the annual value of the land that was

---

[72] Campbell and others, *Medieval capital*, pp. 63–9, 193–8.
[73] Campbell and others, *Medieval capital*, pp. 66–7.
[74] Maddicott, *English peasantry*, p. 301; Campbell, 'Ecology versus economics', pp. 94–6.
[75] *Walter of Henley*, pp. 272–3, 325.

being cultivated. The margin between profit and loss was consequently a very narrow one.

Labour

Labour was an intrinsic component of the costs of manuring and marling, ploughing, harrowing, sowing, weeding, reaping and stacking, carting, threshing, and winnowing. Kosminsky has calculated that since the cash value of labour dues was generally only a fraction of the true labour input required to work the land, 'the demesne economy in England was quite insufficiently provided for by the obligatory labour of villeins'.[76] Employment of waged labour in some form was therefore unavoidable. Waged labour, although generally dearer than customary labour, was more efficient and incurred lower policing costs.[77] It came in various forms – casual, seasonal, and permanent – which were remunerated at different rates and employed in different combinations by different demesnes. Labour was exchanged 'for any of a bewildering variety of payments', including money, grain, land, work done by the demesne plough-teams on their tenements, and rent rebates.[78] Budgeting the true cost of these assorted labour inputs is anything but straightforward.

The mid-thirteenth-century *Fleta* reckoned that it cost 25½d. per acre to prepare and harvest a crop, exclusive of the cost of seed.[79] Walter of Henley endorses this figure, on the basis that three ploughings cost 6d. each, one harrowing 1d., weeding ½d., reaping 5d., and carriage 1d. per acre. These figures are for wheat and are chosen to suit his argument.[80] Kosminsky has argued that this exaggerates cultivation costs, claiming that many demesnes could manage with half this level of expenditure.[81] Nevertheless, Walter's estimate is remarkably close to the actual level of expenditure per sown acre on Merton College's manor of Cuxham in Oxfordshire before 1348. Here the demesne was worked by a combination of customary services and permanent waged labour, augmented by additional casual labour as and when required. On average, the services were worth 8.9d. per sown acre, the wages in grain and cash to the *famuli* a further 8.3d., and payments in grain and cash to casual workers 4.8d., bringing total labour costs on this demesne to approximately 22d. per sown acre.[82] As has been seen, Cuxham obtained above-average yields from a regular three-course rotation. On more intensively worked demesnes, with more frequent ploughings, greater attention to manuring and weeding, and a heavier harvest to reap, cart, stack, and thresh, labour costs would have been correspondingly higher.

---

[76] Kosminsky, *Studies*, p. 289.   [77] Stone, 'Hired and customary labour'.
[78] Farmer, 'Prices and wages', p. 760.   [79] Langdon, *Horses*, p. 267.
[80] *Walter of Henley*, p. 325.   [81] Kosminsky, *Studies*, pp. 288–9.
[82] Calculated from Harvey, *Oxfordshire village*, pp. 75–86, 164–71. Grain liveries have been valued using mean national prices for the period 1290–1347 calculated from Farmer, 'Prices and wages', p. 734.

Like prices, wage rates varied regionally. During the first half of the fourteenth century the manors of the bishop of Winchester typically paid 5d. to 8d. for reaping and binding each acre of grain, whereas the manors of Westminster Abbey closest to London paid two or three times that amount. Agricultural wages were consistently higher in the vicinity of London than those in the provinces, and the effect of the Black Death was to increase that difference.[83] As a general rule, wages were highest where prices were highest and vice versa, with the result that they partially offset the tendency of prices to inflate or deflate land values. Pronounced local variations in the rates and forms of remuneration were, however, superimposed on these broader regional trends.

Draught power

After labour, the single greatest production cost was draught power, primarily for ploughing and harrowing but also for haulage. As with customary labour, employing tenant teams and carts helped subsidise demesne production by keeping direct costs down, although at some sacrifice of efficiency. Certainly, the direct costs of maintaining demesne draught animals, ploughs, and carts could be considerable. Animals had to be managed, fed, and shod, ploughs and carts to be repaired and maintained. All three were subject to depreciation and required regular replacement. At Cuxham *c.* 1300, where the demesne employed two plough-teams each of two horses and six oxen to plough almost 360 acres per year, Langdon has estimated that maintenance of the plough beasts cost 8.5d. per sown acre. Maintaining the demesne's two cart-horses cost a further 3.2d. per sown acre, while maintaining the ploughs and carts, will have added a further 1d. to 2d. per acre, bringing the total to at least 13d. per sown acre per year.[84]

Such costs naturally varied a good deal from demesne to demesne, depending upon the type of plough, size and composition of team, speed of ploughing, and physical wear and tear on ploughs and carts.[85] Wheeled ploughs, largely confined in use to Norfolk and the counties of the extreme south-east, were both the fastest and the costliest to maintain.[86] Horses were similarly both faster and costlier than oxen. Animal for animal they were 40 per cent more expensive to maintain.[87] Savings could nevertheless be obtained by using horses to reduce team size. A team of four horses, for instance, cost 30 per cent less to maintain than one of eight oxen. Light two-horse plough-teams, which made their first appearance on demesnes in the closing decades of the fourteenth century, were no less than 65 per cent cheaper than conventional eight-ox teams.

Wheeled ploughs and small, all-horse teams were well suited to light land.

---

[83] Farmer, 'Prices and wages', p. 766.
[84] Langdon, 'Economics of horses', pp. 32–8; Harvey, *Oxfordshire village*, pp. 57–61.
[85] Chapter 4, pp. 120–34.   [86] Langdon, *Horses*, pp. 127–41.
[87] Langdon, 'Economics of horses', p. 37.

Heavy land required larger, more cumbersome teams which worked more slowly. Wear and tear was obviously greater on heavy land than light. As in later centuries, therefore, light-land farmers enjoyed a real cost advantage over their heavy-land counterparts. It was light-land demesnes which earliest appreciated the efficiency gains which could be obtained from substituting the stronger, faster but dearer horse for the slower and cheaper ox.[88] Moreover, the changeover to horses brought compensatory output gains within the pastoral sector, which was partially liberated from an obligation to breed replacement oxen and able to deploy the pastoral resources thereby released to other more productive activities, such as intensive dairying. Norfolk, in particular, was in the vanguard of this development and its better light soils commanded values well above the national average.[89] From the mid-thirteenth century, as horses were progressively substituted for oxen, so plough-teams shrank in size (Figures 4.13 and 4.14). By the second quarter of the fourteenth century light-soil Sedgeford, in the extreme north-east of the county, was using four teams each of only three horses to prepare an annual sown area of 380–440 acres (i.e. 95–110 sown acres per team), whereas heavy-soil Thorpe Abbotts on the boulder clays of south Norfolk was using two mixed-teams of four horses and four oxen augmented with customary ploughings to prepare an annual sown area of 120–75 acres (i.e. 60–87.5 sown acres per team).[90]

Labour costs apart, unit ploughing costs at Thorpe Abbotts were approximately four times those at Sedgeford. Small wonder, therefore, that it was possible to hire out the demesne plough-teams and ploughs at Thorpe Abbotts for 12d. to 14d. a day in the 1340s (those hiring them getting less than an acre's ploughing per plough per day in return for that payment).[91] Such charges obviously bit deep into the net profit that could be made from arable land and ensured that stiff and heavy clay soils were rarely valued in the *IPM*s at more than 6d. an acre. Yet, paradoxically, the stimulus to improve the cost-effectiveness of ploughing appears to have been greatest in those parts of eastern and south-eastern England where land-values were themselves highest, for it was here that the smallest teams generally worked the largest acreages, helped by the selective use of wheeled ploughs and mixed- or all-horse teams. Elsewhere, and especially in the extreme south, south-west and north, large, slow, ox-dominated plough-teams effectively capped potential profits and depressed the value of land.[92]

Seed and fertiliser

After labour and draught power the other main costs involved in arable production were seed and fertiliser. Since seed costs were a direct function of the

[88] Langdon, *Horses*, pp. 100–1, 159–60.
[89] Campbell, 'Towards an agricultural geography', pp. 91–3; Figures 7.12, 7.13.
[90] NRO, L'Estrange IB 1/4, 3/4; WAL 274×6/479–95.   [91] NRO, WAL 274×6/481.
[92] Langdon, *Horses*, pp. 87–90, 110–11, 118–27.

price of grain they obviously helped counterbalance the effect of the latter upon land values. So, too, did seeding rates, for these tended to be heaviest in areas of high grain prices, such as Kent, Sussex, and Norfolk, owing to the stimulus which those prices provided to the adoption of more intensive methods of production. For instance, during the first half of the fourteenth century it would have cost on average 18d. to sow an acre of wheat at a rate of 2 bushels, but 36d. to sow it at a maximum rate of 4 bushels. Small wonder that commercial incentives had to be strong before demesnes would commit themselves to a policy of heavy seeding rates.

High prices similarly provided the justification for substantial expenditure on the purchase of manure and other fertilisers. Most demesnes relied upon on-the-farm sources of fertiliser in this period, typically, the random droppings of browsing animals, the tathe of folded demesne and tenant sheep, farmyard manure, and marl. Its chief cost was the labour required to obtain and apply it. This was greatest in the case of marl (which was not strictly a fertiliser but was applied to improve the soil's structure and its capacity to retain nutrients). In Norfolk marl cost 3s. 0d. per acre to dig and a further 4d. per acre to spread at Hanworth in 1284–5, the two operations together costing 4s. 0d. an acre at Lessingham in 1290–1 and 6s. 2½d. an acre at Calthorpe in 1323–4.[93] Wherever possible, therefore, marling, like manuring, seems to have been undertaken by the *famuli*, as in the case of the 5 acres marled at Ludham in 1355, or as a labour service by the customary tenants, as was regularly the case at Hanworth, with the result that it does not always appear as a specified item in account rolls.[94] It is this comparative silence which makes it difficult to gauge the actual scale on which marling was undertaken. Even at Martham, where marling is fairly regularly recorded during the second half of the thirteenth and first half of the fourteenth centuries, the acreages involved are rarely specified. Only at Hanworth were the acreages involved recorded as a matter of routine: these reveal an average of 5.2 acres marled each year between 1272 and 1306, representing 4.0 per cent of the acreage sown with cereals, at an approximate cost of 1.6d. per grain acre.[95] The fertility-enhancing qualities of marl were widely known in medieval England but high unit costs limited its application.[96]

Cost constraints applied even more to exploitation of such 'external' sources of fertiliser as sea-sand and urban nightsoil (stable manure, street sweepings, and sewage). The first was usually free for the taking, whereas the second was purchased, often at some expense.[97] Both, however, incurred con-

---

[93] PRO, SC 6/936/30; Eton College Records, vol. 49/242; NRO, Case 24, Shelf C.
[94] NRO, Diocesan Est/10.   [95] PRO, SC 6/936/18–32, 6/937/1–10.
[96] Hallam (ed.), *AHEW*, vol. II, pp. 285–7, 323, 346–8, 388, 404, 435–40.
[97] In Cornwall the right to take sea-sand for use as a fertiliser was protected by royal charter, but this right did not extend to Devon, where barges on the Tamar engaged in the transport of sea-sand were charged 12d. a year: Finberg, *Tavistock Abbey*, p. 89.

siderable carriage and spreading costs and hence were only economic where the carriage was inexpensive and either labour cheap or prices dear. The best attested medieval case of the application of sea-sand is provided by the coastal districts of Cornwall and west Devon. In these damp western areas of limited but vital arable land, sand rich in calcium carbonate served the same function as lime, helping to counteract the natural acidity of the soil and improve grain yields.[98] Its cost lay in the labour required to dig and then spread it and the pack horses and barges used to transport it. On Tavistock Abbey's demesne of Werrington a train of six or seven pack horses under the supervision of a sandman was employed for 24 weeks each year throughout the fifteenth century to bring sand the 14 miles from Bude Bay. The sandman alone was paid 14s. 0d. a year in wages and 3½d. per horse was paid to the lord of Woolston for way-leave across his manor. The yields of wheat and oats obtained from the 50 acres sown on average each year were obviously considered adequate repayment for the expenditure involved, which amounted to at least 3¾d. per sown acre, exclusive of the maintenance costs of the horses and labour involved in applying the sand. The Abbey's home demesne of Hurdwick also used sea-sand – bought and brought up the Tamar by the barge-load – to fertilise its arable and obtained even more impressive yields from a sown area that varied between 90 and 200 acres.[99]

Towns, with their concentration of animals and humans, were the single greatest supplementary source of organic fertiliser.[100] Their demand for provisions provided farmers in the immediately surrounding countryside with a powerful incentive to intensify their output. Purchasing urban manure or 'nightsoil' provided one means of achieving this. Such purchases are particularly well documented on a group of demesnes around Norwich. The aptly named Heigham-by-Norwich, for instance, was purchasing manure, presumably from Norwich, as early as 1239–40, when 7s. ½d. was spent on this item.[101] Other purchases of a similar or greater magnitude are subsequently recorded at Catton, Eaton, Lakenham, Monks Granges, Newton-by-Norwich, and Plumstead, all of them possessions of Norwich Cathedral Priory and within a five-mile radius of the city.[102] At Plumstead the provenance of this manure is not in doubt since the account for 1277–8 refers explicitly to *fimis emptis apud Norwycum*.[103] The sums involved were often quite considerable and appear to have reached a peak at the turn of the thirteenth

---

[98] Hatcher, 'Farming: south-western England', p. 388.

[99] Finberg, *Tavistock Abbey*, pp. 89–115; Fox, 'Farming: Devon and Cornwall', pp. 311–12.

[100] On nightsoil and urban waste see E. L. Sabine, 'City cleaning in medieval London', *Speculum* 12 (1937), 19–43; J. C. Tingey, 'The journals of John Dernell and John Boys, carters at the lathes in Norwich', *NA* 15 (1904), 114–63; D. J. Keene, 'Rubbish in medieval towns', in A. R. Hall and H. K. Kenward (eds.), *Environmental archaeology in the urban context* (London, 1982), pp. 26–30; Slicher Van Bath, *Agrarian history*, pp. 256–7.

[101] NRO, Diocesan Est/1.   [102] NRO, DCN 60/4, 8; DCN 61/42; DCN 26, 28, 29.

[103] NRO, DCN 60/29/4.

and fourteenth centuries. At Monks Granges, immediately outside the city's perimeter to the north, expenditure rose progressively from 8d. in 1265, to 6s. 10¾d. in 1268–9, 22s. 9½d. in 1275–6, and reached 35s. 6d. in 1295–6; thereafter, it declined somewhat to a low of 8s. 6d. in 1318–19, recovered in 1320–1 to an all-time maximum of 42s. 1d., but by 1334–5 was down to a mere 6s. 0d. again.[104] A similar chronology was followed at Eaton, to the south-west of the city, expenditure here reaching a peak of 46s. 2d. in 1297–8 only to cease altogether after 1319.[105] At Newton a maximum of 58s. 4d. plus a further 3s. 4d. for the hire of carts was spent in 1299–1300, while at Lakenham manure to the value of £7 3s. 4d. was purchased in 1296, equivalent to an expenditure of 7½d. per sown acre.[106]

Carriage added considerably to the costs of this exercise, as the accounts for Plumstead make clear. Here the manure was first boated down the Yare as far as Postwick – a boatman was a permanent member of the demesne labour force from 1298 to 1343 – and then loaded onto carts for the last mile or so to the demesne. This combined operation usually accounted for between a third and a half of the final cost and for this reason the rate of manuring at Plumstead was significantly lower than on demesnes closer to Norwich, averaging 0.6d. per sown acre 1277–1332, as compared with 1.2d. at Monks Granges 1255–1335, 2.3d. at Newton 1273–1328, and 2.6d. at Eaton 1263–1318 and Catton 1265–1340.[107] Whenever possible the unpleasant and arduous task of actually spreading the manure was undertaken using labour services and at Newton-by-Norwich in 1288–9 and 1301–2 as well as at Eaton in 1291–2 the acreages involved were equivalent to a fifth of the area sown with grain. This was an impressive achievement and represents a significantly higher rate of manuring than was to prevail in the second half of the fourteenth century. Moreover, at its peak, in the high-price years of the late 1290s and early 1320s, the proportion was undoubtedly even greater. Nevertheless, as the lower rate of manuring at Plumstead plainly demonstrates, at a distance of more than 5 miles from Norwich the costs of carriage began to become prohibitive with the result that on the nitrogen-starved soils at Taverham, another Norwich Cathedral Priory manor 10 miles to the north-west of Norwich, no such purchases are recorded.[108]

As the use of nightsoil in the immediate hinterland of Norwich illustrates, purchasing and applying fertiliser was price dependent. In other words, expenditure rose as prices rose, and diminished as prices fell, both through space and over time. Most other labour-intensive activities – manuring, marling, and weeding – were similarly price dependent. In the main it was higher prices which justified heavier seeding rates and the substitution of horses for oxen in

[104] NRO, DCN 60/26/2; L'Estrange IB 4/4; DCN 60/26/9, 17, 18, 24.
[105] NRO, DCN 60/8/10, 18–25.   [106] NRO, DCN 60/28/3, 61/42.
[107] NRO, DCN 60/29/4–23, 60/26/1–24, 60/28/1–5, 60/8/1–17, 60/4/1–37.
[108] NRO, DCN 60/35.

draught work. There was therefore a positive relationship between costs and prices which meant that one tended to compensate for the other in the determination of land values.

Certain other costs were yield dependent. Larger harvests required more labour to reap, stack, and thresh the grain.[109] Heavier net yields per unit area were themselves usually contingent upon thicker seeding rates. Shortening fallows and lengthening rotations raised the work loads of man and beast and, other things being equal, thereby added substantially to unit costs. Because high rates of output were contingent upon high rates of input high-yielding demesnes were not necessarily profitable. Nor, for that matter, were low-productivity systems necessarily loss making, as M. Bailey's analysis of demesne husbandry in Breckland has amply demonstrated.[110] Low-yielding demesnes could generate worthwhile profits provided that unit costs were kept down. Adopting innovative and intensive methods did not always make economic good sense. Yield-dependent costs thus had the opposite effect from price-dependent costs upon the capacity of unit land values to reflect variations in land productivity.

There is also an important distinction to be drawn between production costs gross and net of any consequent efficiency gains (either within the arable sector or elsewhere). Substituting waged for customary labour is a good example of this. Its cost may ostensibly have been higher but it was more flexible and better motivated. The resultant gain in labour productivity may consequently have reduced total unit costs. Similarly, draught horses may have been costlier per beast than oxen, but they were rarely employed in equal numbers and were capable of delivering higher rates of work, faster journey times, and a fuller and more productive exploitation of pastoral resources. Ironically, it was often higher prices which provided the initial incentive to make the changes necessary to secure these gains, with obvious benefits for unit land values.

Plainly, grain prices and production costs interacted with yields in a variety of ways, sometimes inflating and sometimes deflating the effect of the latter upon unit land values. The net outcome in large measure determined the market rent which the land commanded. This helps explain why the crown's escheators were more interested in recording the financial than the physical yields of the land.[111] Prior Henry de Lakenham was unusual among landlords in taking the trouble to calculate and record both the profit and yield of his arable. He had a good reason for doing so: the precise combination of yields, prices, and production costs prevailing on his estate meant that he could anticipate a healthy profit of at least 8d. an acre from all seventeen of his properties. Yet such a

[109] Clark, 'Labour productivity', pp. 221–31.   [110] Bailey, *A marginal economy?*
[111] Raftis, *Assart data*, pp. 16–17. Raftis cites Oschinsky's observation that 'By value was then understood the annual income from the estate not its capital value': *Walter of Henley*, p. 69.

high rate of return was characteristic of only 4 per cent of England south of the Trent. A further 45 per cent of the country could enjoy a moderate return of 4d. to 8d. an acre but fully 51 per cent of the country could expect no more than a low return of less than 4d. an acre. On this evidence the vast majority of English demesnes in the period 1300–49 were low profit-making and by implication low yielding.

These respective proportions representative of high, moderate, and low rates of return are remarkably close to those predicted by Thünen's land-use model of a closed economy dependent upon animate sources of power and engaged in provisioning a single central city. In such an economy Thünen reckoned that 6 per cent of the arable area would be devoted to the most productive forms of manure-intensive arable husbandry, a further 54 per cent to moderately intensive and productive regimes, and 40 per cent to low-yielding, extensive cropping systems.[112] Artificial though this model may be it suggests that the predominantly low productivity performance of fourteenth-century English demesnes was an essentially rational response to prevailing levels of economic rent which themselves were consistent with the small size of the capital and limited extent of its provisioning hinterland. Across large parts of the country the economic incentives do not yet appear to have been strong enough to have encouraged demesne managers to adopt the measures necessary to secure higher levels of land productivity.[113] When, in the post-medieval centuries, the productivity of arable land eventually rose, that rise was principally achieved by improving the performance of these low-productivity regimes under the twin stimuli of higher prices and lower capital costs so that productivity differentials were progressively narrowed. This in turn was contingent upon a trade-based expansion and concentration of market demand.[114]

7.4   Temporal trends

7.41   Harvest reliability

Medieval harvests were prone to considerable fluctuation from year to year. The natural unpredictability of the weather, periodic depredations of plant pests and diseases, and the ever-present competition of weeds, all ensured that no harvest could be counted upon until it was safely gathered in. On Norfolk demesnes the annual volatility of yields is reflected in mean coefficients of variation calculated on the net yield per seed in the range 36.6 to 51.9 for the principal grain crops (Table 7.10). It is striking that the most dependable crop,

[112] Campbell and others, *Medieval capital*, pp. 5–7.   [113] Campbell, 'Economic rent'.

[114] E. A. Wrigley, 'A simple model of London's importance in changing English society and economy, 1650–1750', *PP* 37 (1967), 44–70; Wrigley, 'Urban growth'; Overton and Campbell, 'Productivity change', pp. 41–2.

namely barley, is that which was grown on the largest scale by the greatest number of demesnes. This is no coincidence: improving the overall reliability of yields was part of the art of raising productivity.[115] Some demesne managers chose to do this by concentrating upon producing those crops which experience showed performed most consistently. Others preferred to diversify their cropping and thereby spread the risk of a poor return. This was one of the merits of dividing the arable between winter- and spring-sown crops, since it was exceptionally rare for both to fare equally badly (in that respect the famine years 1316 and 1317 were a statistical freak).[116] It has been suggested that the scattering of strips in the open fields was another widely employed form of risk aversion.[117] On large estates, spreading production across many different demesnes must have served much the same purpose.

Practical steps could also be taken to reduce hazards by improving drainage, constructing flood defences, containing weed growth, and scaring off pests. Investment in storage similarly allowed surpluses from one year to be carried over as a hedge against shortages the next. The vast and often architecturally ambitious barns erected by many landlords became one of the status symbols of the age.[118] Nevertheless, it has been doubted whether anything more than a fraction of the harvest was ever stored from year to year.[119] Some soils, environments, and locations were more susceptible to risks than others. In Norfolk a handful of fortunate and exceptionally well-managed demesnes enjoyed weighted aggregate coefficients of variation of less than 30.0, whereas corresponding coefficients on those demesnes which yielded least reliably were twice as great (Table 7.11). Most of the latter were on the cold, heavy, boulder-clay soils of the south of the county and several were under the unenlightened management of the abbey of Bury St Edmunds (Figures 7.07 and 7.08). Similar contrasts in harvest reliability are apparent on the demesnes of the bishopric of Winchester.

Grain prices provide the best year-by-year record of harvest fluctuations, with the qualification that they reflect conditions of demand as well as supply. In a well-known study of the period 1480–1619 W. G. Hoskins demonstrated that one harvest in every four could be reckoned as a failure in some degree, while a really bad harvest could be expected every six or seven years.[120] Cold winters were especially likely to trigger poor harvests but so, too, were

[115] See above, p. 336.

[116] Kershaw, 'The Great Famine'; G. H. Dury, 'Crop failures on the Winchester manors, 1232–1349', *TIBG*, new series, 9 (1984), 407; Jordan, *The Great Famine*, pp. 7–39.

[117] McCloskey, 'Open fields'; S. Fenoaltea, 'Transaction costs, Whig history, and the common fields', *Politics and Society* 16 (1988), 171–240.

[118] Brady, 'The gothic barn'.

[119] McCloskey and Nash, 'Corn at interest'; K. G. Persson, 'The seven lean years, elasticity traps, and intervention in grain markets in pre-industrial Europe', *EcHR* 49 (1996), 698–702.

[120] W. G. Hoskins, 'Harvest fluctuations and English economic history, 1480–1610', *AHR* 12 (1964), 28–43.

Table 7.10 *Weighted aggregate coefficient of variation (net yield per seed): Norfolk, 1250–1449 (demesnes with a minimum of ten recorded harvests)*

	Yield per seed (net of seed and tithe)			
	Demesnes		Harvests	
Crop	n	Coefficient of variation	n	Coefficient of variation
Wheat	34	44.4	838	43.6
Rye	28	49.7	673	47.7
Barley	34	36.6	987	36.7
Oats	34	51.1	951	51.9

Notes:
Method: Weighted by share of total grain area.
Source: Norfolk accounts database.

recurrent outbreaks of rust and other parasitic infestations of grain crops.[121] Unfortunately, this did not mean that bad harvests could be forecast with sufficient accuracy to forestall periodic food crises. For Titow outstandingly bad harvests were those in which yields were at least 15 per cent below average. In the late thirteenth and early fourteenth centuries on the incomplete evidence of the Winchester Pipe Rolls such harvests occurred in 1283, 1290, 1310, 1315, 1316, 1339, 1343, 1346, 1349, and 1350. Outstandingly good harvests, in contrast, occurred in 1287, 1298, 1309, 1311, 1313, 1318, 1325, 1326, 1332, 1337, 1338, and 1344.[122]

The effect of plenty and scarcity had, of course, a very different impact on large producers and small.[123] For the bishops of Winchester, with over 8,000 acres under crop, the depressed grain prices of a glutted market made 1288 one of the least profitable years of the thirteenth century whereas the famine prices of 1317 brought bumper revenues from grain sales notwithstanding the meagreness of yields.[124] For their customary tenants, however, the heriots payable whenever holdings changed hands from death or transfer show that where 1288 passed without tremor 1317 represented the climax of a famine-induced crisis with an unprecedented turnover in holdings.[125] This pattern is

---

[121] S. Scott, S. R. Duncan, and C. J. Duncan, 'The origins, interactions and causes of the cycles in grain prices in England, 1450–1812', *AHR* 46 (1998), 1–14.

[122] J. Z. Titow, 'Evidence of weather in the account rolls of the bishopric of Winchester 1209–1350', *EcHR* 12 (1959), 360–407. Cf. Bailey, 'Peasant welfare', p. 235.

[123] Overton, *Agricultural revolution*, pp. 20–1.

[124] Titow, 'Land and population', p. 10. London cornmongers similarly appear to have profited from the unrestrained workings of market forces: Campbell and others, *Medieval capital*, pp. 82–4.

[125] M. M. Postan and J. Z. Titow (with statistical Notes by J. Longden), 'Heriots and prices on Winchester manors', *EcHR* 11 (1958), 392–417.

Table 7.11 *Frequency distribution of weighted aggregate coefficient of variation: Norfolk, 1250–1449 (demesnes with a minimum of ten recorded harvests)*

Weighted aggregate coefficient of variation (net yield per seed)	No. of demesnes %
<30	8.8
30–<40	41.2
40–<50	32.4
50–<60	11.8
60+	5.9
No. of demesnes	34
Min.	22.1
Mean	41.5
Max.	64.5

Source: Norfolk accounts database.

even more apparent in densely crowded east Norfolk where, on the manor of Hakeford Hall in Coltishall, many small-holders were only able to weather deficient harvests by selling off land, a sacrifice which they vainly endeavoured to reverse whenever favourable harvests allowed them to enter the land market as purchasers.[126] These divergent responses to harvest fluctuations serve as a reminder that it was far more in the interests of subsistence producers to maximise their productivity than it was of the lords of vast estates. The former had almost everything to gain from over-production, the latter much to lose.

7.42 Trends in unit land values

To what extent were these pronounced annual fluctuations in output superimposed upon broader temporal trends in productivity? The want, with certain notable exceptions, of consistently detailed data available over the very long run means that it is impossible to return more than a partial and very qualified verdict. By far the largest and geographically most comprehensive body of evidence representative of the widest cross-section of estates derives from the *IPM*s and relates to the changing unit value of arable land. Other things being equal, any change in the productivity of land will be reflected in its value. The problem with the *IPM*s is that the recorded valuations from which trends must

---

[126] Campbell, 'Population pressure', pp. 110–20.

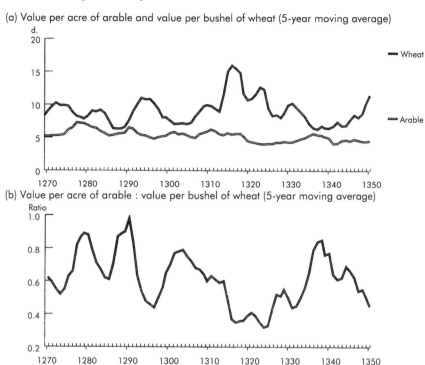

(a) Value per acre of arable and value per bushel of wheat (5-year moving average)

(b) Value per acre of arable : value per bushel of wheat (5-year moving average)

Fig. 7.14. Trends in the mean unit value of demesne arable: fifteen eastern counties, 1270–1349 (source: National and FTC1 *IPM* databases).

be constructed are discontinuous in both time and space. For any given year valuations are available only for certain locations; in subsequent years the locations with equivalent data are different. Reconstructing statistically robust trends from such data is methodologically exceptionally challenging. Simple moving averages nevertheless provide a crude guide to the temporal variations present within the data.

Figure 7.14a illustrates the changing mean value per acre of sown arable over the eighty-year period 1270–1349 within a fifteen-county area comprising Bedfordshire, Berkshire, Buckinghamshire, Cambridgeshire, Essex, Hertfordshire, Huntingdonshire, Kent, Lincolnshire, Middlesex, Norfolk, Northamptonshire, Oxfordshire, Suffolk, and Surrey. As will be noted, arable peaked in value in 1278–9 at over 7d. an acre and registered its lowest value in 1323–4 at barely 4d. an acre. Dividing the value of arable by the corresponding five-year moving average of the price of wheat (with a one-year lag to accommodate the fact that *IPM* valuations reflect recent rather than current values and prices) controls for changing monetary values and at the same time converts those values into their equivalents in physical units of grain. The resultant chronology, as Figure 7.14b shows, is altogether more eventful and sharply etched, with a three-fold differential between the maximum and minimum values (obtained in 1291 and 1324 respectively) and a coefficient of variation of 27.9 per cent. Episodes of good harvests with their low prices and bad harvests with their high stand out clearly but in both cases are superimposed upon an essentially static long-term trend.

There is nothing here to suggest a long-term trend in arable productivity towards either deterioration or improvement during the half-century or so before the Black Death. Levels of physical productivity implied by the valuations may have fallen more or less continuously for the first quarter of the fourteenth century, sinking to an eighty-year minimum at the time of the Great European Famine and ensuing harvest failures of the early 1320s, but in the second quarter of the century that trend was arrested and reversed and by the late 1330s rates of output had returned to levels commensurate with those registered in the late 1270s, late 1280s, and early 1300s. Renewed bad weather coupled with monetary problems and punitive taxation drove productivity down again in the 1340s. The agrarian economy was therefore already under duress when the Black Death struck.[127] What is surprising is that productivity levels should have held up so well during the period from the Great Famine to the Black Death, for this stands out as the most prolonged episode of unstable climatic conditions during the entire second millennium.[128]

Arable valuations relate in the main to the sown area only and hence do not reveal whether lords were raising or lowering the productivity of their arable in any sustained and systematic way by altering the proportion of it under crop. Yet this was one of the most effective ways of producing more from the land, and in the post-medieval centuries was to make a major contribution to the progressive rise in English arable productivity.[129] In 1333 Podimore in Somerset, a property of Glastonbury Abbey, converted from a two- to a three-course system of cropping and thereby brought an extra 16 per cent of its arable under crop. Such conversions were once thought to have been widespread during the thirteenth century but, as Fox has convincingly demonstrated, they were more the exception than the rule.[130] Far more common than the full-scale changeover from one system to the other was the occasional cultivation of *inhoks* from the fallow or pasture, typically with oats or legumes.[131] Outside the area of regular commonfield systems, where cultivators enjoyed greater freedom to modify rotations, temporary and permanent reductions in

---

[127] Maddicott, *English peasantry*; Campbell, 'Population pressure', pp. 116–19; Ormrod, 'Crown and economy', pp. 181–3.

[128] Baillie, 'Dendrochronology provides a background', pp. 107–8; Chapter 1, pp. 22–3.

[129] Overton and Campbell, 'Production et productivité', pp. 273–7, 290, 292.

[130] Fox, 'Alleged transformation to three-field systems'.

[131] E.g. Ravensdale, *Liable to floods*, pp. 116–18; Hallam, 'Farming: southern England', pp. 344–5; Harrison, 'Field systems', pp. 7–12.

fallowing were easier to achieve. Unfortunately, although examples of all these developments can be cited their collective contribution to productivity trends is impossible to quantify. For the cropped area alone can long-term productivity trends be measured and even then only for those demesnes, estates, and localities well represented in the long run by extant accounts.

7.43 Trends in yields of individual crops

No series of manorial accounts is longer and more complete than that for the estates of the bishopric of Winchester. For a period of almost 250 years individual crop yields are precisely documented on up to fifty manors in southern England. The record is by no means continuous, especially for the first half of the thirteenth century, nor are all demesnes consistently represented. Problems of customary measures also mean that temporal trends must mostly be assessed using yields per seed rather than per acre. Nevertheless, the sheer quantity and duration of the Winchester yield statistics place them at the centre of debate about medieval arable productivity.[132]

Table 7.12 summarises the broad trends in yields returned by the principal grain crops – wheat, barley, and oats – averaged across all the Winchester manors from 1225 until 1453.[133] The figures are meaned for successive periods of at least twenty-five years' duration to minimise the impact of individual exceptionally good or bad harvests. The three crops were grown on differing scales, occupied different positions in rotations, and display long-term trends which differ from each other in several significant respects. The trends of wheat and barley correlate most closely ($R = +0.72$). Surprisingly, the correlation between barley and oats is significantly weaker ($R = +0.49$), notwithstanding that both were largely spring sown. Between wheat and oats there is no correlation at all ($R = -0.13$).

The highest mean yields of wheat and barley were obtained in the earliest period. Yields of oats before 1250 were similarly impressive, but were bettered after 1380. In the second half of the thirteenth century yields of all three crops registered a progressive decline, yields of wheat falling least and those of barley most. After 1270 the decline in yields was accompanied by a contraction in the area under crop, which was reduced by over a third across the estate as a whole during the course of the next fifty years.[134] Titow has argued that the bishops were taking the poorest land out of cultivation – much of it sown with oats – and has reasoned that, other things being equal, this should have resulted in an improvement in mean yields as cropping was concentrated onto

[132] W. Beveridge, 'The yield and price of corn in the Middle Ages', *Economic History* (a supplement of *The Economic Journal*) 1 (1927), 155–67; Titow, *Winchester yields*; Farmer, 'Grain yields on Winchester manors'; Thornton, 'Determinants of productivity'.

[133] Titow, *Winchester yields*; Farmer, 'Grain yields on Winchester manors'.

[134] Titow, 'Land and population', pp. 21–2.

Table 7.12 *Trends in mean gross yields per seed on the estates of the bishopric of Winchester and abbey of Westminster, 1225–1453*

| | Indexed mean gross yield per seed | | | | | |
| | Wheat | | Barley | | Oats | |
Years	Winchester	Westminster	Winchester	Westminster	Winchester	Westminster
100=	3.8	3.3	3.3	3.6	2.2	2.4
1225–1249	109		144		123	
1250–1274	103		124		116	
1275–1299[a]	100	100	100	100	100	100
1300–1324	104	87	110	105	101	90
1325–1349	106	91	115	121	103	107
1350–1380	98	87	109	110	111	108
1381–1410	103	99	127	110	134	116
1411–1453	98		112	114	139	

Notes:

[a] Westminster = 1271–1299

Source: B. M. S. Campbell, 'Land, labour, livestock, and productivity trends in English seignorial agriculture, 1208–1450', in Campbell and Overton (eds.), *Land, labour and livestock*, p. 161.

the better soils. The fact that the opposite occurred is in his view evidence that all soils were being progressively sapped of their fertility. Yet if this was the case it is hard to explain why grain yields should thereafter have rallied, the improvement being most marked in the case of barley and least in the case of oats (whose yield should have benefited most from the contraction of cultivation).

In the second half of the fourteenth century yields of oats improved more markedly and during the first half of the fifteenth century gave better returns than ever before. By this period their yield was heavily dependent upon the size of the sheep flocks available for folding.[135] Wheat, in contrast, generally fared worse after 1349 than it had done before. As the lead crop in rotations it should have benefited most from higher stocking densities and increased manure supplies, but the decades following the Black Death brought inferior yields and although there was some recovery in the closing decades of the fourteenth century that recovery was not sustained into the fifteenth century. Barley yields followed the same trend, with the difference that recovery between 1381 and 1410 and decline thereafter were both more pronounced (Table 7.12).

These Winchester yields can be compared with those obtained on the estates of Westminster Abbey over the 135-year period 1275–1410 (Table 7.12).[136] The Westminster demesnes were geographically less focused in distribution, with major groupings both in the vicinity of London (in Middlesex, Essex, and Hertfordshire) and in the west midlands (on the Gloucestershire/Warwickshire border). Their mean yields therefore encompass a greater diversity of experience. They are also less firmly documented than those for Winchester since there are many more gaps in the runs of accounts and more inconsistencies in the representation of individual demesnes. The less complete and focused nature of this yield series may account for the fact that long-term variations in the yields of the individual grains are statistically much more independent of one another than on the Winchester estates. Unlike Winchester there is no correlation between the yields of wheat and barley ($R = -0.19$). Instead, there is a weak association between the yields of wheat and oats ($R = +0.40$), which were both sown in quantity on many of the Westminster demesnes. The closest correlation is, however, between the two spring-sown grains, barley and oats ($R = +0.60$), although even this is by no means strong.

Comparing Westminster with Winchester reveals that barley yields alone followed a roughly similar chronology on both estates ($R = +0.72$), displaying a modest upward trend throughout the period 1275–1349, falling back slightly in the period 1350–80, but then rallying during the decades 1381–1410 (although this late recovery is less marked on the Westminster than the

[135] M. J. Stephenson, 'The productivity of medieval sheep on the great estates, 1100–1500', PhD thesis, University of Cambridge (1987), pp. 176–87.
[136] Farmer, 'Grain yields on Westminster Abbey manors'.

Winchester demesnes). As on the Winchester estates, wheat yields on the Westminster demesnes tended to be slightly better in the second quarter of the fourteenth century than the first, declined somewhat in the period 1350–80, but then improved again in the final decades of the fourteenth century. On both estates oats yields also tended to improve throughout the fourteenth century. Otherwise there is little correlation between their respective yields of wheat and oats. For both estates, the years 1381–1410 were a good period, especially for barley and oats, and arguably the best for over a hundred years; the weather was more favourable, cultivation had been withdrawn from the poorest land, soil nitrogen was benefiting from expanded sowings of legumes, and higher stocking densities were ensuring a better supply of manure.[137] So propitious were these circumstances for a recovery in yields that it is remarkable the recovery was not stronger. Presumably the sharply rising cost of labour in conjunction with depressed land values and prices acted as a powerful countervailing influence (Figure 1.01).

These estate-focused series of yields can be compared with a county series reconstructed for Norfolk (Table 7.13). This county lacks the long and relatively complete individual runs of accounts which enable trends in yields to be reconstructed with the same precision as on the Winchester estate. Even the single best Norfolk series of manorial accounts, that for the prior of Norwich's manor of Sedgeford, is poor by Winchester standards. Over the 176-year period 1256–1431 there are only eighty-seven years with full accounts, many in very poor condition, plus a further nineteen when summary information is recorded in the *Proficuum maneriorum*.[138] More typical are many short or intermittent runs of accounts. Collectively these comprise a substantial if fragmented volume of data amounting to some 1,085 individually recorded harvests from 121 different demesnes over a 185-year period. These data are remarkably representative of the county as a whole and the various cropping types within it and hence make up in geographical focus what they may lack in chronological precision (Figures 7.04 and 7.05).

In Norfolk, rye and oats – the two hardiest crops often grown interchangeably – are the only grains whose trends in yields are highly correlated ($R = +0.89$). As on the Westminster estates, there is also some correlation between the yields of barley and oats ($R = +0.57$). Between these two spring-sown grains and wheat, however, there is no significant correlation ($R = +0.14$ and $+0.04$). Until 1349 mean yields per seed of all three grains displayed a gradual but steady improvement; between 1350 and 1375 all three also registered a setback; thereafter they followed increasingly different paths (Table 7.13). Wheat yields recovered very slightly between 1375 and 1424 but then declined once more. Barley yields recovered strongly and attained a recorded maximum

---

[137] For case studies see Thornton, 'Determinants of productivity'; Stern, 'Hertfordshire manor'.
[138] NRO, DCN 60/33/1–31, 62/1–2; L'Estrange IB 1/4, 3/4, 4/4; DCN R236A.

Table 7.13 Indexed seeding rates, gross yields per seed, gross yields per acre, and gross weighted aggregate grain yields (WAGY): Norfolk, c. 1250–1854

	Wheat			Barley			Oats			WAGY	
	Seeding rate (bus./ac.)	Yield per seed	Yield per acre (bus.)	Seeding rate (bus./ac.)	Yield per seed	Yield per acre (bus.)	Seeding rate (bus./ac.)	Yield per seed	Yield per acre (bus.)	Yield per seed	Yield per acre (bus.)
Index: 100 =	2.8	4.6	14.9	4.5	3.1	15.8	5.8	2.4	13.8	2.7	10.3
pre-1275	109	84	89	99	104	99	103	99	98	92	88
1275–1299	100	100	100	100	100	100	100	100	100	100	100
1300–1324	97	105	100	97	106	102	90	109	96	106	107
1325–1349	99	109	105	101	110	109	95	116	109	114	119
1350–1374	92	86	77	98	101	97	87	99	86	93	82
1375–1399	100	90	87	96	117	109	87	117	101	99	97
1400–1424	95	91	85	93	105	94	86	119	101	91	79
1425–1449	97	82	72	97	105	97	90	123	105	97	86
1584–1599			79			74			112		80
1628–1640			116			75			133		91
1660–1679			86			88			95		80
1680–1709			99			97			145		83
1710–1739			113			139			191		125
1836			156			203			263		201
1854			201			241			333		248

Note:
WAGY = weighted aggregate grain yield. Gross WAGY yield per seed $(RCg) = \sum (RC^i.P^i/Pw.TCg^i/TCg)$ where RC^i is the yield ratio of grain i, P^i is the price of grain i per bushel, Pw is the price of wheat per bushel, TCg^i is the acreage under grain i, and TCg is the total area under grains. Gross WAGY yield per acre $(YCg) = \sum (YCg^i.P^i/Pw.TCg^i/TCg)$ where YCg^i is the net yield in bushels per acre of grain i, P^i is the price of grain i per bushel, Pw is the price of wheat per bushel, TCg^i is the acreage under grain i, and TCg is the total area under grains. See Table 7.01.

Sources: Norfolk accounts database; B. M. S. Campbell, 'Land, labour, livestock, and productivity trends in English seignorial agriculture, 1208–1450', in Campbell and Overton (eds.), Land, labour and livestock, pp. 161, 171.

between 1375 and 1399 before levelling off at close to their average for the period as a whole. Oats yields also recovered strongly after 1375 but unlike wheat and barley continued to make good this gain so that, as on the Winchester estates, higher yields were being returned in the first half of the fifteenth century than at any previously recorded time. As the crop most likely to be sown at the tail end of rotations oats benefited most from the reversion to less intensive systems of cropping.[139]

Trends in the yields of wheat and barley reconstructed for Norfolk correlate well with those recorded on the Winchester estates ($R = +0.69$ for wheat and $+0.91$ for barley). This lends considerable confidence to the Norfolk series, while the broad oscillations which they both display point to the role of such over-arching factors as climatic variation and the reduction in labour supply that arose from the Black Death. Between Norfolk and the Westminster estates the correlations are much weaker ($R = +0.09$ for wheat and $+0.62$ for barley), thereby casting further doubt on the reliability of the Westminster series. Barley alone displays much the same trend in all three series. With oats, although there are some common denominators – notably a marked tendency for yields to be better at the end of the fourteenth century than they had been at the end of the thirteenth – there is no observable statistical correlation between the three series.

7.44 Trends in weighted aggregate grain yields

For Norfolk, unlike the Winchester and Westminster estates, trends in WAGY yields can be calculated. These highlight the productivity watershed represented by the Black Death (Table 7.13). Gross WAGY yields per seed and per acre both rose progressively during the eighty years or so prior to the Black Death, the fortuitous run of good harvests in the 1330s lifting yields to a late final peak. This trend was buoyed up in part by a modest improvement in the prices of oats, dredge, and barley relative to wheat (Table 5.07). By the 1330s and 1340s a bushel of Norfolk malted barley was worth 95 per cent of the value of a bushel of wheat. At least as important was a progressive increase in the intensity of husbandry. By the first half of the fourteenth century fodder cropping, multiple ploughings, systematic manuring, folding and marling, regular weeding, and meticulous harvesting of the crop had become the norm on demesnes in the most economically developed and populous parts of the county.[140] The economic viability of such methods depended upon the combination of high grain prices and cheap labour, preconditions which prevailed until factor costs were transformed by the massive demographic haemorrhage of the Black Death. Few parts of England experienced greater mortality in 1348–9 than Norfolk and the county's population continued to dwindle for the

---

[139] Chapter 6, pp. 286, 290-2, 299.   [140] Campbell, 'Agricultural progress'.

remainder of the fourteenth century.[141] First, labour became scarcer and dearer, then, from the late 1370s, prices fell, those of barley (the county's leading crop) and oats falling most. By the close of the century lower relative prices had knocked a fifth off the unit value of the county's grain acreage (Table 5.07). Demesne managers responded by reducing the intensity of their husbandry as labour became increasingly dear relative to land.

Over the second half of the fourteenth century Norfolk's mean gross WAGY yield per seed declined by 18 per cent. Lower seeding rates magnified this into a 30 per cent reduction in gross output per acre (Table 7.13). Net WAGY yields per seed and per acre fell even more, by 30 per cent and 34 per cent respectively. Many individual demesnes registered far greater declines, particularly those whose high yields were most dependent upon lavish labour inputs (Figure 7.15). On Norwich Cathedral Priory's hitherto top-performing demesne of Martham, for instance, the mean net WAGY yield per acre was 41 per cent lower at the beginning of the fifteenth century than it had been at the beginning of the four-teenth (Table 7.14). Underpinning that decline was a 21 per cent reduction in the number of man-days per cropped acre worked by the demesne's permanent staff of farm servants and an even greater curtailment in the employment of casual labour. Since more land was being left fallow each year the overall reduc-tion in labour inputs per *arable* acre was probably at least 30 per cent.[142]

At Hindolveston, another of the cathedral priory's demesnes, the produc-tivity fall was even more marked and the net WAGY yield per acre was virtu-ally halved over the same period (Table 7.14).[143] Here, too, intensive methods had raised yields to impressive levels by the opening of the fourteenth century, although neither soils nor location were as favourable as at Martham. At Plumstead and Sedgeford, two other priory demesnes, the productivity decline was more modest, in part because yields were lower and had less scope to fall.[144] Moreover, at Sedgeford the fall was partially cushioned by a substan-tial reduction in the sown area as leys were lengthened and cropping was con-centrated onto the better soils. On this light-soil demesne the reduction in yields was reasonably evenly spread across all the principal crops. At Martham, Hindolveston, and Plumstead, in contrast, wheat bore the brunt of the decline. It was invariably the lead crop in rotations and thereby received the maximum benefit from ploughings and manurings; when these were reduced as husbandry became less intensive its harvest was disproportionately affected. Wheat's loss was, however, oats' gain. Sown as the last course when fertility was most depleted, the hardiest and least demanding of the grains responded most positively to the retreat from intensive husbandry. Apart from barley at Plumstead, it was the only crop to benefit in this way.

[141] Shrewsbury, *Bubonic plague*, pp. 94–9; Campbell, 'Population pressure', pp. 95–101.
[142] Campbell, 'Agricultural progress', pp. 38–9.
[143] NRO, L'Estrange IB 4/4; DCN 60/18/7–14, 53–61; DCN R236A.
[144] NRO, DCN 60/29/4–14, 40–46; DCN R236A; L'Estrange IB 3/4, 4/4; DCN 60/33/6–13, 31.

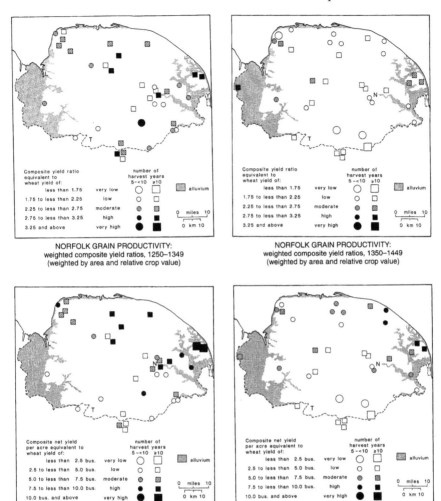

Fig. 7.15. Weighted aggregate grain yields in Norfolk, pre- and post-1350 (L = Lynn; N = Norwich; T = Thetford; Y = Yarmouth) (source: Norfolk accounts database).

Table 7.14 Changes in arable productivity on five demesnes of Norwich Cathedral Priory between 1300–24 and 1400–24

Demesne	National cropping type[a] (with second choice)		Change in sown area %	Percentage change between 1300–24 and 1400–24 in mean net yield per acre						Net WAGY yield per acre 1300–24	
	pre-1350	post-1349		Wheat	Rye	Barley	Oats	Legumes	WAGY	bus.	n
Martham	1(2)	1(4)	−9	−44		−13	+26	−46	−41	10.2	16
Plumstead	1(2)	1(4)	−9	−34	−13	+11	+11	−23	−31	7.3	11
Hindolveston	1(4)	1(4)	−20	−35		−40	+2	−15	−48	9.1	14
Sedgeford	2(1)	1(4)	−30	−12	−16	−17	−12	±0	−25	7.2	16
Taverham	2(1)	1(4)	−37		+9	+76	+71	−55	+16	4.5	14
All			−21	−32	−7	−5	+18	−28	−31	7.7	

Notes:
[a] For definition of national cropping types see Tables 6.01 and 6.02.
Source: B. M. S. Campbell, 'Land, labour, livestock, and productivity trends in English seignorial agriculture, 1208–1450', in Campbell and Overton (eds.), Land, labour and livestock, p. 175.

All four of these demesnes were located at the intensive margin of cultivation and paid a high productivity price when these intensive methods ceased to be economically viable. A fifth priory demesne at Taverham provides an instructive exception to this rule.[145] It applied a cropping system comparable to that at Sedgeford to lighter and poorer soils. In the early fourteenth century it was the least productive of these five demesnes and yet by the early fifteenth century was the only one to have improved its yields, especially those of barley and oats (Table 7.14). Significantly, this improvement occurred in conjunction with a 37 per cent contraction in the cropped area and was almost certainly the result of withdrawing cultivation from the least productive soils. Mean yields rose because production was now concentrated onto the better soils. A similar development can be observed on the bishop of Ely's demesne at Brandon on the Norfolk/Suffolk border in the heart of sandy and infertile Breckland.[146] Here, a 50 per cent reduction in the cropped area between 1340 and 1390 was accompanied by a 45 per cent rise in aggregate productivity, as the yields of rye, barley, and oats all increased substantially.

Taverham and Brandon illustrate productivity trends at the extensive margin of cultivation. Not only did mean yields rise as tillage was withdrawn to the more rewarding soils but a greater share of the reduced acreage was devoted to the more demanding crops with high relative values. This is much as Ricardo predicted and as Postan, Titow, and Farmer have argued, although in neither instance is there anything to suggest that soil exhaustion was a factor in the equation.[147] Low yields before 1350 merely reflected the fact that at the high tide of medieval demographic and economic expansion it was worthwhile to cultivate intrinsically low-yielding soils, to the detriment, in some cases only, of mean yields.

Norfolk was a remarkably dichotomous county, containing the extremes of soil quality and land productivity. In the fourteenth century, however, it was the most densely populated county in the country and stood in the van of many economic developments. It was the experience of the intensive margin of cultivation which therefore dominated mean productivity trends within the county as a whole. This meant that population density and arable land productivity were for the most part positively correlated, just as they were in much of northern France and the Low Countries over the same period.[148] The net

---

[145] NRO, L'Estrange IB 4/4; DCN 60/35/8–13, 43–52; DCN R236A.

[146] Chicago UL, Bacon Roll 644–59; PRO, SC 6/1304/22, 24, 25, 27, 29, 31, 33–5; Elveden Hall, Suffolk, Iveagh Collection 148 (Phillipps 26523).

[147] Chapter 1, pp. 16–20. R. B. Outhwaite, 'Progress and backwardness in English agriculture, 1500–1650', *EcHR* 39 (1986), 1–18, has proposed a similar scenario for the sixteenth century.

[148] H. Van der Wee, 'Introduction – the agricultural development of the Low Countries as revealed by the tithe and rent statistics, 1250–1800', in Van der Wee and Van Cauwenberghe (eds.), *Productivity and agricultural innovation*, pp. 1–23; E. Le Roy Ladurie, 'The end of the Middle Ages: the work of Guy Bois and Hugues Neveux', in E. Le Roy Ladurie and J. Goy, *Tithe and agrarian history from the fourteenth to the nineteenth centuries* (Cambridge, 1982), pp. 71–92.

result, as H. Neveux found in the Cambrésis region of France, was that as population declined, labour inputs were reduced, and productivity levels subsided, so the range of yield differentials narrowed.[149] Productivity at the lower end of the yield spectrum was lifted as the least fertile land was withdrawn from cultivation, while productivity at the top was lowered by a reversion to less intensive methods.

If arable-land productivity in the more intensively farmed areas fell with population in the century or so after the Black Death did it recover with renewed population growth in the sixteenth century? A qualified answer can be obtained by matching the direct evidence of yields provided by manorial accounts against estimated gross yields per acre calculated from the valuations of standing crops contained in probate inventories (since the latter lack information on seeding rates, net yields and yields per seed cannot be estimated). Table 7.13 summarises such a yield series reconstructed for Norfolk. Because the medieval yields relate solely to substantial demesnes, mean yields for the period 1584–1739 have been estimated from probate inventories for farms with at least 20 sown acres. To maximise comparability, county means have been derived using the same system of regional weighting as that applied to the manorial accounts. Gross WAGY yields per acre have also been calculated from the inventories using information on crop proportions and prices internal to them. Finally, corresponding yield statistics – individual and WAGY – derived from the 1836 tithe files and the 1854 agricultural statistics assembled by the Poor Law inspectors provide a nineteenth-century benchmark for comparison.[150]

On the evidence of Table 7.13, neither the high wheat and barley yields nor the high WAGY yields per acre of the first half of the fourteenth century were significantly bettered until the early eighteenth century. This medieval ceiling was breached first by rye in the 1660s and then by oats in the 1680s, but not by wheat and barley (the two leading crops) until after 1710. By early modern standards there was therefore nothing inferior about demesne yield levels in the period 1265–1450. On the contrary, rising yields in the period 1584–1640 represent a recovery to an essentially medieval level of productivity, no doubt using methods of land management which would have been familiar to many a fourteenth-century demesne manager. Dearer land, cheaper labour, and rising prices once more provided the incentive. When, after 1660, that incentive weakened as population pressure slackened, prices stagnated, and wages rose, yields again fell away (a trend reinforced by worsening climatic conditions). An essentially Boserupian relationship thus predominated between population and arable productivity in Norfolk from at least the thirteenth to the eighteenth centuries.[151] There were, however, important deviations from

---

[149] Le Roy Ladurie, 'End of the Middle Ages', p. 85.
[150] Campbell and Overton, 'New perspective', pp. 66–76.   [151] Boserup, *Conditions*.

this trend, exemplified by Taverham and Brandon. These suggest that at the extensive margin of cultivation a more Ricardian relationship may sometimes have prevailed.[152]

Norfolk's early-fourteenth-century yields, like its unit land values (Figure 7.13), were high by national standards and not decisively bettered until four centuries later. Elsewhere medieval yields were lower and hence earlier exceeded. In Hampshire the mean yields obtained on the twenty-four demesnes of the bishopric of Winchester had been matched by the first half of the seventeenth century. By implication the same applied in Hertfordshire and Oxfordshire where yields began to rise several decades before an equivalent trend declared itself in Norfolk.[153] In the Low Countries, too, the early-fourteenth-century productivity peak was exceeded in the seventeenth century, although it was not until the late eighteenth century that the same applied in northern France.[154]

There was a strong spatial dimension to the long-term trend of yields whose dynamic was provided in part by the changing level and configuration of market-determined economic rent and in part by the related diffusion of the technology which made higher productivity attainable. By the early fourteenth century, Norfolk, for a combination of reasons (its superior population density, precocious market involvement, and natural environmental and locational advantages), had established a marked productivity lead over much of the rest of the country, as manifest in the unit value of its arable (Figure 7.13). Over the next five centuries that lead was progressively eroded as the spatial focus of population and economic activity shifted and market expansion coupled with an increase in urbanisation promoted wider and more intense regional specialisation.[155] As different regions became increasingly drawn into this burgeoning commercial nexus, so more farmers intensified and innovated, and thereby secured higher yields. The pattern has yet to be mapped out in detail for either the Middle Ages or succeeding centuries, but already it is clear that from the sixteenth century this process emerged in areas strategically placed to take advantage of the London market before it did in Norfolk.

How do the high Winchester yields of the mid-thirteenth century fit into this chronology? If widely replicated they imply an earlier and higher medieval productivity peak which antedated by over half a century the point when population was at a maximum and wages were at a minimum. Moreover, it is this erosion of mid-thirteenth-century yield levels, taken in conjunction with a general contraction in the scale of demesne cultivation after *c.* 1270, which led Postan and Titow to claim that soils were becoming exhausted on the

---

[152] Grigg, *Dynamics*, pp. 47–67.
[153] Overton and Campbell, 'Productivity change', pp. 38–42.
[154] Van der Wee, 'Introduction', pp. 2, 9–10; Le Roy Ladurie, 'End of the Middle Ages', pp. 81–3.
[155] For an account of these developments see Dodgshon and Butlin (eds.), *Historical geography*, pp. 102–22, 151–222, 323–50.

Winchester estates.[156] Their claim has had wide appeal.[157] Unfortunately, there is no alternative evidence against which that from the Winchester estates might be compared.[158] Nor, with a couple of notable exceptions, has husbandry on the Winchester demesnes been explored in sufficient detail.

The exceptions are the Somerset manors of Rimpton and Taunton which have been the subject of close examination by Thornton.[159] In the case of Rimpton it is the high yields of the mid-thirteenth century that in his view require explanation, not their subsequent decline; all the more so as yields in the opening years of the thirteenth century were unimpressive. At that early stage in direct management stocking levels, labour inputs, and investment in equipment were all comparatively low. That situation was transformed in the second quarter of the thirteenth century. Commencing in the latter years of the episcopate of Peter des Roches (1206–37), demesne productive forces at Rimpton and elsewhere on the estate were developed much more fully and on many manors, where there was scope, more land was brought under the plough. In this way 1,000 acres were added to the sown acreage of the estate as a whole.[160] At Rimpton the cropped acreage was expanded by a fifth and it was during this high-tide of arable exploitation that the demesne secured its peak medieval yields. In part this was fortuitous, for these years coincided with a spell of exceptionally favourable weather conditions, representing the end, perhaps, of the climatic optimum of the early Middle Ages. But the plough-ing up of grassland also released substantial accumulated reserves of soil nitrogen. Far from depressing mean yields, extending the cultivated area ini-tially led to higher rates of output than on established arable land. It was only after some years, as that nitrogen was used up, that mean yields began to fall. Table 7.15 shows that this occurred at Rimpton during the final decades of the thirteenth century. For Shiel this represents the natural productivity sequence wherever land with accumulated reserves of nitrogen was brought into culti-vation.[161] Eventually mean yields stabilised at a new, lower level and in due course much of the assarted land, its nitrogen depleted, reverted to grassland. Hence the progressive contraction in the sown acreage of the Winchester estates from the 1270s, a contraction which is first apparent at Rimpton in the 1280s and became marked from the 1330s.

Thus, although productivity at Rimpton certainly fell back from the 1270s 'a real "crisis" in land productivity had not occurred by 1325'.[162] On the con-

---

[156] Postan, *Economy and society*, pp. 61–71; Titow, *English rural society*, pp. 52–4.

[157] E.g. Miller and Hatcher, *Rural society*, pp. 57, 214–17; Brenner, 'Agrarian class structure', p. 33; Brenner, 'Agrarian roots', p. 308; Farmer, 'Grain yields on Westminster Abbey manors', p. 331.

[158] On extant early manorial accounts see *Manorial records of Cuxham*, pp. 16–26.

[159] Thornton, 'Demesne of Rimpton'; Thornton, 'Determinants of productivity'. See also Biddick, 'Agrarian productivity'.   [160] Titow, 'Land and population', pp. 21–2.

[161] Shiel, 'Improving soil fertility'; Overton, *Agricultural revolution*, p. 117.

[162] Thornton, 'Determinants of productivity', p. 194.

Table 7.15 *Indexed trends in mean net weighted aggregate grain yields (WAGY) per seed and per acre on the bishopric of Winchester's demesne of Rimpton in Somerset, 1209–1403*

Years	n	Net WAGY per seed	Net WAGY per acre (bus.)	Acreage sown	Days' work per sown acre	Net WAGY per acre per '000 man-days
Index: 100 =		2.9	4.9	256	19.6	1.016
1209–1249	12	83	104	87	74	158
1225–1274	12	110	127	104	86	138
1250–1299	18	98	99	102	96	101
1275–1324	31	102	101	98	104	99
1300–1349	41	106	106	90	93	125
1325–1375	46	79	80	81	78	124
1350–1403	42	62	67	69	81	117

Note:
WAGY = weighted aggregate grain yield. Net WAGY yield per seed (RCg^H) = $\sum(RC^{Hi}.P^i/Pw.TCg^i/TCg)$ where RC^{Hi} is the yield ratio of grain i, P^i is the price of grain i per bushel, Pw is the price of wheat per bushel, TCg^i is the acreage under grain i, and TCg is the total area under grains. Net WAGY yield per acre (YCg^H) = $\sum(YCg^{Hi}.P^i/Pw.TCg^i/TCg)$ where YCg^{Hi} is the net yield in bushels per acre of grain i, P^i is the price of grain i per bushel, Pw is the price of wheat per bushel, TCg^i is the acreage under grain i, and TCg is the total area under grains. See Table 7.01.
Source: C. Thornton, 'The determinants of land productivity on the bishop of Winchester's demesne of Rimpton, 1208 to 1403', in Campbell and Overton (eds.), *Land, labour and livestock*, pp. 193, 205.

trary, closer management and increasing labour inputs brought a modest revival in land productivity during the first half of the fourteenth century, albeit at the sacrifice of labour productivity. Had the weather been less adverse it is possible that these gains would have been more conspicuous. That position was maintained until the middle years of the century when, as at Martham, a real crisis in land productivity was precipitated by the dramatic fall in population. No longer could intensive, yield-enhancing methods of production be justified. From mid-century output per unit area fell and this productivity decline persisted notwithstanding a reduction in the cropped area and increased stocking densities.

Rimpton demonstrates that reclamation during the thirteenth century may well have brought temporary windfall gains in land productivity. Nevertheless, these proved impossible to sustain in the long term. Nor, as the post-Black Death decline in yields demonstrates, were generous supplies of animal manure the key to maintaining and improving yields.[163] Rather, keeping land in good heart depended upon the effective deployment of labour. By employing more labour per arable acre, replacing customary labour which was grudgingly performed with hired *famuli* who were better motivated, and through closer supervision of that labour the productivity of Rimpton's core arable was sustained. The price paid for this intensification of husbandry methods was not soil exhaustion. Far from it; there was actually a modest improvement in arable productivity in the first half of the fourteenth century. Rather, it was the productivity and consequently the remuneration of labour that suffered. As inputs per sown acre rose between 1209 and 1324 so there was a concomitant fall in arable output per man-day (Table 7.15). Nor, in the aftermath of the Black Death, did reduced labour inputs restore labour productivity in arable husbandry to its early-thirteenth-century level, for the decline in yields per acre was too severe. Raising land and labour productivity in conjunction with an expansion in the area under cultivation thus appears to have eluded these medieval cultivators. It would require a transformation not of technology but of labour processes and of units of production before this breakthrough – so critical for the economy as a whole – could be achieved, a breakthrough which in Norfolk did not finally come until the eighteenth century.

Norfolk is a single county, Rimpton in Somerset merely one manor of the bishopric of Winchester, and the Winchester estate the sole estate with a long, reasonably complete run of recorded yields. Although each offers valuable insights into long-term trends in arable productivity they do not embody either individually or collectively the experience of the whole country. Only national-level data, for instance, are capable of revealing the productivity contribution of increased farm and regional specialisation. These provisional results are, however, sufficient to convey the complexity and diversity of trends in arable productivity which admit of no simple monocausal explanation. Different crops, different manors, and different localities display different chronologies. More aggregate measures of productivity further qualify the story, although the greater demands which they place on the evidence further restrict the representativeness of the results. To date, trends in WAGY yields are only available for Norfolk, the Winchester demesnes of Rimpton and Taunton in Somerset, and Westminster Abbey's demesne of Kinsbourne in Hertfordshire.[164] In all cases net WAGY yields per acre provide a truer picture than equivalent yields per seed. Net WACY yields per *arable* acre would be even more revealing, but require accurate information on temporary leys and the areas annually left fallow, which are only exceptionally recorded on a systematic annual basis. Comparing productivity levels prevailing on medie-

---

[163] Cf. Farmer, 'Grain yields on Westminster Abbey manors', p. 342.
[164] Table 7.13 and n. 46, above.

val demesnes with those obtained by equivalent farms in later centuries presents even greater methodological and evidential problems.[165] Although few of these problems are entirely insuperable none will be resolved without much painstaking research effort. In this respect, some of the best returns are likely to be obtained from analysis of long-term trends in the unit value of arable land. Not only are valuation data relatively quickly collected, but they are representative of a far wider cross-section of estates and survive for the greater part of the country, including many localities and regions deficient in manorial accounts. They offer the potential of a spatially and temporally more subtle and comprehensive view of aggregate trends in arable productivity over the critical period *c.* 1270–1400. Outside these years reconstructing productivity trends will always be circumscribed by the limited available evidence.

[165] Campbell and Overton, 'New perspective', pp. 50–3, 66–76.

8

Grain output and population: a conundrum

8.1 Total grain output

Medieval English diets were dominated by grain, consumed as pottage, bread, and ale. Grain also bulked large in the diet of man's most powerful working animal, the horse.[1] The aggregate volume of grain output thus determined the size of the population that could be supported and the amount of work it was capable of undertaking. Yet national grain output is not accurately recorded until the advent of official agricultural statistics in 1871. Surrogate sources which cast direct light on this issue are few. They include the 1801 Crop Returns and the *Nonarum inquisitiones* of 1340–1, whose under-exploited returns – part published and part unpublished – constitute a minefield for the unwary.[2] Unfortunately, the utility of both is marred by incomplete survival. For want of a better alternative, indirect methods of estimating grain output have therefore to be relied upon.

By combining estimates of the total grain area with evidence of the composition of that area (Chapter 5) and the per-acre yield of the individual grain crops (Chapter 7) provisional estimates of aggregate grain output can be constructed for the two benchmark dates, *c.* 1300 and 1375 (Tables 8.02 and 8.03). A third and altogether more speculative estimate calculates what the volume of grain output would have been in 1086 if patterns and productivities of cropping roughly equivalent to those in *c.* 1300 had prevailed (Table 8.04).

Fundamental to the accuracy of all three output calculations is the base estimate of the total grain area. Domesday Book is the sole medieval source which casts oblique light upon this.[3] Starting from the recorded number of ploughlands within a sample of twelve counties, and reckoning 120 acres per

---

[1] For grain consumption by horses see Langdon, 'Economics of horses'.
[2] M. E. Turner, 'Arable in England and Wales: estimates from the 1801 crop return', *JHG* 7 (1981), 291–302; *Nonarum inquisitiones in curia scaccarii*; C. R. Erlington, 'Assessments of Gloucestershire: fiscal records in local history', *Transactions of the Bristol and Gloucester Archaeological Society* 103 (1985), 5–15; Masschaele, *Peasants, merchants, and markets*, pp. 83–105.   [3] See the discussion in Darby, *Domesday England*, pp. 127–34.

ploughland and one ploughland per team, F. W. Maitland extrapolated that England south of the Trent contained approximately 9.0 million arable acres in 1086, implying an area not far short of 10 million acres for the country as a whole.[4] Taking the wider evidential base of twenty-eight counties and basing his calculations on recorded plough-teams rather than ploughlands, R. V. Lennard computed that at 100 acres per team England south of the Trent would have contained 7.2 million arable acres.[5] When allowance is made for the omitted counties and under-recorded teams, this implies a national arable area of at least 8 million acres in 1086.[6]

Maitland's and Lennard's estimates are both remarkably high. At late-thirteenth-century rates of yield 8 million arable acres would have been capable of supporting a population of 3.0 million (Table 8.04, estimate 1, column B): i.e. the maximum number conceivably compatible with the evidence of recorded tenant numbers (Table 8.06). This, however, presupposes a much larger class of under-tenants than most Domesday historians are prepared to admit.[7] Even with lower yields and extraction rates (consistent with the conversion of a greater proportion of grain to ale and less to pottage than two centuries later), the 8 million acres of arable projected by Lennard could comfortably have supported a population of 2.3 million (Table 8.04, estimate 3, column B). Those who argue for the positive benefits of a process of commercial growth may have little difficulty with the hypothesis that mean yields rose rather than fell between 1086 and 1300 but it runs directly counter to the Ricardian logic espoused by Postan and his followers. On their reasoning land productivity should have been higher under the less pressurised conditions that prevailed in 1086 than under the ecologically more straitened circumstances of *c.* 1300.[8] Of course, if mean yields were indeed higher in 1086 than *c.* 1300 there would have been ample food for a population of 3.0 million even at relatively low extraction rates. Nor would an arable area of 8 million acres have left much scope for expansion during the twelfth and thirteenth centuries, since it is improbable that the equivalent area *c.* 1300 was more than 10.5 million acres. With 8 million acres already under the plough, a 30 per cent increase over the next two centuries is the most that could have taken place,

[4] F. W. Maitland, *Domesday Book and beyond*, revised edition (Cambridge, 1987), pp. 435–6.

[5] Lennard, *Rural England*, p. 393: his figures omit Cheshire, Cumberland, Derbyshire, Durham, Lancashire, Northumberland, Shropshire, Staffordshire, Westmorland, and Yorkshire. Darby, *Domesday England*, p. 132, accepts the ratio of 100 acres per team.

[6] 8.0m. arable acres is a conservative estimate since Darby counts 81,184 plough-teams: *Domesday England*, p. 336.

[7] Darby, *Domesday England*, pp. 57–61, 88–91; S. P. J. Harvey, 'Domesday England', in Hallam (ed.), *AHEW*, vol. II, pp. 48–9. For a contrary view see A. R. Bridbury, 'The Domesday valuation of manorial income', in Bridbury, *The English economy from Bede to the Reformation* (Woodbridge, 1992), pp. 124–5.

[8] Postan, 'Agrarian society', pp. 556–9; Titow, *Winchester yields*, pp. 30–1; M. M. Postan and J. Hatcher, 'Population and class relations in feudal society', in Aston and Philpin (eds.), *Brenner debate*, pp. 69–70; Chapter 1, pp. 17–19.

notwithstanding the emphasis placed by some historians on the scale and impact of post-Domesday reclamation and colonisation.[9] Perhaps the time has come to recognise that neither the Domesday plough-team numbers nor the ploughland figures provide a convincing basis for estimating the amount of arable land in 1086.[10] Those who have accepted this evidence have been too ready to be impressed by a large Domesday arable area.[11]

These objections vanish if F. Seebohm's alternative method of estimating the Domesday arable area is employed. He calculated the amounts of land held by each recorded category of tenant – freemen, sokemen, villeins, bordars, and cottars – and added to these an allowance for land held in demesne (Table 8.01). This yielded a round total of about 5 million acres within the counties surveyed by the Domesday inquest: barely half the amount estimated by Maitland.[12] Recomputing this estimate using up-to-date counts of tenant numbers and reckoning that 30 per cent of all arable was in demesne raises the total to at least 5.4 million acres.[13] After allowance for under-recording and the omission of the four northernmost counties it would appear that there were approximately 5.75–6.0 million arable acres in 1086 (Table 8.01). Such an area is consistent with a total population of 1.5–2.5 million (depending upon crop combinations, yields, and extraction rates), thereby agreeing with S. Harvey's carefully considered verdict that 'a reasonable estimate would approach two million, and should not exclude a somewhat higher figure'.[14] It dispenses with the need to hypothesise a massive underclass of sub-tenants and allows a 75 per cent expansion of the arable area and two- to two-and-a-half-fold increase in the size of the population over the period 1086–1300, in line with independent evidence of active reclamation and mounting tenant numbers.[15]

Two centuries later the arable area was much enlarged and probably approached in extent the 10.7 million acres attained at the height of the ploughing-up campaign of the Napoleonic War.[16] Although considerable

---

[9] Chapter 1, pp. 11, 18; Donkin, 'Changes', pp. 98–106. Historians have mostly been more content to describe than quantify the expansion of the arable area, although on the estates of the bishopric of Worcester in the west midlands Dyer reckons that it amounted to less than 10 per cent: Dyer, *Lords and peasants*, p. 96.   [10] Darby, *Domesday England*, p. 120.

[11] Postan, *Economy and society*, p. 17; J. Z. Titow, *English rural society 1200–1350* (London, 1969), p. 72.

[12] F. Seebohm, *The English village community* (London, 1883), pp. 102–3.

[13] According to Harvey, 'Demesne agriculture', p. 53, 'the mean average of ratio [*sic*] in Domesday Book of plough-teams belonging to the demesne to plough-teams of the peasantry, county by county, often approximates 1: 2'. The Hundred Rolls of 1279 likewise indicate that within the heavily manorialised, old-settled midland counties demesne arable comprised 30 per cent of the total: Kosminsky, *Studies*, p. 93.   [14] Harvey, 'Domesday England', p. 49.

[15] J. C. Russell, *British medieval population* (Albuquerque, 1948), pp. 55–91; Harley, 'Population and agriculture'; Campbell, 'Commonfields in eastern Norfolk', pp. 18–26; H. E. Hallam, 'Population movements in England, 1086–1350', in Hallam (ed.), *AHEW*, vol. II, pp. 536–93; Smith, 'Human resources', pp. 191–6.

[16] Overton and Campbell, 'Production et productivité', pp. 290–1.

Table 8.01 *Estimated arable acreage of England in 1086*

Type of holding	Mean size of holding (ac.)	According to Seebohm		Based on Darby	
		Number of holdings	Amount of arable (m. ac.)	Number of holdings	Amount of arable (m. ac.)
Villein	22.5	108,407	2.250	109,230	2.458
Bordar, cottar, coscet	3.0	88,500	0.250	88,796	0.266
Sokeman	22.5	23,000	0.500	23,324	0.525
Freeman	40.0	12,000	0.500	13,553	0.542
Sub total		232,000	3.500	234,903	3.791
Demesne	120.0	12,500	1.500	13,500	1.625
Total surveyed counties		244,500	5.000	248,403	5.416
Allowance for omissions		21,250	0.425	20,200	0.440
Grand total for England		265,750	5.425	268,600	5.862

Sources: F. Seebohm, *The English village community* (London, 1883), pp. 102–3; H. C. Darby, *Domesday England* (Cambridge, 1977), p. 337.

tracts of medieval arable had been converted to permanent grass by 1800, these losses need to be set against the gains that had resulted from post-medieval wetland reclamation, the enclosure and conversion to tillage of former common pastures and wastes, and the disaforestment of royal forest. It is thus almost inconceivable that the arable area in 1300 could have exceeded that in 1800. The retreat from this medieval peak of perhaps 10.5 million acres began in 1315. Adverse climatic conditions, sea flooding, shortages of seed-corn, cattle plagues which struck at the plough beasts, and Scottish raids in the north all exacted their toll. By 1340–1 there is firm evidence that thousands of acres of former arable in scores of vills were lying unsown.[17] Then, following the massive mortality crisis of mid-century, there was neither the labour nor the demand to justify keeping so much land in cultivation.[18] By 1375 demesnes, on average, were cropping 24 per cent less land than in 1300 (Table 5.06). Such a contraction, if replicated throughout the economy, would have returned the tillage area to near or just above its probable eleventh-century level. In fact, the cut-back in cultivation was probably more pronounced in the peasant than the demesne sector. It was the peasantry who bore the brunt of the demographic collapse and much peasant arable won by reclamation during previous centuries was undoubtedly converted back to grass or wood, especially as increasing quantities of former demesne arable were transferred on lease to peasant use. An overall reduction in the arable area of approximately one quarter between 1300 and 1375 may therefore be conjectured.

How much of the arable was sown with grain at these three dates (1086, 1300 and 1375) depended upon the frequency of fallows and the scale of legume cultivation. In most of lowland England it was the norm to cultivate at least half of the arable each year, sometimes substantially more, although only in a few exceptional localities did the grain area exceed three-quarters of the total. At the opposite extreme, outfield land, brecks, and some convertible land was subject to only intermittent cropping. As cropping intensified between 1086 and 1300 fallows probably contracted somewhat. If the wholesale conversion of two- to three-field systems may have been less widespread than formerly believed, the cultivation of *inhoks* from the fallow became an increasingly common practice. Where there were fewer constraints upon the flexibility of cropping, additional courses were probably added to rotations. Legumes were necessarily one of the principal beneficiaries of this reduction in fallows. The net outcome was an increase in the grain area from perhaps 56 per cent of the arable in 1086 to 59 per cent in *c.* 1300, by which time it had grown from approximately 3.29 million acres to 6.23 million acres (Table 8.05). No greater area would be sown with grain until the second quarter of the nineteenth century. The subsequent cut-back in grain cropping was pronounced and reinforced in certain localities by the substitution of legumes for grain. By

[17] Baker, 'Contracting arable lands'.   [18] Campbell and others, 'Demesne-farming', pp. 131–7.

c. 1375 barely 4.0 million acres were being sown with grain and it is almost certain that by the middle of the fifteenth century this area had shrunk still further. Never again would grain production be so reduced in extent, nor, until 1830, would grains occupy such a small share of total tillage (Table 8.05).

There is firmer evidence of the actual composition of the grain area in both *c.* 1300 and *c.* 1375. National and regional samples of manorial accounts allow the composition of demesne cropping to be reconstructed with a reasonable degree of accuracy. Tables 8.02 and 8.03 indicate the areas that would have resulted if these demesne crop proportions had been replicated throughout the arable sector as a whole. In fact, non-demesne producers almost certainly grew relatively more of the cheaper grains, especially in those parts of the north and west where demesne cultivation was poorly developed.[19] Consequently, wheat is probably overstated in importance relative to rye and oats. This affects the composition rather than the volume of the final estimates of total grain supply and is primarily of significance because of the different dietary uses to which the various grains were put, with their differing extraction rates. Only a systematic comparison of demesne receipts with tithe receipts and multure payments can clarify the relationship between the product mix of the demesne and non-demesne sectors.[20]

Estimates of the mean yield per acre of each of the grains also perforce derive from the demesne sector alone. Those used here relate to an eleven-county area comprising Norfolk and the ten FTC counties (Table 7.06). The overall range of yields within this extensive area was probably as wide as that within the country as a whole, since it embraced localities of fertile and infer-tile soil and intensive and extensive husbandry. The mean unit value of arable land within these eleven counties was also close to the national average (Figure 7.13), which suggests that these mean yields may be reasonably representative of the rest of the country. On the other hand, the area was climatically better suited to grain production than much of the north and west and the commer-cial incentives to raise productivity were also stronger, owing to the presence of London and ready access to other centres of concentrated demand at home and abroad.[21] These yield figures are consequently more likely to overstate than understate national mean demesne yields. How representative they are of arable husbandry as a whole is another matter.

One influential school of thought argues that peasant yields were inferior to those obtained by demesnes owing to deficiencies of livestock and manure arising from the surplus-extraction relations embodied in serfdom.[22] Much

---

[19] Miller, 'Farming in northern England'.

[20] Miller (ed.), *AHEW*, vol. III, pp. 64–5, 217–18, 228–9, 306; H. S. A. Fox, 'Medieval farming: arable productivity and peasant holdings at Taunton', Economic and Social Research Council, unpublished End of Award Report, B00 23 2203 (1991).

[21] Campbell and others, *Medieval capital*; Nightingale, 'Growth of London'.

[22] Postan, *Economy and society*, p. 124; Titow, *English rural society*, pp. 80–1; Aston and Philpin (eds.), *Brenner debate*, pp. 31–3.

Table 8.02 *Estimated national grain output and the population it was capable of feeding, c. 1300*

		Individual grains			
Variable	Wheat	Rye and rye mixtures	Barley and dredge	Oats	Total grains
1) Total physical output:					
a % national grain area[1]	*36.6*	*7.5*	*19.9*	*36.0*	*100.0*
b Total national grain area (m. ac.)[2]	2.28	0.47	1.24	2.24	6.23
c Net yield per acre (qtrs.)[3]	1.19	1.03	1.63	0.98	7.39
d Total net grain output (m. qtrs.)	2.71	0.48	2.02	2.18	7.39
e Mean price per quarter (s.)[4]	5.92	4.65	4.31	2.35	
f Total value of net grain output (£m.)	0.802	0.113	0.435	0.256	1.606
2) Total unprocessed kilocalorie output:					
g Weight per quarter (lb.)[5]	424	408	368	288	
h Kilocalories per pound[5]	1,520	1,520	1,452	1,676	
i Kilocalories per quarter [*h* × *g*]	644,480	620,160	534,336	482,688	
j Net kilocalorie yield per acre [*i* × *c*]	765,320	635,664	868,296	470,621	670,075
k Total net kilocalorie grain output (m.) [*i* × *d*]	1,744,930	298,762	1,076,687	1,054,191	4,174,570
3) Total processed kilocalorie output:					
l Principal use[6]	bread	bread	ale	²/₃ pottage ¹/₃ fodder	
m Food extraction rate[7]	0.80	0.80	0.30	0.67	0.64
n Total net food output (m. kcal) [*k* × *m*]	1,395,944	239,010	323,006	706,308	2,664,268
o Less 10% wastage (m. kcal) [*n* × 0.9][8]	1,256,350	215,109	290,705	635,677	2,397,841
p Total daily supply of kilocalories (m.) [*o* ÷ 365]	3,442	589	796	1,742	6,569

4.38

4) Total population capable of being fed:

q Total population (m.) capable of being fed at 1,500
kilocalories per person per day [$p \div 1,500$]

Notes and sources:
[1] Calculated from National accounts database for the period 1275–1324.
[2] See Table 8.05.
[3] Calculated from Table 7.06, Norfolk and FTC counties combined.
[4] Calculated from D. L. Farmer, 'Prices and wages', Hallam (ed.), *AHEW*, vol. II, p. 734.
[5] Calculated from Table 5.04.
[6] See Chapter 5, pp. 214–27.
[7] See Table 5.04 and Chapter 5, pp. 214–27.
[8] See B. M. S. Campbell, J. A. Galloway, D. J. Keene, and M. Murphy, *A medieval capital and its grain supply* (n.p., 1993), pp. 34, 43; D. N. McCloskey and J. Nash, 'Corn at interest: the extent and cost of grain storage in medieval England', *American Economic Review* 74 (1984), 182.

Table 8.03 *Estimated national grain output and the population it was capable of feeding, c. 1375*

		Individual grains			Total grains
Variable	Wheat	Rye and rye mixtures	Barley and dredge	Oats	
1) Total physical output:					
a % national grain area[1]	*37.2*	*4.0*	*31.4*	*27.4*	*100.0*
b Total national grain area (m. ac.)[2]	1.49	0.16	1.26	1.10	4.01
c Net yield per acre (qtrs.)[3]	0.98	0.90	1.63	1.03	
d Total net grain output (m. qtrs.)	1.45	0.16	2.05	1.13	4.79
e Mean price per quarter (s.)[4]	6.67	4.57	4.73	2.60	
f Total value of net grain output (£m.)	0.484	0.037	0.485	0.147	1.153
2) Total unprocessed kilocalorie output:					
i Kilocalories per quarter[5]	644,480	620,160	534,336	482,688	
j Net kilocalorie yield per acre [i×c]	628,368	558,144	868,296	494,755	664,303
k Total net kilocalorie grain output (m.) [i×d]	936,268	89,303	1,094,053	544,231	2,663,855
3) Total processed kilocalorie output:					
l Principal use[6]	bread	bread	ale	½ pottage ½ fodder	
m Food extraction rate[7]	0.80	0.80	0.30	0.50	0.53
n Total net food output (m. kcal) [k×m]	749,014	71,442	328,216	272,116	1,420,788
o Less 10% wastage (m. kcal) [n×0.9][8]	674,113	64,298	295,394	244,904	1,278,709
p Total daily supply of kilocalories (m.) [o÷365]	1,847	176	809	671	3,503

4) *Total population capable of being fed:*

q Total population (m.) capable of being fed at 1,500
  kilocalories per person per day [p ÷ 1,500]

2.34

Notes and sources:
[1] Calculated from National accounts database for the period 1350–1399.
[2] See Table 8.05.
[3] Calculated from Table 7.06, Norfolk and FTC counties combined.
[4] Calculated from D. L. Farmer, 'Prices and wages, 1350–1500', in Miller (ed.), *AHEW*, vol. III, p. 444.
[5] See Table 5.04.
[6] See Chapter 5, pp. 214–27.
[7] See Table 5.04 and Chapter 5, pp. 214–27.
[8] See B. M. S. Campbell, J. A. Galloway, D. J. Keene, and M. Murphy, *A medieval capital and its grain supply* (n.p., 1993), pp. 34, 43;
  D. N. McCloskey and J. Nash, 'Corn at interest: the extent and cost of grain storage in medieval England', *American Economic
  Review* 74 (1984), p. 182.

late-reclaimed peasant land was also undoubtedly marginal for grain production. On the other hand, the family labour used to work many peasant holdings was more abundant and better motivated than the forced and hired labour which tilled the demesnes. Depending upon the productivity outcome of the different factor endowments of the demesne and non-demesne sectors total net grain output may have been either under- or over-estimated. Pending resolution of this enigma it has been assumed that demesne yields are broadly representative of grain productivity in general. Combining these independent estimates of the scale, composition, and productivity of the grain sector indicates that approximately 7.4 million quarters of assorted grains were produced in *c.* 1300 and 4.8 million quarters in *c.* 1375.

Estimating total net grain output in 1086 is an altogether more speculative exercise. If, on the evidence of tenant numbers, there were 5.75–6.0 million acres of arable, and if 56 per cent of that arable was devoted to grain, then the national grain area would have been approximately 3.29 million acres. With a lower ratio of labour to land in 1086 than 1300 relative factor costs would have encouraged more land-extensive methods of production. Commercial incentives to specialise according to comparative advantage would also have been weaker. Per-acre yields are therefore likely to have been the same or lower. Allowing at most a 20 per cent improvement (anything more is hard to credit given the actual physical level of yield obtaining in *c.* 1300 and the evidence of only modest gains in yield during the latter years of the thirteenth century) would mean that yields in 1086 were 84–100 per cent of those in 1300. [23] On these assumptions, and if patterns of cropping were broadly equivalent to those in *c.* 1300, approximately 3.4–3.9 million quarters of assorted grains would have been produced in 1086 (i.e. 46–53 per cent of the corresponding grain output in *c.* 1300).

8.2   Total kilocalorie output

When it came to feeding the population it was the kilocalorie rather than the physical volume or cash value of grain production that was all important. Here, the estimated physical volume of each grain's output has been converted into its kilocalorie equivalents (the most convenient standardised measure of their respective food values) using the conversion factors given in Table 5.04. The results are set out in Tables 8.02, 8.03, and 8.04. Of the kilocalories present in raw grain a substantial proportion were unavoidably lost to potential human consumption either in the process of converting them to foodstuffs or because they were fed as fodder to animals. Vermin, rot, accident, negligence, and wilful destruction levied a further tithe upon grain supplies before,

[23] Chapter 7, pp. 370–5.

Table 8.04 *Alternative estimates of national grain output and the population it was capable of feeding in 1086*

Variable	Estimate 1[a]		Estimate 2[b]		Estimate 3[c]	
	A	B	A	B	A	B
Total arable area (m. ac.)	5.875	8.00	5.875	8.00	5.875	8.00
Total grain area at 56% of arable (m. ac.)	3.29	4.50	3.29	4.50	3.29	4.50
Net kilocalorie yield per grain acre	670,075	670,075	670,075	670,075	558,396	558,396
Total net output unprocessed grain (b. kcal)	2.205	3.015	2.205	3.015	1.837	2.513
Total net output processed grain after wastage (b. kcal)	1.235	1.670	1.103	1.508	0.919	1.257
Mean daily supply processed grain (m. kcal)	3,383	4,575	3,022	4,130	2,518	3,444
Total population capable of being fed at 1,500 kilocalories per person per day (m.)	2.26	3.05	2.01	2.75	1.68	2.30
Total value of net output of unprocessed grain (£)	213,000	290,000	213,000	290,000	176,000	242,000

Notes and sources:

[a] Assuming the same net kilocalorie yield per acre as in *c.* 1300 (Table 8.02) and an extraction rate (including wastage) of 0.56.

[b] Assuming the same net kilocalorie yield per acre as in *c.* 1300 (Table 8.02) and an extraction rate (including wastage) of 0.50.

[c] Assuming a 20 per cent improvement in net kilocalorie yield per acre 1086–*c.* 1300 and an extraction rate (including wastage) of 0.50.

A Arable area estimated from the number and size of holdings using Seebohm's method (see pp. 388–9).

B Arable area estimated from the number of recorded plough-teams using Lennard's method (see p. 387).

b. = billion, i.e. 10^{12}.

during, and after processing.[24] As demonstrated in Chapter 5, kilocalorie extraction rates varied with the form of grain consumption. Feeding grain as fodder to livestock for the purposes of producing milk, meat, and lard had the lowest extraction rate of all. Brewing grain into ale of various strengths was also highly extravagant of available kilocalories, approximately 70 per cent of which were lost in the brewing process. Grinding grain into flour and then baking it into bread was far less wasteful, although the efficiency of the conversion varied from grain to grain. All grains required some milling in order to remove their outer husk and render them digestible to humans. Thus processed they could be consumed 'whole' in pottage. This yielded the highest food extraction rates of all.

Not all these forms of grain consumption enjoyed the same dietary status. Nutritious as pottage was, its food status was always low. Nor were all grains equally suited to these various forms of consumption. Wheat was the premier bread grain, barley the most prized brewing grain, and oats the staple fodder grain. Nevertheless, none was completely specialised in its use and there was always much substitution between grains.[25] For transparency of calculation, and because overlapping uses between the grains tended to cancel each other out, estimates of the total processed kilocalorie supply of grains have been based on the principal use made of each of the grains as established from an analysis of patterns of grain disposal on demesnes in the FTC counties (Table 5.05).[26] In the case of wheat, rye, and rye mixtures this was bread, and in the case of barley and dredge it was brewing (although some of each was always consumed as pottage, and barley bread was the staple of the poor in certain regions). Oats are more complicated. They were the most versatile grain, variously consumed as fodder, pottage, ale, and a component of the coarser breads. Which use predominated changed over time and varied from region to region. Different assumptions have therefore been made of the proportions consumed as fodder, pottage, and ale in 1086, *c.* 1300 and *c.* 1375 (Tables 8.02, 8.03 and 8.04). Oats' most abiding human use was as pottage. Initially important as a brewing grain, they were progressively supplanted by dredge and barley. Meanwhile, as the use of working horses grew, they gained steadily in importance as a fodder crop.

Overall, after allowance for wastage, approximately 58 per cent of potential kilocalories were actually retained and consumed after processing into food and drink in *c.* 1300 (Table 8.02). The corresponding proportion in *c.* 1375 was 48 per cent (Table 8.03), the sharp fall in extraction rates reflecting a pronounced dietary shift, with a significantly greater proportion of grain being processed into ale and significantly less into pottage. In both periods very substantial proportions of available kilocalories were lost between pro-

---

[24] Campbell and others, *Medieval capital*, pp. 34, 43; McCloskey and Nash, 'Corn at interest', p. 182.   [25] Chapter 5, pp. 214–27.   [26] Chapter 5, pp. 214–27.

duction and consumption. This was normal. Corresponding extraction rates calculated for the seventeenth and eighteenth centuries (Table 8.05) rarely rose above 50 per cent. Judged by this yardstick the estimated extraction rate of 58 per cent in *c.* 1300 is impressive and may even be too high. If accurate, it is symptomatic of a society attempting to maximise the efficiency with which grain was consumed by throttling back on those forms of grain consumption most extravagant of available kilocalories. The corresponding extraction rate in 1086 is likely to have been lower. Comparison with later centuries suggests that an extraction rate in the range 50 to 56 per cent is most plausible (Table 8.05). Depending upon the total grain area and rate of yield this lower extraction rate would daily have delivered 2,518–3,383 million processed grain kilocalories ($2.518 - 3.383 \times 10^9$ kcal/day) (Table 8.04). Over the next 200 years that supply rose by two- to two-and-a-half-fold, to 6,569 million kilocalories (6.569×10^9 kcal/day). Thereafter, following the dramatic mid-fourteenth-century demographic collapse, supplies were cut back by almost half to approximately 3,503 million kilocalories (3.503×10^9 kcal/day) in *c.* 1375.

8.3   The total population capable of being fed

All the available evidence suggests that by *c.* 1300, whereas a privileged elite lived off the fat of the land, the bulk of the population was dependent upon a heavily grain-based and far from generous diet, with many people subsisting at the minimum nutritional level compatible with survival.[27] The combination of grains grown is consistent with this view, demonstrating as it does the primacy accorded to the food-productive bread and pottage grains at the expense of the land-extravagant brewing grains (Table 8.02). The prominence of the cheaper bread and brewing grains – rye and rye mixtures and dredge – also points to straitened economic and dietary circumstances (Table 5.08). Everything implies the prevalence of production choices intended to minimise the losses incurred by the conversion of grain to food and drink, resulting in the exceptionally high kilocalorie extraction rate of 0.58 (Table 8.05). Once the demographic heat was taken out of the situation diets and production patterns both changed. By 1380 production of pastoral foodstuffs was expanding (Table 4.03) and making an increasing contribution to diets (which themselves were probably improving), while patterns of grain production not only shifted in favour of the higher-quality bread and drink grains but were compatible with a substantial *per capita* increase in ale consumption (Tables 5.08 and 8.03).[28] At 0.48, estimated kilocalorie extraction rates were consequently significantly lower at the close of the fourteenth century than they had been at the beginning.

[27] Dyer, *Standards of living.*   [28] Campbell and others, 'Demesne-farming'.

Table 8.05 *Estimates of population and grain supply: England 1086–1871*

Date	Total grain area (m. ac.)	Grain as % of arable area %	Net kilocalorie grain yield per acre (m. kcal)	Total net kilocalorie grain output (b. kcal)	Total kilocalorie grain imports (b. kcal)	Total net kilocalorie grain supply (b. kcal)	Approximate food extraction rate	Total population (m.)	Daily per person supply of grain kilocalories without imports (kcal)	Daily per person supply of grain kilocalories with imports (kcal)
1086	3.29	56.0	0.56–0.67	1.84–2.21	0.0	1.84–2.21	0.50–0.56	1.68–2.26	1,500	1,500
1300	6.23	59.2	0.67	4.17	0.0	4.17	0.58	4.38	1,500	1,500
1375	4.01	50.3	0.67	2.69	0.0	2.69	0.48	2.34	1,500	1,500
1600	(4.70)	57.1	(0.75)	(3.53)	0.0	(3.53)	(0.52)	4.11	(1,230)	(1,230)
1700	5.22	58.0	1.07	5.59	−0.11	5.48	0.49	5.06	1,480	1,440
1800	5.81	54.4	1.77	10.28	+0.68	10.96	0.45	8.66	1,470	1,560
1830	7.06	49.8	1.83	12.92	+2.43	15.35	0.46	13.28	1,230	1,460
1871	6.79	49.1	2.44	16.57	+10.85	27.42	0.50	21.50	1,060	1,750

Notes:
Figures in brackets are probably under-estimates caused by strong geographical bias in the available data which exaggerate the importance of brewing grains and thereby depress the mean extraction rate.
b. = billion, i.e. 10^{12}.

Source: Table 8.04; M. Overton and B. M. S. Campbell, 'Production et productivité dans l'agriculture anglaise, 1086–1871', *Histoire et Mesure*, 11, 3/4 (1996), 290, 292, 295–6.

Per-capita kilocalorie food requirements varied with age, sex, and occupation. An adult male engaging in hard manual labour, such as winter ploughing or reaping, might need in excess of 4,000 kilocalories per day whereas an elderly and sedentary male would be adequately fed on barely half this. On average, across both sexes and all ages, 'a population which could rely on a normal consumption of 2,000 calories per head would have been, in centuries past, an adequately fed population, at least from the point of view of energy'.[29] Populations could manage on less than this for short periods, and frequently had to during the recurrent seasonal and annual food shortages that were the normal lot of pre-industrial life, but any more prolonged reduction in *per-capita* kilocalorie consumption would have impaired demographic reproduction. An adequate diet was therefore a precondition for the population growth which climaxed *c.* 1300. It was also a precondition for adequate levels of work output in an age which relied upon manual methods of production: keeping body and soul together was one thing, providing sufficient energy to work and earn a living was another. This does not, however, mean that all sections of society were well fed. On the contrary, medieval food resources were so unequally distributed that many of the poorer sections of society are likely to have been seriously malnourished. For instance, Dyer reckons that virgators, who rated among the better-off customary tenants, enjoyed a daily allowance per head of approximately 2,200 kilocalories *c.* 1300.[30] Well over half the population were probably worse off. In dietary terms that did not necessarily mean that they received significantly fewer kilocalories – since life would have become unsupportable had they done so – but, rather, that the forms in which they consumed them were coarser and more monotonous. Those trapped by poverty were forced to trade down to their lowest dietary preferences; their menu, like that of Geoffrey Chaucer's poor widow, was unlikely to include refined wheaten bread and copious quantities of ale.

Grain could contribute up to 75 to 80 per cent of all kilocalories without serious nutritional imbalance occurring, although those who could afford it presumably preferred a more varied diet in which meat was more prominent.[31] Mean dietary dependence upon grains was generally greatest when populations were maximised and living standards minimised, under which circumstances grains might be required to contribute 1,500–1,750 kilocalories per head per day. In Tables 8.02, 8.03, and 8.04 estimations of the populations capable of being fed are made on the assumption that 1,500 grain-derived kilocalories were daily consumed per head. Such quantities are equivalent to three-quarters of a daily diet of 2,000 kilocalories, two-thirds of a diet of 2,250 kilocalories, and 60 per cent of a diet of 2,500 kilocalories. On the whole

---

[29] M. Livi-Bacci, *Population and nutrition*, trans. T. Croft-Murray with C. Ipsen (Cambridge, 1991), p. 27.   [30] Dyer, *Standards of living*, p. 134.
[31] Dyer, *Standards of living*, pp. 55–70, 151–60.

the greater the daily kilocalorie intake the smaller the contribution that grains made to it, for more ample diets are normally associated with greater dietary diversity. Obviously, the assumption of a lower *per-caput* dietary contribution from grains would yield higher population estimates.

With a mean diet of 1,500 grain-derived kilocalories per head per day sufficient grain would have been produced to support total populations of 4.38 million in *c.* 1300 and 2.34 million in *c.* 1375, implying a 46 per cent drop in population between these two dates.[32] If anything, the *c.* 1300 estimate errs on the side of generosity since it rests upon exceptionally high grain acreages and extraction rates. The sub-national sample of crop yields may also lend a slight upward bias to the estimate. In contrast, with a marginally smaller contraction in the arable area and a modest reduction in the dietary contribution of grains it is possible to envisage a slightly larger population in *c.* 1375. The 'safest' population totals on the available output information would therefore be 4.00–4.25 million in *c.* 1300 and 2.25–2.50 million in *c.* 1375.[33] Such revised totals would yield a less pronounced population decline between 1300 and 1375 of roughly 40 per cent. This accords with the scale of the demographic collapse projected by J. C. Russell, long out of favour with most historians but recently reinstated by Blanchard (Table 8.06).

The total of 2.25–2.50 million people estimated for *c.* 1375 is reassuringly close to the normal range of estimates calculated for that date from the poll tax returns of 1377 (Table 8.06). The poll tax is widely regarded as the single most reliable medieval source for projecting the total population.[34] Assumptions about the rate of tax evasion and the age structure of the population are critical to these projections which range from a minimum of 2.20 million to a maximum (contingent upon a 25 per cent rate of evasion and a very youthful age profile) of 3.00–3.40 million (Table 8.06). Few historians would quarrel with an estimate of 2.5 million which relies upon less extreme assumptions and agrees remarkably well with what is now known about demesne land-use and productivity. Such concurrence, if more than coincidental, suggests that the seigniorial sector may indeed, at least at this date, be broadly representative of arable husbandry as a whole.

Independent population estimates for *c.* 1300 are altogether more controversial and a much wider range of possible totals has been proposed (Table 8.06). The highest of these take as their starting point an estimate of 3.0 million or more in 1377 and extrapolate back from that. A 50–60 per cent drop

[32] Cf. Smith, 'Demographic developments', p. 49.

[33] In 1650, when the population was again pressing upon resources but national productive forces were much more fully developed, the total population is estimated to have been 5.23m.: Wrigley and Schofield, *Population history*, pp. 208–9.

[34] 'The poll taxes of 1377 and 1379 can enable us to secure a more accurate picture of the population and society of England than at any time until the first census of 1801': Russell, *British medieval population*, p. 119; Smith, 'Human resources', pp. 190, 198–9.

Table 8.06 *Alternative population estimates of medieval England*

Author (year)	Estimated population (in millions) at given date				
	1086	c. 1300	1377	1520s	1540s
J. C. Russell (1948)	1.10	3.70	2.23		3.22
M. M. Postan (1966)	2.60–3.00	6.00–7.00	3.00–3.40		
J. Cornwall (1970)			2.20	2.30	2.80
H. C. Darby (1977)	1.22–1.48				
J. Hatcher (1977)	1.75–2.25	4.50–≥6.00	2.50–3.00	2.25–2.75	
B. M. S. Campbell (1981)				1.05–2.92	
R. M. Smith (1988)	1.10–2.50	≥6.00	2.50–3.00		2.80–3.10
H. E. Hallam (1989)	≥2.00				
S. Harvey (1989)	≥2.00				
E. A. Wrigley and R. S. Schofield (1989)					2.77
I. S. W. Blanchard (1996)	1.10–1.53	3.40–4.50	2.24		
Grain output method (Tables 8.01–8.03)	1.50–2.50	4.00–4.25	2.25–2.50		

Sources:
J. C. Russell, *British medieval population* (Albuquerque, 1948), pp. 246, 263, 272.
M. M. Postan, 'Medieval agrarian society in its prime: England', in Postan (ed.), *The Cambridge economic history of Europe*, vol. I, *The agrarian life of the Middle Ages*, 2nd edition (Cambridge, 1966), pp. 561–3.
J. Cornwall, 'English population in the early sixteenth century', *EcHR* 23 (1970), 44.
H. C. Darby, *Domesday England* (Cambridge, 1977), p. 89.
J. Hatcher, *Plague, population and the English economy 1348–1530* (London, 1977), pp. 68–9.
B. M. S. Campbell, 'The population of early Tudor England: a re-evaluation of the 1522 Muster Returns and 1524 and 1525 Lay Subsidies', *JHG* 7 (1981), 153.
R. M. Smith, 'Human resources', in Astill and Grant (eds.), *The countryside of medieval England*, pp. 189–91.
H. E. Hallam, 'Population movements in England, 1086–1350', in Hallam (ed.), *AHEW*, vol. II, pp. 536–7.
S. Harvey, 'Domesday England', in Hallam (ed.), *AHEW*, vol. II, p. 49.
E. A. Wrigley and R. S. Schofield, *The population history of England 1541–1871* (Cambridge, 1989), pp. 207–10.
I. S. W. Blanchard, *The Middle Ages: a concept too many?* (Avonbridge, 1996), pp. 36–8.

in population as a result of successive onslaughts of famine and plague between 1315 and 1375 is all that is then required to hypothesise a total of 6.0 million or more *c.* 1300.[35] While there is some compelling local evidence which points to a fall in population of this magnitude it is far from clear that such a decline was replicated at a national level.[36] Indeed, the evidential base used in such extrapolations is far flimsier than that used here to project total grain output. Extrapolating forwards from a high Domesday estimate of population has also been used to produce a similar result.[37]

Those who argue most forcibly for a population *c.* 1300 of 6.0 million or more are also in the main those who place most emphasis upon the low living standards of the bulk of the population and its overwhelming dependence upon a predominantly grain-based diet.[38] For Postan the poverty of the peasantry was expressed in an acute deficiency of livestock resulting in chronic under-manuring and inferior yields on peasant holdings.[39] Such a scenario is clearly irreconcilable with the grain-output estimates set out in Table 8.02. Only if mean yields on peasant holdings matched those on lowland demesnes in the south and east of England could enough grain have been produced to feed a population of at most 4.5 million. With inferior yields on peasant holdings the total must perforce have been smaller; at most 4.0 million and probably less, depending upon the productivity differential. This brings the total remarkably close to the 3.70 million initially proposed by Russell but rejected by Postan.[40]

Postan's preferred population figure of 6.0 million or more in *c.* 1300 would have required an average of 9,000 million processed grain kilocalories per day (9.0×10^9 kcal/day), which, on the most generous extraction rates, translates into a total unprocessed net grain output of 5.5 billion kilocalories a year (5.5×10^{12} kcal/year). If the demesne sector accounted for 25 per cent of all arable, each peasant grain acre would need to have been 40 per cent more productive than its demesne equivalent in order to have secured such an output.[41] A productivity differential of such magnitude in favour of the entire non-demesne sector, although hypothetically possible, would require a radical reassessment of the verdicts most historians have hitherto passed on that sector.[42] Even a more conservative population estimate of 5.0 million in *c.* 1300 would have been contingent upon a 13 per cent productivity advantage in favour of the peasantry. It is also worth noting that no such productivity differential has

---

[35] See the criticisms of I. S. W. Blanchard, *The Middle Ages: a concept too many?* (Avonbridge, 1996), pp. 37–8.

[36] Poos, 'Rural population of Essex'; Smith, 'Demographic developments', pp. 48–9.

[37] Hallam, 'Population movements', pp. 536–7.

[38] An exception is H. E. Hallam, *Rural England 1066–1348* (London, 1981).

[39] Postan, 'Agrarian society', p. 602.   [40] Postan, 'Agrarian society', pp. 561–2.

[41] Chapter 3, pp. 56–60.

[42] E.g. Miller and Hatcher, *Rural society*, pp. 161–4. For a recent reassessment of peasant producers see Raftis, *Peasant economic development*.

to be hypothesised in order to get the output and population estimates to square in *c.* 1375, nor is there any unequivocal evidence to suggest that farm size was subsequently ever a significant determinant of crop yields in English agriculture.[43]

The onus is therefore on those who wish to argue for a medieval population at peak in excess of 4.5 million to demonstrate by what means, on known patterns of land-use and productivity, it could have been fed. Not only does a population of 4.25 million, or thereabouts, fit the available agricultural evidence, it also accords well with recent reassessments of the scale and significance of the urban sector. For instance, the poll tax indicates that in 1377 London contained 1.7 per cent of the country's population.[44] If the same proportion out of a national total of 4.25 million had lived in the capital *c.* 1300 then London would have had approximately 72,000 inhabitants. In fact, the metropolis probably contained a smaller share of England's population at this earlier date: had 1.5 per cent of a national population of 4.25 million lived in London the capital's population would have been 64,000. Significantly, these two estimates more or less split the difference between the 80,000 proposed by D. J. Keene and the 60,000 preferred by P. Nightingale as the likely size of the capital on the eve of the Great European Famine.[45]

London headed England's early-fourteenth-century urban league table. It was one of fourteen or fifteen cities with at least 10,000 inhabitants whose combined populations in *c.* 1300 probably amounted to at least 0.25 million, equivalent to 5.9 per cent of the total population. This corresponds almost exactly to the proportion resident in towns of at least 10,000 inhabitants in 1600.[46] One of the distinctive features of medieval English urbanism was the hundreds of smaller urban places which also existed. Dyer has estimated that towns of all sorts contained approximately 15–20 per cent of the total population *c.* 1300, which, on a figure of 4.25 million, would amount to 0.64–0.85 million people (many of whom, of course, retained a partial or complete dependence upon agriculture for their living).[47] For Dyer the numbers of those who lived in towns and/or engaged in non-agricultural occupations grew faster than the total population during the twelfth and thirteenth centuries, resulting in a process of urbanisation.[48] As R. H. Britnell has emphasised, the

[43] Campbell and Overton (eds.), *Land, labour and livestock*, pp. 27, 246–9, 309–11.
[44] Campbell and others, *Medieval capital*, p. 44.
[45] D. J. Keene, 'A new study of London before the Great Fire', *Urban History Yearbook 1984*, pp. 11–21; Nightingale, 'Growth of London', pp. 95–8.
[46] Campbell and others, *Medieval capital*, p. 11, n. 50.
[47] C. C. Dyer, *Everyday life in medieval England* (London, 1994), p. 302; Masschaele, *Peasants, merchants, and markets*, pp. 86–91. The urban population may have been substantially higher: on the evidence of the 1334 lay subsidy and 1377 poll tax Derek Keene has estimated that 325,000–417,000 may have lived in towns of 10,000 or more inhabitants in *c.* 1300: Campbell and others, *Medieval capital*, pp. 10–11.
[48] Dyer, 'How urbanized was medieval England?', pp. 179–80.

case for such a process is stronger the smaller the total population in *c.* 1300 and the slower its growth during the preceding centuries.[49] The agricultural evidence thus endorses recent work on the non-agricultural sector while conflicting with the views of demographers and some others. In part this conflict arises from the difficulty of reconciling macro and micro estimates and evidence.

8.4   Grain output, population, and GDP

These estimates of grain output and population take on added significance when set against recent provisional estimates of GDP in 1086 and *c.* 1300. Placing radically different interpretations on the manorial valuations given in the Domesday survey and assuming different population totals at that date, Snooks and Mayhew have proposed rival GDP estimates of £137,000 and £300–400,000 in 1086.[50] At late-eleventh-century prices, with an extraction rate of 0.50, and on the assumption that grain contributed 1,500 kilocalories per head per day, the annual *per-capita* requirement of unprocessed grain would have been £0.10. The low population of 1.53 million favoured by Snooks would thus have required a raw grain output of £153,000 to sustain it, a sum in excess of his estimate of total GDP (Table 8.07). Clearly, Snooks's estimate of the latter is too low, all the more so as the population in 1086 is likely to have been at least 10 per cent higher.[51] Two conspicuous sources of under-estimation are a 50 per cent under-evaluation of the cost of peasant subsistence and a failure to allow for the production of a disposable surplus by the unfree.[52]

Mayhew starts with a larger population and offers alternative GDP estimates of £300,000 (which he prefers) and £400,000. The raw grain requirement of his favoured population of 2.25 million would have been £225,000, equivalent to 75 per cent of his lower and 56 per cent of his higher GDP estimate. This suggests that with a relatively large Domesday population of 2.25 million the higher of his two GDP estimates is in fact the more plausible since it is unrealistic to suppose that raw grain could have accounted for as much as three-quarters of all GDP. When allowance is made for the value added to grain when it is traded and processed, for the value of other plant products (legumes, horticultural crops, hay and herbage, wood and timber), for the value of all livestock products (traction, dairy produce, meat, hides, skins, fleeces, and wool), unprocessed and processed, for mineral production (notably iron, lead, and tin), the manufacture of craft goods, and provision of

---

[49] Britnell, 'Commercialisation and economic development', pp. 9–12.
[50] Snooks, 'Dynamic role', pp. 28–35; Mayhew, 'Modelling medieval monetisation', pp. 71–4.
[51] This has been argued most forcibly by Bridbury, 'Domesday valuation', pp. 124–5.
[52] N. J. Mayhew, 'Appendix 2: The calculation of GDP from Domesday Book', in Britnell and Campbell (eds.), *Commercialising economy*, p. 196, n. 10.

Table 8.07 *Alternative estimates of population, GDP, and net grain output as a percentage ...*

Date	Method	Total population (m.)	Total GDP (£m.)	GDP per head (£)	Total annual requirement of unprocessed grains	
					per total population (£m.)	as % of GDP (%)
1086	Snooks 1	1.53	0.137	0.09	0.153	112
1086	Mayhew 1	2.25	0.300	0.13	0.225	75
1086	Output 3B2	2.30	0.400	0.17	0.242	61
1086	Output 3A1	1.68	0.300	0.18	0.176	59
1086	Mayhew 2	2.25	0.400	0.18	0.225	56
1086	Output 1A2	2.26	0.400	0.18	0.213	53
1086	Output 2A2	2.01	0.400	0.20	0.213	53
1086	Output 3A2	1.68	0.400	0.24	0.176	44
c. 1300	Snooks 2	5.75	4.07	0.71	2.185	54
c. 1300	Mayhew 3	6.00	5.00	0.83	2.280	46
c. 1300	Output 4	4.25	3.70	0.87	1.606	43
c. 1300	Output 5	4.25	4.07	0.96	1.606	39

Notes and sources:
Snooks 1 and Snooks 2 as in: G. D. Snooks, 'The dynamic role of the market in the Anglo-Norman economy and beyond, 1086–1300', in Britnell and Campbell (eds.), *Commercialising economy*.
Mayhew 1, Mayhew 2, and Mayhew 3 as in: N. Mayhew, 'Modelling medieval monetisation', in Britnell and Campbell (eds.), *Commercialising economy*.
Output 3B2: arable area of 8.0 m. ac., grain yields lower than in *c.* 1300, extraction rate of 0.50 (Table 8.04), GDP of £400,000.
Output 3A1: arable area of 5.75–6.0 m. ac., grain yields lower than in *c.* 1300, extraction rate of 0.50 (Table 8.04), GDP of £300,000.
Output 1A2: arable area of 5.75–6.0 m. ac., grain cropping and grain yields as in *c.* 1300, extraction rate of 0.56 (Table 8.04), GDP of £400,000.
Output 2A2: arable area of 5.75–6.0 m. ac., grain cropping and grain yields as in *c.* 1300, extraction rate of 0.50 (Table 8.04), GDP of £400,000.
Output 3A2: arable area of 5.75–6.0 m. ac., grain yields lower than in *c.* 1300, extraction rate of 0.50 (Table 8.04), GDP of £400,000.
Output 4: population and grain output derived by grain-output method; GDP estimate derived from Mayhew 3.
Output 5: population and grain output derived by grain-output method; GDP estimate derived from Snooks 2.

urban services, it follows that unprocessed grain alone can hardly have contributed more than 60 per cent of GDP. On this reasoning a low GDP estimate of £137,000 is again ruled out since it is consistently exceeded by the value of raw grain output. So, too, are all output estimates which take Lennard's 8.0 million arable acres as their starting point. Mayhew's lower GDP estimate of £300,000 also proves inadequate. Even with a population of 1.68 million and lower yields and extraction rates in 1086 than *c.* 1300 raw grain output would have comprised 59 per cent of GDP (Table 8.07).

The single most plausible estimate relates to a population of 2.0 million and arable area of 5.75–6.0 million acres, with grain yields much the same as in *c.* 1300 but a markedly lower extraction rate, and raw grain output equivalent to 53 per cent of a GDP of £400,000. At *c.* 1300 prices this is equivalent to a GDP of £1.6 million and a *per-capita* GDP of £0.8. There can be no doubt that with raw grain production contributing roughly half of its GDP this was a poor, low-income economy; even so, the poverty of the population was less abject than envisaged by Snooks.[53] On the evidence assembled in Table 8.07 higher rather than lower estimates of Domesday GDP and GDP *per capita* make the best sense.

Two centuries later grain production still apparently occupied its dominant position within the economy (Table 8.07). For *c.* 1300 Snooks and Mayhew favour similarly high population totals of 5.75 million and 6.0 million and offer broadly similar GDPs of £4.1 million and £5.0 million, albeit derived by fundamentally different methods (Table 8.07).[54] The grain requirements that correspond with these population totals were probably beyond the physical capacity of the grain sector but, if realised, would have amounted to 54 per cent and 46 per cent respectively of GDP as estimated. In fact, Snooks's GDP estimate is probably of the right order of magnitude but as its method of calculation is unclear and it is partially dependent upon his implausible Domesday estimate it has to be set aside. Mayhew's estimate, as he himself acknowledges, is far from robust and is inflated by the large population on which it is based. Recomputing his GDP estimate for the smaller population of 4.25 million implied by the output evidence yields a national income of £3.7 million, approximately 43 per cent of which would have been accounted for by raw grain output (the equivalent proportion is 39 per cent if calculated on Snooks's GDP of £4.1 million). This is still a very large share of GDP but appears nevertheless to have been somewhat smaller than the equivalent share in 1086, betokening growth elsewhere within the economy.

It is normal in low-income economies for agriculture to contribute well over

[53] Snooks, 'Dynamic role', pp. 28, 32.
[54] Snooks, 'Dynamic role', pp. 49–53; Mayhew, 'Modelling medieval monetisation', pp. 57–62.

half of GDP. If grain had comprised three-quarters by value of all agricultural output in *c.* 1300, as certainly appears to have been true of the seigniorial sector within the FTC counties, then agriculture's overall contribution to GDP may have been of the order of 60 per cent.[55] This is consistent with recent views about the size of the urban sector at that date and may be compared with the 37–43 per cent which N. F. R. Crafts reckons agriculture contributed to GDP at the opening of the eighteenth century, by which time manufacturing and commerce were considerably more developed.[56] In 1688 GDP *per capita* – the measure of real economic growth – was approximately two-and-a-half times what it had been in *c.* 1300 (when the purchasing power of agricultural and builders' wages indicates that living standards were approaching a pronounced temporal low: Figure 1.01).[57] Between 1086 and *c.* 1300, in contrast, there appears to have been only a 10–20 per cent improvement.

Although in real terms GDP may have grown by 130–150 per cent between 1086 and *c.* 1300, because the total population grew by perhaps 89–113 per cent over the same period *per-capita* GDP was only marginally better at its end than it had been at the beginning. It is thus difficult to resist Mayhew's conclusion that for all the institutional, commercial, and technological transformations of the age, the principal economic achievement between 1086 and 1300 'seems to have consisted in supporting more people at roughly the same standard of living'.[58] Perhaps there had been a time in the early to mid-thirteenth century when GDP *per capita* had been higher but by *c.* 1300 this had long since passed and there is abundant evidence to show that the fourteenth century brought mounting adversity for many.[59] Probably if GDP *per capita* failed to rise it was because the 'have-nots' increasingly outnumbered the 'haves': material progress was not yet so widely diffused that all, on average, were able to share in it.

Of course, the provisional nature of all these estimates cannot be over-emphasised. As knowledge and information grow revisions will undoubtedly need to be made. The historical challenge is to come up with aggregate estimates of grain output, agricultural output, total population, urban population, and GDP which are internally consistent both within and between each chosen benchmark year. Those offered here possess that merit. They suggest an arable area which expanded from 5.75–6.0 million acres in 1086 to 10.5 million acres or less in *c.* 1300 but which then contracted to *c.* 8.0 million acres in *c.* 1375 in response to a population which grew from 2.0–2.25 million in

[55] Chapter 4, pp. 183–5, and Chapter 5, p. 188.
[56] N. F. R. Crafts, 'British economic growth, 1700–1831: a review of the evidence', *EcHR* 36 (1983), 189.
[57] Phelps Brown and Hopkins, 'Seven centuries of prices'; Farmer, 'Prices and wages', pp. 772–9.
[58] Mayhew, 'Modelling medieval monetisation', p. 74.
[59] Campbell (ed.), *Before the Black Death*.

1086 to 4.0–4.25 million in *c.* 1300 but which thereafter shrank to 2.25–2.5 million in *c.* 1375. During the twelfth and thirteenth centuries grain output grew marginally faster than the arable area, population marginally faster than grain output, and GDP marginally faster than the population. Population growth was thus underpinned by processes of agricultural and economic change which gradually transformed the structure as well as the volume of production.

9

Adapting to change: English seigniorial agriculture, 1250–1450

9.1   Agriculture before the Black Death: constraints versus incentives

On the evidence of the seigniorial sector English agriculture *c.* 1300 presents a paradox. High prices and low real wages coupled with the great physical extent of the agricultural area, strong arable bias to production, growing emphasis upon grains with high food-extraction rates, and impressive value of agriculturally based exports all imply that agriculture was operating at full stretch, yet only limited advantage was taken of the productivity gains which more intensive and innovative forms of husbandry and modes of land-use were capable of delivering. Across the greater part of the country the intensive extremes of coppiced woodland, closely regulated meadows, enclosed grassland, and more or less continuously cropped arable remained the exception rather than the rule. Instead, the predominance of low unit land values indicates that comparatively extensive methods of managing woodland, grassland and arable remained the norm (Table 9.01 and Figure 9.01). Of sampled demesnes, almost 40 per cent practised the more extensive forms of arable husbandry (Table 6.01 and Figures 6.05, 6.06, and 6.07), over 60 per cent the more extensive forms of pastoral husbandry (Table 4.01 and Figures 4.08 and 4.10), and approximately 50 per cent the more extensive forms of mixed husbandry (Table 4.08). Throughout the south, centre, west, and north of the country seigniorial land-use and husbandry remained less differentiated and technologically progressive than in the east midlands, East Anglia, and south-east. The sheer geographical extent of these agriculturally least developed parts of the country lends weight to G. Astill and A. Grant's verdict that 'we are left with a slightly depressing picture of medieval agricultural progress and productivity, which suggests that opportunities for improvement may have existed but were not taken up'.[1]

Yet by the same date a conjunction of favourable environmental, institutional, locational, and economic factors had combined to nurture the

---

[1] Astill and Grant (eds.), *Countryside*, p. 216.

Table 9.01 *Classification of agricultural land by unit value: England south of the Trent, 1300–49*

Unit value category	% of classified area	Mean unit value of arable (d.)	Mean ratio (unit value : unit value) of:		
			grassland to arable	meadow to arable	meadow to pasture
Very high	2.3	16.2	1.1	1.3	2.9
High	23.3	7.0	2.2	2.8	3.3
Average	32.6	4.3	3.6	4.3	3.9
Low	41.8	2.6	5.7	6.5	5.3
All	100.0	100.0	4.5	5.3	4.5

Notes:
Method: K. Bartley and B. M. S. Campbell, '*Inquisitiones Post Mortem*, G.I.S., and the creation of a land-use map of pre Black Death England', *Transactions in G.I.S.* 2 (1997), 333–46.
Variables: (unit value grassland : unit value arable) ×2.0; (unit value meadow : unit value arable) ×1.5; (unit value meadow : unit value pasture) ×1.5; (unit value arable) ×1.2
Source: National *IPM* database.

evolution of uniquely progressive methods in a few key areas. Eastern Kent, eastern Norfolk, and parts of the East Anglian Fenland and Fen edge stand out in this respect, each distinguished by technologically sophisticated agricultural systems that reconciled high productivity with sustainability. The high unit land values which were the concomitant of their advanced agricultural methods highlight the precociousness of these areas (Figure 9.01). In these few privileged parts of the country – amounting to no more than an eighth of the total agricultural area within England south of the River Trent – the best English methods were practically on a par with those to be found in the more developed parts of the continent, especially in northern France and Flanders.[2] Had these methods been more widely adopted English agriculture would have been capable of producing a greater volume of output than in fact pertained at this medieval climax of economic and demographic expansion.

The story of agricultural change during the hundred years or so prior to the Black Death is therefore one of uneven development. Although change of one sort or another can be found almost everywhere, extensification was generally more important than intensification with the result that innovation and specialisation went very much further in some areas than others. The upshot was

[2] B. H. Slicher Van Bath, 'The rise of intensive husbandry in the Low Countries', in J. S. Bromley and E. H. Kossmann (eds.), *Britain and the Netherlands* (London, 1960), vol. I, pp. 130–53; Thoen, 'Agricultural technology'.

N

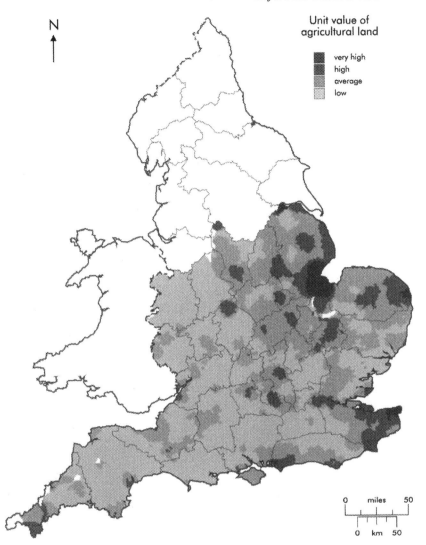

Fig. 9.01. Unit value of agricultural land: England south of the Trent, 1300–49 (see Table 9.01 for explanation) (source: National *IPM* database).

a widening divergence between the least and most progressive and productive husbandry systems. A key puzzle for historians of the pre-Black Death period, as Postan rightly diagnosed in 1972, is therefore 'why the methods, or even the implements, known to medieval men were not employed, or not employed earlier or more widely than they in fact were'.[3] Part of the explanation must

---

[3] Postan, *Economy and society*, p. 42.

obviously lie with the constraints and difficulties – botanical, zoological, biological, institutional, technical, and commercial – against which medieval husbandmen had to struggle and it is these supply-side obstacles to progress which have hitherto received most historical attention. Yet, in a commercialised economy, appropriate incentives needed to be in place before farmers would produce more and the removal of institutional obstacles became worthwhile. Hence, such demand-side factors as the costs of trade, the size and efficiency of the market, and the extent to which demand was concentrated and differentiated were also important.

9.11 Supply-side constraints

Although the agricultural area had probably been stretched further by *c.* 1300 than ever before, a variety of mostly institutional obstacles meant that England nevertheless remained a country with much agriculturally underexploited land. Considerable areas, for example, had been set aside by monarchs and magnates as royal forest and private hunting grounds and as such could not readily be brought into more productive use.[4] There were also substantial amounts of common pasture and waste. True, these did yield a range of agricultural products, but rates of productivity were bound to remain low until more effective means of management and exploitation could be put in place. The medieval solution was to regulate access and exploitation by passing and enforcing bye-laws and imposing stints; later centuries would eventually go further and replace common with private property rights.[5] Enclosure, as the latter process became known, was a costly and socially disruptive undertaking, but it facilitated the substitution of more intensive methods of production, often requiring greater unit inputs of labour and capital. Because of entrenched vested interests and the costs of overcoming them enclosure was not a viable proposition in the early fourteenth century when rural congestion was at its height.[6]

Even when land could be brought into use medieval husbandmen were constrained in what they could produce by the botanical character of the crops that they cultivated and the zoological attributes of the stock that they reared. The grains cultivated on medieval English demesnes seem to have been intrinsically poor reproducers and certainly delivered significantly lower yields per seed than those cultivated by northern French farmers in the same period.[7] Although many different strains of seed must have been sown, there is no con-

---

[4] Young, *Royal forests*; Cantor, 'Forests, chases, parks and warrens', pp. 56–85.
[5] W. O. Ault, *Open-field farming in medieval England* (London, 1972); J. R. Wordie, 'The chronology of English enclosure, 1500–1914', *EcHR* 36 (1983), 483–505; Overton, *Agricultural revolution*, pp. 147–67.   [6] Clark, 'Cost of capital', pp. 276–85.
[7] Chapter 7, pp. 316–17.

temporary English evidence of attempts at seed selection nor of the emergence of specialist seed merchants with a vested interest in a process of seed selection. Not until the seventeenth century would selection upgrade the general quality of seed and thereby deliver improved yield ratios.[8] Medieval livestock were similarly low yielding. Archaeozoological evidence indicates that carcass weights in the fourteenth century were significantly lower than those that had prevailed in Roman and early Saxon times and milk yields and fleece weights were well below those that would prevail in the nineteenth century. Such attributes took time to reverse. Faunal remains indicate a slow but steady growth in the size of cattle and sheep from the fifteenth century as standards of nutrition improved, breeding became more careful and selective, and imports of live animals introduced new genetic strains. Much had been already achieved by the advent of the highly publicised selective breeding of the eighteenth century.[9]

Farmers have always had to work both with and against nature since their efforts could be partially or wholly undone by weeds, pests, and pathogens. Medieval farmers lacked the herbicides, pesticides, and fungicides with which their modern counterparts combat these problems and although they were undoubtedly wise in the ways of nature they lacked scientific knowledge. Getting plants and animals to reproduce successfully was therefore a constant struggle; one, moreover, with the potential always to become tougher should the climate take a malign turn or plant and animal diseases become more virulent. Today it is reckoned that approximately a fifth of the world's grain output is lost to pests and diseases; M. Overton reckons that losses may have reached a third in early modern England and they are likely to have been at least as high in the Middle Ages.[10] Natural predators were dealt with by hawking, hunting, and hiring boys to scare off birds; manorial accounts also record occasional payments to itinerant mole catchers.[11] As if there were not enough pests already, the rabbit which had been introduced for its meat and

---

[8] Allen, *Enclosure and the yeoman*, pp. 206–7.

[9] S. J. M. Davis and J. V. Beckett, 'Animal husbandry and agricultural improvement: the archaeological evidence from animal bones and teeth', *Rural History* 10 (1999), 1–17.

[10] Overton, *Agricultural revolution*, p. 17.

[11] Piers Plowman asks the knight to hunt down hares, foxes, boars and badgers and to set falcons upon wildfowl: *Visions from Piers Plowman*. In the late thirteenth century on the earl of Norfolk's manor of Halvergate in Norfolk a boy was regularly employed to scare off rooks while seed was being sown: PRO, SC 6/936/4–17. Folios 170 and 171 of the East Anglian-produced Luttrell Psalter (BL, Add. MS 42130) similarly depict crows being scared off by a dog and a man with a sling. Canines are not recorded in manorial accounts even though dog bones invariably feature in animal-bone assemblages from archaeological sites: U. Albarella, '"The mystery of husbandry": medieval animals and the problem of integrating historical and archaeological evidence', unpublished paper. On the bishop of Winchester's manor of Downton in 1324–5 2d. was paid to a man who caught sixteen moles in the meadow: Titow, *English rural society*, p. 127.

fur in the twelfth century had become sufficiently acclimatised and numerous by the late fourteenth century to pose a serious nuisance to cereal farmers, especially in the vicinity of warrens.[12]

Crops, once sown, had to compete with weeds.[13] Demesne managers dealt with the problem by preferring tall grain species which could overtop rival plants, by sowing grain thickly (especially at the tail end of rotations when weed infestation was most severe), by ensuring that the seed being sown was clean, by employing casual and customary workers to weed the growing crop, and by repeated ploughing of bare summer fallows to destroy the root systems of perennial weeds.[14] These could be costly measures but they did serve to contain the problem. Against the attacks of insects and pathogens farmers were altogether more impotent. It has recently been suggested that recurrent outbreaks of rust and other parasitic infestations of grain crops were a major reason for cyclically depressed grain yields from at least the fifteenth century on and they may similarly have contributed to the cyclical price variations so apparent in the late thirteenth and early fourteenth centuries (Figure 1.01).[15] Measures adopted to counteract this problem included the exchange of seed corn between manors, the scattering of plots, and the spreading of production across a number of demesnes. Crop rotation should also have provided some protection against pathogens, although few medieval cultivators – including many in otherwise progressive Norfolk – seem to have appreciated that sowing successive grain crops on the same land was more likely to exacerbate than mitigate the problem. This was a particular failing of some of the most intensive rotations. Even when the harvest was safely gathered in, it was vulnerable to fire, mould, vermin attack, and embezzlement. Constructing barns and raising ricks and granaries on staddle stones afforded the securest forms of storage and were forms of investment which lords were best placed to make.[16]

With so many natural adversaries ranged against them, quite apart from the uncertainties of the English climate, it is perhaps small wonder that medieval cultivators succeeded in producing so little from the land. Nor were livestock any more immune to pests and disease. Cattle and especially sheep were prone to a range of diseases collectively referred to by contemporaries as 'murrains'. The more virulent of these outbreaks had the capacity to decimate herds and flocks. The cattle plague of 1319–20, which so reduced the numbers of plough oxen that land had to be taken out of cultivation, was probably an outbreak

[12] Eighty-seven acres belonging to the bishop of Ely at Feltwell in Norfolk could not be sown in 1396–7 because of the depredations of rabbits from the warren of the duke of Lancaster in neighbouring Methwold; by 1422–3 this area had grown to 314½ acres: NRO, Phi/472/1–2 577 ×9.

[13] Straw preserved as thatch confirms how serious weed infestation could be: Hawkes, 'Country thatches'; Letts, *Smoke blackened thatch*, pp. 35–41.

[14] Chapter 6, p. 271; Campbell, 'Agricultural progress', p. 29; Postles, 'Cleaning the arable'.

[15] Scott and others, 'Grain prices', pp. 12–13.

[16] Le Patourel, 'Rural building', pp. 866–74.

of rinderpest.[17] Prevention and cure were beyond the knowledge and powers of contemporaries. Sheep murrains, including outbreaks of scab and liver-fluke, were more prevalent and fatalities in the worst outbreaks could be very high.[18] Shepherds resorted to a variety of medicinal palliatives, including treating infected animals with ointments containing sulphur, copperas, verdi-gris, mercury, or tar, in an endeavour to mitigate infection.[19] The heaviness of the mortalities in the worst outbreaks nevertheless serve as a salutary demon-stration of the limitations of medieval veterinary expertise. By the early four-teenth century sheep murrain seems to have become endemic and it was certainly exacerbated by the extreme weather conditions of the period. Excessive rain in late spring and summer could ruin the hay crop. Severe winters and cold, wet springs left pastures sodden, thereby creating the pre-conditions for outbreaks of liver-fluke to which starving flocks fell easy victim.[20] Excessive folding could also lead to the build up of parasitic infec-tions within the soil. As if these hazards were not enough, livestock were also the rural commodity most prone to theft and were specifically targeted in the repeated Scottish raids upon the north of England which followed the out-break of the War of Independence.[21] In the forests and fastnesses of the north small numbers of sheep and cattle were also regularly killed by wolves.[22] Perhaps it was the king, however, who was the greatest predator of them all, since livestock were an asset he was keen to seize whenever the opportunity arose to take estates into royal hands.[23] Pastoral husbandry was therefore no less risky a business than arable husbandry: indeed, setbacks when they occurred were often greater and were slower and costlier to remedy since it took several years to breed and rear up replacement stock.[24] It is only remark-able that there were so many relatively risk-free runs of years.

Certain environments were plainly more hazardous than others. Sometimes higher returns justified the risks, as was most conspicuously the case of fertile, low-lying, alluvial soils reclaimed for meadow, pasture, and arable and suscep-tible to flooding if ditches and dykes were inadequately maintained and weather and tides proved extreme. By the early fourteenth century, for

---

[17] Kershaw, 'The Great Famine', pp. 106–8; Jordan, *The Great Famine*, pp. 35–9.

[18] In the 1270s an epidemic of scab had such a disastrous effect upon national wool output that prices were driven up; the volume of wool sold by the bishop of Winchester fell by almost two-thirds during these years: Lloyd, *Wool prices*, p. 15. Between 1313 and 1321 murrain reduced Crowland Abbey's Fenland flock from almost 11,000 animals to fewer than 2,000: Kershaw, 'The Great Famine', p. 104.   [19] Trow-Smith, *Livestock husbandry*, p. 156.

[20] E.g. Kershaw, *Bolton Priory*, pp. 83–4: Bolton Priory's flock of 3,027 sheep was reduced by over two-thirds between 1315 and 1317 through the combined effects of bad weather, lack of hay, and epidemic disease.

[21] B. A. Hanawalt, *Crime and conflict in English communities 1300–1348* (Cambridge, Mass., 1979), pp. 70–3; C. McNamee, *The wars of the Bruces: Scotland, England and Ireland, 1306–1328* (East Linton, 1997), pp. 75–7.   [22] Tupling, *Rossendale*, p. 27.

[23] Titow, 'Land and population', pp. 44–6; Biddick, 'Agrarian productivity', pp. 98–104.

[24] Chapter 4, pp. 135, 167–8.

example, reclamation had transformed the low-lying silt fens of East Anglia from a marshy waste into one of the most intensively exploited and prosperous districts in England.[25] Between 1291 and 1341, however, thousands of acres of reclaimed marshland on the east and south coasts were lost to the sea as mounting royal taxation sapped marshland communities of their ability to maintain sea defences at the very time that there was a minor marine transgression and heightened incidence of storm surges.[26]

In lowland mixed-farming contexts where the bulk of working demesnes were located, deep, fertile, relatively easily cultivated loam soils seem to have rewarded those who tilled them with the highest and most reliable yields. Even lighter and inherently less fertile sandy soils could be made to pay since the lower yields which they delivered could be offset against lower costs of cultivation. Ploughing, in particular, required less draught power per unit area than heavier soils, an advantage which was to stand farmers of light-soil in good stead until well into the nineteenth century. Sheep-corn husbandry also proved a particularly cost-effective method of maintaining the fertility of these soils.[27] In contrast, it was the intrinsically more fertile heavy clay soils which appear to have presented the greatest problems to medieval cultivators. Such soils dominated many parts of lowland England and were 'difficult to plough, difficult to break down to a fine tilth, and prone to waterlogging'.[28] In wet years they must have broken the heart as well as the back of many a husbandman.

The distribution of heavy soils is reflected in the distribution of land-use classified as type 3 in Figure 3.14, which is by far the most widely represented of the six demesne land-use types identified in Chapter 3. In East Anglia variants of this land-use type show up on the heavy soils of south-east Norfolk and north central Suffolk, in precisely those locations where demesnes persisted in maintaining the largest and slowest plough-teams and proved most resistant to the wholesale replacement of oxen with horses (Figures 4.13 and 4.14). As a result unit ploughing costs on the abbey of Bury St Edmunds's demesne at Thorpe Abbotts, which employed mixed teams made up of four horses and four oxen, were four times those of Norwich Cathedral Priory's demesne at Sedgeford on the good sands of north-west Norfolk, which by the second quarter of the fourteenth century was operating small three-horse teams. Nor was the effort of ploughing these heavy soils rewarded with superior yields. On the contrary, net weighted aggregate grain yields per acre consistently favoured light-soil Sedgeford over heavy-soil Thorpe Abbotts, whose yields were not only worse but also more variable (Figures 7.08 and 7.09).

[25] Darby, *Medieval Fenland*; R. E. Glasscock, 'The distribution of wealth in East Anglia in the early fourteenth century', *TIBG* 32 (1963), 118–23.

[26] Baker, 'Contracting arable lands'; Bailey, '*Per impetum maris*'.

[27] Bailey, 'Sand into gold', pp. 43–51.

[28] See Overton, *Agricultural revolution*, pp. 58–9, where a map of heavy and light land is given.

The demesne at Thorpe Abbotts may have performed particularly badly, but it was a rare demesne that got significantly better results from these heavy soils. Nationally, it was unusual for stiff and heavy clay soils to command a value of more than 6d. an acre when under tillage, and valuations at or below the national average of 4d. an acre were more typical (Figure 7.13). In the nineteenth century iron ploughs and tiled under-drains would transform clay-land productivity; in the meantime the typical medieval solution to the severe drainage problems presented by these soils was to plough them into steep ridges separated by furrows to carry the water away. From the fifteenth century on, and increasingly during the ensuing early modern period, much clayland was withdrawn from cultivation and converted to grass.[29] This grassing of the heavy clays was a function of a strengthening market for livestock and their products coupled with greater specialisation and inter-regional trade as the structure of demand became more differentiated.[30] Significantly, it has no pre-cursor in the pre-Black Death era of expanding commerce and growing demand for agricultural produce when the indifferent yields mostly obtained from these heavy soils must have depressed mean crop yields within the country as a whole.

The superior performance of Sedgeford compared with Thorpe Abbotts may also have had something to do with the fact that the priors of Norwich Cathedral appear to have been more enterprising landlords than the abbots of Bury St Edmunds. The management policies pursued by landlords could obvi-ously make a material difference to the results that were obtained. Much, too, depended upon the attitude and initiative of the peasant workforce who were charged with so much of the day-to-day operation of a demesne. The coexis-tence of an innovative peasantry could do much to ensure that demesnes within the same locality were cultivated in a progressive manner. Modern studies have found good labour management, linked with the practical and technical ability of the farmer, to be very closely related to overall farm pro-ductivity.[31]

The nature of landlord–tenant relations could also influence the attitude of the workforce: where a landlord was reactionary and repressive or simply inex-perienced and inept his farm servants and other workers were most likely to be negligent, dilatory and even fraudulent.[32] The abbey of Bury St Edmunds was one of the largest and managerially most complex Benedictine landlords in the country and was notorious for its bad relations with its tenantry.[33] The

---

[29] For a case study of land-use trends in Essex over this period see Poos, *Rural society*, pp. 46–51.

[30] Walton, 'Agriculture and rural society', pp. 251–6; Overton and Campbell, 'Norfolk livestock farming', pp. 385, 393–4.

[31] National Economic Development Office, *Farm productivity* (London, 1973).

[32] *Walter of Henley*, p. 317. According to Chaucer, Oswald the reeve of Bawdeswell in Norfolk was adept at deceiving both the auditor and his young master: *General prologue*, pp. 69–70.

[33] H. E. Hallam, 'The life of the people', in Hallam (ed.), *AHEW*, vol. II, pp. 850–2.

rotations reconstructed on its demesnes of Hinderclay, Redgrave, Rickinghall, and Thorpe Abbotts display some of the worst features of medieval cropping, with successive courses of grain on the same land insufficiently relieved by fallow courses or courses of legumes (Figures 6.09, 6.10, and 6.19). The abbey seems to have been indifferent to the poor returns obtained from these demesnes or the fact that in a significant number of years in the fourteenth century Thorpe Abbotts and Tivetshall must have operated at a considerable loss. Here, as on many another great ecclesiastical estate with a strong adherence to feudal socio-property relations, labour motivation and work efficiency offered particular scope for improvement. Clark, for instance, has estimated that *c.* 1300 labour productivity in agriculture was less than a quarter what it was to be in 1850, with much of the subsequent gain coming from increased work intensity.[34]

Unlike the abbot and monks of Bury St Edmunds, the priors of Norwich were clearly intent upon getting the most out of their estates. They were assiduous in refocusing their demesne activities and expanding the overall area under crop, if need be substituting legumes for fallows in order to do so.[35] Significant sums were spent manuring those of their demesnes closest to Norwich and a close interest was taken in the crop yields obtained on all their demesnes.[36] Priors William de Kyrkeby and Henry de Lakenham took a keener interest in whether their arable demesnes were paying their way than almost any other landlords in late-thirteenth- and early-fourteenth-century England, and towards that end developed accountancy procedures that verged on the calculation of profit and loss.[37] In the 1330s and 1350s the priors were quick to lease out demesnes when it appeared expeditious to do so, and were equally prompt at reversing that decision when it proved advisable. Individual priory demesnes took full advantage of the best local knowledge and its two most intensively cultivated demesnes of Martham and Hemsby thereby obtained record yields and estimated profits per acre. Although entitled to customary labour services, all the priory demesnes employed a core staff of paid *famuli*, augmented by casual and customary workers in season and as required. This preference for hired over customary labour may have given them a real productivity edge and may also have delivered significant efficiency gains.[38]

The lessons of the prior of Norwich seem to have been lost on the bishop, who used mostly customary labour to work his demesne at Langham in north Norfolk to the apparent detriment of its productivity.[39] The bishops of Winchester were similarly conservative in their reliance upon customary

---

[34] Clark, 'Labour productivity', pp. 221, 231–5.
[35] Figure 5.03; Chapter 5, pp. 232–5.   [36] Chapter 7, pp. 347, 360–2.
[37] Stone, 'Profit and loss'; Postles, 'Perception of profit'.
[38] Stone, 'Hired and customary labour'.
[39] Raynham Hall, Norfolk, Townshend MSS, Langham accounts; NRO, MS 1307–9 2 B3, 20336 126×3, 1554–5 1 C1; Chapter 7, p. 324.

labour and Thornton has recently suggested that this may be why yields on the bishops' five great demesnes at Taunton in Somerset were not higher in the first half of the fourteenth century.[40] Nevertheless, the depressing effect that customary labour may have had upon the productivity of the seigniorial sector should not be exaggerated. Many demesnes, especially those on small manors belonging to minor lay lords, had always been inadequately provided with customary labour; many, too, where there was an abundant labour supply available for hire and market demand was strong, were commuting their labour services and substituting hired workers.[41] The substitution of hired for customary labour is a subject which would repay more systematic exploration: a cost-benefit analysis of hired versus customary labour in different economic contexts might certainly be revealing. While hired labour was undoubtedly the better motivated and more productive, customary labour, at least in certain situations, may have been the more profitable.

Among landlords, perpetual institutions were potentially the most hidebound by managerial conservatism; on conventual and collegiate estates inertia rather than enterprise could all too easily rule. Nevertheless, such landlords were also in a uniquely privileged position to develop the management of their estates on a long-term sustainable basis. This applied particularly to pastoral husbandry which was spared the periodic asset stripping experienced by most lay and episcopal estates whenever there was a wardship or vacancy. It is therefore no coincidence that mean stocking densities tended to be higher on demesnes in conventual and collegiate ownership than on other categories of estate where capital accumulation was subject to recurrent setbacks. Predictably, demesnes taken into the king's hands tended to be the most understocked of all, nor do royal demesnes appear to have fared much better.[42] In matters of agricultural management the crown appears to have been the greatest of the negligent landlords of England: an illustration, no doubt, of the diseconomies of scale.

Most demesne managers undoubtedly acquired their practical and technical knowledge and experience on the job. Much information and advice must also have been exchanged between manors belonging to the same estate and estates belonging to the same religious order. In addition, from the mid-thirteenth century written treatises on agriculture and accounting began to be

---

[40] C. Thornton, 'The level of land productivity at Taunton, Somerset, 1283–1348', unpublished paper.

[41] On the evidence of the 1279 Hundred Rolls 46 per cent of all demesne land was unprovided with labour services: Kosminsky, *Studies*, p. 287. On manors with labour services these accounted for only about a third of all the labour employed: Postan, *The famulus*, p. 5. Britnell, 'Commerce and capitalism', p. 364, estimates that wage labour 'may have accounted for about a fifth to a quarter of the total labour expended in producing goods and services'.

[42] Within the FTC counties *c.* 1300 stocking densities on demesnes under crown management were 25 per cent below average, whereas those on conventual, collegiate, and episcopal demesnes were 17 per cent above average: FTC1 accounts database.

produced with the aim of helping landlords and their officials get the most out of their estates by providing a guide to sound administrative and agricultural practices. The earliest of these, Robert Grosseteste's *Rules*, was written in the early 1240s specifically to advise the newly widowed countess of Lincoln on the management of her property, and was subsequently copied for use on other estates. It appears to have been followed in the 1250s or 1260s by the *Seneschaucy*, intended to assist men of legal background who found themselves taking up posts that involved a knowledge of estate management. Then, in perhaps the 1270s or 1280s, came Walter of Henley's *Husbandry*, the most influential and widely read of these treatises. It provided relatively minor landlords with practical advice on how to extract, honourably and honestly, the greatest profit from their estates. Finally, the anonymous *Husbandry*, composed at the very end of the thirteenth century, gave advice on presenting and auditing manorial accounts. These treatises were far from revolutionary in their effects; but they are symptomatic of a changing mentality and, in a modest way, engendered a more professional approach to the managerial problems to which the era of estate high farming gave rise.[43] Against this must be set their emphasis upon tried and tested methods, which may have discouraged some landlords from experimentation and innovation.[44] Too slavishly followed, they fostered a conservative turn of mind.[45]

There is ample evidence, however, to demonstrate that lords were perfectly capable of investing in new technology and forms of organisation when it suited them. Widespread windmill construction in the period 1180–1300, made possible by a breakthrough in woodworking technology, provides a striking example of the rapid diffusion of an important innovation.[46] Moreover, because of the substantial capital sums involved – a new windmill cost about £10 to erect in the late thirteenth century – it was from the start an almost exclusively seigniorial innovation.[47] The first recorded reference to a windmill is in 1185 and within a century of its introduction it had become a common sight in many parts of England.[48] The spread of the windmill appears to have followed a classic diffusion pattern, with a slow period of initial adoption giving way to a spate of intense windmill-building activity in the 1230s and 1240s, after which the rate of diffusion gradually slackened. Adoption of improved record keeping and accounting is an equally striking example of the seigniorial capacity for innovation, creating in the manorial account the single most vital source for the analysis of technological change in medieval agriculture.[49] Lords, too, were behind the introduction of the

[43] D. Oschinsky, 'Medieval treatises on estate accounting', *Economic History Review* 17 (1947), 52–61; *Walter of Henley*; Harvey, 'Agricultural treatises'; Miller and Hatcher, *Rural society*, p. 214.   [44] Langdon, 'Agricultural equipment', p. 106.

[45] Langdon, 'Was England a technological backwater?', p. 277.

[46] R. Holt, *The mills of medieval England* (Oxford, 1988), pp. 17–35; Astill, 'Archaeological approach', pp. 212–13.   [47] Holt, *Mills*, pp. 86–7.   [48] Holt, *Mills*, pp. 17–35.

[49] Chapter 2, pp. 26–37; *Manorial records of Cuxham*, pp. 12–71; Clanchy, *From memory to written record*.

rabbit to England from France at some stage in the twelfth century.[50] Here, however, the diffusion process was slower. Notwithstanding that rabbits were highly prized for their meat and fur, they were slow to acclimatise to their new surroundings and required careful rearing and cosseting inside specially created warrens. Establishing and exploiting warrens was therefore a skilled business. Although many were established during the second quarter of the thirteenth century it was not until the fourteenth century that rabbits had become sufficiently acclimatised to support a steadily rising annual cull: a clear illustration of demand running up against a supply constraint which it took time to resolve.[51]

Building up an adequate stock of horses similarly took time and was an important determinant of the speed with which horses replaced oxen for farm work. Horse haulage never became universally adopted during the Middle Ages while all-horse traction always remained confined to the Chiltern Hills, parts of Kent and Norfolk, and a few other isolated locations. In part, the wider adoption of horses for haulage than for traction reflects the fact that the former was essentially a 'bolt on' innovation whereas the latter was more integral to the prevailing system of husbandry and therefore contingent upon a host of associated agronomic adjustments and changes. The same was true of the diffusion of vetches whose adoption was often closely related to the adoption of horse power since they were commonly grown as a fodder substitute for hay.[52] Vetches were a relatively cheap innovation but relied for their spread upon a supply of seed and an appreciation of the advantages they offered over other better-established legumes.

As J. Mokyr points out, much agricultural technology, especially that to do with cropping, tends to be highly site specific. Hence new techniques often have to be modified and adapted to suit the particular factor endowments of individual producers. A significant part of the development cost is thus imposed on the user of an innovation, and the additional experimentation slows down the process.[53] Consequently, farming systems could never be transferred lock, stock, and barrel from one context to another. Because so much trial and error was involved the diffusion of improved agricultural systems invariably proceeded at a pace that was more evolutionary than revolutionary.[54] The very slowness and difficulty of the process meant that none of the key agricultural innovations of the age was therefore universal in its adoption. It was those populous parts of the east and south-east most strongly exposed to concentrated demand and the quickening pace of commerce that

---

[50] E. M. Veale, 'The rabbit in England', *AHR* 5 (1957), 85–90.

[51] M. Bailey, 'The rabbit and the medieval East Anglian economy', *AHR* 36 (1988), 1–20.

[52] Campbell, 'Diffusion of vetches'.

[53] J. Mokyr, *The lever of riches: technological creativity and economic progress* (Oxford, 1990), p. 32.

[54] For a case study of a demesne in the process of changing its farming system see Hogan, 'Clays, *culturae*'.

consistently proved themselves to be most receptive to the adoption of new technology. It was here, by the early fourteenth century, that windmills, written accounting, rabbit warrens, horse traction as well as horse haulage, vetches, continuous cropping, and substantial barns of sophisticated construction were all most likely to be encountered either separately or in combination. It was here, too, that medieval agricultural systems attained their fullest development.

Lords and their managers were hardly a uniform group. Variations in age, ability and aptitude will have compounded the more obvious differences in the scale and type of estate, location and composition of household, and command over land, labour, and capital. Standards of management are therefore bound to have varied considerably from demesne to demesne and estate to estate. To presume that all lords were equally willing to make the most of available agricultural techniques and opportunities would be a mistake. For many, the exercise of lordship rather than enlightened land management was the priority. Even the most enterprising often had to contend with entrenched conservatism, inflexible commonfield regulations, reluctant workers, high costs of capital, and low unit land values. Such disincentives may, however, have had as much if not more to do with the limitations of demand as with the shortcomings of supply.

9.12 Demand-side incentives

Landlords knew what their estates were for, and that was to yield income.[55] Few would have been so irrational as to pass by profit opportunities that were to be had for the taking. If lords were not prepared to take advantage of available economic opportunities demesnes could always be leased to those who were. Direct management only made sense if the pecuniary advantage that it offered meant that it gave better returns than leasing while at the same time supplying lords with provisions they could not conveniently obtain by alternative means. During the thirteenth century market demand had grown at all levels, with the result that the production decisions of demesne managers were increasingly influenced by the relative price of agricultural products, rent of land, and cost of labour.[56] As the market grew so, too, did the incentive to specialise; indeed, with few remaining opportunities for extending the agricultural area, specialisation and exchange provided the best prospect of averting diminishing returns and off-setting environmental limitations.[57] The extent of that specialisation was however determined by the size of the market.[58]

[55] Bolton, *English economy*, pp. 101–2.
[56] Persson, *Pre-industrial growth*; Britnell, *Commercialisation*; Miller and Hatcher, *Towns, commerce and crafts*; Campbell, 'Matching supply to demand'.
[57] Persson, *Pre-industrial growth*, pp. 71–3.
[58] Campbell, 'Ecology versus economics', pp. 91–4; Campbell, 'Economic rent', pp. 235–42; Campbell and others, *Medieval capital*, pp. 46–75.

Proliferating markets, fairs, and boroughs during the twelfth and thirteenth centuries testify to the vigour with which internal trade was then expanding.[59] Towns not only increased in number but grew in size and, on the most generous estimates, doubled their share of the population from perhaps 10 per cent in 1086 to 20 per cent in 1300.[60] Greater economic and market integration delivered disproportionate benefits to the cities at the top of the urban hierarchy. Between 1086 and 1300 London consolidated its primacy and more than trebled its size.[61] Where in 1086 possibly one in ninety English people had lived in London, by 1300 this had narrowed to approximately one in sixty. London grew in part because during the thirteenth century its role as a capital city was enhanced. The king and his court spent more and more of their time in London, and Westminster became in effect the permanent seat for the Exchequer, Chancery, and Royal Courts of Law, attracting governmental, administrative, and legal business to the city. The metropolis benefited even more from the enhancement of its role as a regional, national, and international centre of commerce. It was the one English port frequented by merchants from all the countries with which England traded and denizen merchants were more active in its overseas trade than in that of any other domestic port.[62] Owing to their initiative, London captured an increasing share of the expanding volume of overseas trade. Specialist groups emerged within the city dependent upon trade and commerce for a living and to service that trade London became a manufacturing and distribution centre, with the finest London-made goods commanding a national market.[63]

Great cities like London were the forcing ground of agricultural change as levels and contours of economic rent within their hinterlands were reconfigured in response to expanding urban demand. Economic rent, as Ricardo was the first to realise, is partly a function of land quality and population density but more particularly, as Thünen demonstrated in 1826, of distance from the market as determined by transport costs.[64] Since the transportation of goods to the market involves costs, the distance that they have to be transported determines the type of goods produced and the manner of their production. Economic rent also varies according to the facility with which a commodity can be transported and its perishability. Thus both bulky goods such as oats and faggots and perishable goods such as fresh milk and

[59] Miller and Hatcher, *Towns, commerce and crafts*, pp. 135–80.
[60] Dyer, 'How urbanized was medieval England?'.
[61] D. J. Keene, 'Medieval London and its region', *London Journal* 14 (1989), 99–111.
[62] Carus-Wilson and Coleman, *Export trade*, pp. 41–6; Miller and Hatcher, *Towns, commerce and crafts*, pp. 214–15, 228–9.
[63] Keene, 'London and its region'; Nightingale, 'Growth of London'; Miller and Hatcher, *Towns, commerce and crafts*, pp. 181–254.
[64] D. Ricardo, *The principles of political economy and taxation*, ed. O. St Clair (London, 1957), pp. 33–45; Grigg, *Dynamics*, pp. 50–1, 135–40; Thünen, *Isolated state*; Chisholm, *Rural settlement*, pp. 20–32.

fat animals yield an economic rent that declines very sharply with distance from the market. Economic rent further affects production costs and thereby determines the type of farming system which is employed, so that although the crop being grown and livestock kept may remain the same, the manner of their cultivation and management may vary. This particularly affects the intensity of their production, which tends to decrease with distance from the market as lower production costs are necessary to offset higher transportation costs.

Analysis of land-use and farming systems within London's hinterland *c.* 1300 reveals that they were strongly influenced by the provisioning needs of the capital.[65] Nevertheless, at this early stage in the city's history London's 'Thünen field of force' embraced – for all commodities – no more than perhaps a fifth of the total national land area. Of this, less than half was engaged in the regular supply of grain to the city. Even within that area of regular grain supply relatively low levels of economic rent prevailed over high, with the result that extensive systems of production prevailed over intensive.[66] This is consistent with the Thünen model which predicts that within the provisioning hinterland of a major city high-yielding, intensive systems will occupy 6 per cent of the arable, moderately productive and semi-intensive systems a further 54 per cent, and low-yielding extensive systems 40 per cent of the arable.[67] The provisioning hinterlands of lesser urban centres were even smaller, few drawing their grain from beyond a radius of 25 miles.[68]

Little wonder, therefore, that, at a national scale, low economic rent and extensive farming systems remained very much the order of the day. Low unit land values accounted for half of the total agricultural area south of the River Trent and were predominant throughout those southern and western parts of the country most remote from London and the leading entrepots of the east coast (Figure 9.01). For producers in these areas 'the local markets and the communities around them were the more important outlets for the produce of the countryside'.[69] Such markets were, however, incapable of stimulating economic rent and agricultural intensification to the same extent as major urban concentrations of demand. Consequently, low economic rent lay like a shadow across the land, discouraging intensification, innovation, and specialisation and causing extensive systems of production to predominate. Only where the economic rent for subsistence production exceeded that for com-

---

[65] Campbell and others, *Medieval capital*; Galloway and others, 'Fuelling the city'.

[66] Campbell, 'Economic rent', pp. 240–3.

[67] By comparison, south of the Trent high unit land values accounted for 12 per cent, average unit land values for 35 per cent, and low unit land values for 52 per cent of the agricultural area: Figure 9.01.

[68] Thünen, *Isolated state*; Campbell and others, *Medieval capital*, pp. 172–4; Campbell, 'Economic rent', pp. 241–4; G. W. Grantham, 'Espaces privilégiés: productivité agraire et zones d'approvisionnement des villes dans l'Europe préindustrielle', *Annales* 52 (1997), 695–725.

[69] Farmer, 'Marketing', p. 329.

mercial production may the intensity of production on peasant holdings have exceeded that on demesnes in the same locality.[70]

On this analysis the prevalence throughout much of northern, central, western, and southern England of extensive farming systems, relatively unprogressive in technology and characterised by low output per unit area, is symptomatic less of ignorance, backwardness or under-investment than of the limited economic opportunities with which most producers had to contend. In England's land-locked interior, especially, major markets were mostly too remote and transport costs too high to justify investment in the kinds of intensive and productive methods practised in more favourably situated parts of the country. This may help to explain the low yield levels with which a number of medieval English landlords evidently acquiesced. Certainly, as described by Biddick, the management policy pursued by the bishop of Winchester on his demesnes in southern England in the early thirteenth century is redolent of what Thünen's theory would predict for an area of low economic rent. Arable crops were produced for predominantly local consumption by the bishop and his household using relatively extensive methods, while wool and dairy products – high in value relative to their bulk and thereby better able to withstand the costs of carriage to distant markets – were relied upon as the principal agricultural sources of cash.[71]

For most producers penalised by distance either from major urban centres or cheap water transport, producing grain for local and regional markets and live animals and wool for national and international markets may well have been the best commercial options; hence the pre-eminence of wool and wool products among English exports, notwithstanding wool's relatively modest share of gross agricultural output.[72] It had the twin merits of a high value relative to its bulk, which ensured its transportability, and an ability to withstand prolonged storage without significant deterioration. As such it provided a commercial life-line for peasants and lords alike. An alternative strategy in areas of low land values and therefore cheap provisions and low wages would have been to turn low labour costs to advantage by producing manufactured goods for sale in distant markets. This was not to occur on a significant scale until the close of the Middle Ages, when it was to transform the fortunes of many hitherto under-developed areas in the midlands, the north, and the south-west of the country.[73]

J. Langton and G. Hoppe have stressed the mutually beneficial tripartite

---

[70] Campbell, 'Economic rent', pp. 235–8.
[71] Biddick, 'Agrarian productivity'. See also Farmer, 'Marketing', pp. 349, 360–2, 397. Cf. D. L. Farmer, 'Two Wiltshire manors and their markets', *AHR* 37 (1989), 1–11.
[72] Chapter 4, pp. 155–6; Miller and Hatcher, *Towns, commerce and crafts*, p. 213. On demesnes in the London region *c.* 1300 wool accounted for only an estimated 9 per cent of gross agricultural output by value: Campbell, 'Measuring commercialisation', p. 149.
[73] Baker, 'Changes', pp. 218–36.

relationship which subsequently developed during the early modern period between expanding metropolitan demand, the evolution of capitalist agriculture, and the growth of proto-industrialisation.[74] But in the thirteenth and fourteenth centuries the corresponding relationship remained bipartite rather than tripartite, since proto-industrialisation existed only in embryo, and the capacity of urban demand alone to stimulate agricultural development was further qualified by the lesser scale of the metropolis and other leading urban centres. The central problem of medieval agricultural production was therefore as much a deficiency of demand as an inelasticity of supply. Farm specialisation may have represented one of the most effective ways of achieving a greater division of labour and making best comparative use of a range of pedological and environmental types, with all the productivity benefits which this implies, but so long as the demand of the urban and industrial sectors for food and raw materials remained circumscribed the agricultural sector would be unable to reap the full rewards of greater specialisation.[75]

The crucial historical lesson to be drawn from this is that technology *per se* was not the limiting factor upon agricultural productivity and output which it has so often been represented as being. Structural, institutional, and economic disincentives to greater specialisation and the wider adoption of available technology were more important. In the increasingly commercialised world of the thirteenth century it was factor costs and economic rent that determined in the main how land was used, what was produced, and whether it was worth adopting more intensive and innovative methods. When and where circumstances were propitious medieval English farmers were as innovative and productive as their seventeenth-century successors, although change in neither period was rapid.[76] In fact, much seventeenth-century progress was based upon the wider adoption of essentially medieval methods and it is the diffusion and progressive improvement of those methods which constitutes a central theme of the post-medieval story of English agriculture.[77]

For agriculture, the vital difference between the respective demographic and economic high-tide marks of the fourteenth and the seventeenth centuries was therefore less the available technology than the size, composition, and concentration of the market. Thus, where by the late seventeenth century London drew selectively upon most of England to provision its population of 400,000, in the early fourteenth century the capital relied in the main upon the Thames

---

[74] J. Langton and G. Hoppe, *Town and country in the development of early modern Western Europe* (Norwich, 1983).

[75] Cf. P. K. O'Brien and G. Toniolo, 'The poverty of Italy and the backwardness of its agriculture before 1914', in Campbell and Overton (eds.), *Land, labour and livestock*, pp. 405–9.

[76] P. Glennie, 'Continuity and change in Hertfordshire agriculture, 1550–1700: ii – trends in crop yields and their determinants', *AHR* 36 (1988), 145–61; Campbell and Overton, 'New perspective', pp. 74, 92–3; Langdon and others, 'Introduction', pp. 2–3.

[77] Glennie, 'Continuity and change', pp. 155–6; Campbell and Overton, 'New perspective', pp. 88–95; Overton, *Agricultural revolution*, pp. 197–203.

basin and adjoining counties of the south midlands and East Anglia to support a much smaller population of 60–80,000.[78] At this earlier date the capital and other major urban centres at home and abroad were too small to raise economic rent over a sufficiently wide area to encourage agricultural intensification by more than a minority of farmers in a few favoured localities. Across much of the country the impulse to change remained weak with the result that limited technological development and low productivity remained the order of the day on the majority of demesnes. Rather than invest costly capital in the land, where the value of that land was low, landlords quite sensibly preferred to invest in better administration (written accounting), enhanced food processing (windmills), improved transportation (bridges and horse haulage), and the development of a commercial infrastructure (markets, fairs, and boroughs).

The key question for historians of the early fourteenth century is therefore less why agriculture was not more productive than why London and the leading provincial cities were not larger. Certainly, there is no reason to suppose that the provisioning of a larger medieval metropolis would have presented any insuperable organisational problems, for a well-developed commercial infrastructure was already in place and medieval supply systems bear every sign of having been elastic.[79] Nor, in the long run, was the rising agricultural productivity of its hinterland the source of the capital's growth. Rather, London grew because its political and commercial status was enhanced. Such enhancement took time to bring about and arose from processes essentially exogenous to agriculture. Yet until the metropolis gained in gravitational force and the economy became more developed and diversified surplus labour would remain entrapped on the land, heightening rural congestion, depressing labour productivity in agriculture, and ensuring that demand remained more dispersed than concentrated. While this was the case the incentives to specialise, intensify, invest, and innovate were bound to remain selective in nature and geographically circumscribed in impact.[80] This is not to belittle the amount of urban and commercial development that had already taken place by *c.* 1300, which was certainly greater than anything that had occurred before, but rather to emphasise how much still remained to be achieved. Creating a more open and agriculturally less self-sufficient economy was here of paramount importance. It was this which eventually provided the precondition for

---

[78] F. J. Fisher, 'The development of the London food market, 1540–1640', *Economic History Review* 5 (1935), 46–64; Wrigley, 'London's importance'; Wrigley, 'Urban growth'; Campbell and others, *Medieval capital*.

[79] On marketing links and supply networks see Campbell and others, *Medieval capital*, pp. 78–110, 182–3. Cf. the more specialised and intensive husbandry of relatively highly urbanised Flanders: Thoen, 'Agricultural technology'.

[80] The timing of the post-medieval yield rise in Hertfordshire, Oxfordshire, and Norfolk is consistent with a spatially selective rise in economic rent as the metropolitan market expanded: Overton and Campbell, 'Productivity change', pp. 40–2.

a fuller articulation of agricultural productive forces. This symbiotic process of commercial and agrarian development would nevertheless take another five centuries to come to full fruition.

9.2   Agriculture after the Black Death: adapting to a major demand-side shock

Until *c.* 1300 strengthening centripetal market forces within the north European economy as a whole, focusing upon the core economic region subtended by the cities of Paris, Ghent, and London, had exercised a growing influence upon English agriculture. The impact of those forces is apparent in an increasingly hierarchical differentiation of husbandry types, whose spatial configuration reflects the growing primacy of economic over purely environmental or institutional factors.[81] A more or less national market for wool had developed, a long-distance, inter-regional trade in livestock had become established, and grain was being traded over ever greater distances, especially wherever water transport was available and the gravitational pull of major cities was strong. Had change continued in this vein it seems certain that further agricultural specialisation would have resulted and the intensity of production would have risen.

Yet at the same time there were several more ominous developments. Demand was becoming increasingly weighted towards low-income consumers, who had been obliged to trade down to the cheapest grains. Food prices were becoming more volatile, heightening the risks of market participation and dependence. Crown interference was undermining the once lucrative wool trade. War, taxation, and purveyancing were becoming ever more disruptive and burdensome, threatening the peaceful pursuit of trade and agriculture, as territorially ambitious monarchs drew upon their enlarged economic and demographic resources to engage in wars of domination and conquest.[82] Shortages of draught animals and seed were forcing land to be withdrawn from cultivation.[83] To an agrarian economy thus constructed the demographic collapse of mid-century was a catastrophe with a silver lining. The dramatic demographic reduction in the size of the market may have meant that there was less scope for agricultural specialisation and intensification; but it also made possible a radical restructuring of demand to the dietary advantage of the poorer social groups. Moreover, market forces now favoured tenants rather than lords thereby reinforcing the existing trend towards the substitution of hired for customary labour.

---

[81] Power and Campbell, 'Cluster analysis'.

[82] J. H. Munro, 'Industrial transformations in the north-west European textile trades, *c.* 1290– *c.* 1340: economic progress or economic crisis?', in B. M. S. Campbell (ed.), *Before the Black Death* (Manchester, 1991), pp. 110–48; R. R. Davies, *Domination and conquest. The experience of Ireland, Scotland and Wales 1100–1300* (Cambridge, 1990).

[83] Baker, 'Contracting arable lands'; Campbell, 'Ecology versus economics', pp. 94–6.

Lords responded to these altered economic circumstances in a number of mostly predictable ways. Many gave up direct demesne management and leased out some or all of their demesnes, in the process becoming less reliant upon direct consumption for the provisioning of their own households. The seigniorial sector thus began to contract absolutely as well as relatively until by the mid-fifteenth century it was only a few mostly home farms that remained in hand. Arable production was scaled down and altered in composition as demand for the cheaper bread and brewing grains dwindled and *per-capita* ale consumption grew. Pasture expanded at the expense of arable and on some open-field demesnes more of the arable was sown with fodder crops, enabling more animals to be kept.[84] Fuller use was made of cart-horses, especially in those parts of the country which serviced London, and – except where the adoption of horse-ploughing was already well advanced – ox-ploughing began to make a come-back. As arable husbandry contracted so, too, did its draught requirement, with the result that the pastoral sector became less subservient to the arable and the proportion of non-working animals expanded from 49 per cent to 58 per cent of livestock units. Sheep benefited most, especially wherever lords began to expropriate the fold rights of their tenants, but more cattle were also kept, initially for milk but increasingly for meat. Together, a 13 per cent reduction in the mean grain acreage and 40 per cent increase in the mean number of livestock units yielded a 56 per cent increase in mean stocking density per demesne; the corresponding rise in the mean stocking density of non-working livestock was an impressive 90 per cent. But for the farming out of dairy herds that rise in mean stocking densities would undoubtedly have been greater.[85]

Modifying and developing existing husbandry systems largely accommodated these pronounced production shifts. Five rather than six pastoral husbandry types may be distinguished in the post-Black Death period (Table 4.01), six rather than seven arable husbandry types (Tables 6.01 and 6.02), and seven rather than eight mixed husbandry types (Table 4.08).[86] Apart from the emergence of a new arable husbandry type distinguished by the large-scale fodder cropping of legumes, no fundamentally new farming systems came into being. On the contrary, differences between farming types became less rather than more marked. There was a pronounced diminution in the intensity of the more intensive systems and seigniorial husbandry almost everywhere became more mixed. A continuum of mixed-farming systems – intensive mixed farming, mixed farming with sheep, and extensive mixed farming – emerged as the most characteristic demesne husbandry type of the post-Black Death period. Of these three systems extensive mixed farming was the most widely represented, emerging as a truly national system with important

---

[84] Chapter 5, p. 246; Chapter 6, pp. 276–85.   [85] Chapter 4, pp. 150–1, 172–83, 186.
[86] Campbell and others, 'Demesne-farming', pp. 138–43.

regional variations. Demesnes operating this farming type registered some of the most notable gains in stocking densities no doubt as a consequence of the active conversion of arable to grassland. Sheep-corn husbandry, which proliferated on southern downlands and woldlands, registered even more striking gains in stocking densities, reflecting the general tendency for the more extensive, grass-based forms of mixed husbandry to become both more numerous and more pastoral.

It was in central and southern England that these mixed-farming systems made their most striking gains. These areas stood in the vanguard of the shift from corn to horn – as they were to do again in comparable periods of agrarian change – for the simple reason that they were best placed environmentally and economically to switch resources from arable to pastoral production.[87] Above all that meant converting arable land to pasture. Further east, and especially on the lighter soils of East Anglia, the mixed-farming systems that had developed were traditionally both more intensive and more arable based, insofar as they were underpinned by fodder cropping and temporary grazing/folding on the arable. Demesne producers here lacked an equivalent comparative advantage when it came to substituting more extensive and, consequently, more grass-based mixed-farming systems. They had prospered when the demand for grain was strong and now, as in all subsequent agricultural recessions, suffered as the terms of trade shifted in favour of lower-cost forms of pastoral production.[88] Demesne-farming systems in the west and north were similarly slow to change.[89] Here methods of pastoral farming had always been relatively extensive and an inherent land-use bias towards grass meant that good arable land remained at something of a premium. There was therefore no great incentive to alter the established balance of production, at least for the time being.[90]

By the opening of the fifteenth century, therefore, the country's agricultural geography had been subtly but profoundly transformed and changes were set in train which would work themselves out over the course of that century. Underpinning that transformation was a reconfiguration of economic rent as land suited to pastoral production gained relative to that suited to arable, as the relative distribution of population and therefore the demand for land

---

[87] For the expansion of grass at the expense of arable after 1349 in the midlands see Dyer, 'Occupation of land: west midlands', pp. 78–80; Dyer, *Warwickshire farming*, pp. 9–12. For the shifting frontier between corn and horn in the seventeenth century see Kussmaul, 'Agrarian change'.

[88] Cf. J. D. Chambers and G. E. Mingay, *The agricultural revolution 1750–1880* (London, 1966), p. 181; P. J. Perry, 'Where was the "Great Agricultural Depression"? A geography of agricultural bankruptcy in late Victorian England and Wales', *AHR* 20 (1972), 30–45.

[89] E. Miller, 'The occupation of the land: Yorkshire and Lancashire', in Miller (ed.), *AHEW*, vol. III, pp. 45–8; Fox, 'Occupation of land: Devon and Cornwall', pp. 152–63.

[90] For evidence of a late fifteenth-century swing to grassland in the south-west see Fox, 'Occupation of land: Devon and Cornwall', pp. 152–4.

shifted, and as the hinterlands of major urban markets were redrawn.[91] Changing factor and commodity prices played to the comparative advantage of some regions more than others. Population densities fell but the rate of fall was spatially uneven as heightened migration compounded local and regional differentials in fertility and mortality.[92] The thinning population of much of the countryside was matched by the dwindling size of the greater towns and cities.

By 1500 only five towns (Bristol, Exeter, London, Newcastle-upon-Tyne, and Norwich) remained with 10,000 or more inhabitants and London – the primate city – had shrunk to perhaps 40–50,000 inhabitants.[93] The urban share of the population may have held up reasonably well but this was small consolation for the loss of the kind of concentrated demand capable of driving up economic rent and generating highly differentiated patterns of land-use within a wide provisioning hinterland.[94] After 1380 the selective impact of concentrated urban demand upon the countryside diminished and the lengthening shadow of low economic rent upon the land dissuaded farmers from becoming too specialised or intensive in their production.[95] Not only were townsmen fewer but *per capita*, like the rest of the population, they were also consuming less pottage and bread and more ale and meat than at the beginning of the century, with the result that their provisioning hinterlands for grain shrank by more than their hinterlands for pastoral products.[96] The demise of Henley, 68 miles upstream from London, as a major grain entrepot for that city is symptomatic. The urban incentive for arable farmers to specialise was therefore weaker and more restricted than at the beginning of the fourteenth century when most urban populations were at their medieval peak. Moreover, in a thirstier age the grain that was demanded was as likely to be for brewing as for baking and sometimes therefore had a different agricultural provenance.[97] Livestock producers fared somewhat better. Urban demand for meat was more buoyant than that for bread and animals could be driven to market over considerable distances. The only real problem was the increasing tendency towards over-production of animals with the result that growing numbers of

---

[91] Thünen, *Isolated state*; Chisholm, *Rural settlement*, pp. 20–32; Grigg, *Dynamics*, pp. 135–40; Campbell and others, *Medieval capital*, pp. 4–7.

[92] The changing distribution of population is implicit in the changing distribution of taxable wealth: Schofield, 'Geographical distribution of wealth'; Darby and others, 'Distribution of wealth', pp. 257–61; A. Dyer, *Decline and growth in English towns, 1400–1640* (London, 1991), pp. 40–2. On migration see Poos, *Rural society*, pp. 159–79.

[93] J. de Vries, *European urbanization 1500–1800* (London, 1984), pp. 64, 279.

[94] Campbell, 'Economic rent'.   [95] Campbell and others, 'Demesne-farming', pp. 177–9.

[96] Dyer, *Decline and growth*, pp. 20–4; Keene, *Cheapside*, pp. 19–20; Dyer, *Standards of living*, pp. 199–202; Galloway, 'London's grain supply'; Farmer, 'Marketing', pp. 372–3.

[97] Campbell, 'Matching supply to demand', pp. 835–6; J. A. Galloway, 'London's grain supply: changes in production, distribution and consumption during the fourteenth century', *Franco-British Studies: Journal of the British Institute in Paris* 20 (1995), 23–34.

producers found themselves vying for the same limited markets. Even within pastoral husbandry there were no easy profits to be made.[98]

The changing geography of seigniorial agriculture therefore represents a reversion to more land-extensive forms of agriculture in which grain production was increasingly counterbalanced by animal husbandry. As a result, localities and regions became better able to meet more of their consumption requirements internally from within their own resources. Whereas at the opening of the fourteenth century it is possible to recognise the clear impact of growing centres of concentrated urban demand, both at home and overseas, upon the pattern of seigniorial husbandry, by the opening of the following century such centripetal influences had diminished in scale and are less self-evident in their impact.[99] The very fact that demesne agriculture became less differentiated suggests that incentives to specialise and intensify had weakened. The new agricultural landscape that emerged was influenced less by distance from major markets and more by land quality and local demand for the land and its products. The latent patterns of specialisation and intensification that may be detected at the climax of medieval economic expansion *c.* 1300 had mostly fallen into abeyance in an agrarian world seemingly reoriented along more local and intra-regional lines. This, however, is almost certainly to under-estimate the commercial potential of animals and animal products, both of which were capable of being marketed at a far greater range than grain.[100]

If the orbit of grain markets became more circumscribed it does not necessarily follow that the same applied to the markets for live animals, dairy products, hides, skins, and wool.[101] In fact, an active trade in live animals undoubtedly helped sustain the expansion of flocks and herds that was such a feature of this period (Table 4.05). Most demesnes relied upon the market for replacement work-horses, for the plough and especially for the cart, and the same often applied to oxen. The conduct of pastoral husbandry also regularly generated surplus animals for sale: redundant, decrepit, and sickly animals requiring replacement, surplus calves and lambs from specialist herds and flocks, animals purpose-bred for sale and others fattened for meat (Table 4.06). On what scale that trade was conducted and over what distances

---

[98] M. Overton and B. M. S. Campbell, 'Norfolk livestock farming 1250–1740: a comparative study of manorial accounts and probate inventories', *JHG* 18 (1992), 377–96; M. Mate, 'Pastoral farming in south-east England in the fifteenth century', *EcHR* 40 (1987), 535–6; Campbell, 'Fair field', p. 64.

[99] Campbell and others, *Medieval capital*, pp. 172–83; Power and Campbell, 'Cluster analysis', p. 242.

[100] Lloyd, *English wool trade*; Farmer, 'Marketing', pp. 377–408; Overton and Campbell, 'Norfolk livestock farming', pp. 377–8, 393–4.

[101] Lloyd, *English wool trade*; M. Kowaleski, 'Town and country in late medieval England: the hide and leather trade', in P. J. Corfield and D. J. Keene (eds.), *Work in towns 850–1850* (Leicester, 1990), pp. 57–73.

remains to be established, but neither is likely to have been inconsiderable.[102]

Does this mean that demesne husbandry became more or less commercialised in the post-Black Death period?[103] There can be no simple answer since much depends upon how commercialisation is defined and measured.[104] Within the FTC counties sales of agricultural produce per demesne generated marginally less income at the close of the fourteenth century than they had done at the beginning, although the decline from £21.00 to £19.10 is largely attributable to the post-1375 fall in prices. On this crude criterion agricultural sales were no more important at the end of the period than they had been at the beginning. In fact, sales of crop and crop products, which had accounted for almost two-thirds of all income *c.* 1300, declined in value by 30 per cent. There are proportionately fewer examples of highly commercialised crop production in the later sample of demesnes and more examples of weak commercialisation. In part, of course, this was because it was animals rather than crops that now offered the better commercial opportunities. Over the course of the fourteenth century the proportion of agricultural sales income contributed by animals and animal products (including game and fish) rose from 36 per cent to 50 per cent. By *c.* 1390 sales of live animals were contributing 19 per cent of sales income and sales of animal products – wool, hides, dairy produce, and lactage – 30 per cent. The single greatest gain came from the leasing of dairy herds, since when they had been managed directly a proportion of the dairy produce had always been retained for consumption on the manor or estate. By becoming more pastoral, seigniorial husbandry therefore also became more commercial: indeed, this was doubly the case for, whereas lords were mainly sellers of grain, they entered the market both to buy and sell animals.

Seigniorial agriculture also became more commercialised in another important but less conspicuous sense. The alternative to purchase and sale – intra-estate transfer of crops and animals – declined in both relative and absolute significance (Tables 4.04, 4.05, 4.06, 5.05). Two-and-a-half times as much grain was sold as transferred *c.* 1390 compared with twice as much *c.* 1300.[105] Purchase and sale similarly gained relative to transfer in the replacement and disposal of livestock, especially equines and bovines. In this very important respect demesnes were becoming more dependent upon the market, although paradoxically at a time when concentrated demand was of waning rather than waxing significance. This change is most marked on conventual and collegiate estates which had hitherto made greatest use of intra-estate transfers as a means of restocking manors and provisioning the central household. By

[102] Campbell, 'Measuring commercialisation', pp. 142–3, 148–9, 152–3, 163–74; Blanchard, 'European cattle trade', pp. 428–31; M. K. McIntosh, *Autonomy and community: the royal manor of Havering, 1200–1500* (Cambridge, 1986), pp. 141–3; Mate, 'Pastoral farming'.

[103] Campbell, 'Fair field', pp. 68–9.   [104] Campbell, 'Measuring commercialisation'.

[105] Chapter 5, pp. 193–203.

implication, therefore, households in their turn must have purchased an enlarged proportion of their provisions, presumably because with so much excess production capacity the market was now better able to satisfy their needs. Nevertheless, the direct consumption of estate produce did not end entirely. Many demesnes were retained in hand precisely because they were managed as home farms intended to provision the household. A reduced but nonetheless significant proportion of production continued to bypass the market. The agricultural economy remained only partially commercialised.

Autarky therefore lingered on, as did the traditional reliance upon customary labour by the more manorialised estates within the seigniorial sector. Lords were especially reluctant in an age of mounting labour scarcity to relinquish their right to levy harvest works. Customary labour faded away gradually and did not finally disappear until the last demesnes were set at farm in the second half of the fifteenth century. In all these respects the post-Black Death period witnessed no radical break with the past. Changes took place but the forms which change took were invariably shaped by pre-Black Death practice. Even direct demesne management itself was a legacy of the past. It lingered on long after its original rationale had passed until, by the middle years of the fifteenth century, it had become something of an anachronism. Thenceforth, lords are primarily of agricultural interest for their activities as lessors rather than as direct producers. It is to other sources and categories of producer that it is necessary to turn in order to reconstruct the on-going story of English agricultural development.

9.3 The medieval antecedents of English agricultural progress

The story to be told of medieval agriculture, at least as far as the minority seigniorial sector is concerned, anticipates in many respects that told of later centuries. Whether in the thirteenth century, the sixteenth century, or the eighteenth century, producers coping with rising demand were faced by many of the same production choices and responded in the same kinds of way.[106] There are parallels, too, between the late fourteenth and late seventeenth centuries, when demand contracted and producers found themselves faced by a potential surfeit of supply. Agricultural technology would, of course, improve over time as knowledge advanced (although practical experience of individual practices could lapse) but until well into the nineteenth century the constraints of a basically animate and organic technology would apply. Improvement was therefore always an uneven and gradual process, contingent upon associated structural and institutional changes and involving many inter-related developments across a broad front. Its pace only quickened as the pace of economic life itself quickened in the eighteenth and nineteenth centuries. Productivity

---

[106] Overton, *Agricultural revolution*, pp. 88–105.

gains, when they eventually came, were rarely dramatic and were greater within the pastoral than the arable sector. By 1800 crop yields per cropped acre were only twice what they had been in 1300, while carcass weights and fleece weights had trebled and milk yields quadrupled.[107] Compared with the productivity gains being won elsewhere in the economy, those within agriculture were unimpressive.

Had national economic progress over these five centuries depended solely upon the productivity of domestic agriculture it could not have been as great. Much of the problem *c.* 1300 was that England was too exclusively reliant upon domestic agriculture for food, drink, fuel, raw materials, and export earnings. More than at any subsequent point in time it was output per unit of land that determined the size and density of the population that could be supported, while upon output per agricultural worker hinged the proportion of that population which could engage in non-agricultural activities.[108] The prominence of primary products in the country's already significant export trade highlights the relatively under-developed state of the economy and demonstrates that at this time England's greatest international comparative advantage appears to have lain in the commercial production of high-quality wool. This is symptomatic of a land-extensive, grass-rich, agrarian economy, locationally peripheral to Europe's then core manufacturing regions of Flanders and northern Italy. To compound matters, the export trade was still largely dominated by alien merchants who thereby appropriated the bulk of the profits of that trade. Nor was much value added to the bulk of the commodities that were exported: wool worked up as cloth constituted only a small and probably declining proportion of all the wool and woollen goods that were exported *c.* 1300.

Until the national economy was released from its overwhelming reliance upon domestic agriculture it would be difficult to effect a fuller development of productive forces, including, paradoxically, those of agriculture. Such development came in later centuries. As knowledge and technology advanced, so inanimate were increasingly substituted for animate sources of power and inorganic for organic raw materials, thereby lessening the tyranny of biological reproduction.[109] More importantly, through fuller and less passive participation in Pirenne's 'great commerce' significantly enhanced opportunities were offered for wealth creation. Via international trade England broke out of a narrow and exclusive dependence upon its own agricultural sector, gained

---

[107] Clark, 'Labour productivity', pp. 215–16.

[108] J. P. Gibbs and W. T. Martin, 'Urbanization, technology and the division of labour', *American Sociological Review* 27 (1962), 667–77; J. Merrington, 'Town and country in the transition to capitalism', *New Left Review* 93 (1975), 452–506.

[109] E. A. Wrigley, 'The supply of raw materials in the industrial revolution', *EcHR* 15 (1962), 1–16; E. A. Wrigley, *Continuity, chance and change: the character of the industrial revolution in England* (Cambridge, 1988), pp. 34–50.

access to the produce of other lands, and thereby achieved superior rates of economic and demographic growth.

Already in the Middle Ages land-scarce and highly urbanised regions such as Flanders, with large manufacturing populations to support, had used trade to obtain such land-extensive products as timber, pitch, wax, furs, wool, hides, live animals on the hoof, and low-yielding grain from regions of relative land abundance. The latter, in return, gained access to the bullion, specialised manufactures, and quality produce of these more developed regions, typically cloth, craft wares of many sorts, and wine. A series of positive feedbacks ensured that this Smithian process of reciprocal exchange, once initiated, had the capacity to become self-perpetuating. Trade reinforced market growth, which in turn stimulated a greater division of labour thereby lowering unit production costs, especially in mass-produced manufactured goods. Expanding trade flows reduced unit transaction costs through the significant scale economies that they offered. Together, cheaper produce and cheaper trade enabled a progressive widening in the geographical extent and deepening in the intensity of commercial interaction from which further market growth ensued.[110] The process was not without limits. Marginal costs could rise and diminishing returns set in as the trading system grew in scale and complexity. Expanding demand could encounter major difficulties in overcoming supply-side rigidities, especially within agriculture. Transaction costs could rise and commerce contract if warfare became too sustained and widespread. Once commercial recession set in excess population tended to become entrapped on the land, thereby promoting production strategies geared towards self-sufficiency rather than exchange as the economic rent for subsistence goods rose above that of those destined for the market. Specialisation and the labour-productivity growth that it was capable of bestowing were thereby negated. Latent tensions between commercial and subsistence interests also became greatly aggravated. Such processes were undoubtedly at work in the early fourteenth century.[111]

It was nevertheless in the fourteenth century that the wider commercial context of English agricultural production began to be transformed, for this was when denizen merchants successfully captured an increasing share of the export trade. As a result more of the profits of trade accrued to England, thereby facilitating the relative growth of mercantile cities of which London was by far the most notable. It was strongly reinforced in the fifteenth century by the growth of proto-industry as manufacturing activity migrated to least-cost locations, which added value to agricultural and mineral products and supplied a widening range of domestically produced trade goods. This

---

[110] K. G. Persson, *Pre-industrial economic growth, social organization and technological progress in Europe* (Oxford, 1988), pp. 71–3; G. W. Grantham, '*Contra* Ricardo: on the macroeconomics of pre-industrial economies', *European Review of Economic History* 3 (1999).

[111] Munro, 'Industrial transformations'; Fischer, *Great wave*, pp. 30–45.

brought real economic benefits to remote regions of cheap land, cheap victuals, and cheap labour. Meanwhile, several of the old supply-side constraints began to be resolved: serfdom decayed, disaforestment took place, interest rates fell, private began haltingly to replace common property rights, and greater use was made of iron rather than wooden implements. The stage was thereby set for the eventual emergence from the late sixteenth century of that fruitful triumvirate – large-scale, concentrated, metropolitan demand, proto-industry, and capitalist agriculture – from whose interaction so many economic dividends were in due course to flow.[112] Not the least of these was the greater differentiation of domestic demand for foodstuffs, which promoted new and more developed forms of agricultural specialisation. As important was enhanced commercial access to the agricultural products of other nations – tea, coffee, sugar, timber, silk, cotton, linen, young animals, and grain as required – whose range had been greatly extended since the Middle Ages by the discovery of new lands and the opening up of new sea routes. It also seems fairly certain that the unit costs of domestic and overseas trade must have fallen as cargo vessels grew in scale and the sea-borne trade grew relative to the old overland trade.

These commercial developments were propitious for farmers. International trade broke down the self-sufficiency of individual regions and countries and offered those able and willing to produce the goods demanded by overseas markets the opportunity to tap lucrative fresh sources of wealth. External trade also underpinned much domestic urban growth, which in turn provided farmers with further incentives to invest, specialise, and innovate.[113] Market growth on a whole range of scales and levels broke down old ways and encouraged farmers to capitalise upon their comparative advantages, thereby counteracting the opposing tendency towards diminishing returns and raising the mean productivity of agriculture as a whole.[114]

English national income grew by almost two-and-a-half-fold between 1086 and 1300, it grew by at least as much again between 1300 and 1688, and by a further two-and-a-half-fold between 1688 and 1800.[115] As the economy grew and demand-side incentives strengthened, so economic rent rose and was reconfigured. Specialisation, intensification, and investment were thereby encouraged and, as structural and institutional rigidities were progressively overcome, farmers responded with a quickening pace of innovation and rising

---

[112] Langton and Hoppe, *Town and country*.

[113] B. M. S. Campbell, 'The sources of tradable surpluses: English agricultural exports 1250–1350', in L. Berggren, N. Hybel, and A. Landen (eds.), *Trade and transport in northern Europe 1150–1400* (Toronto, forthcoming); B. M. S. Campbell, J. A. Galloway, D. J. Keene, and M. Murphy, *A medieval capital and its grain supply* (n.p., 1993).

[114] Persson, *Pre-industrial growth*, pp. 63–103; Overton and Campbell, 'Productivity change', pp. 9–22.

[115] Table 8.07; Snooks, 'Dynamic role', p. 50; Deane and Cole, *British economic growth*, p. 78.

productivity which further impelled the economy forward.[116] From a medieval perspective there was nothing intrinsically new or different about this, except the heightened pace of change and the strong tendency by the eighteenth century for agricultural output and the productivities of land and labour to rise together.[117] Probably something similar had happened during the most vigorous phase of medieval expansion and growth in the late twelfth and early thirteenth centuries. At that earlier and less mature stage of economic development there had, however, been far less prospect of progress fructifying and becoming self-sustaining in more than the medium term. Even had it done so, it could not have withstood the massive demand shock inflicted by the succession of exogenous environmental setbacks which began with the Great European Famine and culminated with the Black Death.

---

[116] For the wider economic and demographic implications of agricultural change from the late seventeenth century on, see A. H. John, 'Agricultural productivity and economic growth in England, 1700–1760', *JEH* 25 (1965), 19–34; E. L. Jones, *Agriculture and the Industrial Revolution* (Oxford, 1974); D. B. Grigg, 'Breaking out: England in the eighteenth and nineteenth centuries', in *Population growth*, pp. 163–89; Wrigley, 'Urban growth'.

[117] Mokyr, *Lever of riches*, pp. 31–56; Overton, *Agricultural revolution*, pp. 1–9; Overton and Campbell, 'Production et productivité'.

Appendix 1

Demesne-level classification of husbandry types

The following listing classifies individual manors according to their respective pastoral (P), cropping (C) and mixed-farming (M) types, pre-1350 (1) and post-1349 (2). Thus, 'C1 1' means 'pre-1350 cropping-type 1'. The characteristics of each husbandry type are specified in Tables 4.01, 6.01, 6.02, and 4.08 and mapped in Figures 4.01–11, 6.01–07, and 6.13–18. References are indicated by superscript numbers and listed at the end of this appendix (pp. 451–2).

Bedfordshire: *Barton in the Clay*[1] (P2 4); *Clapham Bayeux*[2] (P1 2); *Cranfield*[1] (P2 3); *Eaton Bray*[1] (P1 3; C1 3; M1 5); *Edworth*[2] (P1 6; C1 6; M1 7); *Grovebury*[2] (P1 3); *Grove in Leighton Buzzard*[2] (P1 3); *Harrold*[1] (P2 3); *Higham Gobion*[1 & 4] (P2 6; C2 3; M2 7); *Houghton*[1] (P1 5; C1 3; M1 7); *Little Staughton*[1] (P1 4); *Millow*[3] (P1 3; C1 3; M 1); *Pegsdon*[3] (C1 3); *Shillington*[1, 3, & 4] (P1 2; P2 3; C2 3; M2 3); *Sundon*[1] (P1 4; C1 3; M1 5); *Swanton in Harrold*[1] (P1 3; C1 3; M1 1).

Berkshire: *Ashbury*[6] (P1 3; C1 5; M1 6); *Avington*[3] (P1 4; C1 4; M1 5); *Billingbear*[1,3,4,&9] (P1 2; P2 3; C1 3; C2 5; M1 3; M2 6); *Bray*[3] (P1 2; C1 2; M1 3); *Brightwalton*[4] (P2 3; C2 4; M2 3); *Brightwell*[3, 4, & 9] (P1 2; P2 3; C1 4; C2 4; M1 3; M2 3); *Coleshill*[4] (P2 3; C2 4; M2 3); *Coombe*[3] (P1 3; C1 3; M1 6); *Coxwell*[1] (P1 2); *Cresswell in Bray*[4] (P2 3; C2 4; M2 4); *Culham*[1, 3, & 4] (P1 2; P2 3; C1 3; C2 3; M1 2; M2 1); *Didcot*[4] (P2 6; C2 4; M2 7); *Drayton*[4] (P2 4; C2 8; M2 5); *Eaton Hastings*[1] (P1 6; P2 5; C2 4; M2 7); *Hampstead Marshall*[3] (P1 2; C1 4; M1 3); *Hampstead Norreys*[3] (P1 3; C1 3; M1 1); *Harwell*[1, 3, 4, & 9] (P1 3; P2 4; C1 6; C2 4; M1 5; M2 5); *Hinton Waldrist*[4] (P2 3; C2 4; M2 5); *Inkpen*[3 & 4] (P1 3; P2 3; C1 4; C2 4; M1 6; M2 3); *Kennington*[3] (P1 2; C1 2; M1 7); *Letcombe Regis*[1 & 3] (P1 2; C1 5); *Long Wittenham*[4] (P2 4; C2 1; M2 5); *Shallingford and Newbury*[2] (P2 3); *Southcot*[1] (P2 4; C2 4; M2 3); *Speen*[1] (P1 4; P2 4; C1 3; C2 1; M1 5; M2 5); *Templeton*[3] (P1 4; C1 3; M1 5); *Upton with Blewbury*[1] (P1 3; C1 3; M1 5); *Waltham St Lawrence*[1, 3, 4, & 9] (P1 5; P2 4; C1 3; C2 3; M1 7; M2 5); *Wantage*[3] (C1 6); *Wargrave*[1,3,4,&9] (P1 2; P2 3; C1 3; C2 3; M1 3; M2 2); *Woodspeen*[1] (P1 4; C1 4; M1 5); *Woolstone*[3 & 4] (P1 3; P2 3; C1 4; C2 4; M1 6; M2 3); *Wyke near Westbrook*[1] (P1 2).

Buckinghamshire: *Aylesbury*[1 & 4] (P2 3; C2 3; M2 3); *Bledlow*[3] (P1 2; C1 3; M1 3); *Brill*[1] (P1 6); *Cheddington*[3 & 4] (P1 3; C1 1; M1 1); *Cippenham*[3] (P1 2; C1 4; M1 3); *Cuddington*[4] (P2 3); *Denham*[1 & 4] (P1 1; P2 3; C1 1; C2 3; M1 1; M2 6); *Fulmer*[2] (P1 4;

C1 3; M1 5); *Halton*[1 & 4] (P1 3; P2 4; C1 5; C2 4; M1 3; M2 3); *Holmer*[3] (P1 1; C1 5; M1 5); *Horsenden*[4] (P2 4; C2 4; M2 3); *Ibstone*[3] (P1 1; C1 3; M1 5); *Iver*[3 & 4] (P1 6; P2 3; C1 7; C2 5; M1 7; M2 6); *Ivinghoe*[1, 3, 4, & 9] (P1 3; P2 3; C1 5; C2 3; M1 6; M2 3); *Langley Marish*[1 & 3] (P1 3; C1 2; M1 2); *Little Missenden*[1] (P1 2); *Middle Claydon*[2] (P1 3); *Monks Risborough*[4] (P2 3; C2 4; M2 4); *Moreton*[1, 3, 4, & 9] (P1 2; P2 3; C1 6; C2 3; M1 1; M2 6); *Quainton*[4] (P2 3; C2 4; M2 3); *Quarrendon*[4] (P2 4; C2 8; M2 4); *Steeple Claydon*[1] (P1 3; P2 4; C2 3; M2 3); *Stone*[1] (P1 3); *Temple Bulstrode*[3] (P1 3; C1 5; M1 6); *Tingewick*[3 & 4] (P1 3; P2 5); *Turweston*[3 & 4] (P1 3; P2 4; C1 3; C2 4; M1 1; M2 3); *Twyford*[2] (P2 3); *Water Eaton*[4] (P2 3; C2 8; M2 6); *Weedon in the Vale*[4] (P2 3; C2 8; M2 3); *Wendover*[3] (P1 6; C1 5; M1 7); *West Wycombe*[1, 2, & 9] (P1 3; P2 1; C1 3; C2 4; M1 3; M2 1); *Westcott*[1 & 3] (P1 6; C1 5; M1 7); *Whaddon*[1] (P2 3; C2 5; M2 6).

Cambridgeshire: *Burwell*[1] (P2 1; C2 8; M2 3); *Cottenham*[1] (P1 2); *Ditton Valence*[1] (P1 6; P2 3; C2 4; M2 3); *Downham in the Isle*[1 & 2] (P1 2; P2 3; C1 1; C2 4; M1 1; M2 6); *Dry Drayton*[1] (P1 4); *Elsworth*[1] (P2 3; C2 3; M2 3); *Grantchester*[1 & 8] (P1 2); *Harston*[1] (P1 2; C1 3; M1 1); *Kennett*[1] (P1 3; C1 2; M1 2); *Knapwell*[1] (P2 4); *Melbourne*[1] (P1 4; C1 3; M1 1); *Meldreth*[1 & 2] (P1 2; P2 6; C1 3; C2 3; M1 1; M2 7); *Oakington*[1] (P1 4; P2 4; C2 3; M2 3); *Soham*[1] (P1 6; P2 3; C2 1; M2 1); *Uphall and Cherry Hinton*[1] (P2 1; C2 3; M2 4); *Wisbech Barton*[1 & 2] (P1 4; P2 3; C1 4; C2 5; M1 6; M2 6).

Cheshire: *Drakelow*[1] (P2 6; C2 5; M2 6); *Frodsham*[1] (P1 2; P2 3; C2 5; M2 6); *Wrenbury*[2] (C1 5).

Cornwall: *Cargoll in Newlyn East*[1] (P1 3); *Egloshayle*[1] (P1 3); *Lawhitton*[1] (P1 3); *Pawton*[1] (P1 3); *Penheale*[1] (P2 3; C2 5; M2 6); *Penryn*[1] (P1 3); *St Germans*[1] (P1 3); *St Keverne in the Lizard*[1] (P1 2); *Tregear*[1] (P1 3); *Whaleborough*[1] (P2 3; C2 5; M2 6).

Cumberland: *Birkby*[1] (C1 5); *Bolton*[1] (C1 7); *Cockermouth*[1] (C1 7); *Cockermouth and Birkby*[1] (P1 3; M1 6).

Derbyshire: *Belper*[1] (P1 2; C1 7; M1 7); *Chapel-en-le-Frith*[2] (P1 6); *Melbourne*[1 & 2] (P1 5; C1 1).

Devon: *Ashbury*[16] (C1 4); *Barton St Mary*[1] (P1 3; C1 7; M1 6); *Bishops Nympton*[1] (P1 3); *Bishopsteignton*[1] (P1 3); *Chudleigh*[1] (P1 3); *Clyst*[1 & 12] (P1 2; P2 3; C2 5; M2 6); *Crediton*[1] (P1 3); *Exminster*[1] (P1 2; C1 7; M1 8); *Goodrington*[1] (P2 3; C2 5; M2 3); *Hemyock*[1] (P1 3; C1 5; M1 6); *Honiton*[1] (P1 3; C1 5; M1 6); *Hurdwick*[14] (C1 7; C2 2); *Kenn*[2] (C1 7); *Langtree*[1] (P2 3); *Otterton*[1] (P2 4); *Paignton*[1] (P1 3); *Pinhoe*[1] (P2 4); *Plympton*[1] (P1 3; C1 4; M1 6); *Tawton*[1] (P1 3); *Tiverton*[1] (P1 2; C1 7; M1 3); *Topsham*[1] (P1 4; C1 7; M1 5); *Uplyme*[2 & 16] (P1 3; C1 5; M1 6); *Werrington*[14] (P1 1; C1 7; C2 5; M1 1); *Yealmpton*[1] (P2 3; C2 5; M2 6).

Dorset: *Ashley*[2] (P2 1; C2 4; M2 3); *Buckland Newton*[6] (P1 4; C1 5; M1 6); *Canford*[1] (P1 2; C1 4; M1 1); *Colbeare*[16] (C1 4); *Cranborne*[2] (P1 4; C1 4; M1 6); *Kingston Lacy*[1] (P1 3; C1 4; M1 6); *Marnhull*[16] (C1 6); *Pimperne*[10] (C1 4); *Plush*[16] (C1 4); *Portland*[2] (P1 4; C1 4; M1 5); *Steeple and Creech*[1 & 2] (P1 3; P2 3; C1 4; C2 5; M1 6; M2 5); *Stoke*[10] (C1 6); *Sturminster Newton*[6] (P1 2; C1 6; M1 3); *Sutton Waldron*[2] (P2 4; C2 4; M2 5); *Tarrant*[1 & 2] (P2 4; C1 4; C2 4; M2 5); *Walterstone*[2] (P2 1; C2 5; M2 5); *Wivelsford*[10] (C1 4); *Wyke*[1 & 2] (P1 5; C1 4; M1 7).

Durham: *Bearpark*[1 & 2] (P1 6; P2 6); *Bellasis*[1] (P1 5); *Bewley*[1] (P1 2; P2 5; C2 8; M2 7); *Billingham*[1] (P1 5); *Burnhopshall*[2] (P2 1); *Cotum*[2] (P1 5; C1 5; M1 7); *Coundon*[2] (P1 4); *Dalton*[1] (P1 6); *Elvethall*[1 & 2] (P1 3; P2 6; C1 4; C2 4; M1 6; M2 6); *Ferryhill*[1 & 2] (P1 5; P2 6; C2 5; M2 7); *Finchale*[1 & 2] (P2 3); *Fulwell*[1] (P2 5; C2 3; M2 4); *Heighington*[2]

(P1 6); *Houghall*[1] (P1 6; P2 6); *Jarrow*[1 & 2] (P1 3; P2 3; C1 6; C2 8; M1 6; M2 3); *Ketton*[1] (P1 5; P2 5); *Maudelyns*[1 & 2] (P1 6; P2 6; C1 5; C2 5; M1 7; M2 7); *Merrington*[1] (P2 5); *Middleham*[2] (P1 5); *Middridge*[2] (P1 4); *Monkwearmouth*[1 & 2] (P1 5; P2 5; C1 4; C2 8; M1 3; M2 4); *Muggleswick*[1] (P1 6); *Pittington*[1 & 2] (P1 5; P2 5); *Quarrington*[2] (P1 5; C1 6; M1 7); *Rainton*[1] (P1 6); *Ricknall*[2] (P1 5); *Wardley* [1] (P1 6; P2 6); *Wersclule*[2] (P1 3); *Westoe*[1] (P1 6; P2 5; C2 8; M2 7); *Wingate*[1 & 2] (P2 3); *Witton*[1 & 2] (P1 4; P2 3; C1 5; C2 2; M1 6; M2 6).

Essex: *Bardfield*[10] (C1 5); *Bekeswell*[4] (P2 1; C2 5; M2 6); *Berners Roding*[4] (P2 3; C2 5; M2 6); *Berwick Berners*[4] (P2 3; C2 5; M2 6); *Birchanger*[4] (P2 1; C2 5; M2 6); *Birdbrook*[1, 3, & 4] (P1 2; P2 3; C1 5; C2 5; M1 3; M2 5); *Bocking*[1, 3, & 4] (P1 2; P2 3; C1 5; C2 5; M1 3; M2 6); *Boreham*[4] (P2 3; C2 5; M2 6); *Borley*[1 & 4] (P1 2; P2 3; C1 4; C2 4; M1 3; M2 1); *Bradwell*[2] (P1 1); *Bulmer*[4] (P2 3; C2 4; M2 5); *Bures*[4] (P2 3; C2 5; M2 6); *Chelmsford*[1] (P1 2); *Chesterford*[3] (P1 2; C1 4; M1 3); *Childerditch*[4] (P2 3; C2 5; M2 6); *Chingford*[3 & 1] (P1 3; C1 5; M1 5); *Clacton*[1] (P1 3); *Claret*[10] (C1 5); *Copford*[1] (P1 4); *Cressing Temple*[3] (P1 3; C1 5; M1 6); *Crondon*[1] (P1 2); *Dagenham*[2] (P1 4); *Dovercourt*[1 & 3] (P1 2; C1 4; M1 3); *Eastwood*[1 & 3] (P1 2; P2 1; C1 3; C2 5; M1 3; M2 6); *Fanton*[3] (P1 2; C1 5; M1 3); *Faulkbourne*[1] (P2 4); *Feering*[1, 3, & 4] (P1 3; P2 3; C1 5; C2 5; M1 3; M2 6); *Felsted*[3] (P1 2; C1 5; M1 3); *Grays Thurrock*[3] (P1 3; C1 1; M1 1); *Great Bardfield*[3] (P1 2; C1 5; M1 3); *Great Hallingbury*[1] (P1 3; C1 5; M1 6); *Hadleigh*[1 & 3] (P1 3; P2 3; C1 4; C2 4; M1 5; M2 6); *Hanningfield*[3] (P1 3; C1 5; M1 5); *Hatfield Broadoak*[4] (P2 3; C2 5; M2 6); *High Easter*[4] (P2 1; C2 3; M2 5); *Hornchurch*[4] (P2 3; C2 2; M2 6); *Hutton*[4] (P2 3; C2 5; M2 6); *Kelvedon*[1, 3, & 4] (P1 3; P2 3; C1 5; C2 5; M1 5; M2 6); *Laindon*[1] (P1 4); *Langenhoe*[4] (P2 4; C2 5; M2 5); *Lawling*[1 & 4] (P1 2; P2 1; C1 7; C2 5; M1 3; M2 6); *Little Baddow*[3] (C1 3); *Little Maldon*[4] (P2 3; C2 5; M2 6); *Mersea*[3] (P1 4; C1 5; M1 5); *Messing Hall*[4] (P2 6; C2 5; M2 6); *Middleton*[3 & 4] (P1 3; P2 3; C1 5; C2 5; M1 3; M2 6); *Milton Hall*[1] (P1 2; P2 1; C1 5; C2 4; M1 3; M2 6); *Moulsham*[4] (P2 3; C2 5; M2 6); *Newport*[3] (P1 2; C1 5; M1 3); *Orsett*[1] (P1 3); *Quickbury in Sheering*[1] (P1 3; C1 5; M1 3); *Rayne*[1] (P1 4); *Roydon*[3] (P1 2; C1 5; M1 4); *Smeetham in Bulmer*[1] (P2 3); *Southchurch*[4] (P2 1; C2 5; M2 5); *Southminster*[1] (P1 3); *Stapleford Abbots*[4] (P2 3; C2 5; M2 6); *Stebbing*[4] (P2 6; C2 5; M2 7); *Sutton*[1] (P1 3); *Takeley*[4] (P2 3; C2 5; M2 6); *Tendring*[1] (P2 4); *Theydon*[3] (P1 3; C1 5; C2 5; M1 5); *Thundersley*[1] (P1 2); *Tolleshunt Major*[4] (P2 3; C2 5; M2 6); *Ugley*[4] (P2 3; C2 5; M2 6); *Walthambury*[4] (C2 5); *West Hanningfield*[1] (P1 2; C1 6; M1 3); *West Thurrock*[3] (P1 1; C1 2; M1 5); *Wicken Bonhunt*[3] (P1 4; C1 1; M1 5); *Wickham Bishops*[1] (P1 3); *Widford*[1] (P1 3; C1 5; M1 6); *Witham*[3] (P1 2; C1 5; M1 3); *Wix*[4] (P2 6; C2 2; M2 6); *Woodham*[1] (P2 3); *Wrabness*[1] (P2 3; C2 2; M2 2); *Writtle*[1] (P2 3; C2 4; M2 4); *Writtle Rectory*[4] (P2 5; C2 4; M2 4).

Gloucestershire: *Alveston*[1] (P1 6); *Avening*[1] (P2 4; C2 3; M2 5); *Awre*[1 & 2] (P1 2; P2 3; C1 4; C2 4; M1 7; M2 7); *Benynton*[2] (C1 5); *Berkeley*[1] (P1 2; C1 5; M1 6); *Bibury*[1] (P1 4; P2 4; C2 3; M2 5); *Bourton on the Hill*[1] (P1 4; P2 4; C2 4; M2 3); *Bradel*[2] (C1 5); *Brimpsfield*[1] (P2 4; C2 3; M2 5); *Cam and Coaley*[1] (P1 3); *Chaceley*[1] (P2 3; C2 8; M2 3); *Egeton*[2] (C1 5); *Hardwicke*[1] (P1 4; P2 5; C2 8; M2 4); *Hawkesbury*[1 & 2] (P1 2; P2 3); *Horsley*[1] (P2 4; C1 4; C2 3; M2 3); *Horton*[1] (P2 3); *'Langebr and Gosynton'*[2] (C1 5); *Minchinhampton*[1] (P1 3; P2 4); *Oldem*[2] (C1 5); *Pucklechurch*[1] (C2 5); *Stone*[2] (C1 5); *Temple Guiting*[2] (C1 3); *Todenham*[1] (P1 3); *Wotton under Edge*[1] (P1 4; C1 5; M1 6).

Hampshire, Isle of Wight: *Appleford*[1] (P1 6; C1 4; M1 7); *Bowcombe*[1] (P1 3; C1 1; M1 1); *Chillerton*[1] (P1 6; C1 1; M1 7); *Niton*[1] (P1 4; C1 1; M1 5); *Pan*[1 & 2] (P1 6; C1 1; M1 7); *Shorwell*[1] (P1 5; C1 1; M1 7); *Whitefield*[2] (P1 4); *Wootton*[1] (P1 2; C1 4); *Wroxall*[1] (P2 4; C2 4; M2 5).

Hampshire: *Anstey*[1] (P1 4); *Ashmansworth*[1 & 9] (P2 4; C2 5; M2 5); *Beaulieu*[1] (P1 2); *Beauworth*[1 & 9] (P1 6; P2 4; C1 7; C2 5; M1 7; M2 5); *Bentley*[1 & 9] (P1 2; P2 4; C1 5; C2 5; M1 7; M2 5); '*Bergerie*'[2] (P1 1); *Bishops Sutton*[1 & 9] (P1 3; P2 3; C1 5; C2 4; M1 6; M2 3); *Bishops Waltham*[1 & 9] (P1 3; P2 3; C1 4; C2 4; M1 6; M2 3); *Bishopstoke*[1 & 9] (P1 2; P2 3; C2 4; M2 6); *Bitterne*[1 & 9] (P1 2; C1 4; M1 6); *Brockhampton*[1] (P2 6; C2 4; M2 7); *Burgate*[1] (P1 2); *Burghclere*[1 & 9] (P1 6; P2 4; C1 5; C2 5; M1 7; M2 5); *Cheriton*[1 & 9] (P1 4; P2 3; C1 5; C2 4; M1 6; M2 3); *Colbury*[1] (P1 2); *Cold Henley*[1] (P2 4; C2 4; M2 5); *Crawley*[1 & 9] (P1 3; P2 4; C1 4; C2 4; M1 6; M2 5); *Droxford*[1] (P2 4; C2 4; M2 5); *East Meon*[1 & 9] (P1 4; P2 4; C1 7; C2 5; M1 4; M2 5); *East Woodhay*[1 & 9] (P1 6; P2 4; C1 3; C2 4; M1 7; M2 5); *Fareham*[1 & 9] (P1 4; P2 3; C1 5; C2 5; M1 5; M2 3); *Farringdon in Bishops Waltham*[1] (P1 3); *Hambledon*[1 & 9] (P1 3; P2 3; C1 5; C2 1; M1 6; M2 3); *Hartford*[1] (P1 2); *Hickley*[7] (P1 2; P2 3; C1 1; C2 4; M1 8; M2 6); *High Clere*[1 & 9] (P1 2; P2 3; C1 3; C2 4; M1 6; M2 6); *Holbury*[1] (P1 2); *Honnington*[2] (P2 4); *Itchingswell*[1 & 9] (P1 4; P2 4; C1 5; C2 5; M1 5; M2 5); *Long Sutton*[2] (P1 4; C1 5); *Mardon*[1 & 9] (P1 2; P2 3; C1 5; C2 4; M1 3; M2 6); *Marwell*[1] (P2 3; C2 4; M2 3); *North Waltham*[1 & 9] (P1 6; P2 4; C1 3; C2 5; M1 7; M2 5); *Odiham*[1] (P1 2; C1 3; M1 3); *Old Alresford*[1 & 9] (P1 6; P2 3; C1 4; C2 4; M1 7; M2 3); *Otterwood*[1] (P1 2); *Overton*[1 & 9] (P1 3; P2 3; C1 7; C2 4; M1 5; M2 3); *Soberton*[1] (P1 2); *Sowley*[1] (P1 2); *St Leonards*[1] (P1 2); *Tichborne*[1] (P2 3; C2 4; M2 3); *Twyford*[1 & 9] (P2 4; C2 4; M2 5); *West Meon*[2 & 9] (P1 3; P2 3; C1 7; C2 5; M1 6; M2 3); *Wield*[1 & 9] (P1 6; P2 3; C1 5; C2 5; M1 7; M2 6); *Wolvesey*[1 & 9] (P2 6; C2 5; M2 7); *Wootton*[1 & 2] (P1 3; P2 3; C1 4; C2 4; M1 4; M2 3).

Herefordshire: *Aconbury*[1] (P2 3); *Bridge Sollers*[1] (P2 6); *Bunshill*[2] (P1 2); *Clifford*[1] (P1 2); *Durneford*[1] (P2 6; C2 4; M2 7); *Garway*[1] (P1 3); *Huntington*[1] (P2 4; C2 5; M2 5); *Kilpeck*[1] (P2 3); *Leinthall*[2] (P1 6); *Mansell Lacy*[1] (P2 6); *Mathon*[1] (P2 6; C2 5; M2 6).

Hertfordshire: *Albury*[4] (P2 3; C2 5; M2 5); *Aldenham*[1, 3, & 4] (P1 3; P2 3; C1 5; C2 5; M1 3; M2 5); *Amwell*[1 & 3] (P1 2; C1 2; M1 3); *Ashwell*[1, 3, & 4] (P1 1; P2 5; C1 3; C2 3; M1 4; M2 4); *Berkhamsted*[3] (C1 5); *Bishops Stortford*[1] (P1 2); *Broxbourne*[2] (P1 6); *Great Gaddesden*[2 & 4] (P1 1; P2 1; C1 5; C2 5; M1 3; M2 2); *Kings Langley*[3] (P1 2; C1 5; M1 3); *Kinsbourne*[3 & 4] (P1 1; P2 1; C1 5; C2 5; M1 6; M2 5); *Knebworth*[1] (P2 3; C2 5; M2 6); *Meesden*[1] (P1 2; P2 3; C1 5; C2 5; M1 3; M2 6); *Much Hadham*[1] (P1 3); *Pelham Furneaux*[1] (P1 4); *Sawbridgeworth*[4] (P2 3; C2 5; M2 6); *Shenley*[1] (P1 3; C1 5; M1 5); *Standon*[3] (P1 2; C1 5; M1 3); *Stevenage*[1, 3, & 4] (P1 2; P2 4; C1 5; C2 5; M1 3; M2 5); *Temple Dinsley*[3] (P1 1); *Therfield*[3] (P1 3; C1 4; M1 3); *Walkern*[1, 2, & 4] (P1 2; P2 3; C1 5; C2 4; M1 3; M2 6); *Weston*[1 & 3] (P1 3; C1 5; M1 6); *Wheathampstead*[1, 2, 3, & 4] (P1 1; P2 1; C1 5; C2 5; M1 3; M2 5); *Wymondley*[1 & 3] (P1 2; P2 3; C1 5; C2 4; M1 3; M2 3).

Huntingdonshire: *Abbots Ripton*[1] (P1 2; P2 3); *Alyngton*[2] (P1 2; P2 3; C2 8; M2 3); *Broughton*[2 & 18] (P1 3; P2 3; C1 1; C2 8; M1 3; M2 3); *Holywell*[1] (P1 4; P2 3; C1 5; C2 8; M1 7; M2 3); *Houghton*[1, 2, & 18] (P1 3; P2 3; C2 4; M2 3); *Morborne*[1] (P1 2); *Slepe*[1 & 18] (P1 2; P2 3; C2 8; M2 3); *Upwood*[1, 2, & 18] (P2 3; C2 8; M2 3); *Warboys*[18] (P2 3; C2 8; M2 3); *Weston*[1] (P1 2; P2 3); *Wistow*[1 & 18] (P1 2; P2 3; C2 8; M2 3).

Kent: *Adisham*[1 & 4] (P1 3; P2 4; C1 1; C2 4; M1 1; M2 4); *Agney*[1, 3, & 4] (P1 2; P2 1; C1 4; C2 4; M1 1; M2 3); *Aldington*[1] (P1 5); *Appledore*[1, 2, 3, & 4] (P1 1; P2 3; C1 7; C2 5; M1 5;

M2 6); *Barksore*[1 & 3] (P1 3; P2 3; C1 4; C2 5; M1 6; M2 6); *Barton*[4] (P2 3; C2 1; M2 3); *Bekesbourne*[4] (P2 3; C2 1; M2 3); *Bexley*[1 & 3] (P1 3; C1 4; M1 6); *Bishopsbourne*[1] (P1 1); Blean[1 & 3] (P1 2; C1 7); *Boughton under Blean*[1] (P1 4; C1 4; M1 5); *Brook*[1, 2, 3, & 4] (P1 2; P2 3; C1 4; C2 4; M1 1; M2 6); *Charing*[1] (P1 4); *Chartham*[3 & 4] (P1 2; P2 1; C1 1; M1 1; M2 3); *Chingley*[1] (P2 3; C2 5; M2 6); *Cliffe*[1 & 3] (P1 3; P2 3; C1 1; C2 4; M1 1; M2 3); *Cobham*[3] (P1 3; C1 4; M1 1); *Copton*[3 & 4] (P2 1; C1 1; C2 4; M2 4); *Cosington*[1] (P2 3; C2 8; M2 3); *Dartford*[3] (P1 3; C1 4; M1 5); *Denge Marsh*[1 & 4] (P2 1; C2 8; M2 6); *East Farleigh*[1, 2, & 3] (P1 3; P2 4; C1 5; C2 8; M1 1; M2 5); *East Peckham*[3] (P1 2; C1 4; C2 5; M1 3); *Eastry*[1, 3, & 4] (P1 2; P2 1; C1 1; C2 4; M1 1; M2 4); *Ebony*[1, 2, & 3] (P1 2; P2 3; C1 7; C2 5; M1 6; M2 6); *Eltham*[4] (P2 3; C2 2; M2 5); *Elverton*[1, 3, & 4] (P1 3; P2 1; C1 1; C2 8; M1 1; M2 3); *Fairfield*[3] (C1 4); *Gillingham*[1] (P1 4; C1 4; M1 5); *Great Chart*[1, 3, & 4] (P1 2; P2 3; C1 4; C2 8; M1 3; M2 6); *Ham*[3 & 4] (P1 3; P2 1; C1 4; C2 1; M1 5; M2 5); *Hollingbourne*[1 & 3] (P1 2; P2 4; C1 4; C2 4; M1 1; M2 5); *Ickham*[1, 3, & 4] (P1 3; P2 3; C1 1; C2 4; M1 1; M2 3); *Lamberhurst*[4] (P2 3; C2 5; M2 6); *Leeds Manor*[3] (P1 6; C1 4; M1 7); *Lessness*[3] (P1 2; C1 3; M1 4); *Leysdown*[3] (P1 6; C1 7; M1 7); *Little Chart*[3] (P1 2; C1 4; M1 6); *Loose*[1 & 3] (P1 3; C1 4; M1 6); *Lydden*[1 & 3] (P1 3; P2 3; C1 1; M1 1); *Lyminge*[1] (P1 4; C1 1; M1 4); *Maidstone*[1 & 3] (P1 4; C1 4; M1 5); *Meopham*[1, 3, & 4] (P1 3; P2 3; C1 4; C2 8; M1 3; M2 5); *Mersham*[1] (P1 3; P2 4; C1 5; C2 5; M1 6; M2 5); *Milton*[3] (P1 2; C1 4; M1 3); *Monkton*[1 & 4] (P1 3; P2 1; C1 1; C2 1; M1 1; M2 3); *Newnham Court*[4] (P2 4; C2 4; M2 3); *Northfleet*[1 & 3] (P1 4; C1 4; M1 4); *Northstead*[2] (P1 6); *Orgarswick*[4] (P1 2; P2 1; C2 8; M1 1; M2 2); *Orpington*[1 & 3] (P1 3; P2 4; C1 4; C2 4; M1 1; M2 4); *Ospringe*[1 & 3] (P1 2; C1 1; M1 4); *Otford*[1, 3, & 4] (P1 2; P2 3; C1 5; C2 4; M1 3; M2 3); *Peckham*[1 & 2] (P1 2; P2 4; C1 4; C2 4; M1 3; M2 5); *Reculver*[2] (P1 2); *Ruckinge*[1 & 3] (P1 2; P2 3; C1 4; C2 5; M1 1; M2 6); *Saltwood*[1] (P1 2); *Scotney in Lydd*[4] (P2 1; C2 8; M2 3); *Sharpness*[1 & 4] (P1 3; P2 1; C2 1; M2 5); *Strood*[3] (P1 4; C1 1; M1 5); *Teynham*[1] (P1 3); *West Cliffe*[3] (P1 5; C1 4; M1 7); *West Farleigh*[1 & 3] (P1 3; P2 3; C1 4; C2 4; M1 3; M2 3); *West Peckham*[1] (P1 6; C1 1; M1 7); *Westerham*[1, 3, & 4] (P1 3; P2 3; C1 3; C2 5; M1 1; M2 6); *Westgate*[1] (P1 1); *Westwell*[1, 2, 3, & 4] (P1 3; P2 4; C1 4; C2 4; M1 1; M2 3); *Willop in Aldington*[1] (P1 2); *Wingham*[1] (P1 4); *Wingham Barton*[1] (P1 2); *Wrotham*[3 & 4] (P2 4; C1 4; C2 4; M2 5); *Yalding*[1] (P1 6).

Lancashire: *Accrington*[1] (P1 6); *Ightenhill*[1] (P1 6); *Lytham St Annes*[1] (P2 3; C2 8; M2 3); *Standen near Clitheroe*[1] (P1 6); *Swinehurst*[1] (P1 6); *West Derby*[1] (P1 6; C1 7; M1 8); *Widnes*[2] (P1 2).

Leicestershire: *Barton*[1] (P1 6; C1 5; M1 7); *Beaumanor near Woodhouse*[1 & 2] (P1 4; C1 4; M1 5); *Castle Donington*[1 & 8] (P1 3; C1 5; M1 6); *Cold Overton*[1] (P1 6; C1 5; M1 7); *Diseworth*[1] (P1 6; C1 5; M1 7); *Great Easton*[1] (P1 3; C1 4; M1 5); *Kings Norton*[1] (P2 3); *Kirby Bellars*[1] (P2 4; C2 8; M2 3); *Lutterworth*[1 & 2] (P2 3; C1 4); *Nailstone*[2] (C1 3); *Newbold Vernon*[10] (C1 3); *Owston and Knossington*[1] (P2 3); *Stretton*[1] (P1 6; C1 5; M1 7); *Withcote*[1] (P1 6).

Lincolnshire: *Aswick Grange*[1] (P1 2); *Baston*[1] (P1 6); *Bolingbroke*[1 & 8] (P1 3); *Bowthorpe*[1] (P1 2); *Brocklesby*[1 & 8] (P1 3; C1 1; M1 6); *Bucknall*[1] (P1 6); *Caythorpe*[1] (P2 4); *Dowdyke*[1] (P1 2); *Fiskerton*[1] (P1 3; C1 1; M1 1); *Frampton*[1] (P1 2); *Fulstow*[1 & 2] (P1 2; P2 3; C1 3; M1 2); *Gedney*[1] (P2 5; C2 3; M2 6); *Greetham*[1 & 8] (P1 3); *Harrington*[1] (P2 3; C2 4; M2 5); *Hildick*[2] (P1 2); *Holywell*[1] (P1 6; C1 5); *Kirkton in Lindsey*[1] (P1 6; C1 3; M1 7); *Langtoft*[1] (P1 2); *Long Bennington*[1] (P1 6; C1 3; M1 7); *Martin by Horncastle*[1] (P2 4); *Munkelode*[2] (P1 1); *Nomansland*[2] (P1 1); *Rippingale*[1] (P1 6; C1 6; M1 7); *Scotter and Scotterthorpe*[1] (P1 3; C1 2; M1 1); *Sedgebrook*[1 & 8] (P1 2; C1 3;

M1 1); *Somerton Castle*[1] (P2 3; C2 3; M2 3); *Stallingborough*[1 & 2] (P1 5); *Sutton on Sea*[1, 2, & 8] (P1 3; C1 7; M1 6); *Swaton*[1 & 8] (P1 5; C1 1; M1 1); *Temple-Bruer*[1] (P1 6); *Thoresby*[1 & 8] (P1 2; C1 4; M1 1); *Thurlby*[1] (P2 3); *Walcot near Alkborough*[1] (P1 4; C1 2; M1 5); *Wathall*[1] (P1 3; C1 1; M1 1); *Whaplode*[1] (P1 2); *Wrangle*[1 & 8] (P1 2; C1 3; M1 1).

Middlesex: *Acton*[3] (P1 4; C1 5; M1 5); *Ashford*[1, 3, & 4] (P1 3; P2 4; C1 2; C2 2; M1 2; M2 5); *Colham* [3, 4, & 8] (P1 2; P2 3; C1 7; C2 3; M1 2; M2 5); *Ebury alias Eye*[3 & 4] (P1 2; P2 3; C1 2; C2 1; M1 2; M2 1); *Edgware*[1 & 3] (P1 2; C1 5; M1 3); *Fulham*[1] (P1 3); *Halliford*[1, 3, & 4] (P1 2; P2 4; C1 2; C2 1; M1 2; M2 5); *Hampstead*[3] (P1 2; C1 7; M1 3); *Haringay*[1] (P1 2); *Harmondsworth*[4] (P2 3; C2 4; M2 5); *Harrow*[1] (P1 2; C1 7; M1 8); *Hayes*[1] (P1 6; C1 7; M1 3); *Hyde*[4] (P2 3; C2 4; M2 5); *Isleworth*[1 & 3] (P1 2; C1 4; M1 2); *Knightsbridge*[1 & 3] (P1 3; P2 3; C1 3; M1 2); *Laleham*[1, 3, & 4] (P1 3; P2 4; C1 1; C2 2; M1 1; M2 5); *Ruislip*[3] (C1 5); *Staines*[3] (P1 3; C1 1; M1 5); *Stepney*[1] (P1 5); *Sunbury*[1] (P1 4); *Teddington*[3] (P1 3; C1 2; M1 2); *Tottenham Court*[4] (P2 6; C2 5; M2 7); *Yeoveney*[3 & 4] (P1 2; P2 3; C1 4; C2 4; M1 1; M2 6).

Norfolk (Archival references for Norfolk demesnes are all to the Norfolk accounts database, listed in Appendix 2 below.): *Acle* (P1 3; C1 1; M1 1); *Aldborough* (P2 1; C2 2; M2 2); *Aldeby* (P2 3; C2 1; M2 1); *Alderford* (P1 1; C1 1; M1 1); *Arminghall* (P1 2; C1 1; M1 1); *Ashby* (P2 3; C2 1; M2 1); *Ashill* (P1 3; P2 1; C1 1; C2 1; M1 2; M2 1); *Attleborough* (P1 2; C1 2; M1 2); *Attlebridge* (P1 3; C1 2; M1 2); *Aylmerton* (P1 1; C1 1; M1 1); *Barton Bendish* (P1 6; P2 4; C1 1; C2 2; M1 7; M2 2); *Bauburgh* (P1 2; C1 2; M1 2); *Beetley* (P1 3; C1 2; M1 2); *Bintree* (P2 5; C2 1; M2 3); *Bircham* (P1 1; P2 1; C1 2; C2 2; M1 5; M2 2); *Blickling* (P2 1; C2 1; M2 2); *Boughton* (P2 3; C2 8; M2 1); *Bradenham* (P1 2; C1 4; M1 3); *Brancaster* (P2 3; C2 1; M2 1); *Brandiston* (P1 2; C1 1; M1 1); *Bressingham* (P1 2; P2 6; C1 4; C2 4; M1 1; M2 3); *Briston* (P1 6; C1 4; M1 7); *Bunwell* (P1 2; C1 1; M1 1); *Burgh in Flegg* (P1 2; P2 1; C1 1; C2 1; M1 1; M2 1); *Burnham Thorpe* (P1 3; P2 3; C1 1; C2 1; M1 2; M2 1); *Burston* (P2 4; C2 4; M2 1); *Caister cum Markshall* (P1 2; C1 2; M1 2); *Calthorpe* (P2 3; C2 1; M2 1); *Catton* (P1 3; C1 2; M1 2); *Cawston* (P1 3; C1 2; M1 2); *Costessey* (P1 2; P2 1; C1 2; C2 1; M1 2; M2 2); *Crimplesham* (P1 6; C1 2; M1 7); *Cringleford* (P1 2; P2 3; C1 2; C2 1; M1 2; M2 1); *Crownthorpe* (P1 2; C1 1; M1 1); *Deopham* (P1 4; P2 5; C1 1; C2 1; M1 2; M2 3); *Diss* (P2 6; C2 1; M2 3); *Ditchingham* (P1 2; C1 4; M1 3); *Earsham* (P1 2; C1 4; M1 3); *East Carleton* (P1 2; P2 6; C1 1; C2 4; M1 1; M2 3); *East Lexham* (P2 1; C2 2; M2 2); *East Wretham* (P1 3; C1 2; C2 2; M1 2); *Eaton* (P1 2; P2 3; C1 2; C2 1; M1 2; M2 1); *Eccles* (P1 3; P2 1; C1 2; C2 2; M1 2; M2 2); *Felbrigg* (P2 1; C2 1; M2 1); *Feltwell* (P1 3; P2 1; C1 1; C2 1; M1 2; M2 2); *Fincham* (P1 1; P2 4; C1 1; C2 1; M1 5; M2 1); *Flegg* (P1 2; P2 3; C1 1; C2 1; M1 1; M2 1); *Forncett* (P1 2; C1 1; M1 1); *Foxley* (P1 2; C1 1; M1 1); *Framingham* (P1 2; C1 1; M1 1); *Fring* (P1 1; C1 2; M1 2); *Gateley* (P1 2; C1 4; M1 1); *Gaywood* (P1 3; C1 4; M1 6); *Gimingham* (P2 3; C2 1; M2 1); *Gnatingdon* (P1 1; C1 2; M1 2); *Great Cressingham* (P1 2; P2 1; C1 2; C2 1; M1 2; M2 1); *Gresham* (P1 6; C1 2; M1 1); *Gressenhall* (P2 3; C2 2; M2 1); *Haddeston* (P1 2; C1 1; M1 1); *Hainford* (P2 1; C2 1; M2 1); *Halvergate* (P1 4; P2 3; C1 1; C2 4; M1 6; M2 1); *Hanworth* (P1 2; C1 1; M1 1); *Hardley* (P2 3; C2 1; M2 1); *Hargham* (P1 1; C1 2; M1 3); *Harpley* (P1 1; C1 1; M1 1); *Hautbois* (P1 2; P2 3; C1 2; C2 1; M1 2; M2 1); *Haveringland* (P2 3; C2 1; M2 1); *Heacham* (P1 3; P2 3; C1 2; C2 1; M1 2; M2 1); *Heigham by Norwich* (P1 2; P2 5; C1 1; C2 1; M1 2; M2 3); *Helhoughton* (P1 1; C1 1; M1 3); *Hempnal* (P1 2; C1 4; M1 1); *Hemsby* (P1 2; P2 3; C1 1; C2 1;

M1 1; M2 1); *Hethel* (P2 6; C2 4; M2 3); *Hevingham* (P1 2; P2 1; C1 1; C2 1; M1 1; M2 1); *Hevingham with Marsham* (P1 3; P2 3; C1 1; C2 1; M1 2; M2 1); *'Heythe' near Plumstead* (P1 3; C1 2; M1 2); *Hilborough* (P2 1; C2 1; M2 2); *Hilgay* (P2 3; C2 2; M2 1); *Hindolveston* (P1 2; P2 3; C1 1; C2 1; M1 1; M2 1); *Hindringham* (P1 2; P2 3; C1 1; C2 1; M1 1; M2 1); *Hingham* (P1 3; C1 5; M1 3); *Hockham* (P2 1; C2 1; M2 2); *Holkham* (P2 1; C2 1; M2 2); *Horning* (P2 3; C2 1; M2 1); *Hoveton* (P2 1; C2 1; M2 1); *Howardes Manor* (P2 6; C2 1; M2 3); *Hunstanton* (P1 1; P2 1; C1 1; C2 1; M1 1; M2 1); *Intwood* (P1 3; P2 1; C1 2; C2 1; M1 2; M2 1); *Kempstone* (P1 2; P2 1; C1 4; C2 4; M1 3; M2 1); *Kerdiston* (P1 2; C1 1; M1 1); *Keswick* (P1 2; P2 3; C1 2; C2 1; M1 2; M2 1); *Knapton* (P1 1; C1 1; M1 1); *Lakenham* (P1 2; P2 1; C1 2; C2 1; M1 2; M2 1); *Langham* (P1 3; P2 1; C1 1; C2 1; M1 2; M2 1); *Lessingham* (P1 2; C1 1; M1 1); *Litcham* (P2 6; C2 1; M2 3); *Little Ellingham* (P1 2; C1 1; M1 2); *Loddon* (P1 5; C1 1; M1 4); *Lopham* (P1 2; C1 4; M1 1); *Ludham* (P2 3; C2 1; M2 1); *Marham* (P2 3; C2 1; M2 1); *Martham* (P1 2; P2 3; C1 1; C2 1; M1 1; M2 1); *Melton* (P1 2; P2 1; C1 2; C2 2; M1 2; M2 1); *Methwold* (P2 1; C2 2; M2 2); *Mileham* (P2 3; C2 5; M2 1); *Monks Grange* (P1 2; C1 2; M1 2); *Mundham* (P2 3; C2 4; M2 1); *Newton* (P1 3; P2 3; C1 4; C2 5; M1 6; M2 1); *Newton by Norwich* (P1 2; P2 3; C1 2; C2 1; M1 2; M2 1); *North Creake* (P1 1; C1 2; M1 1); *North Elmham* (P1 2; P2 6; C1 2; C2 1; M1 2; M2 1); *North Walsham* (P2 3; C2 1; M2 1); *Ormesby* (P1 2; P2 3; C1 1; C2 1; M1 1; M2 1); *Osmundiston* (P1 3; C1 1; M1 2); *Plumstead* (P1 2; P2 3; C1 1; C2 1; M1 2; M2 1); *Popenhoe* (P1 3; P2 3; C1 3; C2 3; M1 3; M2 2); *Potter Heigham* (P2 5; C2 1; M2 2); *Quidenham* (P2 1; C2 1; M2 2); *Raynham* (P1 2; P2 1; C1 1; C2 1; M1 2; M2 1); *Reedham* (P2 3; C2 1; M2 1); *Ringstead* (P2 1; C2 1; M2 1); *Rougham* (P2 3; C2 4; M2 1); *Saxthorpe* (P1 2; P2 1; C1 1; C2 1; M1 1; M2 1); *Scottow* (P2 1; C2 1; M2 1); *Scratby* (P1 2; P2 3; C1 1; C2 1; M1 1; M2 1); *Sedgeford* (P1 3; P2 3; C1 2; C2 1; M1 2; M2 1); *Seething* (P1 2; P2 3; C1 4; C2 1; M1 1; M2 1); *Shotesham* (P2 6; C2 4; M2 3); *Shropham* (P2 1; C2 1; M2 2); *Sloley* (P1 2; C1 1; M1 1); *South Walsham* (P1 3; C1 1; M1 5); *Southery* (P1 2; C1 2; M1 1); *Sporle* (P1 1; C1 1; M1 6); *Stanhoe* (P1 1; P2 6; C1 2; C2 1; M1 2; M2 3); *Stiffkey* (P1 1; C1 1; M1 4); *Stradsett* (P2 3; C2 3; M2 1); *Suffield* (P1 2; C1 1; M1 1); *Syderstone* (P2 1; C2 1; M2 2); *Tacolneston* (P2 3; C2 4; M2 1); *Taverham* (P1 2; P2 3; C1 2; C2 1; M1 2; M2 1); *Thornage* (P1 6; P2 1; C1 1; C2 1; M1 7; M2 1); *Thornham* (P1 1; P2 3; C1 1; C2 1; M1 1; M2 3); *Thorpe Abbotts* (P1 2; P2 3; C1 1; C2 4; M1 1; M2 1); *Thurning* (P1 2; P2 1; C1 1; C2 1; M1 1; M2 1); *Thwaite* (P2 5; C2 1; M2 3); *Titchwell* (P1 1; C1 1; M1 1); *Tivetshall* (P1 2; P2 3; C1 4; C2 4; M1 1; M2 1); *Topcroft* (P2 3; C2 4; M2 1); *Tunstead* (P2 1; C2 1; M2 1); *Wattisfield* (P2 1; C2 4; M2 1); *West Harling* (P1 2; P2 1; C1 2; C2 2; M1 2; M2 2); *West Lexham* (P1 6; C1 2; M1 7); *West Newton* (P1 1; C1 2; M1 2); *West Walton* (P1 3; P2 3; C1 4; C2 8; M1 6; M2 1); *Wicklewood* (P1 2; C1 1; M1 1); *Wiggenhall* (P1 5; C1 4; M1 7); *Wilton* (P2 1; C2 1; M2 1); *Wimbotsham* (P1 2; C1 2; M1 1); *Worstead* (P1 1; C1 1; M1 4); *Wroxham* (P1 1; C1 2; M1 2); *Wymondham* (P1 2; P2 3; C1 1; C2 4; M1 1; M2 1).

Northamptonshire: *Addington*[1 & 3] (P1 5; C1 4; M1 4); *Aldwincle*[3] (P1 3; C1 4; M1 5); *Ashby St Ledgers*[1] (P2 3; C2 5; M2 5); *Ashton*[3 & 4] (P1 4; P2 4; C1 4; C2 4; M1 5; M2 5); *Biggin*[3 & 4] (P1 3; P2 3; C1 5; C2 8; M1 6; M2 3); *Boroughbury*[1, 3, & 4] (P1 2; P2 3; C1 4; C2 3; M1 1; M2 1); *Castor*[3] (P1 2; C1 4; M1 3); *Cottingham*[3] (P1 3; C1 5; M1 6); *Culworth*[4] (P2 5); *Elmington*[1 & 3] (P1 5; C1 4; M1 4); *Eye*[3 & 4] (P1 2; P2 3; C1 4; C2 3; M1 1; M2 3); *Glinton*[3] (P1 2; C1 3; M1 1); *Higham Ferrers*[1, 3, & 4] (P1 5; P2 4; C1 3;

C2 3; M1 7; M2 7); *Irthlingborough*[1 & 3] (P1 3; C1 3; M1 5); *Kettering*[3] (P1 2; C1 2; M1 2); *Lolham*[3] (P1 6; C1 4; M1 7); *Long Buckby*[1 & 3] (P1 3; P2 4; C1 5; C2 8; M1 6; M2 3); *Longthorpe*[1 & 3] (P1 2; P2 3; C1 4; C2 1; M1 3; M2 1); *Maidwell*[3 & 4] (P1 2; P2 3; C1 5; C2 8; M1 3; M2 3); *Manor of Park*[1] (P1 2; C1 3); *Naseby*[1] (P1 3; C1 7; M1 3); *Oundle*[3] (P1 5; C1 4; M1 4); *Overstone*[1] (P1 2); *Passenham*[4] (P2 6); *Pury*[1] (P1 6; C1 5; M1 7); *Radstone*[1 & 3] (P1 2; C1 5; M1 3); *Raunds*[4] (P2 6; C2 3; M2 7); *Rushden and Higham Park*[3] (P1 2; C1 3; M1 3); *Silverstone*[1] (P1 2); *Stanwick*[3] (P1 4; C1 3; M1 5); *Thorp Waterville*[3] (P1 3; C1 4; M1 5); *Thrupp and Norton*[1] (P2 6; C2 4; M2 6); *Torpel*[3] (P1 2; C1 5; M1 3); *Upton*[3] (P1 5; C1 5; M1 4); *Walton*[3] (P1 4; C1 4; M1 4); *Warmington*[3] (P1 3; C1 4; M1 6); *Weedon Bec*[3] (C1 5); *Wellingborough*[1 & 3] (P1 2; C1 4; M1 3); *Werrington*[3] (P1 2; C1 4; M1 1); *Wootton*[2] (P1 2; C1 1).

Northumberland: *Bamburgh*[1] (P1 6; C1 5; M1 7); *Embleton*[1] (P1 5); *Holy Island*[1 & 2] (P1 2; P2 3; C2 4; M2 3); *Stamford*[1] (P1 6).

Nottinghamshire: *Bingham*[1] (P1 6; C1 5; M1 7); *Collingham*[1] (P1 3; C1 2; M1 6); *Gringley on the Hill*[1] (P1 6); *Kneesall*[1] (P1 2); *Laneham*[1] (P2 6); *Scrooby*[1] (P2 6); *Southwell*[1] (P2 6); *Wheatley*[1] (P1 2; C1 5; M1 6).

Oxfordshire: *Adderbury*[1, 3, 4, & 9] (P1 3; P2 3; C1 3; C2 3; M1 6; M2 1); *Bicester*[3] (P1 4; C1 3; M1 6); *Black Bourton*[1] (P1 3); *Broadwell*[3] (P1 6; C1 3; M1 7); *Caversfield*[2 & 3] (P1 5); *Checkendon*[1] (P1 1); *Clifton*[3] (P1 2; C1 3; M1 3); *Combe*[1 & 3] (P1 4; C1 6; M1 7); *Cottisford*[3] (P1 3); *Crowmarsh*[2 & 4] (P1 3; P2 3; C1 3; C2 4; M1 4; M2 3); *Cuxham*[1, 2, & 3] (P1 2; P2 3; C1 5; C2 4; M1 3; M2 6); *Forest Hill*[1, 3, & 17] (P1 3; C1 3; M1 3); *Fritwell*[1] (P2 3); *Hampton Gay*[1] (P1 3); *Handborough*[1] (P1 2); *Henton*[3] (C1 4); *Holywell*[1, 3, & 4] (P1 2; P2 3; C1 1; M1 2); *Islip*[1, 3, & 4] (P1 2; P2 4; C1 3; C2 8; M1 3; M2 3); *Kidlington*[3] (P1 4); *Launton*[1 & 3] (P1 3; P2 3; C1 5; C2 8; M1 6; M2 6); *Little Faringdon and Langford*[1] (P1 6); *Little Tew*[1 & 3] (P1 4; C1 3; M1 3); *Merton*[3] (P1 3; C1 3; M1 6); *Middleton Stoney*[3] (P1 4; C1 3; M1 5); *Milton*[1] (P2 3; C2 3; M2 1); *North Leigh*[1] (P2 3; C2 3; M2 3); *Nuneham*[3] (P1 2; C1 7; M1 6); *Oddington*[3] (C1 5); *Sandford-on-Thames*[3] (P1 6; C1 5; M1 7); *Shifford*[1] (P2 6; C2 3; M2 7); *Shilton*[1] (P1 2); *Sibford Gower*[3] (P1 3; C1 3; M1 5); *South Stoke*[4] (P2 1; C2 3; M2 4); *South Weston*[1] (P1 3); *Steeple Barton*[1] (P1 3); *Stone*[17] (C1 3); *Stratton*[3] (P1 2; C1 3; M1 1); *Temple Cowley*[3] (P1 4; C1 4; M1 5); *Upper Heyford*[3 & 4] (P1 2; P2 3; C1 3; C2 4; M1 3; M2 3); *Warpsgrove*[3] (P1 3; C1 5; M1 6); *Water Eaton*[1] (P1 3); *Waterperry*[2 & 17] (P1 3; C1 4; M1 6); *Watlington*[1, 3, & 17] (P1 2; C1 4; M1 7); *Whitchurch*[1 & 3] (P1 2; C1 5; M1 7); *Witney*[1, 3, 4, & 9] (P1 3; P2 3; C1 4; C2 4; M1 6; M2 3); *Wootton*[3] (P1 6; C1 3; M1 7).

Rutland: *Market Overton*[1 & 2] (P1 3; P2 3; C1 3; C2 2; M1 6; M2 6); *Oakham*[1 & 2] (P1 2; P2 5; C1 1; C2 8; M1 1; M2 4); *Stretton*[1] (P1 6; C1 5); *Tinwell*[1] (P1 4; C1 4; M1 6).

Shropshire: *Adderley*[1] (P1 2; C1 5; M1 6); *Cleobury Barnes*[1] (P2 3; C2 5; M2 6); *Ludlow*[2] (P1 6); *Stanton Lacy*[1] (P2 6; C2 5; M2 7).

Somerset: *Ashcott*[6] (P1 2; C1 6; M1 7); *Baltonsborough*[6] (P1 2; C1 3; M1 6); *Barwick*[1] (P1 2; C1 4; M1 7); *Batcombe*[6] (P1 4; C1 5; M1 6); *Beckington*[1] (P2 6); *Bedminster*[2] (P1 2; C1 5; M1 7); *Bishops Hull in Taunton*[1 & 9] (P2 5; C2 5; M2 7); *Brent*[6] (P1 2; C1 4; M1 3); *Bridgwater*[1] (P1 6; C1 4; M1 7); *Butleigh*[6] (P1 2; C1 6; M1 3); *Charlton*[1] (P1 4; C1 4; M1 5); *Ditcheat*[6] (P1 2; C1 6; M1 3); *Doulting*[6] (P1 4; C1 7; M1 8); *East Pennard*[6] (P1 5; C1 6; M1 3); *Farleigh Hungerford*[1] (P2 3; C2 5; M2 6); *Glastonbury*[6] (P1 2; C1 6; M1 3); *Godney*[6] (P1 2; C1 6; M1 6); *Greinton*[6] (P1 5; C1 6; M1 7); *Hatch Beauchamp*[2] (P1 6; C1 5; M1 3); *Hentsridge*[1] (P1 2; C1 5; M1 3); *High Ham*[6] (P1 5; C1 6; M1 3); *Holway in Taunton*[1 & 9] (P1 6; P2 5; C1 5; C2 5; M1 7; M2 7); *Houndstreet*[16] (C1 4);

Hurcott[1] (P1 6; C1 6; M1 7); *Kingsbury*[1] (P1 4; C1 4; M1 5); *Kingston and Nailsbourne*[1 & 9] (P1 4; P2 3; C1 7; C2 5; M1 5; M2 6); *Marksbury*[6] (P1 3; C1 6; M1 6); *Marston*[1] (P1 6; C1 5; M1 7); *Meare*[6] (P1 2; C1 6; M1 3); *Mells*[6] (P1 4; C1 5; M1 6); *Pilton*[6] (P1 2; C1 5; M1 3); *Podimore Milton*[6] (P1 3; C1 5; M1 5); *Porlock*[1 & 2] (P2 3; C2 5; M2 3); *Portbury*[2] (P1 6; C1 1; M1 7); *Poundisford in Taunton*[1 & 9] (P1 2; P2 3; C1 5; C2 5; M1 3; M2 6); *Queen Camel*[1] (P1 2; C1 5; M1 3); *Rimpton*[1 & 9] (P1 2; P2 3; C1 5; C2 5; M1 3; M2 6); *Shapwick*[6] (P1 4; C1 6; M1 7); *Staplegrove in Taunton*[1 & 9] (P1 6; P2 5; C1 5; C2 5; M1 7; M2 7); *Stockwood*[1] (P1 6; C1 5; M1 7); *Stogursey*[1] (P1 6; C1 5; M1 7); *Street*[6] (P1 6; C1 6; M1 3); *Trendle in Pitminster*[1] (P1 6); *Walton*[6] (P1 5; C1 6; M1 3); *Wellow*[1 & 2] (P2 4; C2 5; M2 5); *West Hatch*[1] (P1 6; P2 6; C1 5; C2 4; M1 7; M2 6); *Westonzoyland*[6] (P1 2; C1 1; M1 1); *Winscombe*[1] (P1 3); *Wrington*[6] (P1 3; C1 6; M1 3).

Staffordshire: *Baswich*[1] (P1 3); *Brewood*[1] (P1 3); *Eccleshall*[1] (P1 2); *Keele*[1] (P1 3; C1 5; M1 6); *Longdon*[1] (P1 3; C1 3; M1 7); *Marchington*[1] (P1 5; C1 5; M1 4); *Rolleston*[1] (P1 6; C1 5; M1 3); *Scropton*[2] (P1 6; C1 5; M1 7); *Sedgley*[1] (P2 3); *Tutbury*[1] (P1 5; C1 1; M1 4).

Suffolk: *Acton*[1] (P2 3; C2 8; M2 3); *Berlawe near Stradbrook*[11] (P2 5; C2 5; M2 5); *Blakenham*[1] (P1 2; C1 2; M1 3); *Brandon*[5] (P1 3; P2 1; C1 2; C2 1; M1 2; M2 2); *Bungay*[5] (P1 2; C1 4; M1 1); *Chevington*[1] (P1 5; P2 3; C2 5; M2 6); *Clare*[10] (C1 5); *Clopton*[1] (P1 6; P2 3; C1 4; C2 5; M1 7; M2 5); *Clopton Kingshall*[1] (P2 3; C2 5; M2 5); *Cratfield*[1] (P1 2; C1 5; M1 3); *Dalham*[1] (P1 4); *Denham*[1] (P1 2; C1 1); *Dunningworth*[1] (P1 3; C1 2; M1 2); *Earl Soham*[1] (P1 2; C1 5; M1 3); *Earl Stonham*[1] (P1 2; C1 4; M1 1); *Erbury*[1] (P2 3; C2 5; M2 6); *Exning*[1] (P2 5; C2 3; M2 4); *Framlingham*[1] (P1 2; C1 5; M1 3); *Great Saxham*[1] (P2 3; C2 4; M2 6); *Hargrave*[1] (P2 3; C1 5; C2 5; M2 6); *Henley*[1 & 2] (P1 2; C1 4; M1 1); *Hinderclay*[5] (P1 2; P2 3; C1 1; C2 1; M1 1; M2 1); *Hollesley*[1] (P1 4); *Hoo near Kettleburgh*[1] (P1 2; C1 4; M1 3); *Horham*[1 & 2] (P1 2; P2 3; C1 5; M1 3); *Hoxne*[5] (P1 6; C1 1; M1 8); *Hundon*[1 & 10] (P2 3; C2 5; M2 6); *Kelsale*[1] (P1 6; C1 5; M1 7); *Lackford*[1] (P2 1); *Lakenheath*[1] (P1 3; P2 1; C1 2; C2 2; M1 2; M2 2); *Lawshall*[1] (P1 2; P2 3); *Little Ashfield*[1] (P1 2; P2 3; C2 4; M2 6); *Melton*[1 & 2] (P1 2; P2 3; C1 2; C2 2; M2 1); *Monks Eleigh*[1] (P1 2; P2 3; C1 5; C2 5; M1 4; M2 6); *Nayland*[1] (P1 2); *Peasenhall*[1] (P1 2); *Redgrave*[5] (P1 3; P2 3; C1 4; C2 4; M1 1; M2 1); *Reydon*[1] (P2 3); *Rickinghall*[5] (P1 2; P2 3; C1 4; C2 4; M1 1; M2 1); *Risby*[1] (P2 4; C2 4; M2 3); *Staverton*[1] (P1 6; C1 2; M1 7); *Syleham*[1] (P1 2); *Walton*[1] (P1 2; C1 1); *Wattisfield*[5] (P2 3); *Westley*[1] (P1 3; C1 2; M1 5); *Wicklow in Hacheston*[1] (P1 2; C1 7; M1 3); *Woodhall*[10] (C1 4).

Surrey: *Banstead*[1] (P1 3; P2 4; C1 3; C2 4; M1 5; M2 3); *Battersea*[1 & 3] (P1 2; P2 3; C1 2; C2 2; M1 3; M2 6); *Beddington*[4] (C2 4); *Bensham in Croydon*[1] (P1 4); *Betchworth*[4] (P2 3; C2 4; M2 6); *Byfleet*[3] (P1 3; C1 2; M1 6); *Cheam*[1, 2, & 3] (P1 3; C1 1; M1 1); *Chessington*[3] (P1 3; C1 7; M1 6); *Claygate*[1] (P1 3); *Croydon and Cheam*[1] (P1 6; C1 2; M1 7); *Dorking*[4] (P2 3; C2 1; M2 2); *East Gomshall*[1] (P2 4); *East Horsley*[3] (P1 4; C1 4; M1 5); *Esher*[1, 3, & 9] (P1 3; C1 7; M1 8); *Farleigh*[1, 2, 3, & 4] (P1 3; P2 3; C1 5; M1 3); *Farnham*[1, 3, 4, & 9] (P1 3; P2 3; C1 4; C2 4; M1 6; M2 3); *Lambeth*[1] (P1 2; C1 7; M1 8); *Leatherhead*[3 & 4] (P1 3; P2 4; C1 4; M1 6); *Malden*[3 & 4] (P1 3; P2 3; C1 5; C2 5; M1 6; M2 6); *Merstham*[3 & 4] (P1 3; P2 4; C1 5; C2 5; M1 6; M2 5); *Morden*[1 & 3] (P1 2; P2 4; C1 5; C2 5; M1 3; M2 5); *Pyrford*[1, 3, & 4] (P1 3; P2 4; C1 7; C2 2; M1 3; M2 2); *Sheen*[3] (P1 3; C1 7; M1 2); *Shere and Vachery*[2] (P2 3); *Stoke by Guildford*[1] (P1 3); *Thorncroft*[3 & 4] (P1 3; P2 4; C1 4; C2 4; M1 6; M2 5); *Tyting*[1] (P1 3); *Walton-on-*

Thames[3] (P1 4; C1 7; M1 5); *Walworth*[1 & 3] (P1 5; P2 5; C1 1; M1 4); *Wandsworth*[4] (P2 3; C2 2; M2 5); *West Gomshall*[1] (P2 4); *West Horsley*[4] (P2 3; C2 1; M2 3); *Wimbledon*[1] (P1 2; C1 7; M1 3).

Sussex: *Alciston*[1, 2, & 13] (P1 3; P2 3; C2 4; M2 3); *Apuldram*[1] (P1 3; C1 1; M1 1); *Barnhorn Manor in Bexhill*[1] (P2 3); *Beddingham*[1 & 13] (P1 4; P2 4; C1 1; C2 4; M1 5; M2 5); *Bersted*[1] (C2 8); *Bishopstone*[2 & 13] (C2 4); *Bosham*[1, 2, & 13] (P1 3; P2 3; C1 4; C2 4; M1 6; M2 3); *Chalvington*[1] (P1 6; P2 3; C1 3; C2 5; M1 7; M2 6); *Chidham*[1] (P1 3); *Duncton*[1] (P2 3; C2 5; M2 6); *East Lavant*[1] (P1 3); *Eastbourne*[13] (C1 1); *Ecclesdon in Angmering*[1] (P2 5; C2 8; M2 4); *Ferring*[13] (C2 4); *Funtington*[1] (P1 4; P2 4; C1 4; M1 5); *Glynde*[1 & 2] (P1 3; P2 3; C1 4; C2 4; M1 6; M2 3); *Heighton*[1] (P1 4; P2 4; C1 4; C2 4; M1 5; M2 3); *Heyshott*[1] (P2 3; C2 5; M2 6); *Lodsworth*[1] (P1 3); *Marley*[1] (P1 2; P2 3; C1 7; C2 5; M1 3; M2 6); *Pagham*[1] (P1 3); *Peppering*[13] (C2 4); *Petworth*[1 & 2] (P1 3; P2 3; C2 2; M2 2); *Preston*[13] (C2 4); *Sidlesham and Greatham*[13] (C2 8); *Slindon*[1] (P1 3); *Steyning*[2] (P1 3; C1 1; M1 6); *Stoneham in South Malling*[1] (P2 3); *Stoughton*[1] (P1 4; C1 4; M1 5); *Streat*[1] (P2 5); *Sutton*[1] (P2 4; C2 5; M2 5); *Tangmere*[1] (P1 6; P2 3; C2 4; M2 3); *Warminghurst*[1] (P1 3; C1 4; M1 4); *West Stoke*[1] (P1 4; C1 1; M1 5); *West Thorney*[1, 2, & 13] (P1 4; P2 3; C1 1; C2 8; M1 5; M2 7); *Westdean*[1] (P2 4; C2 4; M2 5); *Wiston*[1 & 2] (P1 4; P2 3; C2 4; M2 3).

Warwickshire: *Barton*[2] (P1 6; C1 5); *Braundon*[10] (P1 2; C1 2; M1 2); *Caluden in Wyken*[1] (P1 6; C1 7; M1 7); *Chilvers Coton*[1] (P1 5; C1 3; M1 3); *Compton*[1] (P1 4); *Cubbington*[1] (P1 2; C1 3; M1 3); *Fillongley*[2] (P1 5; C1 6; M1 4); *Fletchamstead*[1] (P1 6; C1 3; M1 3); *Grafton*[2] (P1 4; C1 3; M1 5); *Harbury*[1] (P1 2; C1 3; M1 7); *Itchington*[1] (P2 5; C2 3; M2 4); *Kington Brailes*[1] (C1 6); *Knowle*[1] (P1 2; P2 3; C2 5; M2 6); *Ladbroke*[1] (P2 3; C2 3; M2 6); *Long Compton*[1] (P2 4); *Sherbourne*[1] (P1 6; C1 3; M1 3); *Stretton*[2] (P1 6; C1 5); *Studley*[1] (P1 6; C1 5; M1 3); *Sutton under Brailes*[1] (P1 5); *Talton in Tredington*[1] (P1 6); *Temple Balsall*[1] (P1 2; C1 5; M1 3); *Temple Cretton?*[2] (P1 5; C1 3; M1 5); *Tysoe*[1] (P1 2; C1 4; M1 3); *Warwick*[1] (P1 2; C1 3; M1 3); *Wolvey*[1] (P1 3; C1 4; M1 5).

Westmorland: *Maulds Maeburn*[1] (P2 3).

Wiltshire: *Aldbourne*[1] (P1 4); *Amesbury*[1] (P1 4; C1 4; M1 6); *Badbury*[6] (P1 3; C1 6; M1 6); *Bishops Fonthill*[1 & 9] (P1 4; C1 4; M1 5); *Bishopstone*[1 & 9] (P1 3; P2 4; C1 4; C2 4; M1 6; M2 3); *Bromham*[15] (P1 3; P2 3; C1 2; C2 2; M1 2; M2 5); *Calne*[1] (P1 4; C1 4; M1 5); *Calstone*[2] (P1 3; C1 6; M1 6); *Chippenham*[1 & 2] (P2 3); *Christian Malford*[6] (P1 5; C1 5; M1 3); *Collingbourne*[1] (P1 3; C1 4; M1 6); *Cowesfield*[1] (P1 6; C1 5; M1 7); *Damerham*[6] (P1 3; C1 5; M1 6); *Downton*[1 & 9] (P2 3; C1 4; C2 4; M2 3); *Ebbesbourne Wake*[1] (P2 5; C2 4; M2 7); *Edington*[1] (P1 4; P2 4; C1 4; C2 4; M1 6; M2 5); *Everleigh*[1] (P1 4; C1 4; M1 6); *Grittleton*[6] (P1 3; C1 6; M1 3); *Heytesbury*[1] (P2 3; C2 4; M2 5); *Idmiston*[6] (P1 3; C1 4; M1 5); *Inglesham*[1] (P1 2); *Keevil*[2] (P1 6; C1 3; M1 7); *Kingston Deverill*[1 & 2] (P2 4; C2 3; M2 5); *Kington St Michael*[6] (P1 4; C1 4; M1 3); *Knoyle*[1 & 9] (P1 4; P2 3; C1 4; C2 4; M1 6; M2 3); *Mere*[1] (P1 4; C1 5; M1 5); *Monkton Deverill*[16] (C1 4); *Nettleton*[6] (P1 4; C1 5; M1 3); *Sevenhampton*[1] (P1 3; C1 6; M1 6); *Sutton Veny*[1] (P1 3); *Trowbridge*[1] (P1 2); *Upton Knoyle*[1] (P1 6; P2 4; C1 4; C2 4; M1 7; M2 5); *Winterbourne*[1] (P1 3; C1 4; M1 6); *Winterbourne Monkton*[6] (P1 4; C1 4; M1 3).

Worcestershire: *Broadway*[1] (P2 6; C2 4; M2 6); *Hanley Castle*[2] (P1 2); *Hewell Grange*[2] (P2 3; C2 5; M2 6); *Leigh*[2] (P2 3; C2 5; M2 6); *Longdon*[2] (P1 6; C1 3); *Oldington*[1] (P1 4); *Peachley*[1 & 2] (P1 3; P2 4); *Pensham*[1] (P2 3; C2 3; M2 3); *Pershore*[1] (P1 3; P2 4); *Pinvin*[1] (P2 6); *Wadborough*[1] (P2 3; C2 8; M2 3); *Wickhamford*[1] (P1 2).

Yorkshire, East Riding: *Beverley*[1] (P2 6); *Burstwick*[1] (P1 6; P2 3; C1 5; C2 8; M1 7;

M2 6); *Burton Constable*[1] (Pl 6; Cl 5; M1 7); *Cleton*[1] (Pl 6; Cl 6; M1 7); *Easington*[1] (Pl 6; Cl 5; M1 7); *Faxfleet*[1] (Pl 3); *Halsham*[2] (Pl 2); *Howsham*[1] (P2 3; C2 5; M2 6); *Keyingham*[1] (Pl 2; P2 6; Cl 5; C2 8; M1 3; M2 7); *Little Humber*[1] (Pl 5; Cl 4; M1 7); *Metham*[1] (Pl 6); *Ringborough*[1] (Pl 6); *Sherburn*[1] (P2 6; C2 5; M2 7); *Skidby*[1] (P2 6; C2 8; M2 7); *Skipsea Church*[2] (Pl 5); *South Burton*[1] (P2 4; C2 3; M2 5); *Wetwang*[1] (P2 6; C2 4; M2 7).

Yorkshire, North Riding: *Carthorpe*[2] (Pl 5); *East Cowton*[1] (Pl 5; Cl 5; M1 3); *Helmsley*[1] (P2 5; C2 5; M2 7); *Holme*[1] (Pl 6; Cl 5; M1 7); *Little Langton*[1] (Pl 2); *Stanghow*[1] (Pl 6; Cl 5; M1 7); *West Tanfield*[1 & 2] (Pl 5; P2 5; C2 4; M2 4).

Yorkshire, West Riding: *Ackworth*[1] (Pl 4; Cl 7; M1 5); *Acomb*[1] (Pl 2); *Altofts*[1] (Pl 6); *Broughton*[1] (Pl 6; Cl 7); *Campsall*[1] (Pl 4; Cl 5; M1 5); *Cawood*[1] (P2 6; C2 2; M2 7); *Conisbrough*[1] (Pl 6; Cl 5; M1 7); *Cowick*[2] (Pl 2; Cl 5; M1 3); *Cowick with Snaith*[2] (Pl 6; Cl 5; M1 3); *East Haddesley*[2] (Pl 2; Cl 5; M1 3); *Elmsall*[1] (Pl 4); *Great Sandal*[1] (Pl 2; Cl 7; M1 8); *Harewood*[1] (Pl 3; Cl 5; M1 6); *Kippax*[1 & 8] (Pl 6; Cl 5; M1 7); *Methley*[2] (P2 3; C2 5; M2 6); *North Deighton*[1] (Pl 6; Cl 5; M1 7); *Owston*[1] (Pl 5); *Paddockthorpe in Newton*[1] (P2 3); *Pollington*[2] (Pl 4; P2 6; Cl 7; M1 5); *Rockley and Stainbrogh*[1] (P2 3); *Roecliffe*[1] (Pl 6; Cl 5; M1 7); *Roundhay*[1] (Pl 2); *Skipton*[1] (Pl 2; Cl 7; M1 6); *Soothill*[1] (Pl 6; Cl 7; M1 8); *Sowerby*[2] (Pl 2); *Tanshelf*[1 & 8] (Pl 6; Cl 4; M1 7); *Temple Hirst*[1] (Pl 3; Cl 7; M1 6); *Thorner*[6] (P2 4; C2 5; M2 5); *Tickhill*[1] (Pl 6; Cl 5; M1 7); *Thorpe in Balne*[1] (Pl 6; Cl 5; M1 7); *Vernoyl in Balne*[2] (Pl 2); *Whitgift*[1 & 2] (Pl 4; Cl 5; M1 5).

Yorkshire (Riding uncertain): *Couhouse*[1] (C2 5); *Porterlawe*[2] (Pl 1; Cl 2).

Scotland (Berwickshire): *Coldingham*[2] (Pl 3; P2 3; Cl 7; C2 5; M1 6; M2 3).

Wales (Monmouthshire): *Bergbeven*[2] (Pl 2); *Grosmont*[2] (Pl 6); *Llangathney*[2] (Pl 4); *Llangwm*[10] (Cl 5); *Llantrissent*[10] (Cl 6); *Llanvihangel*[2] (Pl 6); *New Grange*[10] (Cl 5); *Tregoythel*[2] (Pl 5); *Treigruk*[10] (Cl 5); *Trelleck*[10] (Cl 7); *Troy*[10] (Cl 5); *Usk*[10] (Cl 7); *White Castle*[2] (Pl 2).

Wales (county uncertain): *'Coidemor'*[2] (P2 3).

References

1 Data supplied by John Langdon and listed in Appendix C of his PhD thesis, 'Horses, oxen and technological innovation: the use of draught animals in English farming from 1066 to 1500', University of Birmingham, 1983, pp. 416–56.

2 Data collected and supplied by John Langdon but not listed in his PhD thesis.

3 'Feeding the city 1' accounts database, listed in Campbell and others, *Medieval capital*, Appendix 1, pp. 184–90.

4 'Feeding the city 2' accounts database, listed in Appendix 3 below.

5 Norfolk accounts database, listed in Appendix 2 below.

6 Data supplied by Martin Ecclestone and collected in conjunction with his MA thesis, 'Dairy production on the Glastonbury Abbey demesnes 1258–1334', University of Bristol, 1996.

7 Data supplied by David Postles.

8 Grouped De Lacy manorial account: Notts. RO, DD FJ 6/1/1.

9 Data extracted from selected Winchester Pipe Rolls: Hants. RO, 11M59 B1/38, 43, 45, 53, 58, 76, 97.

10 PRO SC 6/1109/24.

11 NRO Phillipps 26517.
12 Alcock, 'East Devon manor'.
13 Brandon, 'Demesne arable farming'.
14 Finberg, *Tavistock Abbey*.
15 Hare, 'Lord and tenant in Wiltshire'.
16 Keil, 'The estates of Glastonbury Abbey'.
17 D. Postles, 'Grain issues from some properties of Oseney Abbey, 1274–1348', *Oxoniensia* 44 (1979), 100–16.
18 Raftis, *Estates of Ramsey Abbey*.

Appendix 2

Demesnes represented in the Norfolk accounts database

The information is presented according to the following template:

> *Manor name*: lord; number of accounts; year date of accounts (* denotes account recording demesne at farm); archive and documentary references.

Archives and manuscript references are those pertaining at the time that the manorial accounts were examined and data extracted (a task largely complete by the mid-1980s). No attempt has been made to take account of subsequent changes in the location and calendaring of documents. Square brackets denote accounts which have come to light since the database was created. Virtually all the documents listed are manorial accounts; the one main exception is the chartulary of Norwich Cathedral Priory known as *Proficuum maneriorum* (NRO DCN 40/13) which contains information abstracted by the priory from its original accounts, many of which no longer survive. For some years there are both accounts and entries in the *Proficuum maneriorum*; for others (marked thus °) there are only the latter. A ? indicates that the provenance, lord, or date is uncertain or unknown.

See Figure A2.01 for the distribution of the Norfolk demesnes.

Norfolk

Acle: Bigod, earls of Norfolk; 7 acnts.; 1268–9, 1270–1, 1271–2, 1272–3, 1277–8, 1278–9, 1279–80; PRO, SC 6/929/1–7. *Aldborough*: ?; 1 acnt.; 1430–1; PRO, SC 6/929/8. *Aldeby*: ?; 1 acnt.; 1312–13; NRO, DCN 60/2/1. *Aldeby*: Aldeby Priory; 15 acnts.; 1399–1400, 1401–2*, 1403–4, 1406–7, 1407–8, 1409–10, 1410–11, 1412–13, 1413–14, 1416–17*, 1419–20, 1420–1, 1423–4, 1434–5, 1446–7*; NRO, MS 21065–79 34 E5. *Alderford/Witchingham*: manor of Cleyhalle; ?; 1 acnt.; 1344–5; NRO, Phi/465 577 × 9. *Antingham*: De Antingham; 2 acnts.; 1409–10, 1423–4*; NRO, MS 6031 16 B8, MS 6242 16 D4. *Arminghall*: Norwich Cathedral Priory (chamberlain); 1 acnt. + *Proficuum maneriorum*° × 9; 1294–5°, 1297–8°, 1298–9°, 1299–1300°, 1300–1°, 1301–2°, 1302–3°, 1305–6°, 1306–7°, 1347–8; NRO, DCN 40/13°, DCN 61/7. *Ashby*: abbey of St Benet at Holme; 5 acnts.; 1238–9, 1239–40, 1245–6, *c.* 1379, *c.* 1392; NRO, Diocesan Est/1, Est/2/1, Est/9. *Ashill*: manor of Panworth; De Nerford; 21 acnts.; 1320–1, 1325–6, 1326–7, 1336–7, 1342–3, 1354–5, 1355–6, 1357–8, 1358–9, 1359–60,

1360–1, 1361–2, 1376–7, 1377–8, 1378–9, 1379–80, 1380–1, 1381–2, 1384–5, 1385–6, 1389–90; NRO, MS 21086 34 E6, NRS 21161–3 45 A4. *Ashill*: manor of Uphall; De Hastyngs, earls of Pembroke; 3 acnts.; 1365–6, 1372–3, 1373–4; PRO, SC 6/929/9, SC 6/944/19–20. *Attleborough*: Bigod, earls of Norfolk; 8 acnts.; 1274–5, 1275–6, 1277–8, 1278–9, 1279–80, 1281–2, 1292–3, 1294–5; PRO, SC 6/929/14–21. *Attlebridge*: Norwich Cathedral Priory (almoner); 3 acnts.; 1307–8, 1314–15, *c.* 1330; NRO, DCN 61/11–13. *Aylmerton*: De Becham; 1 acnt.; 1345–6; NRO, WKC 2/24 394×3. *Barton Bendish*: De Scalis; 2 acnts.; 1333–4, 1356–7; NRO, Hare 185×4/175–6. *Bastwick*: abbey of St Benet at Holme; 1 acnt.; 1239–40; NRO, Diocesan Est/1. *Bauburgh*: Norwich Cathedral Priory (sacrist); 8 acnts.; 1274–5, 1296–7, 1305–6, 1307–8, 1313–14, 1324–5, 1326–7, 1337–8; NRO, DCN 61/14–15, DCN 60/8/12, DCN 61/16–20. *Beaudesert* (in the Fens): bishop of Ely; 2 acnts.; 1315–16, *temp.* Ed. II/III; PRO, SC 6/1132/13; SC 6/1135/7. *Beeston*: earl of Arundel; 2 acnts.; 1285–6, 1286–7; Holkham Estate Records, Tittleshall bundle 3. *Beetley* with *North Elmham*: bishop of Norwich; 2 acnts.; 1327–8, 1329–30; NRO, DCN 61/24–5. *Bintree*: ?; 1 acnt.; 1378–9; Holkham Estate Records, vol. 6, bundle 3, no. 108. *Bircham*: De Clare, earls of Gloucester and Hertford; 33 acnts.; 1310–11, 1311–12, 1322–3, 1323–4, 1324–5, 1325–6, 1326–7, 1327–8, 1330–1, 1331–2, pre-1335, 1333–4, 1335–6, 1336–7, 1337–8, 1339–40, 1340–1, 1341–2, 1342–3, 1343–4, 1344–5, 1346–7, 1347–8, 1348–9, 1350–1, 1351–2, 1352–3, 1355–6, 1356–7, 1357–8, 1360–1, 1361–2, 1372–3*; PRO, SC 6/930/1–10, 33, 11–31. *Blickling*: Erpingham; 3 acnts.; 1410–11, 1411–12, 1420–1*; NRO, NRS 10196 25 A1, NRS 10535 25 B5. *Boughton*: De Causton; 1 acnt.; 1350–1; NRO, Hare 187×1/411. *Bradenham*: De Lacy, earls of Lincoln; 1 acnt.; 1276–7; Nottinghamshire RO, DD FJ Manorial VI 1(1), Accounts iii 3, membrane 7. *Brancaster*: Ramsey Abbey; 12 acnts.; 1253–4; *c.* 1262; 1303, 1324–5, 1351–2, 1352–3, 1359–60, 1362–3, 1367–8, 1368–9, 1369–70, 1378–9*; BL, Add. Charter 39669; NRO, L'Estrange EG 1; PRO, SC 6/931/1–11. *Brandiston*: manor of Guton Hall; De Gyney; 4 acnts.; 1316–17, 1320–1, 1337–8, 1347–8; Magdalen College Archives, Estate records 166/10, 3, 12, 7. [*Braydeston*: Carbonel; 1 acnt.; 1424–5; NRO, MC 495/1, 747×7]. *Bressingham*: Bigod, earls of Norfolk; 3 acnts.; 1269–70, 1272–3, 1276–7; PRO, SC 6/931/21–3. *Bressingham*: De Verdoun; 7 acnts.; 1327, 1336–7, 1341–2; 1368–9*, 1396–7; 1401–2; 1405–6*; BL, Add. Charter 16535–7; NRO, Phi/468 577×9; BL, Add. Charter 16538; NRO, Phi/468 577×9. *Bridgham*: bishop of Ely; 1 acnt.; *temp.* Ed. I/II; PRO, SC 6/931/20. *Briston*: ?; 1 acnt.; 1300–1; PRO, SC 6/931/24. *Bromehill*: Bromhill Priory; 1 acnt.; 1461–2; Christ's College Archives, Bromhill Priory Account 1. *Bunwell area*: ?; 1 acnt.; 1343–4; NRO, MS 1960 2 C4. *Burgh in Flegg*: the Queen; 3 acnts.; 1296–7, 1330–1, 1390–1; PRO, SC 6/1090/4, SC 6/931/27–8. *Burnham Thorpe*: De Calthorpe; 7 acnts.; 1314–15; 1316–17; 1318–19, 1350–1, 1383–4, 1415–16, 1424–5*; NRO, WAL 269×1/63–5; Raynham Hall, Townshend MSS; NRO, NRS 10228 25 A3, WAL 269×1/66–7. *Burston*: ?; 1 acnt.; 1384–5; BL, Add. Charter 26530. *Caister-cum-Markshall*: Bigod, earls of Norfolk; 16 acnts.; 1269–70, 1270–1, 1272–3, 1274–5, 1275–6, 1277–8, 1279–80, 1280–1, 1281–2, 1283–4, 1284–5, 1289–90, 1292–3, 1295–6, 1296–7, 1299–1300; PRO, SC 6/932/11–26. *Calthorpe*: rector of Calthorpe, Great Hospital, Norwich; 37 acnts.; 1315–16, 1320–1, 1323–4?, 1329–30, 1330–1, 1335–6, 1336–7, 1339–40, 1345–6, 1346–7, 1356–7, 1358–9, 1360–1, 1364–5, 1365–6, 1366–7, 1367–8, 1368–9, 1369–70, 1370–1, 1373–4, 1374–5, 1375–6, 1376–7, 1377–8, 1379–80, 1381–2, 1382–3, 1383–4, 1386–7, 1387–8, 1389–90, 1390–1, 1391–2, 1393–4, 1394–5,

1400–1*; N.R.O, Case 24, Shelf C. *Calthorpe*: manor of Hookhall; ?; 1 acnt.; 1413–14; Raynham Hall, Townshend, MSS, box 24. *Catton*: Norwich Cathedral Priory (prior); 23 acnts.; 1265–6, 1268–9, 1272–3, 1273–4, 1275–6, 1280–1, 1282–3, 1295–6, 1301–2, 1302–3, 1308–9, 1309–10, 1311–12, 1312–13, 1320–1, 1322–3, 1323–4, 1324–5, 1334–5, 1338–9, 1339–40, 1340–1, 1343–4; NRO, DCN 60/4/1, DCN 60/26/26, DCN 60/4/2, 5, L'Estrange IB 4/4, DCN 60/4/6, 7, 11, 14, 15, 17, 18, 19, 21, 24, 25, 27, 28, 34, 36, 39, 40, 43. *Catton*: glebe; Norwich Cathedral Priory (communar); 2 acnts.; 1318–19, 1325–6; NRO, DCN 60/4/22, 31. *Cawston*: the Queen; 3 acnts.; 1275–6, 1296–7, 1331–2; PRO, SC 6/1089/7, SC 6/1090/4, SC 6/933/11. *Coltishall*: manor of Hakeford Hall; De Hakeford; 2 acnts.; 1293, 1315 (both, entries in the manor court rolls); King's College Archives, E 29 and 30. *Costessey*: De Clare, earls of Gloucester and Hertford; 8 acnts.; 1277–8, 1278–9, *c.* 1280, 1281, 1282–3, 1283–4, 1290–1, 1291–2; PRO, SC 6/933/13 membrane 16, membrane 14, membranes 1, 5, and 15, membrane 4, membrane 3, membrane 2, SC 6/933/14. *Costessey*: rector of Costessey, Great Hospital, Norwich; 31 acnts.; 1374–5, 1377–8, 1378–9, 1380–1, 1383–4, 1384–5, 1388–9, 1390–1, 1391–2, 1392–3, 1393–4, 1394–5, 1396–7, 1398–9, 1400–1, 1408–9, 1411–12, 1412–13, 1413–14, 1414–15, 1415–16, 1416–17, 1421–2, 1422–3, 1423–4, 1424–5, 1425–6, 1428–9, 1429–30, 1430–1, 1431–2*; NRO, Case 24, Shelf C. *Creake, North?*: De Thorpe; 1 acnt.; 1336–7; NRO, Phi/471 577×9. *Creake*: Creake Abbey; 1 acnt.; 1441–2; Raynham Hall, Townshend MSS. *Crimplesham*: De Clare, earls of Gloucester and Hertford; 1 acnt.; 1304–5; PRO, SC 6/933/18. *Cringleford*: Great Hospital, Norwich; 10 acnts.; 1337–8, 1344–5, 1346–7, 1393–4, 1411–12, 1412–13, 1413–14, 1427–8, 1428–9, 1433–4*; NRO, Case 24, Shelf D. *Crownthorp*: De Crungethorpe; 4 acnts.; 1319–20, 1327–8, 1334, 1347–8; NRO, Kimberley MAC/D/1. *Deopham*: Canterbury Cathedral Priory; 8 acnts.; 1286–7, 1307–8, 1311–12, 1326–7, 1334–5, 1348–9, 1377–8, 1400–1*; CCA, Deopham Beadles' Acnts. *Dereham*: bishop of Ely; 1 acnt.; *temp*. Ed. I/II; PRO, SC 6/931/20. *Diss*: ?; 1 acnt.; 1351–2; PRO, SC 6/935/1. *Ditchingham*: Bigod, earls of Norfolk; 18 acnts.; 1269–70, 1271–2, 1272–3, 1274–5, 1275–6, 1276–7, 1278–9, 1279–80, 1281–2, 1282–3, 1283–4, 1289–90, 1290–1, 1292–3, 1294–5, 1299–1300, 1305–6, 1398–9*; PRO, SC 6/933/20–29, SC 6/934/1–9, 11. *Earsham*: Bigod, earls of Norfolk; 18 acnts.; 1269–70, 1271–2, 1272–3, 1273–4, 1276–7*, 1279–80*, 1281–2, 1283–4, 1284–5, 1289–90, 1290–1, 1292–3?, 1294–5, 1298–9, 1299–1300, 1304–5, 1305–6, 1390–1*; PRO, SC 6/934/12–14, 16, 19, 23–4, 26–8, 31–2, 34–6, 38–9; NRO, Phi/505/3. *East Carleton*: De Curszon; 1 acnt.; 1277–8; John Rylands Library, Manchester, Phillipps charter 17. *East Carleton*: Appylierd; 1 acnt.; 1405–6; John Rylands Library, Manchester, Phillipps charter 18. *East Lexham*: Folyot; 1 acnt.; 1315–16; NRO, Kimberley MAC/C/1. *East Lexham*: De Camoys and Hastings; 7 acnts.; 1361–2, 1364–5, 1366–7, 1368–9, 1376–7, 1385–6, 1428–9*; NRO, Kimberley MAC/C/2–8. *Easton*: abbey of St Benet at Holme; 3 acnts.; 1239–40, 1240–1, 1245–6; NRO, Diocesan Est/1, Est/2/1. *East Wretham*: Hockeburn Priory; 12 acnts.; 1303–4, 1304–5, 1305–6, 1306–7, 1307–8, 1336–7, 1337–8, 1338–9, 1339–40, 1341–2, 1350–1, 1440–1; Eton College Records, vol. 30, no. 43 membrane 3, membranes 1–2, membranes 4–7, membranes 8–10, vol. 30, nos. 44–9. *Eaton*: Norwich Cathedral Priory (prior); 44 acnts. + *Proficuum maneriorum*° ×17; 1263–4, 1265–6, 1272–3, 1273–4, 1275–6; 1282–3; 1287–8; 1288–9; 1291–2, 1292–3°; 1294–5, 1295–6, 1296–7°, 1297–8, 1298–9°, 1299–1300°, 1301–2°, 1302–3°, 1303–4°, 1305–6, 1306–7°; 1308–9; 1309–10, 1311–12, 1312–13, 1317–18, 1318–19, 1320–1, 1322–3, 1324–5,

1325–6, 1326–7, 1327–8, 1328–9°, 1330–1°, 1331–2°, 1332–3°, 1333–4, 1334–5, 1335–6°, 1336–7°, 1337–8°, 1338–9°, 1339–40°, 1349–50; 1358–9, 1361–2, 1366–7, 1369–70, 1373–4, 1384–5, 1394–5, 1395–6*, 1400–1, 1401–2, 1405–6, 1406–7, 1409–10, 1410–11, 1411–12*, 1422–3; NRO, DCN 60/8/1–4, L'Estrange IB 4/4; BLO, MS rolls, Norfolk 20; NRO, DCN 60/8/5; BLO, MS rolls, Norfolk 21; NRO, DCN 60/8/6, DCN 40/13°, DCN 60/8/7–9, DCN 40/13°, DCN 60/8/10, DCN 40/13°, DCN 60/8/12–13, DCN 40/13°; BLO, MS rolls, Norfolk 23; NRO, DCN 60/8/14–24, DCN 40/13°, DCN 62/2, DCN 60/8/25, DCN 40/13°, DCN 60/8/28; BLO, MS rolls, Norfolk 29–33, 35–8, 40–5. *Eccles*: bishop of Norwich; 14 acnts.; 1342–3, 1344–5, 1346–7, 1349, 1350–1, 1353–4, 1354–5, 1355–6, 1356–7, 1357–8*, 1392–3*, 1408–9, 1411–12, 1427–8*; Raynham Hall, Townshend MSS. *Felbrigg*: De Felbrygge; 21 acnts.; 1399–1400, 1400–1, 1401–2, 1402–3, 1403–4, 1404–5, 1405–6, 1406–7, 1407–8, 1408–9, 1409–10, 1410–11, 1411–12, 1412–13, 1414–15, 1415–16, 1416–17, 1417–18, 1418–19, 1420–1, 1421–2; NRO, WKC 2/130–31/398 × 6. *Feltwell*: bishop of Ely; 6 acnts.; *temp.* Ed. I/II; 1337; 1346–7, 1396–7, 1422–3, 1426–7*; PRO, SC 6/931/20; NRO, Deeds various, box T, 155D and BL, Add. Charter 67812; NRO, Bradfer-Lawrence V 10; NRO, Phi/472/1–3 577 × 9. *Feltwell*: manor of Easthall; De Playz; 2 acnts.; early 14th C., 1349–50; Christ's College Archives, Estate records. *Fincham*: De Grantcurt; 3 acnts.; 1268, 1279–80, 1351–2; NRO, Hare 189 × 5/780; PRO, SC 6/935/1. *Fincham*: manor of New Hall and Neleshall; ?; 3 acnts.; 1278, 1279, 1280; NRO, Hare 189 × 5/781. *Fincham*: De Fyncham; 10 acnts.; 1352–3, 1353–4, 1354–5, 1356–7, 1357–8, 1361–2, 1362–3, 1363–4, 1364–5, 1365–6; NRO, Hare 189 × 5/782–5, Hare 189 × 6/785–7, Hare 189 × 6/789–90. *Flegg*: abbey of St Benet at Holme; 21 acnts.; 1341, *c.* 1351, *c.* 1353, *c.* 1355, *c.* 1363, 1368–9, 1369–70, 1372, 1374–5, *c.* 1380, 1407, 1409, 1416–17, 1420–1, 1422–3, 1425–6, 1426–7, 1427–8, 1430, 1432, 1444*; NRO, Diocesan Est/9, Est/58/8, Est/9–10. *Fordham*: Norwich Cathedral Priory; 1 acnt.; 1343–4; NRO, DCN 61/28. *Forncett*: Bigod, earls of Norfolk; 16 acnts.; 1270, 1272–3, 1274–5, 1277–8, 1278–9, 1279–80, 1281–2, 1283–4, 1285–6, 1289–90, 1292–3, 1299–1300, 1302–3, 1303–4, 1305–6, 1308–9; PRO, SC 6/935/2–17, SC 6/1121/1. *Foxley*: ?, 1 acnt.; 1305–6; PRO, SC 6/935/19. *Framingham*: Bigod, earls of Norfolk; 16 acnts.; 1269–70, 1271–2, 1272–3, 1274–5, 1276–7, 1277–8, 1278–9, 1279–80, 1280–1, 1283–4, 1289–90, 1290–1, 1295–6, 1299–1300, 1302–3, 1307–8; PRO, SC 6/935/20, 22–3, 25–9, 31–7, SC 6/1121/1. *Fring*: De Hakeford; 1 acnt.; 1307–8; NRO, MS 126 × 6/20404. *Gasthorpe*: ?; 1 acnt.; 1417–18, John Rylands Library, Manchester, Phillipps charter 19. *Gateley*: Norwich Cathedral Priory (prior); 31 acnts. + *Proficuum maneriorum*° × 18; 1263–4, 1265–6, 1272–3, 1273–4, 1274–5, 1275–6, 1277–8, 1285–6, 1287–8, 1293–4°, 1294–5, 1295–6, 1296–7°, 1297–8, 1298–9°, 1299–1300, 1300–1°, 1302–3°, 1303–4°, 1304–5°, 1305–6, 1306–7°, 1309–10, 1311–12, 1312–13, 1317–18, 1318–19, 1319–20, 1320–1, 1322–3, 1324–5, 1325–6, 1326–7, 1327–8, 1328–9°, 1329–30°, 1330–1°, 1331–2, 1332–3°, 1333–4, 1334–5°, 1335–6°, 1336–7°, 1337–8°, 1338–9°, 1339–40°, 1344–5, 1350–1, 1375–6*?; NRO, DCN 60/13/1–5, L'Estrange IB 4/4, DCN 60/13/6–8, DCN 40/13°, DCN 60/13/9–10, DCN 40/13°, DCN 60/13/11, DCN 40/13°, DCN 60/13/12, DCN 40/13°, DCN 60/13/13, DCN 60/13/14–23, DCN 62/1, DCN 60/13/24, DCN 40/13°, DCN 60/13/25, DCN 40/13°, DCN 62/2, DCN 40/13°, DCN 60/13/26–7. *Gaywood*: bishop of Norwich; 1 acnt.; 1331–2; NRO, DCN 61/30. *Gimingham*: duke of Lancaster; 12 acnts.; 1358–9, 1359–60; 1367–8, 1381–2, 1384–5; 1391–2; 1392–3, 1393–4, 1395–6; 1397–8, 1401–2, 1412–13*; PRO, DL 29/288/4719–20; NRO, MS 6001

16 A6, NRS 11331 26 B6, NRS 11060 25 E2; PRO, DL 29/288/4734; NRO, NRS 11058 25 E2, NRS 11332 26 B6, NRS 11069 25 E3; PRO, DL 29/289/4744, 4752, DL 29/290/4765. *Gnatingdon*: Norwich Cathedral Priory (prior); 27 acnts. + *Proficuum maneriorum*° × 22; 1255–6, 1263–4, 1265–6, 1272–3, 1273–4, 1275–6, 1287–8, 1292–3°, 1293–4°, 1294–5, 1295–6, 1296–7°, 1297–8, 1298–9°, 1299–1300, 1300–1°, 1301–2°, 1302–3°, 1303–4°, 1304–5°, 1305–6, 1306–7°, 1307–8°, 1308–9, 1309–10, 1311–12, 1312–13, 1317–18, 1318–19, 1319–20, 1320–1, 1322–3, 1324–5, 1325–6, 1326–7, 1327–8, 1328–9°, 1329–30°, 1330–1°, 1331–2°, 1332–3°, 1333–4, 1334–5°, 1335–6°, 1336–7°, 1337–8°, 1338–9°, 1339–40°, 1349–50; NRO, DCN 60/14/1–5, L'Estrange IB 4/4, DCN 60/14/6, DCN 40/13°, DCN 60/14/7–8, DCN 40/13°, DCN 60/14/9, DCN 40/13°, DCN 60/14/10, DCN 40/13°, DCN 60/14/12, DCN 40/13°, L'Estrange IB 1/4, DCN 60/14/13–22, DCN 62/1, DCN 60/14/23, DCN 40/13°, DCN 62/2, DCN 40/13°, DCN 60/14/24. *Great Cressingham*: Norwich Cathedral Priory (cellarer); 15 acnts. + *Proficuum maneriorum*° × 11; 1294–5°, 1295–6°, 1296–7°, 1297–8°, 1298–9°, 1299–1300°, 1300–1°, 1301–2°, 1302–3°, 1305–6°, 1306–7°, 1308–9, 1322–3; 1326–7; 1362–3, 1363–4, 1365–6, 1373–4, 1375–6, 1376–7, 1379–80, 1380–1, 1412–13, 1415–16, 1416–17, 1428–9*; NRO, DCN 40/13°, Supp. 10/12/1982 (R187A); Harvard Law Library, MS 85 and NRO, DCN 40/13°; NRO, Supp. 10/12/1982 (R187A). *Great Snoring*: De Burgelion; 1 acnt.; 1327–8; NRO, Phi/499 578 × 1. *Gresham*: De Stutevill; 1 acnt.; 1306–7; PRO, SC 6/936/1. *Gressenhall*: De Hastings; 21 acnts.; 1279–80, 1361–2, 1362–3, 1363–4, 1364–5, 1365–6, 1366–7, 1367–8, 1368–9, 1370–1, 1371–2, 1372–3, 1375–6, 1376–7, 1377–8, 1378–9, 1379–80, 1380–1, 1381–2, 1382–3, 1386–7*; NRO, ING 245 × 5/186, L'Estrange G1–6, G10. *Grimston?*: ?; 1 acnt.; 1380–1; Holkham Estate Records, Tittleshall bundle 23. *Haddiscoe*: Knights Templar; 1 acnt.; 1311–12; PRO, E 358/18. *Hainford*: ?; 1 acnt.; 1363–4; BL, Add. Roll 26060. *Halvergate*: Bigod, earls of Norfolk; 17 acnts.; 1268–9, 1269–70, 1270–1, 1273–4, 1276–7, 1279–80, 1280–1, 1281–2, 1283–4, 1284–5, 1285–6, 1289–90, 1292–3, 1299–1300, 1303–4, 1305–6; 1361–2; PRO, SC 6/936/2–17; NRO, Phi/477 577 × 9. *Hanworth*: Bigod, earls of Norfolk; 24 acnts.; 1272–3, 1273–4, 1274–5, 1276–7, 1277–8, 1278–9, 1279–80, 1280–1, 1282–3, 1283–4, 1284–5, 1285–6, 1289–90, 1290–1, 1292–3, 1295–6, 1298–9, 1299–1300, 1300–1, 1302–3, 1303–4, 1305–6, 1307, 1307–8; PRO, SC 6/936/18–25, 27–32, SC 6/937/1–10, SC 6/1121/1. *Happisburgh*: ecclesiastical manor; 1 acnt.; 1409; PRO, SC 6/937/13. *Hardley*: abbey of St Benet at Holme; 5 acnts.; 1296, 1329–30, 1336; 1356–7; *c.* 1361; NRO, Diocesan Est/2 2/5, Ch. Comm. 101426 5/13, Diocesan Est/2 2/9; Lambeth Palace Library, ED 476; NRO, Ch. Comm. 101426 7/13. *Hardley*: rector of Hardley, Great Hospital, Norwich; 3 acnts.; 1374, 1398–9, 1427–8*; NRO, Case 24, Shelf D. *Hargham*: De Lavenham; 4 acnts.; early 14th C.; 1329–30, 1330–1, 1333–4; NRO, Gates 16/9/62 T 192E; CUL, Buxton MS, box 78, bundle 59. *Harpley*: De Gurney (rector of Harpley); 1 acnt.; 1305–6; NRO, MS 3205 4 A3. *Hautbois*: abbey of St Benet at Holme; 5 acnts.; 1329; 1358–9; 1363, 1367–8, 1372; NRO, Diocesan Est/2 2/3; Lambeth Palace Library, ED 479; NRO, Diocesan Est/2 2/15–17. *Haveringland*: priory of Horsham St Faith; 5 acnts.; 1356–7, 1358–9, 1364–5, 1376–7, 1413–14*; BL, Add. Charter 15199–15202, Add. Charter 9327. *Heacham*: Lewes Priory; 17 acnts.; 1296–7, 1300–1, 1303–4, *temp.* Ed. II, 1330–1, 1333–4, 1359–60, *c.* 1362, 1369–70, 1371–2, 1372–3, 1373–4, 1386–7, 1389–90, 1393–4, *temp.* Hen. V, 1422–3*; NRO, L'Estrange DG 1–8. *Heigham by Norwich*: abbey of St Benet at Holme; 6 acnts.; 1239–40, 1240–1, 1245–6, 1301–2, 1305–6, 1381; NRO, Diocesan Est/1, Est/2 1, Est/2

2/6, Est2 2/7, Est/2 2/20. *Helhoughton*: De Shardelowe; 1 acnt.; 1333–4; Raynham Hall, Townshend MSS, box 13. *Hempnall*: Fitzwalter; 1 acnt.; 1287–8; PRO, SC 6/1118/13. *Hemsby*: Norwich Cathedral Priory (prior); 19 acnts. + *Proficuum maneriorum*° × 21; 1265–6, 1272–3, 1275–6; 1278–9; 1287–8, 1292–3°, 1293–4°, 1294–5, 1295–6, 1296–7°, 1297–8, 1298–9°, 1299–1300, 1300–1°, 1301–2°, 1302–3°, 1303–4°, 1304–5°, 1305–6, 1306–7°, 1312–13, 1317–18, 1318–19, 1320–1, 1322–3, 1324–5, 1327–8, 1328–9°, 1329–30°, 1330–1°, 1331–2°, 1332–3°, 1333–4°, 1334–5, 1335–6°, 1336–7°, 1337–8°, 1338–9°, 1339–40°; 1366–7; NRO, DCN 60/15/1–2; L'Estrange IB 4/4; BLO, MS rolls, Norfolk, 47; NRO, DCN 60/15/3, DCN 40/13°, DCN 60/15/4–5, DCN 40/13°, DCN 60/15/6, DCN 40/13°, DCN 60/15/7, DCN 40/13°, DCN 60/15/8, DCN 40/13°, DCN 60/15/9–15, DCN 40/13°, DCN 60/15/16, DCN 40/13°; Raynham Hall, Townshend MSS. *Hethel*: Appelierd; 1 acnt.; 1404–5; NRO, Gurney Collection and Mills and Reeve RQG 151. *Hevingham*: Le Cat; 4 acnts.; 1287–8, 1327–8, 1346–7, 1417–18; NRO, NRS 14750A 29 D4, NRS 14751 29 D4, NRS 14748 29 D4, NRS 14750 29 D4. *Hevingham with Marsham*: bishop of Norwich; 5 acnts.; 1331–2, 1344, 1353–4, 1357–8, 1361–2*; NRO, DCN 61/34, NRS 14664 29 D2, NRS 14762 29 D4, NRS 13996 28 F3, NRS 14749 29 D4. *'Heythe'* (near Great Plumstead): Norwich Cathedral Priory (prior); 3 acnts. + *Proficuum maneriorum*° × 4; 1333–4, 1334–5, 1336–7°, 1337–8°, 1338–9°, 1339–40°, 1345–6; NRO, DCN 62/2, DCN 61/35, DCN 40/13°, DCN 61/36. *Hilborough*: De Clifton; 6 acnts.; 1367–8, 1375–6, 1376–7, 1409–10*, 1411–12, 1412–13, 1416–17; NRO, box T Daleth vii T 73E. *Hilgay*: Ramsey Abbey; 3 acnts.; early 14th C., 1359–60; 1421–2*; PRO, SC 6/937/14, 15; BL, Add. Charter 39933. *Hindolveston*: Norwich Cathedral Priory (prior); 57 acnts. + *Proficuum maneriorum*° × 15; 1255–6, 1261–2, 1263–4, 1265–6, 1272–3, 1273–4, 1275–6, 1277–8, 1282–3, 1287–8, 1293–4°, 1294–5, 1295–6, 1296–7°, 1297–8, 1298–9°, 1299–1300, 1300–1°, 1301–2°, 1302–3°, 1303–4°, 1304–5°, 1305–6°, 1306–7°, 1308–9, 1309–10, 1311–12, 1312–13, 1313–14, 1317–18, 1318–19, 1320–1, 1322–3, 1324–5, 1325–6, 1326–7, 1327–8, 1328–9°, 1329–30°, 1330–1°, 1331–2°, 1333–4*, 1334–5*, 1339–40°, 1344–5, 1348–9, 1350–1, 1352–3, 1353–4, 1358–9, 1360–1, 1361–2, 1362–3, 1363–4, 1367–8, 1373–4, 1379–80, 1384–5*, 1391–2*, 1395–6, 1396–7, 1397–98, 1400–1, 1402–3, 1404–5, 1405–6, 1406–7, 1408–9, 1411–12, 1414–15, 1415–16, 1417–18*; NRO, DCN 60/18/1–6, L'Estrange IB 4/4, DCN 60/18/7–9, DCN 40/13°, DCN 60/18/11–12, DCN 40/13°, DCN 60/18/13, DCN 40/13°, DCN 60/18/14, DCN 40/13°, DCN 60/18/15–25, 27–8, DCN 40/13°, DCN 62/2, DCN 60/18/29, DCN 40/13°, DCN 60/18/30–1, 34–44, 49–62. *Hindringham*: Norwich Cathedral Priory (prior); 36 acnts. + *Proficuum maneriorum*° × 11; 1255–6, 1263–4, 1265–6, 1272–3, 1273–4, 1275–6, 1277–8, 1287–8, 1291–2, 1292–3, 1293–4°, 1294–5, 1295–6, 1296–7°, 1297–8, 1298–9°, 1299–1300, 1300–1°, 1301–2°, 1302–3°, 1305–6, 1306–7°, 1309–10, 1311–12, 1312–13, 1317–18, 1318–19, 1320–1, 1322–3, 1324–5, 1326–7°, 1327–8, 1328–9°, 1329–30°, 1339–40°, 1341–2, 1343–4, 1349–50, 1363–4, 1376–7, 1381–2, 1387–8*, 1392–3*, 1400–1, 1415–16, 1422–3, 1425–6*; NRO, DCN 60/20/1–5, L'Estrange IB 4/4, DCN 60/20/6–9, DCN 40/13°, DCN 60/20/10–11, DCN 40/13°, DCN 60/20/12, DCN 40/13°, DCN 60/20/13, DCN 40/13°, DCN 60/20/14, DCN 40/13°, DCN 60/20/15–22, DCN 40/13°, DCN 60/20/23, DCN 40/13°, DCN 60/20/24–6, 30–3, 35–9. *Hingham*: ?; 2 acnts.; 1271–2; 1302–3; BL, Campb. IX 8; NRO, Kimberley MAC/B/1. *Hockham*: ?; 2 acnts.; 1380–1, 1383–4; NRO, MS 13853–4 16 F7. *Holkham*: Dereham Abbey; 2 acnts.; 1366–7, 1385–6; Holkham Estate Records, vol. 1, bundle 4, nos. 53 and 73. *Horning*:

abbey of St Benet at Holme; 1 acnt.; 1239–40; NRO, Diocesan Est/1. *Horning*: St James' Hospital, Horning; 6 acnts.; *c.* 1356, *c.* 1363, 1367–8, *c.* 1372, *c.* 1378, *c.* 1382; NRO, Diocesan Est/13, Est/2 2/15–17, Est/13. *Horsham St Faith*: Horsham Priory; 1 acnt.; 1407–8; NRO, NRS 19517 42 C6. *Hoveton*: abbey of St Benet at Holme; 10 acnts.; 1239–40, 1246, 1296, 1329–30, 1336, 1343, *c.* 1361, 1394, 1422, 1439*; NRO, Diocesan Est/1, Ch. Comm. 101426 3/13, Diocesan Est/2 2/5, Ch. Comm. 101426 5/13, Diocesan Est/2 2/9, 2/11, Ch. Comm. 101426 7/13, 2/13, 11/13, 8/13. *Howardes Manor*: Brews; 5 acnts.; 1385–6, 1386–7, 1387–8, 1388–9, 1419–20*; Raynham Hall, Townshend MSS. *Hudeston*: ?; 2 acnts.; 1342–2, 1395–6*; CUL, Buxton MS, box 74, bundle 11, and box 77, bundle 50. *Hunstanton*: De Holm; 3 acnts.; 1330–1, 1332–3, 1333–4; NRO, L'Estrange BG 2–5. *Hunstanton*: L'Estrange; 12 acnts.; *c.* 1336–7, 1338–9, 1339–40, 1340–1, 1342–3, 1345–6, 1347–8, 1367–8, 1368–9, 1370–1, 1408–9, 1438–9*; NRO, L'Estrange BG 6, BG 1, 7 and 9, 8, 10–19. *Ingham*: the countess marshall of England; 1 acnt.; 1344–5; NRO, Phi/487 578×1. *Intwood*: De Hedersete; 5 acnts.; 1308–9, 1325–6 or 1345–6, 1333–4, 1417–18, 1425–6; NRO, NRS 23349–52 Z 97. *Kelling*: ?; 1 acnt.; 1345; PRO, SC 6/937/18. *Kempstone*: Castle Acre Priory; 31 acnts.; 1315–16, 1323–4, 1325–6, 1326–7, 1330–1, 1353–4, 1357–8, 1366–7, 1367–8, 1377–8, 1388–9, 1393–4, 1399–1400, 1400–1, 1406–7, 1409–10, 1414–15, 1416–17, 1418–19, 1424–5, 1425–6, 1427–8, 1428–9, 1434–5, 1435–6, 1439–40, 1441–2, 1444–5, 1445–6, 1448–9, 1452–3*; NRO, WIS 163×1/2–6, 8, 10, 12–15, 17–18, WIS 163×2/19, 21–34, 36–8. *Kerdiston*: L'Amysel; 2 acnts.; 1297–8; 1331–2; W. Suffolk RO, E 18/900/1; NRO, Phi/488 578×1. *Keswick*: De Vallibus; 4 acnts.; 1274–5, *c.* 1302–3, 1312–13, 1319–20; NRO, NRS 23357 Z 98. *Keswick*: Clere; 9 acnts.; 1366–7, 1367–8, 1370–1, 1371–2, 1372–3, 1373–4, 1374–5, 1375–6, 1376–7; NRO, NRS 23358 Z 98. *Knapton*: De Playz; 2 acnts.; 1345–6, 1347–8; St George's Chapel, Windsor, MS XV 53 98–9. *Lakenham*: Norwich Cathedral Priory (chamberlain); 3 acnts. + *Proficuum maneriorum*° ×1; 1295–6, 1306–7°, 1366–7, 1463–4; NRO, DCN 61/42; DCN 40/13°, DCN 61/43–4. [*Langford*: ?; 1 acnt.; NRO, Ncc (Petre) box 8/22]. *Langham*: bishop of Norwich; 17 acnts.; 1326–7; 1327–8; 1329–30; 1330–1, 1344, 1348–9, 1349–50, 1350–1; 1352–3, 1353–4; 1354, 1355, 1364–5, 1365–6, 1366–7, 1368–9, 1381–2*; NRO, MS 1303 2 B3; Raynham Hall, Townshend MSS, box 48; NRO, MS 1554 1 C1; Raynham Hall, Townshend MSS, box 48, 34; NRO, MS 1555 1 C1, MS 1308 2 B3; Raynham Hall, Townshend MSS, box 48. *Lessingham*: abbess of Bec; 2 acnts.; 1290–1, 1297–8; Eton College Records, vol. 49, nos. 242–3. *Litcham*: ?; 1 acnt.; 1383–4; NRO, Kimberley MAC/E/1. *Little Ellingham*: De Wisham; 5 acnts.; 1342–3, 1343–4, 1344–5, 1349–50, 1350–1; Nottingham UL, Manvers collection 24–8. *Little Fransham*: ?; 2 acnts.; 1383–4*, 1402–3; NRO, MS 13122 40 A5, MS 13127 40 A5. *Loddon*: Bigod, earls of Norfolk; 5 acnts.; 1282–3, 1284–5, 1289–90, 1292–3, 1295–6; PRO, SC 6/937/22–6. *Long Stratton*: ?; 1 acnt.; 1410–11; BL, Add. Charter 18554. *Lopham*: Bigod, earls of Norfolk; 18 acnts.; 1268–9, 1270, 1270–1, 1271–2, 1272–3, 1273–4, 1275–6, 1277–8, 1279–80, 1281–2, 1282–3, 1283–4, 1289–90, 1290–1, 1292–3, 1295–6, 1298–9, 1305–6; PRO, SC 6/937/27–33, SC 6/938/1–11. *Ludham*: abbey of St Benet at Holme; 3 acnts.; 1239–40, 1245–6, *c.* 1355; NRO, Diocesan Est/1, Est/2 1, Est/10. *Marham*: De Brews; 2 acnts.; 1355–6; 1356–7; NRO, Hare 194×5/2200; Raynham Hall, Townshend MSS. *Marham*: Marham Abbey; 4 acnts.; 1405–6, 1408–9, 1419–20, 1426–7; NRO, Hare 194 ×5/2201–4. *Martham*: Norwich Cathedral Priory (prior); 53 acnts. + *Proficuum maneriorum*° ×19; 1261–2, 1263–4, 1265–6, 1272–3, 1273–4, 1275–6, 1287–8, 1292–3°,

1293–4°, 1294–5, 1295–6, 1296–7°, 1297–8, 1298–9°, 1299–1300, 1300–1°, 1301–2°, 1302–3°, 1303–4°, 1304–5°, 1305–6, 1306–7°, 1309–10, 1311–12, 1312–13, 1317–18, 1318–19, 1319–20, 1320–1, 1322–3, 1324–5, 1325–6, 1326–7, 1327–8, 1328–9°, 1329–30°, 1330–1°, 1331–2°, 1332–3°, 1333–4, 1334–5, 1335–6°, 1336–7°, 1337–8°, 1338–9°, 1339–40, 1349–50, 1355–6, 1363–4, 1377–8, 1379–80, 1387–8, 1388–9, 1390–1, 1392–3, 1393–4, 1396–7, 1397–8, 1400–1, 1402–3, 1406–7, 1407–8, 1412–13, 1413–14, 1414–15, 1415–16, 1418–19, 1419–20, 1421–2, 1422–3, 1423–4, 1424–5*; NRO, DCN 60/23/1–5, L'Estrange IB 4/4, DCN 60/23/6, DCN 40/13°, DCN 60/23/7–8, DCN 40/13°, DCN 60/23/9, DCN 40/13°, DCN 60/23/10, DCN 40/13°, DCN 60/23/11, DCN 40/13°, DCN 60/23/12–21, DCN 62/1, DCN 60/23/22, DCN 40/13°, DCN 62/2, DCN 60/23/23, DCN 40/13°, NNAS 5890 20 D1, DCN 60/23/25, NNAS 5892 20 D1, NNAS 5894–6 20 D1, NNAS 5898–903 20 D1, NNAS 5904–18 20 D2. *Martham*: glebe, Norwich Cathedral Priory (cellarer); 5 acnts.; 1323–4, 1337–8, 1355, 1359–60, 1386–7; NRO, NNAS 5889 20 D1, DCN 60/23/24, NNAS 5891 20 D1, NNAS 5893 20 D1, NNAS 5897 20 D1. *Mautby*: De Mauteby; 2 acnts.; 1336–7, late 14th C.; NRO, Phi/490–1 578 × 1. *Melton*: Norwich Cathedral Priory; 4 acnts.; 1332–3, 1366–7, 1369–70, 1395–6*; NRO, DCN 60/25/1–4. *Methwold*: duke of Lancaster; 2 acnts.; 1365–6, 1444–5; PRO, DL 29 288/4720, DL 29 293/4811. *Mileham*: earl of Arundel; 2 acnts.; 1349–50, 1350–1; Holkham Estate Records, Tittleshall bundle 17 and 16. *Monks Granges*: Norwich Cathedral Priory (prior); 27 acnts. + *Proficuum maneriorum*° × 19; 1255–6, 1264–5, 1265–6, 1268–9, 1272–3, 1273–4, 1275–6, 1287–8, 1292–3°, 1293–4°, 1294–5, 1295–6, 1296–7°, 1297–8°, 1298–9°, 1299–1300, 1300–1°, 1301–2°, 1302–3°, 1303–4°, 1304–5°, 1305–6, 1306–7°, 1309–10, 1311–12, 1312–13, 1313–14, 1317–18, 1318–19, 1320–1, 1322–3, 1324–5, 1325–6, 1326–7, 1327–8, 1328–9°, 1329–30°, 1331–2, 1332–3°, 1333–4, 1334–5, 1335–6°, 1336–7°, 1337–8°, 1338–9°, 1339–40°; NRO, DCN 60/26/1–3, DCN 60/26/26, DCN 60/26/4–6, L'Estrange IB 4/4, DCN 60/26/7, DCN 40/13°, DCN 60/26/8–9, DCN 40/13°, DCN 60/26/10, DCN 40/13°, DCN 60/26/11, DCN 40/13°, DCN 60/26/12–21, DCN 62/1, DCN 60/26/22, DCN 40/13°, DCN 60/26/23, DCN 40/13°, DCN 62/2, DCN 60/26/24–5, DCN 40/13°. *Mundham*: Great Hospital, Norwich; 9 acnts.; *c.* 1343, 1352–3, 1366–7, 1371–2, 1399–1400, 1428–9, 1429–30, 1430–1, 1432–3*; NRO, Case 24, Shelf F. *Neatishead*: abbey of St Benet at Holme; 1 acnt.; 1239–40; NRO, Diocesan Est/1. *Newton*: lord Guydone; 2 acnts.; 1281–2, 1327; CUL, Cholmondeley (Houghton) MS, reeves' and bailiffs' acnts., 30–1. *Newton by Norwich*: Norwich Cathedral Priory (prior); 10 acnts. + *Proficuum maneriorum*° × 26; 1273–4, 1288–9, 1293–4°, 1294–5°, 1295–6°, 1296–7°, 1297–8°, 1298–9°, 1299–1300, 1300–1°, 1301–2, 1302–3°, 1303–4°, 1304–5°, 1305–6°, 1306–7°, 1324–5°, 1326–7°, 1327–8, 1328–9°, 1329–30°, 1330–1°, 1331–2°, 1332–3°, 1333–4°, 1334–5°, 1335–6°, 1336–7°, 1337–8°, 1338–9°, 1339–40°, 1366–7, 1377–8, 1409–10, 1417–18, 1425–6; NRO, DCN 60/28/1–2, DCN 40/13°, DCN 60/28/3, DCN 40/13°, DCN 60/28/4, DCN 40/13°, DCN 60/28/5, DCN 40/13°, DCN 60/28/6–10. *North Elmham*: Norwich Cathedral Priory (prior); 29 acnts. + *Proficuum maneriorum*° × 10; 1255–6, *c.* 1260, *c.* 1264–5, 1272–3, 1273–4, 1275–6, 1282–3, 1287–8, 1288–9, 1295–6, 1296–7°, 1297–8, 1298–9°, 1305–6, 1306–7°, 1309–10, 1311–12, 1312–13, 1317–18, 1319–20, 1320–1, 1322–3, 1324–5, 1325–6, 1326–7, 1327–8, 1331–2°, 1332–3°, 1333–4, 1334–5, 1335–6°, 1336–7°, 1337–8°, 1338–9°, 1339–40°, 1340–1, 1356–7, 1373–4, 1383–4*; NRO, DCN 60/10/1–5, L'Estrange IB 4/4, DCN 60/10/6–9, DCN 40/13°, DCN 60/10/10, DCN 40/13°, DCN

60/10/11, DCN 40/13°, DCN 60/10/12–21, DCN 62/1, DCN 60/10/22, DCN 40/13°, DCN 62/2, DCN 60/10/23, DCN 40/13°, DCN 60/10/24–7. *North Walsham*: abbey of St Benet at Holme; 8 acnts.; 1239–40, 1245–6, 1354–5, 1367–8, 1368–9, 1389, 1427, 1451*; NRO, Diocesan Est/1, Est/2/1, Est/12. *Northwold*: bishop of Norwich; 1 acnt.; *temp.* Ed. I/II; PRO, SC 6/931/20. *Ormesby*: glebe, Hospital of St Paul; 3 acnts.; 1294–5, 1303–4, 1337–8; NRO, DCN 61/52–4. *Ormesby*: Clere and Rothenhale; 28 acnts.; *temp.* Ed. III, 1408–9, 1423–4, 1424–5, 1425–6, 1427–8, 1428–9, 1429–30, 1430–1, 1432–3, *c.* 1435, 1436–7, 1437–8, 1438–9, 1439–40, 1440–1, 1441–2, 1443–4, 1444–5, 1445–6, 1446–7, 1447–8, 1449–50, 1451–2, 1452–3, 1454–5, 1457–8, 1458–9*; PRO, SC 6/941/7, SC 6/938/26, SC 6/939/1–6, SC 6/939/8A and B, SC 6/939/10–13, SC 6/940/1–6, SC 6/941/6, SC 6/940/7–11, SC 6/941/1–3. *Osmundiston*: De Scheltone; 1 acnt.; 1288–9; Elveden Hall, Suffolk, Iveagh Collection, Cornwallis (Bateman) MS, box 47, no. 8. [*Palgrave*: ?; 1 acnt.; 1383–4; Raynham Hall, Townshend MSS]. *Plumstead*: Norwich Cathedral Priory (prior); 45 acnts. + *Proficuum maneriorum*° × 15; 1263–4, 1265–6, 1272–3, 1277–8, 1287–8, 1288–9, 1292–3°, 1293–4°, 1294–5, 1295–6, 1296–7, 1297–8, 1298–9, 1299–1300, 1300–1°, 1301–2°, 1302–3, 1303–4°, 1304–5°, 1312–13, 1319–20, 1320–1, 1324–5, 1325–6, 1326–7, 1327–8, 1328–9°, 1329–30°, 1330–1°, 1331–2, 1332–3°, 1333–4, 1334–5, 1335–6°, 1336–7°, 1337–8°, 1338–9°, 1339–40°, 1342–3, 1349–50, 1351–2, 1353–4, 1354–5, 1359–60, 1369–70, 1370–1, 1371–2, 1375–6, 1381–2, 1382–3, 1391–2, 1395–6, 1398–9, 1402–3, 1404–5, 1409–10, 1415–16, 1416–17, 1418–19, 1419–20; NRO, DCN 60/29/1–6; DCN 40/13°, DCN 60/29/7–9, DCN 60/29/11–14, DCN 40/13°, DCN 60/29/15, DCN 40/13°, DCN 60/29/16, DCN 60/29/18–21, DCN 62/1, DCN 60/29/22, DCN 40/13°, DCN 60/29/23, DCN 40/13°, DCN 62/2, DCN 60/29/24, DCN 40/13°, DCN 60/29/25–46. *Plumstead*: Norwich Cathedral Priory (precentor); 2 acnts.; 1295–6, 1312–13; NRO, DCN 60/29/10 and 17. *Popenhoe*: Ramsey Abbey; 6 acnts.; 1260–1; 1284, 1291–2, 1324–5, 1337–8, 1390–1; BL, Add. Charter 39934; PRO, SC 6/942/12–13, SC 6/943/5, SC 6/942/16–17. *Potter Heigham*: abbey of St Benet at Holme; 2 acnts.; 1245–6, 1390; NRO, Diocesan Est/2/1, Est/11. *Pulham*: bishop of Ely; 1 acnt.; *temp.* Ed. I/II; PRO, SC 6/931/20. *Quidenham*: ?; 1 acnt.; 1388–9; NRO, Phi/493 578 × 1. *Raynham*: De Ingaldesthorpe; 10 acnts.; 1284–5, 1286–7, 1287–8, 1315–16, 1339–40, 1342–3, 1345–6, 1348–9, 1350–1, 1395–6*; Raynham Hall, Townshend MSS. *Raynham*: De Scales; 4 acnts.; 1304–5, 1372–3, 1373–4, 1402–3*; NRO, MS 1455 1 B1. *Reedham*: Berneye; 12 acnts.; 1377–8, 1378–9, 1379–80, 1380–1, 1381–2, 1383–4, 1384–5, 1386–7, 1392–3, 1393–4, 1394–5, 1444–5*; BL, Add. Charter 26852–63. *Ringstead*: Ramsey Abbey; 16 acnts.; 1253–4; *c.* 1263, *c.* 1311–12; 1324–5; *c.* 1325–6, 1336–7, 1390–1, 1394–5, 1395–6, 1398–9, 1399–1400, 1402–3, 1404–5, 1406–7, 1407–8, 1408–9*; BL, Add. Charter 39669; NRO, L'Estrange EG 1; PRO, SC 6/942/15; NRO, L'Estrange EG 2–9. *Rougham*: De Yelverton; 12 acnts.; 1397–8, 1422–3, 1440–1, 1442–3, 1451–2, 1454–5, 1456–7, 1457–8, 1458–9, 1460–1, 1461–2, 1463–4; NRO, MS 21483/3 NRS 7421, MS 21483/1 NRS 6492, MS 21483/3 NRS 7422, MS 21483/1. *Saxthorpe*: manor of Loundhall; ?; 12 acnts.; late 13th C., 1296–7?, early 14th C., 1320–1 or 1340–1, 1336–7, 1350–1, 1355–6, 1357–8, 1385–6, 1404–5*, 1432–3*, 1438–9; NRO, NRS 19659 42 D7, NRS 19692 42 E4, NRS 19660 42 D7, NRS 19652 42 D7, NRS 19691 42 E4, NRS 19657 42 D7, NRS 19654–5 42 D7, NRS 19658 42 D7, NRS 19656 42 D7, NRS 19690 42 E4, NRS 19677 42 E3, NRS 19650 42 D7. *Scottow*: abbey of St Benet at Holme; 1 acnt.; *c.* 1365–6; NRO, Diocesan Est/11. *Scratby*: Norwich Cathedral Priory (sacrist); 12 acnts.; 1295–6, 1297–8, 1301–2,

1314–15, 1315–16, 1317–18, 1319–20, 1328–9, 1342–3, 1356–7, 1360–1, 1362–3; NRO, DCN 60/30/1–12. *Sedgeford*: Norwich Cathedral Priory (prior); 88 acnts. + *Proficuum maneriorum°* × 20; 1255–6, 1263–4, 1265–6, 1269, 1272–3, 1273–4, 1275–6, 1278–9, 1285–6, 1287–8, 1291–2, 1292–3°, 1293–4, 1294–5, 1295–6, 1296–7°, 1297–8°, 1298–9°, 1299–1300, 1300–1, 1301–2°, 1302–3°, 1303–4°, 1304–5°, 1305–6, 1306–7°, 1307–8°, 1309–10, 1311–12, 1312–13, 1317–18, 1318–19, 1319–20, 1320–1, 1322–3, 1324–5, 1325–6, 1326–7, 1327–8, 1328–9°, 1329–30°, 1330–1°, 1331–2°, 1332–3°, 1333–4, 1334–5°, 1335–6°, 1336–7°, 1337–8°, 1338–9°, 1339–40, 1340–1, 1342–3, 1344–5, 1349–50, 1351–2, 1352–3, 1353–4, 1355–6, 1356–7, 1361–2, 1363–4, 1365–6, 1366–7, 1367–8, 1368–9, 1369–70, 1371–2, 1373–4, 1374–5, 1377–8, 1379–80, 1380–1, 1381–2, 1382–3, 1383–4, 1384–5, 1386–7, 1388–9, 1389–90, 1391–2, 1392–3, 1393–4, 1394–5, 1398–9, 1400–1, 1401–2, 1402–3, 1405–6, 1406–7, 1407–8, 1408–9, 1409–10, 1411–12, 1412–13, 1413–14, 1415–16, 1416–17, 1420–1, 1421–2, 1422–3, 1423–4, 1424–5, 1425–6, 1428–9, 1429–30, 1430–1, 1432–3*; NRO, DCN 60/33/1–3, L'Estrange IB 1/4, DCN 60/33/4–5, L'Estrange IB 4/4, DCN 60/33/6–9, DCN 40/13°, DCN 60/33/10–12, DCN 40/13°, DCN 60/33/13, L'Estrange IB 1/4, DCN 40/13°, DCN 60/33/14, DCN 40/13°, DCN 60/33/15–16, DCN 60/33/18–25, DCN 62/1, DCN 60/33/27, DCN 40/13°, DCN 62/2, DCN 40/13°, L'Estrange IB 1/4, DCN 60/33/29, L'Estrange IB 1/4, DCN 60/33/30, L'Estrange IB 1/4, L'Estrange IB 3/4, DCN 60/33/31, L'Estrange IB 3/4. *Seething*: Bigod, earls of Norfolk; 2 acnts.; 1283–4, 1289–90; PRO, SC 6/943/10–11. *Seething*: Great Hospital, Norwich; 7 acnts.; 1310–11, 1392–3, 1393–4, 1395–6, 1398–9, 1400–1, 1412*; NRO, Case 24, Shelf F. *Shipdham*: bishop of Ely; 1 acnt.; *temp.* Ed. I/II; PRO, SC 6/931/20. *Shotesham*: abbey of St Benet at Holme; 5 acnts.; 1238–9, 1239–40, 1246, *c.* 1369–70, *c.* 1370–1; NRO, Diocesan Est/1, Est/2/1, Est/11, Est/2 2/13. *Shropham*: manor of Bradcar Hall; De Coggessale; 2 acnts.; 1351, 1383–4; NRO, Case 24, Shelf G. *Sloley*: ?; 2 acnts.; 1347–8, 1454–5; CUL, Cholmondeley (Houghton) MS, reeves' and bailiffs' acnts., 33–4. *Southery with Hilgay*: abbot of Bury St Edmunds; 1 acnt.; 1308–9; N. Yorkshire RO, 2JX: 3/14. *South Walsham*: abbey of St Benet at Holme; 1 acnt.; 1239–40; NRO, Diocesan Est/1. *South Walsham*: Bigod, earls of Norfolk; 11 acnts.; 1268–9, 1270, 1270–1, 1276–7, 1277–8, 1281–2, 1282–3, 1283–4, 1290–1, 1292–3, 1296–7; PRO, SC 6/944/21–31. *Sporle*: Sporle Priory; 1 acnt.; 1345–6; PRO, SC 6/1126/9. *Stanhoe*: De Calthorp; 7 acnts.; 1314–15, 1329–30, 1336–7, 1337–8, 1338–9, 1339–40, 1375–6; Raynham Hall, Townshend MSS, box 21. *Stiffkey*: De Hemgham and Manny; 2 acnts.; 1294; 1341–2; NRO, Case 24, Shelf I; Raynham Hall, Townshend MSS, box 24. *Stradsett*: Haukyn; 1 acnt.; 1365–6; PRO, SC 6/943/16. *Suffield*: Bigod, earls of Norfolk; 9 acnts.; 1272–3, 1278–9, 1281–2, 1283–4, 1284–5, *c.* 1290, 1292–3, 1294–5, 1299–1300; PRO, SC 6/944/1–2, 4–6, 10, 7–9. *Swanton Abbot*: abbey of St Benet at Holme; 2 acnts.; 1239–40, 1245–6; NRO, Diocesan Est/1, Est/2/1. *Syderstone*: Cokesford Priory?; 1 acnt.; 1376–7; CUL, Cholmondeley (Houghton) MS, reeves' and bailiffs' acnts., 29. *Tacolneston*: Wylliams manor, De Unedale; 2 acnts.; 1327–8, 1354–5; Pomeroy and Sons, Wymondham. *Taverham*: Norwich Cathedral Priory (prior); 55 acnts. + *Proficuum maneriorum°* × 19; 1255–6, 1261–2, 1263–4, 1265–6, 1268–9, 1272–3, 1273–4, 1275–6, 1277–8, 1282–3, 1287–8, 1291–2, 1292–3, 1293–4°, 1294–5°, 1295–6, 1296–7°, 1297–8°, 1298–9°, 1299–1300°, 1300–1°, 1301–2°, 1303–4°, 1304–5°, 1305–6, 1306–7°, 1309–10, 1311–12, 1312–13, 1317–18, 1318–19, 1320–1, 1322–3, 1324–5, 1325–6, 1326–7, 1327–8, 1328–9°, 1329–30°, 1330–1°, 1331–2°, 1332–3, 1333–4, 1334–5, 1335–6°, 1336–7°,

1337–8°, 1338–9, 1339–40°, 1342–3, 1344–5, 1349–50, 1351–2, 1353–4, 1362–3, 1364–5, 1365–6, 1366–7, 1367–8, 1369–70, 1370–1, 1371–2, 1372–3, 1373–4, 1413–14, 1414–15, 1415–16, 1416–17, 1417–18, 1418–19, 1419–20, 1420–1, 1422–3, 1423–4; NRO, DCN 60/35/1–7, L'Estrange IB 4/4, DCN 60/35/8–12, DCN 40/13°, DCN 60/35/13, DCN 40/13°, DCN 60/35/14, DCN 40/13°, DCN 60/35/15–23, DCN 62/1, DCN 60/35/24, DCN 40/13°, DCN 60/35/25, DCN 62/2, DCN 60/35/26, DCN 40/13°, DCN 60/35/27, DCN 40/13°, DCN 60/35/28–52. *Terrington*: bishop of Ely; 2 acnts.; 1315–16, *temp.* Ed. II/III; PRO, SC 6/1132/13, SC 6/1135/7. *Thornage*: bishop of Norwich; 12 acnts.; 1326–7; 1370–1, 1373–4, 1375–6, 1376–7, 1377–8, 1380–1, 1383–4, 1386–7*, 1394–5*, 1409–10, 1413–14*; NRO, DCN 61/60; Chicago UL, Bacon Rolls 52936, 540–2. *Thornham*: Norwich Cathedral Priory (prior); 24 acnts. + *Proficuum maneriorum*° × 12; 1255–6, 1263–4, 1265–6, 1273–4, 1275–6, 1277–8, 1287–8, 1292–3°, 1294–5, 1295–6, 1296–7°, 1297–8, 1298–9°, 1299–1300, 1300–1°, 1301–2°, 1302–3°, 1303–4°, 1304–5°, 1305–6, 1306–7°, 1307–8°, 1309–10, 1317–18, 1318–19, 1319–20, 1320–1, 1322–3, 1324–5°, 1325–6, 1326–7, 1327–8, 1329–30°, 1349–50, 1350–1, 1351–2; NRO, DCN 60/14/1, DCN 60/37/1–3, L'Estrange IB 4/4, DCN 60/37/4–5, DCN 40/13°, DCN 60/37/6–7, DCN 40/13°, DCN 60/37/8, DCN 40/13°, DCN 60/37/9, DCN 40/13°, DCN 60/37/10, DCN 40/13°, DCN 60/37/11–16, DCN 40/13°, DCN 60/37/17, DCN 62/1, DCN 60/37/18, DCN 40/13°, DCN 60/37/19–21. *Thorpe Abbotts*: abbey of Bury St Edmunds; 19 acnts.; 1336–7, 1339–40, 1342–3, 1345–6, 1347–8, 1349–50, 1350–1, 1351–2, 1353–4, 1356–7; 1357–8; 1358–9, 1359–60, 1361–2, 1362–3, 1367–8, 1371–2, 1375–6, 1378–9; NRO, WAL 274×6/478, 480–8; Elveden Hall, Suffolk, Iveagh Collection, Cornwallis (Bateman) MS, box 60 no. 4; NRO, WAL 274 ×6/479, 489–95. *Thurgarton*: abbey of St Benet at Holme; 1 acnt.; 1239–40; NRO, Diocesan Est/1. *Thurne*: abbey of St Benet at Holme; 2 acnts.; 1239–40, 1245–6; NRO, Diocesan Est/1, Est/2/1. *Thurning*: Burnel; 4 acnts.; 1319–20, 1371–2, 1374–5, 1452–3*; NRO, NRS 2796–9 12 E2. *Thwaite*: abbey of St Benet at Holme; 1 acnt.; 1388; NRO, Diocesan Est/2 2/21. *Tibenham*: abbey of St Benet at Holme; 1 acnt.; 1246; NRO, Diocesan Est/2 1. *Titchwell*: Lovel; 6 acnts.; 1337–8, 1338–9, 1341–2, 1343–4, 1346–7, 1433–4; Magdalen College Archives, 166/1, 166/6, 166/4, 166/8, 166/9, 177/4. *Tivetshall*: abbey of Bury St Edmunds; 23 acnts.; 1335–6, 1340–1; 1344–5, 1345–6, 1349–50; 1350–1, 1352–3, 1354–5, 1356–7, 1361–2, 1363–4; 1364–5; 1365–6, 1367–8; 1368–9; 1369–70; 1372–3, 1374–5, 1375–6, 1376–7; 1377–8, 1379–80; 1393–4*; NRO, WAL 288×1/1245–6; Raynham Hall, Townshend MSS; NRO, WAL 288×1/1247, WAL 288×2/1249, 1248, 1250–1, WAL 274×3/451; Raynham Hall, Townshend MSS; NRO, WAL 274×3/452–3; Raynham Hall, Townshend MSS; NRO, WAL 274×3/454; Raynham Hall, Townshend MSS; NRO, WAL 274×3/455, WAL 288×2/1252; Raynham Hall, Townshend MSS. *Topcroft*: De Clyftone; 1 acnt.; 1353–4; NRO, WIS 163×1/9. *Tunstead*: duke of Lancaster; 4 acnts.; 1358–9, 1359–60, 1365–6, 1381–2*; PRO, DL 29/288/4719–20, 4722, 4724. *Walpole*: bishop of Ely; 2 acnts.; 1315–16, *temp.* Ed. II/III; PRO, SC 6/1132/13, SC 6/1135/7. *Walsingham*: De Clare, earls of Gloucester and Hertford; 3 acnts.; *temp.* Hen. III, 1332, 1336; PRO, SC 6/1109/10, 24, SC 6/1110/3. *Walton*: bishop of Ely; 2 acnts.; 1315–16, *temp.* Ed. II/III; PRO, SC 6/1132/13, SC 6/1135/7. *West Harling*: De Sekford; 8 acnts.; 1328–9, 1329–30, 1332–3, 1335–6, 1357–8, 1368–9, 1372–3, 1377–8; John Rylands Library, Manchester, Phillips Charters 12, 10, 11, 13–16, 9. *West Lexham*: ?; 1 acnt.; 1278–9; PRO, SC 6/937/21. *West Newton*: De Parker; 1 acnt.; 1294–5; BL, Add. Charter 9151. *West Tofts*: ?; 1 acnt.; *c.* 1317–18;

NRO, Phi/500 578×1. *West Walton*: Lewes Priory; 18 acnts.; 1332–3, 1334–5, 1335–6, 1336–7, 1347–8, 1361–2, 1364–5, 1371–2, 1372–3, 1377–8, 1395–6, 1397–8, 1398–9, 1400–1, 1401–2, 1416–17, 1420–1, 1435–6, 1458–9*; NRO, Hare 210×2/4012–23, Hare 210×3/4024–31. *West Winch*: Bardolf; 2 acnts.; 1336–7, 1379–80*; CUL, Cholmondeley (Houghton) MS, reeves' and bailiffs' acnts., 36–7. *Wicklewood*: Norwich Cathedral Priory (almoner); 1 acnt.; 1337–8; NRO, DCN 61/61. *Wiggenhall*: glebe, Norwich Cathedral Priory (cellarer); 1 acnt.; 1301–2; NRO, DCN 61/62. *Wilton*: De Ponyngges; 1 acnt.; 1357–8; BL, Add. Charter 67873. *Wimbotsham*: Ramsey Abbey; 4 acnts.; 1249–50, 1331, 1337–8, *c.* 1390*; NRO, Hare 212×1/4207–9, 4212. *Wimbotsham*: De Ingaldesthorp; 14 acnts.; 1276–7, 1284–5, 1287–8, 1301–2, 1302–3, 1313–14, 1325–6, 1331–2, 1332–3, 1333–4, 1335–6, 1337–8, 1344–5, 1357–8*; NRO, Hare 213×1/4272–8, Hare 213×4/4372–3, Hare 213×1/4279–83. *Witchingham*?: manor of Hithburg; ?; 1 acnt.; 1366–7; Raynham Hall, Townshend MSS. *Worstead*: glebe, Norwich Cathedral Priory (cellarer); 4 acnts.; 1273–4, 1276–7, 1320–1, 1357–8*; NRO, DCN 60/39/1, L'Estrange IB 4/4, DCN 60/39/2–3. *Wroxham*: prioress of Carrow; 1 acnt.; 1342–3; NRO, NRS 2848 12 F1. *Wymondham*: various manors, including Barnakes, Cromwell, Gresehaugh, and Randolf, plus those of Lord Peter de Unedale and Wymondham Abbey; at least 34 acnts.; 1280–1, 1285–6, 1290–1, 1293–4, 1294–5, 1295–6, 1312–13, 1313–14, 1316–17, 1321–2, 1325–6, 1327–8, 1329–30, 1330–1, 1331–2, 1332–3, 1333–4, 1334–5, 1337–8, 1338–9, 1340–1, 1341–2, 1344–5, 1345–6, 1349–50, 1350–1, 1352–3, 1353–4, 1358–9, 1360–1, 1362–3, 1363–4, 1367–8*, 1369–70*; NRO, NRS 10108 22 F5, NRS 11277 26 B1, NRS 10107 22 F5, NRS 11277 26 B1, NRS 14038 28 F6, NRS 18516 and 18517 33 D3, NRS 18523 33 D4, NRS 8811 21 E4, NRS 18524 33 D4, NRS 18518 33 D3, NRS 18519–23 33 D3, NRS 18525–6 33 D4, NRS 18523 33 D4, NRS 11278 26 B1, NRS 18527 33 D4, NRS 18534 33 D5, NRS 18523 33 D4, NRS 11279 26 B1, NRS 18528 and 18530 33 D4, NRS 18544 33 D5, NRS 18529 33 D4, NRS 8812 21 E4, NRS 18529 33 D4, NRS 11280 26 B1, NRS 18566 33 D6, NRS 18545 33 D5, NRS 18565 33 D6, NRS 18523 33 D4, NRS 11277 26 B1.

Suffolk

Brandon: bishop of Ely; 38 acnts.; *temp.* Ed. I/II; 1302, 1337; 1338–9, 1341–2; 1343–4; 1345–6, 1346–7, 1347–8, 1349–50; 1351–2; 1353–4; 1354–5; 1361–2; 1362–3, 1364–5, 1365–6; 1367, 1367–8, 1368–9, 1369–70, 1370–1, 1371–2; 1372–3, 1373–4; 1374–5; 1379–80, 1382–3, 1385–6; 1386–7; 1388–9, 1389–90, 1390–1, 1391–2, 1392–3, 1393–4, 1394–5, 1395–6*; PRO, SC 6/931/20; Chicago UL, Bacon Roll 643; PRO, SC 6/1304/22–3; Chicago UL, Bacon Roll 644; PRO, SC 6/1304/24–7; Chicago UL, Bacon Roll 645; PRO, SC 6/1304/28; Chicago UL, Bacon Roll 646; PRO, SC 6/1304/29; Chicago UL, Bacon Rolls 647–9; PRO, SC 6/1304/30–5; Chicago UL, Bacon Roll 643; PRO, SC 6/1304/36; Chicago UL, Bacon Rolls 650–2; Elveden Hall, Suffolk, Iveagh Collection 148 (Phillipps 26523); Chicago UL, Bacon Rolls 653–60. [A microfilm of the Bacon Rolls is available in the W. Suffolk RO: no. J529/2.] *Bungay*: Bigod, earls of Norfolk; 13 acnts.; 1269–70, 1274–5, 1275–6, 1279–80, 1282–3, 1287–8, 1288–9, 1293–4, 1294–5, 1300–1, 1302–3, 1304–5, 1305–6; PRO, SC 6/991/16–28. *Denham*: Norwich Cathedral Priory (prior); *Proficuum maneriorum*°×11; 1295–6°, 1296–7°, 1297–8°, 1298–9°, 1299–1300°, 1300–1°, 1302–3°, 1303–4°, 1304–5°, 1305–6°, 1306–7°; NRO, DCN 40/13. *Hinderclay*: abbey of Bury St Edmunds; 103 acnts.; *c.* 1251, 1256,

1262–3, 1264–5, 1267–8, 1268–9, 1271–2, 1272–3, 1273–4, 1276–7, 1277–8, 1279–80, 1281–2, 1282–3, 1283–4, 1284–5, 1286–7, 1288–9, 1289–90, 1290–1, 1291–2, 1292–3, 1294–5, 1295–6, 1296–7, 1297–8, 1298–9, 1299–1300, 1301–2, 1302–3, 1303–4, 1304–5, 1305–6, 1306–7, 1307–8, 1308–9, 1309–10, 1310–11, 1312–13, 1313–14, 1315–16, 1317–18, 1318–19, 1319–20, 1320–1, 1321–2, 1323–4, 1324–5, 1325–6, 1326–7, 1327–8, 1329–30, 1330–1, 1331–2, 1332–3, 1333–4, 1334–5, 1335–6, 1336–7, 1337–8, 1343–4, 1346–7, 1348–9, 1349–50, 1350–1, 1351–2, 1352–3?, 1353–4, 1354–5, 1355–6, 1356–7, 1357–8, 1360–1, 1361–2, 1364–5, 1366–7, 1367–8, 1368–9, 1369–70, 1372–3, 1373–4, 1375–6, 1376–7, 1377–8, 1378–9, 1379–80, 1380–1, 1381–2, 1383–4, 1384–5, 1385–6, 1386–7, 1387–8, 1389–90, 1391–2, 1392–3, 1393–4, 1394–5, 1395–6, 1400–1, 1401–2, 1404, 1405–6; Chicago UL, Bacon Rolls 405–30, 432–44, 416, 445–501, 503, 505–10. *Hoxne*: bishop of Norwich; 1 acnt.; 1326–7; NRO, DCN 61/69. *Redgrave*: abbey of Bury St Edmunds; 56 acnts.; 1323–4, 1324–5, 1330–1, 1336–7; 1338–9; 1339–40; 1340–1; 1341–2, 1342–3; 1343–4, 1344–5; 1345–6, 1347–8, 1348–9?, 1349–50, 1350–1, 1351–2, 1353–4, 1355–6, 1356–7, 1358–9, 1359–60, 1360–1, 1361–2, 1362–3, 1363–4, 1364–5, 1366–7, 1367–8, 1368–9, 1369–70, 1370–1; 1371–2; 1372–3, 1373–4, 1375–6, 1376–7, 1378–9, 1379–80, 1380–1, 1381–2; 1382–3; 1383–4, 1384–5, 1385–6, 1386–7, 1387–8, 1388–9, 1390–1, 1391–2, 1392–3, 1394–5, 1396–7, 1398–9, 1402, 1412–13*; Chicago UL, Bacon Rolls 325–8; BL, Add. Roll 63372; Chicago UL, Bacon Roll 329; BL, Add. Roll 63373; Chicago UL, Bacon Rolls 330–1; BL, Add. Rolls 63374–5; Chicago UL, Bacon Rolls 332–52; BL, Add. Roll 63376; Chicago UL, Bacon Rolls 353–60; Chicago UL, Bacon Roll 361 and BL, Add. Roll 63377; Chicago UL, Bacon Rolls 363–76. *Rickinghall*: abbey of Bury St Edmunds; 62 acnts.; 1312, 1327–8, 1332–3, 1334–5, 1335–6, 1336–7, 1337–8, 1338–9, 1339–40, 1340–1, 1341–2, 1342–3, 1343–4, 1344–5, 1345–6, 1347–8, 1348–9, 1349–50, 1350–1, 1351–2, 1353–4, 1354–5, 1355–6; 1356–7; 1357–8, 1358–9, 1359–60, 1360–1, 1361–2, 1362–3, 1363–4, 1364–5, 1367–8, 1368–9, 1369–70, 1370–1, 1371–2, 1372–3, 1373–4, 1375–6, 1377–8*, 1378–9*, 1379–80*, 1381–2, 1382–3, 1384–5, 1385–6, 1386–7, 1387–8, 1388–9, 1390–1, 1391–2, 1392–3, 1393–4, 1394–5, 1395–6, 1396–7, 1398–9, 1399–1400, 1400–1, 1401–2, 1402–3*; BL, Add. Rolls 63512, 63440, 63445, 63513–32; Chicago UL, Bacon Roll 517; BL, Add. Rolls 63533–52, 63554, 63553, 63555–70. *Wattisfield*: abbey of Bury St Edmunds; 1 acnt.; 1364–5; Chicago UL; Bacon Roll 481. *Unidentified*: bishop of Norwich; 1 acnt.; c. 1381–2; NRO, DCN 61/78.

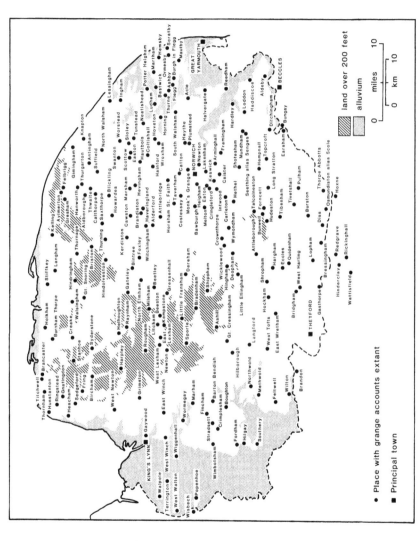

DISTRIBUTION OF DEMESNES UTILIZED IN THE ANALYSIS

land over 200 feet

alluvium

0   miles   10

0   km   10

● Place with grange accounts extant

■ Principal town

GREAT YARMOUTH

KING'S LYNN

NORWICH

THETFORD

BECCLES

Appendix 3

Demesnes represented in the FTC accounts databases

The FTC2 accounts database is presented according to the following template:

Manor name: lord (? denotes that the lord is unknown); year date of accounts; archive and documentary references.

Corresponding information for the demesnes represented in the FTC1 accounts database is listed in B. M. S. Campbell, J. A. Galloway, D. J. Keene, and M. Murphy, *A medieval capital and its grain supply: agrarian production and its distribution in the London region* c. *1300,* Historical Geography Research Series, 30 (1993), Appendix 1, pp. 184–90.

Bedfordshire: *Grovebury:* Worship/Wisthepe; 1389–90; St George's Chapel, Windsor, MS XV 61 32. *Higham Gobion:* Butler; 1379–80, 1380–1, 1381–2; Beds. RO, BS 1175. *Shillington:* Ramsey Abbey; 1377–8, 1380–1, 1383–4; PRO, SC 6/741/22–4.
Berkshire: *Billingbear:* bishop of Winchester; 1377–8, 1381–2, 1383–4, 1389–90; Hants. RO, 11M59/B1/130, 133, 135, 141. *Brightwalton:* Battle Abbey; 1386–7, 1388–9, 1391–2, 1393–4; PRO, SC 6/742/26–9. *Brightwell:* bishop of Winchester; 1377–8, 1381–2, 1384–5, 1389–90; Hants. RO, 11M59/B1/130, 133, 136, 141. *Coleshill:* Edington Priory; 1385–6, 1390–1, 1391–2, 1395–6; PRO, SC 6/743/7–10. *Cresswell in Bray:* Windsor College; 1379–80; St George's Chapel, Windsor, MS XV 61 29. *Culham:* bishop of Winchester; 1375–6, 1376–7, 1377–8, 1381–2; Hants. RO, 11M59 B1/128–30, 133. *Didcot:* De Stonor; 1383–4; PRO, SC 6/748/4. *Drayton:* New College, Oxford; 1392–3, 1393–4, 1394–5, 1395–6; New College Archives, 5971–4. *Harwell:* bishop of Winchester; 1377–8, 1381–2, 1383–4, 1389–90; Hants. RO, 11M59/B1/130, 133, 135, 141. *Hinton Waldrist:* De Bohun; 1376–7, 1380–1, 1388–9; PRO, DL 29/652/10535–6, 10539. *Inkpen:* Titchfield Priory; 1383–4, 1385–6 Berks. RO, D/EC/M88 and M90. *Long Wittenham:* New College, Oxford; 1386–7, 1388–9, 1390–1; New College Archives, CA, 9145–7. *Waltham St Lawrence:* bishop of Winchester; 1377–8, 1381–2, 1383–4, 1389–90; Hants. RO, 11M59/B1/130, 133, 135, 141. *Wargrave:* bishop of Winchester; 1377–8, 1381–2, 1384–5, 1389–90; Hants. RO, 11M59/B1/130, 133, 136, 141. *Woolstone:* Winchester Cathedral Priory; 1381–2, 1383–4, 1387–8, 1391–2; PRO, SC 6/757/8, 9, 12, 14.
Buckinghamshire: *Aylesbury:* earl of Ormond; 1375–6, 1377–8, 1382–3; Birmingham Reference Library, Hampton 1802–4. *Cheddington:* Merton College, Oxford;

1375–6, 1376–7, 1377–8; Merton College Archives, MM 5601–3. *Cuddington*: Rochester Cathedral Priory; 1380–1; PRO, SC 6/760/14. *Denham*: abbot of Westminster; 1387–8, 1390–1; WAM, 3412, 3409. *Halton*: Canterbury Cathedral Priory; 1379–80, 1384–5; CCA, DCc/Halton 4 and 6. *Horsenden*: Braybrook; 1379–80; Bucks. RO, Box H/11a. *Iver*: Windsor College; 1375–6, 1381–2; Bucks. RO, BASM 110/53; St George's Chapel, Windsor, MS XV 53 65. *Ivinghoe*: bishop of Winchester; 1377–8, 1381–2, 1384–5, 1389–90; Hants. RO, 11M59/B1/130, 133, 136, 141. *Monks Risborough*: Canterbury Cathedral Priory; 1378–9, 1379–80, 1384–5; CCA, DCc/Risborough 3–5. *Moreton*: bishop of Winchester; 1375–6, 1377–8, 1378–9, 1379–80; Hants. RO, 11M59 B1/128, 130, 131, 132. *Quainton*: De Missenden; 1379–80, 1383–4, 1392–3; Bucks. RO, D/BASM/9/24, 14, 31. *Quarrendon*: Beauchamp; 1390–1, 1391–2, 1392–3, 1393–4; Oxon. RO, DIL X/D/2–5. *Tingewick*: St Catherine's Abbey, Rouen; 1378–9, 1379–80; New College Archives, 7087–8. *Turweston*: Westminster Abbey; 1377–8, 1383–4, 1386–7, 1390–1; WAM, 7818, 7824, 7827, 7831. *Water Eaton*: De Grey (De Wilton); 1381–2, 1391–2, 1394–5; Bucks. RO, D/BASM/9/13, 16, 21. *Weedon in the Vale*: ?; 1377–8, 1380–1, 1383–4, 1390–1; New College Archives, 6058–9, 6063, 6069. *West Wycombe*: bishop of Winchester; 1377–8, 1381–2, 1384–5, 1389–90, Hants. RO, 11M59/B1/130, 133, 136, 141.

Essex: *Bekeswell*: Westminster Abbey; 1377–8, 1378–9, 1381–2, 1387–8; Essex RO, D/DM M84–5, 87, 91. *Berners Roding*: De Berners; 1383–4, 1384–5, 1385–6; Essex RO, D/DGe M250, D/DU 497/15–16. *Berwick Berners*: De Gildeburgh; 1381–2; Essex RO, D/DHf M45. *Birchanger*: New College, Oxford; 1393–4, 1394–5; New College Archives, 6387, 5785. *Birdbrook*: Westminster Abbey; 1377–8, 1380–1, 1383–4, 1390–1; WAM, 25473, 25476, 25479, 25486. *Bocking*: Canterbury Cathedral Priory; 1375–6, 1376–7; CCA, DCc/Bocking 39 and 40. *Boreham*: De Burnell; 1378–9; PRO, SC 6/837/1. *Borley*: Canterbury Cathedral Priory; 1384–5; CCA, DCc/Borley 8. *Bulmer*: De Sutton; 1392–3; PRO, SC 6/1245/9. *Bures*: De La Pole; 1384–5; PRO, SC 6/1245/10. *Childerditch*: Coggeshall Abbey; 1387–8, 1396–7; Essex RO, D/DP M1113–4. *Cressing Temple*: Knights Hospitallers; 1386; BL, Add. Roll 41476b. *Feering*: Westminster Abbey; 1382–3, 1393–4, 1395–6; WAM, 25712, 25727, 25732. *Hatfield Broadoak*: De Bohun; 1377–8; Essex RO, D/DQ 18. *High Easter*: duchess of Gloucester; 1398–9; PRO, DL 29/42/817. *Hornchurch*: New College, Oxford; 1392–3, 1393–4, 1394–5; New College Archives, 6386–8. *Hutton*: Battle Abbey; 1388–9, 1389–90; PRO, SC 6/844/30–1. *Kelvedon*: Westminster Abbey; 1377–8, 1380–1, 1381–2, 1383–4; WAM, 25833, 25839, 25841, 25843. *Langenhoe*: De Sutton; 1380–1, 1381–2, 1383–4, 1396–7; Essex RO, D/DGe M200, 224–5, D/DC 2/16. *Lawling*: Canterbury Cathedral Priory; 1380–1, 1387–8; CCA, DCc/Lawling 17 and 20. *Little Maldon*: ?; 1376–7, 1378–9, 1379–80, 1380–1; Essex RO, D/DMb M13–16. *Messing Hall*: ?; 1396–7, 1400–1; Essex RO, D/DH X19 and 21. *Middleton*: Canterbury Cathedral Priory; 1375–6, 1384–5, 1386–7, 1388–9; CCA, DCc/Middleton 65–8. *Moulsham*: Westminster Abbey; 1377–8, 1378–9; Essex RO, D/DM M84 and 86. *Southchurch*: Canterbury Cathedral Priory; 1384–5; CCA, DCc/Southchurch 5. *Stapleford Abbots*: De Sutton; 1383–4; PRO, SC 6/847/10. *Stebbing*: Wanton; 1377–8; BL, Add. roll, LL 66016. *Takeley*: New College, Oxford; 1396–7, 1398–9, 1399–1400; New College Archives, 7006–8. *Tolleshunt Major*: Coggeshall Abbey; 1397–8; PRO, SC 6/848/13. *Ugley*: De Waterton; 1391–2, 1392–3,

1393–4, 1394–5; PRO, DL 29/42/809–11, 813. *Walthambury*: duchess of Gloucester; 1397–8; PRO, DL 29/42/815. *Wix*: Wix Priory; 1381–2, 1389–90; PRO, SC 6/849/15 and 19. *Writtle Rectory*: New College, Oxford; 1392–3, 1393–4, 1395–6, 1396–7; New College Archives, 7312–14, 7323.

Hertfordshire: *Albury*: De La Lee; 1376–7, 1380–1, 1384–5, 1385–6; Herts. RO, D/EAp M35–6, 38–9. *Aldenham*: Westminster Abbey; 1377–8, 1380–1, 1383–4, 1390–1; WAM, 26100, 26103, 26106, 26113. *Ashwell*: Westminster Abbey; 1380–1, 1394–5, 1396–7; WAM, 26269, 26285, 26288. *Great Gaddesden*: De Holand; 1383–4; Herts. RO, 2632. *Kinsbourne*: Westminster Abbey; 1377–8, 1380–1, 1383–4, 1390–1; WAM, 8842, 8845, 8848, 8855. *Oddingselles in Pirton*: Oddyngsels; 1378–9; Herts. RO, DE 24. *Sawbridgeworth*: Westminster Abbey; 1377–8, 1380–1, 1383–4, 1390–1; WAM, 26309, 26313, 26317, 26324. *Stevenage*: Westminster Abbey; 1376–7, 1380–1, 1383–4; Guildhall Library, London, 25404/44; WAM, 26355, 26358. *Walkern*: De Morle; 1390–1; Herts. RO, 9357. *Wheathampstead*: Westminster Abbey; 1377–8, 1380–1, 1387–8, 1390–1; Herts. RO, D/ELw M183, M185, M192–3.

Kent: *Adisham*: Canterbury Cathedral Priory; 1376–7; CCA, DCc/Adisham 39. *Agney*: Canterbury Cathedral Priory; 1384–5; CCA, DCc/Agney 57–8. *Appledore*: Canterbury Cathedral Priory; 1377–8, 1379–80, 1384–5, 1385–6; CCA, DCc/Appledore 51, 54, 56; Lambeth Palace Library, ED 194. *Barton*: Canterbury Cathedral Priory; 1377–8, 1389–90; CCA, DCc/Barton Carucate 14 and 15. *Bekesbourne*: Doget; 1391–2, 1396–7; BL, Harl. Rolls Z3 and 4. *Brook*: Canterbury Cathedral Priory; 1377–8, 1378–9; CCA, DCc/Brook 22–3. *Chartham*: Canterbury Cathedral Priory; 1375–6, 1376–7, 1381–2, 1383–4; CCA, DCc/Chartham 39–42. *Copton*: Canterbury Cathedral Priory; 1379–80, 1380–1, 1383–4; CCA, DCc/Copton 33–4, 36. *Denge Marsh*: Battle Abbey; 1376–7, 1380–1, 1383–4; PRO, SC 6/889/27, SC 6/890/4–5. *Eastry*: Canterbury Cathedral Priory; 1382–3; Lambeth Palace Library, ED 386. *Eltham*: crown; 1383–4; PRO, SC 6/890/20. *Elverton*: Canterbury Cathedral Priory; 1376–7, 1377–8, 1380–1; CCA, DCc/Elverton 38–40. *Great Chart*: Canterbury Cathedral Priory; 1375–6, 1377–8; CCA, DCc/Great Chart 77–8. *Ham*: Boxley Abbey; 1393–4, 1397–8; PRO, SC 6/892/1–2. *Ickham*: Canterbury Cathedral Priory; 1374–5, 1376–7; CCA, DCc/Ickham 56–7. *Lamberhurst*: Robertsbridge Abbey; 1376–7; Centre for Kentish Studies, Maidstone, U442 M89. *Meopham*: Canterbury Cathedral Priory; 1374–5, 1383–4; CCA, DCc/Meopham 101–2. *Monkton*: Canterbury Cathedral Priory; 1382–3, 1383–4; CCA, DCc/Monkton 96–7. *Newnham Court*: Boxley Abbey; 1375–6, 1388–9; PRO, SC 6/893/27–8. *Orgarswick*: Canterbury Cathedral Priory; 1380–1; CCA, DCc/Agney 56. *Otford*: archbishop of Canterbury; 1382–3, 1391–2; Lambeth Palace Library, ED 835–6. *Scotney in Lydd*: ?; 1393–4; BLO, MS DD All Souls C183 SC. *Sharpness*: Boxley Abbey; 1377–8, 1383–4, 1388–9, 1390–1; PRO, SC 6/897/7, 9, 11, 12. *Westerham*: Westminster Abbey; 1376–7, 1380–1, 1383–4, 1388–9; WAM, 26474, 26482, 26488, 26505. *Westwell*: Canterbury Cathedral Priory; 1385–6; Lambeth Palace Library, ED 1110. *Wrotham*: archbishop of Canterbury; 1393–4, 1394–5; Centre for Kentish Studies, Maidstone, U55 M64 and 67.

Middlesex: *Ashford*: Westminster Abbey; 1388–9, 1389–90, 1390–1, 1392–3; WAM, 26804–5, 26807, 26813. *Colham*: Le Strange; 1371–2, 1375–6, 1379–80, 1387–8; Lancs., RO, DDK 1746/1–4. *Ebury*: abbot of Westminster; 1377–8, 1380–1, 1381–2, 1393–4; WAM, 26933, 26936–7, 26947. *Halliford*: Westminster Abbey; 1377–8,

1381–2, 1383–4, 1390–1; WAM, 27034, 27038, 27040, 27044. *Harmondsworth*: Winchester College; 1386–7, 1397–8; Winchester College Archives, 11501–2. *Hyde*: Westminster Abbey; 1377–8, 1381–2, 1383–4, 1391–2; WAM, 27084, 27090, 27094, 27102. *Laleham*: abbot of Westminster; 1375–6, 1381–2, 1382–3; WAM, 27156–8. *Tottenham Court*: ?; 1376–7; Guildhall Library, London, 25347/2. *Yeoveney*: abbot of Westminster; 1376–7, 1377–8, 1378–9; WAM, 16887–9.

Northamptonshire: *Ashton*: Colepeper; 1395–6; BL, Harl. Roll R5. *Biggin*: Peterborough Abbey; 1372–3; N'hants. RO, Fitzwilliam Roll 267. *Boroughbury*: Peterborough Abbey; 1378–9, 1389–90; N'hants. RO, PDC.AR.1.10. *Culworth*: Windsor; 1385–6, 1386–7; N'hants. RO, Aa 66 and 38. *Eye*: Peterborough Abbey; 1393–4; N'hants. RO, Fitzwilliam Roll 261. *Higham Ferrers*: ?; 1380–1, 1382–3; PRO, DL 29/324/5305 and 5308. *Maidwell*: Latimer and De Seyton; 1383–4, 1386–7; N'hants. RO, Finch Hatton 482 and 475. *Passenham*: duke of Lancaster; 1380–1; PRO, DL 29/324/5306. *Raunds*: duke of Lancaster; 1380–1; PRO, DL 29/324/5305.

Oxfordshire: *Adderbury*: bishop of Winchester; 1377–8, 1381–2, 1384–5, 1389–90; Hants. RO, 11M59/B1/130, 133, 136, 141. *Crowmarsh*: Battle Abbey; 1391–2; PRO, SC 6/958/16. *Holywell*: Merton College, Oxford; 1377–8, 1380–1, 1383–4, 1391–2; Merton College Archives, MM 4526, 4528, 4530, 4533. *Islip*: abbot of Westminster; 1390–1, 1393–4, 1395–6; WAM, 14828–30. *South Stoke*: Eynsham Abbey; 1396–7; BLO, MSS DD CH CH M93. *Upper Heyford*: New College, Oxford; 1380–1, 1383–4, 1384–5, 1390–1; New College Archives, 6281, 6284–5, 6290. *Witney*: bishop of Winchester; 1377–8, 1381–2, 1384–5, 1389–90; Hants. RO, 11M59/B1/130, 133, 136, 141.

Surrey: *Beddington*: Carew; 1385–6; Surrey RO, 2163/1/11. *Betchworth*: earl of Arundel; 1380–1, 1385–6, 1391–2; Surrey RO, 260/1/16–18. *Dorking*: earl of Arundel; 1381–2, 1385–6, 1386–7, 1388–9; Arundel Castle Archives, A1778–80, 1782. *Farleigh*: Merton College, Oxford; 1374–5, 1375–6; Merton College Archives, MM 4862, 4864. *Farnham*: bishop of Winchester; 1377–8, 1381–2, 1384–5, 1389–90; Hants. RO, 11M59/B1/130, 133, 136, 141. *Leatherhead*: Merton College, Oxford; 1376–7; Merton College Archives, MM 5730. *Malden*: Merton College, Oxford; 1379–80, 1380–1; Merton College Archives, MM 4678–9. *Merstham*: Canterbury Cathedral Priory; 1387–8, 1388–9; Surrey RO, 7/1/14; CCA, DCc/Merstham 50. *Pyrford*: abbot of Westminster; 1377–8, 1380–1, 1385–6, 1390–1; WAM, 27423–4, 27427–8. *Thorncroft*: Merton College, Oxford; 1374–5, 1375–6; Merton College Archives, MM 5767–8. *Wandsworth*: Westminster Abbey; 1376–7, 1377–8, 1378–9, 1390–1; WAM, 27539, 27541, 27544, 27570. *West Horsley*: Berners; 1382–3, 1389–90, 1390–1; PRO, SC 6/1013/12, 15, 16.

Consolidated bibliography

Printed sources

The account-book of Beaulieu Abbey, ed. S. F. Hockey, Camden Society 4th series 16, London, 1975.

The archives of the abbey of Bury St Edmunds, ed. R. M. Thomson, Suffolk Records Society 21, Woodbridge, 1980.

The Black Death, trans. and ed. R. Horrox, Manchester, 1994.

Chaucer, Geoffrey, *The general prologue to the Canterbury tales*, ed. J. Winny, Cambridge, 1966.

Household accounts from medieval England, ed. C. M. Woolgar, 2 vols., British Academy, Records of Social and Economic History, new series, 17 and 18, Oxford, 1992, 1993.

The inventories and account rolls of the Benedictine houses or cells of Jarrow and Monk-Wearmouth, ed. J. Raine, Surtees Society 29, Newcastle upon Tyne, 1854.

Lordship and landscape in Norfolk 1250–1350: the early records of Holkham, ed. W. Hassall and J. Beauroy, British Academy, Records of Social and Economic History, new series, 20, Oxford, 1993.

Manorial records of Cuxham, Oxfordshire circa 1200–1359, ed. P. D. A. Harvey, Oxfordshire Record Society 50, London, 1976.

Ministers' accounts of the earldom of Cornwall 1296–7, ed. M. Midgley, 2 vols., Camden Society 3rd series 66, London, 1942.

Nonarum inquisitiones in curia scaccarii, ed. Record Commissioners, London, 1807.

'Norwich Cathedral Priory gardeners' accounts, 1329–1530', ed. C. Nobel, in *Farming and gardening in late medieval Norfolk*, Norfolk Record Society 61, Norwich, 1997, pp. 1–93.

Paston letters and papers of the fifteenth century, ed. N. Davis, 3 vols., Oxford, 1971.

The Peasants' Revolt of 1381, ed. R. B. Dobson, 2nd edition, London, 1983.

The Pinchbeck Register of the abbey of Bury St Edmunds etc., ed. F. Hervey, 2 vols., Brighton, 1925.

The Pipe Roll of the bishopric of Winchester 1301–2, ed. M. Page, Hampshire Record Series 14, Winchester, 1996.

The prior's manor-houses: inventories of eleven of the manor-houses of the prior of Norwich made in the year 1352 A.D., trans. and ed. D. Yaxley, Dereham, 1988.

Rotuli Hundredorum, ed. Record Commissioners, 2 vols., London, 1812, 1818.

Select documents of the English lands of the Abbey of Bec, ed. M. Chibnall, Camden Society 3rd series 73, London, 1951.

A Suffolk hundred in the year 1283. The assessment of the Hundred of Blackbourne for a tax of one thirtieth, and a return showing the land tenure there, ed. E. Powell, Cambridge, 1910.

Visions from Piers Plowman taken from the poem of William Langland, trans. N. Coghill, London, 1949.

Walter of Henley and other treatises on estate management and accounting, ed. D. Oschinsky, Oxford, 1971.

The Warwickshire Hundred Rolls of 1279–80: Stoneleigh and Kineton Hundreds, ed. T. John, British Academy, Records of Social and Economic History, new series, 19, Oxford, 1992.

Secondary works (including unpublished theses and papers)

Abel, W., *Agrarkrisen und Agrarkonjunktur in Mitteleuropa vom 13. bis zum 19. Jahrhundert*, Berlin, 1935, trans. O. Ordish, *Agricultural fluctuations in Europe from the thirteenth to the twentieth centuries*, London, 1980.

Die Wüstungen des Ausgehenden Mittelalters, 2nd edition, Stuttgart, 1955.

Adams, I., *Agrarian landscape terms: a glossary for historical geography*, Institute of British Geographers, Special Publication 9, London, 1976.

The agrarian history of England and Wales, 8 vols., general eds. H. P. R. Finberg and J. Thirsk, Cambridge, 1967 – in progress.

Albarella, U., 'Size, power, wool and veal: zooarchaeological evidence for late medieval innovations', in G. De Boe and F. Verhaeghe (eds.), *Environment and subsistence in medieval Europe: papers of the 'Medieval Europe Brugge 1997' conference*, I.A.P. Rapporten 9, Zellik, 1997, pp. 19–30.

'"The mystery of husbandry": medieval animals and the problem of integrating historical and archaeological evidence', unpublished paper.

Alcock, N. W., 'An east Devon manor in the later Middle Ages. Part I: 1374–1420. The manor farm', *Reports and Transactions of the Devonshire Association* 102 (1970), 141–87.

Allen, R. C., *The 'capital intensive farmer' and the English agricultural revolution: a reassessment*, Discussion Paper 87–11, Department of Economics, University of British Columbia, Vancouver, 1987.

'The two English agricultural revolutions, 1459–1850', in Campbell and Overton (eds.), *Land, labour and livestock*, pp. 236–54.

Enclosure and the yeoman: the agricultural development of the south midlands 1450–1850, Oxford, 1992.

Allison, K. J., 'The sheep-corn husbandry of Norfolk in the sixteenth and seventeenth centuries', *AHR* 5 (1957), 12–30.

'Flock management in the sixteenth and seventeenth centuries', *EcHR* 11 (1958), 98–112.

Ambrosioli, M., *The wild and the sown. Botany and agriculture in Western Europe: 1350–1850*, trans. M. M. Salvatorelli, Cambridge, 1997.

Ashley, W., *The bread of our forefathers: an inquiry in economic history*, Oxford, 1928.

Astill, G., 'An archaeological approach to the development of agricultural technologies in medieval England', in Astill and Langdon (eds.), *Medieval farming*, pp. 193–224.

Astill, G. and Grant, A. (eds.), *The countryside of medieval England*, Oxford, 1988.

Astill, G. and Langdon, J. L. (eds.), *Medieval farming and technology: the impact of agricultural change in north-west Europe in the Middle Ages*, Leiden, 1997.

Aston, T. H., Coss, P. R., Dyer, C. C., and Thirsk, J. (eds.), *Social relations and ideas: essays in honour of R. H. Hilton*, Cambridge, 1983.

Aston, T. H. and Philpin, C. H. E. (eds.), *The Brenner debate: agrarian class structure and economic development in pre-industrial Europe*, Cambridge, 1985.

Atkin, M. A., 'Land use and management in the upland demesne of the de Lacy estate of Blackburnshire c. 1300', *AHR* 42 (1994), 1–19.

Ault, W. O., *Open-field farming in medieval England: a study of village by-laws*, London, 1972.

Backhouse, J., *The Luttrell Psalter*, London, 1989.

Bailey, M., 'The rabbit and the medieval East Anglian economy', *AHR* 36 (1988), 1–20.
A marginal economy? East-Anglian Breckland in the later Middle Ages, Cambridge, 1989.
'The concept of the margin in the medieval English economy', *EcHR* 42 (1989), 1–17.
'Sand into gold: the evolution of the foldcourse system in west Suffolk, 1200–1600', *AHR* 38 (1990), 40–57.
'*Per impetum maris*: natural disaster and economic decline in eastern England, 1275–1350', in Campbell (ed.), *Before the Black Death*, pp. 184–208.
'Demographic decline in late medieval England: some thoughts on recent research', *EcHR* 49 (1996), 1–19.
'Peasant welfare in England, 1290–1348', *EcHR* 51 (1998), 223–51.

Baillie, M. G. L., 'Dendrochronology raises questions about the nature of the AD 536 dust-veil event', *The Holocene* 4 (1994), 212–17.
'Dendrochronology provides an independent background for studies of the human past', in D. Cavaciocchi (ed.), *L'uomo e la foresta secc. XIII–XVIII*, Prato, 1995, pp. 99–119.
'Gas hydrate hazards: have human populations been affected?', unpublished manuscript.

Baker, A. R. H., 'Evidence in the *Nonarum inquisitiones* of contracting arable lands in England during the early fourteenth century', *EcHR* 19 (1966), 518–32; reprinted with a supplementary note in A. R. H. Baker, J. D. Hamshere and J. Langton (eds.), *Geographical interpretations of historical sources: readings in historical geography*, Newton Abbot, 1970, pp. 85–102.
'Changes in the later Middle Ages', in Darby (ed.), *New historical geography*, pp. 186–247.
'Field systems of south-east England', in Baker and Butlin (eds.), *Field systems*, pp. 377–429.

Baker, A. R. H. and Butlin, R. A. (eds.), *Studies of field systems in the British Isles*, Cambridge, 1973.

Bartley, K. C., 'Classifying the past: discriminant analysis and its application to medieval farming systems', *History and Computing* 8.1 (1996), 1–10.

'Mapping medieval England', *Mapping Awareness* 10, 6 (1996), 34–6.

Bartley, K. C. and Campbell, B. M. S., '*Inquisitiones post mortem*, G.I.S., and the creation of a land-use map of pre Black Death England', *Transactions in G.I.S.* 2
(1997), 333–46.

Bennett, J., *Ale, beer, and brewsters in England: women's work in a changing world,
1300–1600*, Oxford, 1996.

Beresford, M. W., 'The poll tax and the census of sheep, 1549', *AHR* 1 and 2 (1953 and
1954), 9–15 and 15–29.

The lost villages of England, Lutterworth, 1963.

Beresford, M. W. and Hurst, J. G., *Deserted medieval villages: studies*, London, 1971.

Beveridge, W., 'The yield and price of corn in the Middle Ages', *Economic History*, a
supplement of *The Economic Journal*, 1 (1927), 155–67.

Biddick, K., 'Animal husbandry and pastoral land-use on the fen-edge, Peterborough,
England: an archaeological and historical reconstruction, 2500 BC–1350 AD',
PhD thesis, University of Toronto, 1982.

'Pig husbandry on the Peterborough Abbey estate from the twelfth to the fourteenth
century A.D.', in J. Clutton-Brock and C. Grigson (eds.), *Animals and archaeology*, vol. IV, *Husbandry in Europe*, BAR International Series 227, Oxford, 1985,
pp. 161–77.

The other economy: pastoral husbandry on a medieval estate, Berkeley and Los
Angeles, 1989.

Biddick, K. with Bijleveld, C. C. J. H., 'Agrarian productivity on the estates of the bishopric of Winchester in the early thirteenth century: a managerial perspective', in
Campbell and Overton (eds.), *Land, labour and livestock*, pp. 95–123.

Bieleman, J., 'Dutch agriculture in the Golden Age, 1570–1660', in K. Davids and
L. Noordegraaf (eds.), *The Dutch economy in the Golden Age: nine studies*,
Amsterdam, 1993, pp. 159–85.

Birrell, J. R., 'The forest economy of the Honour of Tutbury in the fourteenth and
fifteenth centuries', *University of Birmingham Historical Journal* 8 (1962),
114–34.

Bischoff, J. P., '"I cannot do't without counters": fleece weights and sheep breeds in late
thirteenth and early fourteenth century England', *Agricultural History* 57 (1983),
142–60.

Bishop, T. A. M., 'The distribution of manorial demesne in the vale of Yorkshire',
EHR 49 (1934), 386–406.

'Monastic granges in Yorkshire', *EHR* 51 (1936), 193–214.

'The rotation of crops at Westerham, 1297–1350', *Economic History Review* 9
(1938), 38–44.

Blake, W. J., 'Norfolk manorial lords in 1316', *NA* 30 (1952), 261.

Blanchard, I. S. W., 'Economic change in Derbyshire in the late Middle Ages,
1272–1540', PhD thesis, University of London, 1967.

'Population change, enclosure and the early Tudor economy', *EcHR* 23 (1970),
427–45.

'The continental European cattle trade, 1400–1600', *EcHR* 39 (1986), 427–60.

The Middle Ages: a concept too many?, Avonbridge, 1996.

Bois, G., *The crisis of feudalism: economy and society in eastern Normandy c. 1300–1550*, Cambridge, 1984.

Bökönyi, S., 'The development of stockbreeding and herding in medieval Europe', in Sweeney (ed.), *Agriculture*, pp. 41–61.

Bolton, J. L., *The medieval English economy, 1150–1500*, London, 1980.

Booth, P. H. W., *The financial administration of the lordship and county of Chester 1272–1377*, Chetham Society, Manchester, 1981.

Boserup, E., *The conditions of agricultural growth: the economics of agrarian change under population pressure*, Chicago, 1965.

Bowden, P. J., 'Agricultural prices, wages, farm profits, and rents', in Thirsk (ed.), *AHEW*, vol. V.ii, pp. 1–118.

Brady, N., 'The gothic barn of England: icon of prestige and authority', in E. Smith and M. Wolfe (eds.), *Technology and resource use in medieval Europe: cathedrals, mills and mines*, Aldershot, 1997, pp. 76–105.

Brandon, P. F., 'Arable farming in a Sussex scarp-foot parish during the late Middle Ages', *Sussex Archaeological Collections* 100 (1962), 60–72.

'Medieval clearances in the East Sussex Weald', *TIBG* 48 (1969), 135–53.

'Agriculture and the effects of floods and weather at Barnhorne, Sussex, during the late Middle Ages', *Sussex Archaeological Collections* 109 (1971), 69–93.

'Demesne arable farming in coastal Sussex during the later Middle Ages', *AHR* 19 (1971), 113–34.

'Cereal yields on the Sussex estates of Battle Abbey during the later Middle Ages', *EcHR* 25 (1972), 403–29.

'Farming techniques: south-eastern England', in Hallam (ed.), *AHEW*, vol. II, pp. 312–25.

'New settlement: south-eastern England', in Miller (ed.), *AHEW*, vol. III, pp. 174–89.

Brenner, R., 'Agrarian class structure and economic development in pre-industrial Europe', *PP* 70 (1976), 30–75; reprinted in Aston and Philpin (eds.), *Brenner debate*, pp. 10–63.

'The agrarian roots of European capitalism', *PP* 97 (1982), 16–113; reprinted in Aston and Philpin (eds.), *Brenner debate*, pp. 213–327.

Bridbury, A. R., 'The Domesday valuation of manorial income', in Bridbury, *The English economy from Bede to the Reformation*, Woodbridge, 1992, pp. 111–32.

Briffa, K. R., Jones, P. D., Bartholin, T. S., Eckstein, D., Schweingruber, F. H., Karlen, W., Zetterberg, P. and Eronen, M., 'Fennoscandian summers from AD 500: temperature changes on short and long timescales', *Climate Dynamics* 7 (1992), 111–19.

Britnell, R. H., 'Minor landlords in England and medieval agrarian capitalism', *PP* 89 (1980), 3–22.

'Agriculture in a region of ancient enclosure, 1185–1500', *Nottingham Medieval Studies* 27 (1983), 37–55.

'The Pastons and their Norfolk', *AHR* 36 (1988), 132–44.

'Farming practice and techniques: eastern England', in Miller (ed.), *AHEW*, vol. III, pp. 194–210.

'The occupation of the land: eastern England', in Miller (ed.), *AHEW*, vol. III, pp. 53–67.

'Commerce and capitalism in late medieval England: problems of description and theory', *Journal of Historical Sociology* 6 (1993), 359–76.

The commercialisation of English society 1000–1500, Cambridge, 1993.

'Commercialisation and economic development in England, 1000–1300', in Britnell and Campbell (eds.), *Commercialising economy*, pp. 7–26.

Britnell, R. H. and Campbell, B. M. S. (eds.), *A commercialising economy: England 1086 to c. 1300*, Manchester, 1995.

Campbell, B. M. S., 'Field systems in eastern Norfolk during the Middle Ages: a study with particular reference to the demographic and agrarian changes of the four-teenth century', PhD thesis, University of Cambridge, 1975.

'Commonfield origins – the regional dimension', in T. Rowley (ed.), *The origins of open-field agriculture*, London, 1981, pp. 112–29.

'The extent and layout of commonfields in eastern Norfolk', *NA* 38 (1981), 5–32.

'The population of early Tudor England: a re-evaluation of the 1522 muster returns and 1524 and 1525 lay subsidies', *JHG* 7 (1981), 145–54.

'The regional uniqueness of English field systems? Some evidence from eastern Norfolk', *AHR* 29 (1981), 16–28.

'Agricultural progress in medieval England: some evidence from eastern Norfolk', *EcHR* 36 (1983), 26–46.

'Arable productivity in medieval England: some evidence from Norfolk', *JEH* 43 (1983), 379–404.

'Population pressure, inheritance and the land market in a fourteenth-century peasant community', in R. M. Smith (ed.), *Land, kinship and life-cycle*, Cambridge, 1984, pp. 87–134.

'The complexity of manorial structure in medieval Norfolk: a case study', *NA* 39 (1986), 225–61.

'The diffusion of vetches in medieval England', *EcHR* 41 (1988), 193–208.

'Towards an agricultural geography of medieval England', *AHR* 36 (1988), 87–98.

'Laying foundations: the agrarian history of England and Wales, 1042–1350', *AHR* 37 (1989), 188–92.

'People and land in the Middle Ages, 1066–1500', in R. A. Dodgshon and R. A. Butlin (eds.), *An historical geography of England and Wales*, 2nd edition, London, 1990, pp. 69–121.

'Land, labour, livestock, and productivity trends in English seignorial agriculture, 1208–1450', in Campbell and Overton (eds.), *Land, labour and livestock*, pp. 144–82.

'Commercial dairy production on medieval English demesnes: the case of Norfolk', *Anthropozoologica* 16 (1992), 107–18.

'A fair field once full of folk: agrarian change in an era of population decline, 1348–1500', *AHR* 41 (1993), 60–70.

'Medieval land use and land values', in Wade-Martins (ed.), *Historical atlas of Norfolk*, pp. 48–9.

'Medieval manorial structure', in Wade-Martins (ed.), *Historical atlas of Norfolk*, pp. 52–3.

'Ecology versus economics in late thirteenth- and early fourteenth-century English agriculture', in Sweeney (ed.), *Agriculture*, pp. 76–108.

'Measuring the commercialisation of seigneurial agriculture *c.* 1300', in Britnell and Campbell (eds.), *Commercialising economy*, pp. 132–93.

'The livestock of Chaucer's reeve: fact or fiction?', in DeWindt (ed.), *The salt of common life*, pp. 271–305.

'Economic rent and the intensification of English agriculture, 1086–1350', in Astill and Langdon (eds.), *Medieval farming*, pp. 225–50.

'Matching supply to demand: crop production and disposal by English demesnes in the century of the Black Death', *JEH* 57 (1997), 827–58.

'Constraint or constrained? Changing perspectives on medieval English agriculture', *Neha-Jaarboek voor economische, bedrijfs- en techniekgeschiedenis*, 61 (1998), 15–35.

'The sources of tradable surpluses: English agricultural exports 1250–1350', in L. Berggren, N. Hybel, and A. Landen (eds.), *Trade and transport in northern Europe 1150–1400*, Toronto, forthcoming.

Campbell, B. M. S. (ed.), *Before the Black Death: studies in the 'crisis' of the early fourteenth century*, Manchester, 1991.

Campbell, B. M. S. and Bartley, K. C., *Lay lordship, land, and wealth: the human geography of early fourteenth-century England*, Manchester, forthcoming.

Campbell, B. M. S., Bartley, K. C., and Power, J. P., 'The demesne-farming systems of post Black Death England: a classification', *AHR* 44 (1996), 131–79.

Campbell, B. M. S., Galloway, J. A., Keene, D. J., and Murphy, M., *A medieval capital and its grain supply: agrarian production and its distribution in the London region c. 1300,* Historical Geography Research Series 30, n.p., 1993.

Campbell, B. M. S., Galloway, J. A., and Murphy, M., 'Rural landuse in the metropolitan hinterland, 1270–1339: the evidence of *inquisitiones post mortem*', *AHR* 40 (1992), 1–22.

Campbell, B. M. S. and Overton, M. (eds.), *Land, labour and livestock: historical studies in European agricultural productivity*, Manchester, 1991.

'A new perspective on medieval and early modern agriculture: six centuries of Norfolk farming *c.* 1250–*c.* 1850', *PP* 141 (1993), 38–105.

Campbell, B. M. S. and Power, J. P., 'Mapping the agricultural geography of medieval England', *JHG* 15 (1989), 24–39.

Cantor, L., 'Forests, chases, parks and warrens', in L. Cantor (ed.), *The English medieval landscape*, London, 1982, pp. 56–85.

Carus-Wilson, E. M. (ed.), *Essays in economic history*, 3 vols., London, 1954, 1962.

Carus-Wilson, E. M. and Coleman, O., *England's export trade, 1275–1547*, Oxford, 1963.

Chambers, J. D., *Population, economy, and society in pre-industrial England*, London, 1972.

Chambers, J. D. and Mingay, G. E., *The agricultural revolution 1750–1880*, London, 1966.

Chédeville, A., *Chartres et ses campagnes (XIe–XIIIe siècles)*, Paris, 1973.

Chisholm, M., *Rural settlement and land-use: an essay on location*, London, 1962.

Chorley, G. P. H., 'The agricultural revolution in northern Europe, 1750–1880: nitrogen, legumes and crop productivity', *EcHR* 34 (1981), 71–93.

Clanchy, M. T., *From memory to written record: England 1066 to 1307*, London, 1979.

Clark, G., 'The cost of capital and medieval agricultural technique', *Explorations in Economic History* 25 (1988), 265–94.

'Labour productivity in English agriculture, 1300–1860', in Campbell and Overton (eds.), *Land, labour and livestock*, pp. 211–35.

'The economics of exhaustion, the Postan thesis, and the agricultural revolution', *JEH* 52 (1992), 61–84.

'A revolution too many: the agricultural revolution, 1700–1850', Agricultural History Center, University of California at Davis, Working Paper Series 91, 1997.

Cooter, W. S., 'Ecological dimensions of medieval agrarian systems', *Agricultural History* 52 (1978), 458–77.

Coppock, T. R., *An agricultural atlas of England and Wales*, London, 1964.

Cornwall, J., 'English population in the early sixteenth century', *EcHR* 23 (1970), 32–44.

Court, W. H. B., *The rise of the midland industries, 1600–1838*, London, 1938.

Cracknell, B. E., *Canvey Island: the history of a marshland community*, Department of English Local History, University of Leicester, Occasional Paper 12, Leicester, 1959.

Crafts, N. F. R., 'British economic growth, 1700–1831: a review of the evidence', *EcHR* 36 (1983), 177–99.

Darby, H. C. *The medieval Fenland*, Cambridge, 1940.

'The changing English landscape', *Geographical Journal* 117 (1951), 377–94.

The Domesday geography of eastern England, Cambridge, 1971.

Domesday England, Cambridge, 1977.

Darby, H. C. (ed.), *An historical geography of England before A.D. 1800*, Cambridge, 1936.

(ed.), *A new historical geography of England*, Cambridge, 1973.

Darby, H. C., Glasscock, R. E., Sheail, J., and Versey, G. R., 'The changing geographical distribution of wealth in England 1086–1334–1525', *JHG* 5 (1979), 247–62.

Davenport, F. J., *The economic development of a Norfolk manor 1086–1565*, Cambridge, 1906.

Davies, R. R., *Lordship and society in the March of Wales 1282–1400*, Oxford, 1978.

Domination and conquest. The experience of Ireland, Scotland and Wales 1100–1300, Cambridge, 1990.

Davis, R. H. C., *The medieval warhorse: origin, development and redevelopment*, London, 1989.

Davis, S. J. M. and Beckett, J. V., 'Animal husbandry and agricultural improvement: the archaeological evidence from animal bones and teeth', *Rural History* 10 (1999), 1–17.

Deane, P. and Cole, W. A., *British economic growth 1699–1959*, 2nd edition, Cambridge, 1969.

Denholm-Young, N., *Seignorial administration in England*, Oxford, 1937.

Denton, W., *England in the fifteenth century*, London, 1888.

Derville, A., 'Le Rendement du blé dans la région lilloise (1285–1541)', *Bulletin de la commission historique du Département du Nord* 40 (1975–6), 23–39.

DeWindt, E. B. (ed.), *The salt of common life: individuality and choice in the medieval town, countryside and church. Essays presented to J. Ambrose Raftis on the occasion of his 70th birthday*, Kalamazoo, 1995.

Dickson, D., *New foundations: Ireland 1660–1800*, Dublin, 1987.

Dobson, R. B., *Durham Priory, 1400–1450*, Cambridge, 1973.

Dodwell, B., 'The free peasantry of East Anglia in Domesday', *NA* 27 (1939), 145–57.

Donkin, R. A., 'Cattle on the estates of medieval Cistercian monasteries in England and Wales', *EcHR* 15 (1962), 31–53.

'Changes in the early Middle Ages', in Darby (ed.), *New historical geography*, pp. 75–135.

The Cistercians: studies in the geography of medieval England and Wales, Toronto, 1978.

Drew, J. S., 'Manorial accounts of St Swithun's Priory, Winchester', in Carus-Wilson (ed.), *Essays*, vol. II, pp. 12–30.

Du Boulay, F. R. H., 'Who were farming the English demesnes at the end of the Middle Ages?', *EcHR* 17 (1964–5), 443–55.

The lordship of Canterbury: an essay on medieval society, London, 1966.

Dunford, M. and Perrons, D., *The arena of capital*, Basingstoke and London, 1983.

Dury, G. H., 'Crop failures on the Winchester manors, 1232–1349', *TIBG*, new series, 9 (1984), 401–18.

Dyer, A., *Decline and growth in English towns, 1400–1640*, London, 1991.

Dyer, C. C., *Lords and peasants in a changing society: the estates of the bishopric of Worcester, 650–1540*, Cambridge, 1980.

Warwickshire farming 1349–c. 1520: preparations for agricultural revolution, Dugdale Society Occasional Papers 27, Oxford, 1981.

'English diet in the later Middle Ages', in Aston and others (eds.), *Social relations*, pp. 191–216.

'The social and economic background to the rural revolt of 1381', in R. H. Hilton and T. H. Aston (eds.), *The English rising of 1381*, Cambridge, 1984, pp. 9–24; reprinted in C. C. Dyer, *Everyday life in medieval England*, London, 1994, pp. 191–220.

'Changes in diet in the late Middle Ages: the case of harvest workers', *AHR* 36 (1988), 21–37.

Standards of living in the later Middle Ages: social change in England c. 1200–1520, Cambridge, 1989.

'The consumer and the market in the later Middle Ages', *EcHR* 42 (1989), 305–27.

'The past, the present and the future in medieval rural history', *Rural History* 1 (1990), 37–49.

'Farming practice and technology: the west midlands', in Miller (ed.), *AHEW*, vol. III, pp. 222–38.

'The occupation of the land: the west midlands', in Miller (ed.), *AHEW*, vol. III, pp. 77–92.

'The hidden trade of the Middle Ages: evidence from the west midlands of England', *JHG* 18 (1992), 141–57.

Everyday life in medieval England, London, 1994.

'How urbanized was medieval England?' in J.-M. Duvosquel and E. Thoen (eds.), *Peasants and townsmen in medieval Europe: studia in honorem Adriaan Verhulst*, Ghent, 1995, pp. 169–83.

'Sheepcotes: evidence for medieval sheepfarming', *Medieval Archaeology* 39 (1995), 136–64.

Dymond, D., *The Norfolk landscape*, London, 1985.

Ecclestone, M., 'Dairy production on the Glastonbury Abbey demesnes 1258–1334', MA thesis, University of Bristol, 1996.

Edwards, J. F. and Hindle, B. P., 'The transportation system of medieval England and Wales', *JHG* 17 (1991), 123–34.

Edwards, P., 'The horse trade in the midlands in the seventeenth century', *AHR* 27 (1979), 90–100.

Epstein, S. R., 'Regional fairs, institutional innovation, and economic growth in late medieval Europe', *EcHR* 47 (1994), 459–82.

Erlington, C. R., 'Assessments of Gloucestershire: fiscal records in local history', *Transactions of the Bristol and Gloucester Archaeological Society* 103 (1985), 5–15.

Ernle, Lord (formerly R. Prothero), 'The enclosures of open-field farms', *Journal of the Ministry of Agriculture* 27 (1920), 831–41.

English farming past and present, London, 1912; 3rd edition, 1922.

Evans, N., *The East Anglian linen industry: rural industry and local economy 1500–1850*, Aldershot, 1985.

Farmer, D. L., 'Grain yields on the Winchester manors in the later Middle Ages', *EcHR* 30 (1977), 555–66.

'Grain yields on Westminster Abbey manors, 1271–1410', *Canadian Journal of History* 18 (1983), 331–47.

'Prices and wages', in Hallam (ed.), *AHEW*, vol. II, pp. 716–817.

'Two Wiltshire manors and their markets', *AHR* 37 (1989), 1–11.

'Marketing the produce of the countryside, 1200–1500', in Miller (ed.), *AHEW*, vol. III, pp. 324–430.

'Prices and wages, 1350–1500', in Miller (ed.), *AHEW*, vol. III, pp. 431–525.

'Woodland and pasture sales on the Winchester manors in the thirteenth century: disposing of a surplus or producing for the market?', in Britnell and Campbell (eds.), *Commercialising economy*, pp. 102–31.

Fenoaltea, S., 'Authority, efficiency and agricultural organization in medieval England and beyond: a hypothesis', *JEH* 35 (1975), 693–718.

'Transaction costs, Whig history, and the common fields', *Politics and Society* 16 (1988), 171–240.

Finberg, H. P. R., *Tavistock Abbey: a study in the social and economic history of Devon*, Cambridge, 1951; 2nd edition, 1969.

'An early reference to the Welsh cattle trade', *AHR* 2 (1954), 12–14.

'An agrarian history of England', *AHR* 4 (1956), 2–3.

Fischer, D. H., *The great wave: price revolutions and the rhythm of history*, Oxford, 1996.

Fisher, F. J., 'The development of the London food market, 1540–1640', *Economic History Review* 5 (1935), 46–64.

Fourquin, G., *Les Campagnes de la région Parisienne à la fin du Moyen Age (du milieu du XIIIe au début du XVIe siècle)*, Paris, 1964.

Fox, H. S. A., 'Some ecological dimensions of medieval field systems', in K. Biddick (ed.), *Archaeological approaches to medieval Europe*, Kalamazoo, 1984, pp. 119–58.

'The alleged transformation from two-field to three-field systems in medieval England', *EcHR* 39 (1986), 526–48.

'The people of the wolds in English settlement history', in M. Aston, D. Austin and C. C. Dyer (eds.), *The rural settlements of medieval England: studies dedicated to Maurice Beresford and John Hurst*, Oxford, 1989, pp. 77–101.

'Peasant farmers, patterns of settlement and pays: transformations in the landscapes of Devon and Cornwall during the later Middle Ages', in R. Higham (ed.), *Landscape and townscape in the south-west*, Exeter, 1990, pp. 41–73.

'Farming practice and techniques: Devon and Cornwall', in Miller (ed.), *AHEW*, vol. III, pp. 303–23.

'The occupation of the land: Devon and Cornwall', in Miller (ed.), *AHEW*, vol. III, pp. 152–74.

'Medieval farming: arable productivity and peasant holdings at Taunton', Economic and Social Research Council, unpublished End of Award Report, B00 23 2203, 1991.

'Medieval Dartmoor as seen through its account rolls', in *The archaeology of Dartmoor: perspectives from the 1990s*, Devon Archaeological Society Proceedings 52, Exeter, 1994, pp. 149–71.

Frank Jr., R. W., 'The "hungry gap", crop failure, and famine: the fourteenth-century agricultural crisis and *Piers Plowman*', in Sweeney (ed.), *Agriculture in the Middle Ages*, pp. 227–43.

Fryde, E. B. and Fryde, N., 'Peasant rebellion and peasant discontents', in Miller (ed.), *AHEW*, vol. III, pp. 744–819.

Galloway, J. A., 'London's grain supply: changes in production, distribution and consumption during the fourteenth century', *Franco-British Studies: Journal of the British Institute in Paris* 20 (1995), 23–34.

'Driven by drink? Ale consumption and the agrarian economy of the London region, c. 1300–1400', in M. Carlin and J. T. Rosenthal (eds.), *Food and eating in medieval Europe*, London, 1998, pp. 87–100.

Galloway, J. A., Keene, D. J., and Murphy, M., 'Fuelling the city: production and distribution of firewood and fuel in London's region, 1290–1400', *EcHR* 49 (1996), 447–72.

Galloway, J. A. and Murphy, M., 'Feeding the city: medieval London and its agrarian hinterland', *London Journal* 16 (1991), 3–14.

Galloway, J. A., Murphy, M., and Myhill, O., *Kentish demesne accounts up to 1350: a catalogue*, London, 1993.

Gardiner, M., 'Medieval farming and flooding in the Brede valley', in J. Eddison (ed.), *Romney Marsh: the debatable ground*, Oxford, 1995, pp. 127–37.

Gatrell, A. C., *Distance and space: a geographical perspective*, Oxford, 1983.

Geertz, C., *Agricultural involution: the process of ecological change in Indonesia*, Berkeley, 1963.

Gibbs, J. P. and Martin, W. T., 'Urbanization, technology and the division of labour', *American Sociological Review* 27 (1962), 667–77.

Glasscock, R. E., 'The distribution of wealth in East Anglia in the early fourteenth century', *TIBG* 32 (1963), 118–23.

'England circa 1334', in Darby (ed.), *New historical geography*, pp. 136–85.

Glennie, P., 'Continuity and change in Hertfordshire agriculture 1550–1700: i – patterns of agricultural production', *AHR* 36 (1988), 55–75.

'Continuity and change in Hertfordshire agriculture, 1550–1700: ii – trends in crop yields and their determinants', *AHR* 36 (1988), 145–61.

'Measuring crop yields in early modern England', in Campbell and Overton (eds.), *Land, labour and livestock*, pp. 255–83.

Grant, A., 'Animal resources', in Astill and Grant (eds.), *Countryside*, pp. 149–87.

Grantham, G. W., 'Jean Meuvret and the subsistence problem in early modern France', *JEH* 49 (1989), 184–200.

'*Contra* Ricardo: on the macroeconomics of pre-industrial economies', *European Review of Economic History* 3 (1999), 199–233.

'Time's arrow and time's cycle in the medieval economy: the significance of recent developments in economic theory for the history of medieval economic growth', unpublished paper presented at the 'Fifth Anglo-American seminar on the medieval economy and society', Cardiff, 1995.

'Espaces privilégiés: productivité agraire et zones d'approvisionnement des villes dans l'Europe préindustrielle', *Annales* 52 (1997), 695–725.

Gras, N. S. B., *The evolution of the English corn market from the twelfth to the eighteenth century*, Cambridge, Mass., 1915.

Gray, H. L., 'The commutation of villein services in England before the Black Death', *EHR* 29, 116 (1914), 625–56.

English field systems, Cambridge, Mass., 1915.

Greenway, D. E., 'A newly discovered fragment of the Hundred Rolls of 1279–80', *Journal of the Society of Archivists* 7 (1982), 73–7.

Greig, J., 'Plant resources', in Astill and Grant (eds.), *Countryside*, pp. 108–27.

Grigg, D. B., *The agricultural systems of the world: an evolutionary approach*, Cambridge, 1974.

Population growth and agrarian change: an historical perspective, Cambridge, 1980.

'Breaking out: England in the eighteenth and nineteenth centuries', in *Population growth and agrarian change*, pp. 163–89.

'Western Europe in the thirteenth and fourteenth centuries: a case of overpopulation?', in *Population growth and agrarian change*, pp. 64–82.

The dynamics of agricultural change: the historical experience, London, 1982.

Habakkuk, H. J., 'The agrarian history of England and Wales: regional farming systems and agrarian change, 1640–1750', *EcHR* 40 (1987), 281–96.

Halcrow, E. M., 'The decline of demesne farming on the estates of Durham Cathedral Priory', *EcHR* 7 (1954), 345–56.

'The administration and agrarian policy of the manors of Durham Cathedral Priory', BLitt thesis, University of Oxford, 1959.

Hall, D. N., *Medieval fields*, Princes Risborough, 1982.

Hallam, H. E., *Settlement and society: a study of the early agrarian history of south Lincolnshire*, Cambridge, 1965.

'The Postan thesis', *Historical Studies* 15 (1972), 203–22.

Rural England 1066–1348, London, 1981.

'Farming techniques: eastern England', in Hallam (ed.), *AHEW*, vol. II, pp. 272–312.

'Farming techniques: southern England', in Hallam (ed.), *AHEW*, vol. II, pp. 341–68.

'Population movements in England, 1086–1350', in Hallam (ed.), *AHEW*, vol. II, pp. 508–93.

'The life of the people', in Hallam (ed.), *AHEW*, vol. II, pp. 818–53.

Hallam, H. E. (ed.), *The agrarian history of England and Wales*, vol. II, *1042–1350*, Cambridge, 1988.

Hanawalt, B. A., *Crime and conflict in English communities 1300–1348*, Cambridge, Mass., 1979.

Hare, J. N., 'Lord and tenant in Wiltshire, *c.* 1380–*c.* 1520, with particular reference to regional and seigneurial variations', PhD thesis, University of London, 1976.

'Change and continuity in Wiltshire agriculture: the later Middle Ages', in W. Minchinton (ed.), *Agricultural improvement: medieval and modern*, Exeter Papers in Economic History 14, Exeter, 1981, pp. 1–18.

'The demesne lessees of fifteenth-century Wiltshire', *AHR* 29 (1981), 1–15.

Harley, J. B., 'Population and agriculture from the Warwickshire Hundred Rolls of 1279', *EcHR* 11 (1958), 8–18.

'The Hundred Rolls of 1279', *Amateur Historian* 5 (1961), 9–16.

Harris, A., 'Changes in the early railway age: 1800–1850', in Darby (ed.), *New historical geography*, pp. 465–526.

Harrison, B., 'Field systems and demesne farming on the Wiltshire estates of Saint Swithun's Priory, Winchester, 1248–1340', *AHR* 43 (1995), 1–18.

'Demesne husbandry and field systems on the north Hampshire estates of Saint Swithun's Priory, Winchester, 1248–1340', unpublished paper.

Harrison, D., 'Bridges and economic development, 1300–1800', *EcHR* 45 (1992), 240–61.

Harrod, H., 'Some details of a murrain of the fourteenth century; from the court rolls of a Norfolk manor', *Archaeologia* 41 (1866), 1–14.

Harvey, B. F., 'The leasing of the abbot of Westminster's demesnes in the later Middle Ages', *EcHR* 22 (1969), 17–27.

Westminster Abbey and its estates in the Middle Ages, Oxford, 1977.

'Introduction: the "crisis" of the early fourteenth century', in Campbell (ed.), *Before the Black Death*, pp. 1–24.

'The aristocratic consumer in England in the long thirteenth century', in M. Prestwich, R. H. Britnell and R. Frame (eds.), *Thirteenth century England*, vol. VI, *Proceedings of the sixth conference held at St. Aidan's College, Durham, September 1995*, Woodbridge, 1997, pp. 19–37.

Harvey, N., 'The inning and winning of the Romney marshes', *Agriculture* 62 (1955), 334–8.

Harvey, P. D. A., *A medieval Oxfordshire village: Cuxham 1240–1400*, London, 1965.

'Agricultural treatises and manorial accounting in medieval England', *AHR* 20 (1972), 170–82.

'The Pipe Rolls and the adoption of demesne farming in England', *EcHR* 27 (1974), 345–59.

Manorial records, British Records Association, Archives and the User 5, London, 1984.

'Farming practice and techniques: the Home Counties', in Miller (ed.), *AHEW*, vol. III, pp. 254–68.

Harvey, S. P. J., 'The extent and profitability of demesne agriculture in England in the later eleventh century', in Aston and others (eds.), *Social relations*, pp. 45–72.

'Domesday England', in Hallam (ed.), *AHEW*, vol. II, pp. 45–138.

Harwood Long, W., 'The low yields of corn in medieval England', *EcHR* 32 (1979), 459–69.

Hatcher, J., *Rural economy and society in the Duchy of Cornwall, 1300–1500*, Cambridge, 1970.

Plague, population and the English economy 1348–1530, Studies in Economic and Social History, London, 1977.

'English serfdom and villeinage: towards a reassessment', *PP* 90 (1981), 3–39.

'Farming techniques: south-western England', in Hallam (ed.), *AHEW*, vol. II, pp. 383–99.

Hawkes, N., 'Secrets of medieval life entwined in country thatches', *The Times*, 12 August 1995, p. 3.

Hewitt, H. J., *Mediaeval Cheshire: an economic and social history of Cheshire in the reigns of the three Edwards*, Chetham Society, Manchester, 1929.

Hilton, R. H., *The economic development of some Leicestershire estates in the fourteenth and fifteenth centuries*, Oxford, 1947.

'Peasant movements in England before 1381', *EcHR* 2 (1949), 117–36.

'Winchcombe Abbey and the manor of Sherborne', *University of Birmingham Historical Journal* 2 (1949–50), 50–52.

'Medieval agrarian history', in W. G. Hoskins (ed.), *The Victoria history of the counties of England: a history of Leicestershire*, London, 1954, vol. II, pp. 145–98.

Bond men made free: medieval peasant movements and the English rising of 1381, London, 1973.

The English peasantry in the later Middle Ages: the Ford Lectures for 1973 and related studies, Oxford, 1975.

'Rent and capital formation in feudal society', in *English peasantry*, pp. 174–214.

The decline of serfdom in medieval England, 2nd edition, London and Basingstoke, 1983.

Hodgson, R. I., 'Medieval colonization in northern Ryedale, Yorkshire', *Geographical Journal* 135 (1969), 44–54.

Hogan, M. P., 'Clays, *culturae* and the cultivator's wisdom: management efficiency at fourteenth-century Wistow', *AHR* 36 (1988), 117–31.

Holt, R., *The mills of medieval England*, Oxford, 1988.

Hopcroft, R. L., 'The social origins of agrarian change in late medieval England', *American Journal of Sociology* 99 (1994), 1,559–95.

Hoskins, W. G., 'Harvest fluctuations and English economic history, 1480–1619', *AHR* 12 (1964), 28–46.

Howell, C., *Land, family and inheritance in transition: Kibworth Harcourt 1280–1700*, Cambridge, 1983.

Hunnisett, R. F., 'The reliability of inquisitions as historical evidence', in D. A. Bullough and R. L. Storey (eds.), *The study of medieval records: essays in honour of Kathleen Major*, Oxford, 1971, pp. 206–35.

Hurst, J. G., 'Rural building in England and Wales: England', in Hallam (ed.), *AHEW*, vol. II, pp. 854–932.

Hybel, N., *Crisis or change. The concept of crisis in the light of agrarian structural reorganization in late medieval England*, trans. J. Manley, Aarhus, 1989.

Jack, R. I., 'Farming techniques: Wales and the Marches', in Hallam (ed.), *AHEW*, vol. II, pp. 412–96.

John, A. H., 'Agricultural productivity and economic growth in England, 1700–1760', *JEH* 25 (1965), 19–34.

Jones, E. L., 'Agriculture and economic growth in England, 1660–1750: agricultural change', *JEH* 25 (1965), 1–18; reprinted in *Agriculture and the Industrial Revolution*, pp. 1–18.

'The condition of English agriculture, 1500–1640', *EcHR* 21 (1968), 614–19.

Agriculture and the Industrial Revolution, Oxford, 1974.

Jordan, W. C., *The Great Famine: northern Europe in the early fourteenth century*, Princeton, 1996.

Keene, D. J., 'Rubbish in medieval towns', in A. R. Hall and H. K. Kenward (eds.), *Environmental archaeology in the urban context*, Council for British Archaeology, Research Report 43, London, 1982, pp. 26–30.

'A new study of London before the Great Fire', *Urban History Yearbook 1984*, pp. 11–21.

Cheapside before the Great Fire, London, 1985.

'Medieval London and its region', *London Journal* 14 (1989), 99–111.

Keene, D. J., Campbell, B. M. S., Galloway, J. A., and Murphy, M., 'Feeding the city 2: London and its hinterland *c.* 1300–1400', Economic and Social Research Council, unpublished End of Award Report, 1994.

Keil, I., 'The estates of Glastonbury Abbey in the later Middle Ages: a study in administration and economic change', PhD thesis, University of Bristol, 1964.

'Farming on the Dorset estates of Glastonbury Abbey in the early fourteenth century', *Proceedings of the Dorset Natural History and Archaeology Society* 87 (1966), 234–50.

Kerridge, E., *The farmers of old England*, London, 1973.

Kershaw, I., *Bolton Priory: the economy of a northern monastery 1286–1325*, Oxford, 1973.

'The Great Famine and agrarian crisis in England 1315–22', *PP* 59 (1973), 3–50; reprinted in R. H. Hilton (ed.), *Peasants, knights and heretics*, Cambridge, 1976, pp. 85–132.

King, E., 'Farming practice and techniques: the east midlands', in Miller (ed.), *AHEW*, vol. III, pp. 210–22.

'The occupation of the land: the east midlands', in Miller (ed.), *AHEW*, vol. III, pp. 67–77.

Knowles, D. and Hadcock, R. N., *Medieval religious houses: England and Wales*, London, 1953.

Kosminsky, E. A., 'The Hundred Rolls of 1279–80 as a source for English agrarian history', *Economic History Review* 3 (1931–2), 16–44.

Studies in the agrarian history of England in the thirteenth century, trans. R. Kisch, ed. R. H. Hilton, Oxford, 1956.

Kowaleski, M., 'Town and country in late medieval England: the hide and leather trade', in P. J. Corfield and D. J. Keene (eds.), *Work in towns 850–1850*, Leicester, 1990, pp. 57–73.

Local markets and regional trade in medieval Exeter, Cambridge, 1995.

'The grain trade in fourteenth-century Devon', in DeWindt (ed.), *The salt of common life*, pp. 1–52.

Kussmaul, A., *Servants in husbandry in early modern England* (Cambridge, 1981).

'Agrarian change in seventeenth-century England: the economic historian as pale-ontologist', *JEH* 45 (1985), 1–30.

Lambert, J. M., Jennings, J. N., Smith, C. T., Green, C., and Hutchinson, J. N., *The making of the Broads: a reconsideration of their origin in the light of new evidence*, Royal Geographical Society Research Series 3, London, 1960.

Langdon, J. L., 'The economics of horses and oxen in medieval England', *AHR* 30 (1982), 31–40.

'Horses, oxen and technological innovation: the use of draught animals in English farming from 1066 to 1500', PhD thesis, University of Birmingham, 1983.

'Horse hauling: a revolution in vehicle transport in twelfth- and thirteenth-century England?', *PP* 103 (1984), 37–66.

Horses, oxen and technological innovation: the use of draught animals in English farming from 1066–1500, Cambridge, 1986.

'Agricultural equipment', in Astill and Grant (eds.), *Countryside*, pp. 86–107.

'Inland water transport in medieval England', *JHG* 19 (1993), 1–11.

'Was England a technological backwater in the Middle Ages?', in Astill and Langdon (eds.), *Medieval farming*, pp. 275–92.

Langdon, J. L., Astill, G., and Myrdal, J., 'Introduction', in Astill and Langdon (eds.), *Medieval farming*, pp. 1–10.

Langton, J. and Hoppe, G., *Town and country in the development of early modern Western Europe*, Historical Geography Research Series 11, Norwich, 1983.

Le Patourel, H. E. J., 'Rural building in England and Wales', in Miller (ed.), *AHEW*, vol. III, pp. 820–93.

Le Roy Ladurie, E., 'The end of the Middle Ages: the work of Guy Bois and Hugues Neveux', in E. Le Roy Ladurie and J. Goy, *Tithe and agrarian history from the fourteenth to the nineteenth centuries: an essay in comparative history*, Cambridge, 1982, pp. 71–92.

Lennard, R. V., 'The alleged exhaustion of the soil in medieval England', *Economic Journal* 32 (1922), 12–27.

'Statistics of corn yields in medieval England: some critical questions', *Economic History* 3, 11 (1936), 173–92.

Rural England 1086–1135, Oxford, 1959.

Letts, J. B., *Smoke blackened thatch: a unique source of late medieval plant remains from southern England*, Reading, 1999.

Link, A. N., *Technological change and productivity growth*, London, 1987.

Lipsey, R. G., *An introduction to positive economics*, London, 1972.

Livi-Bacci, M., *Population and nutrition: an essay on European demographic history*, trans. T. Croft-Murray with C. Ipsen, Cambridge, 1991.

Livingstone, M. R., 'Sir John Pulteney's landed estates: the acquisition and management of land by a London merchant', MA thesis, University of London, 1991.

Livingstone, M. R. and Bartley, K. C., 'Historical problems, GIS solutions? Spatio-temporal patterns in medieval data', in A. Barnual (ed.), *Mapping historical data*, Moscow, forthcoming.

Lloyd, T. H., *The movement of wool prices in medieval England*, Economic History Review Supplement 6, Cambridge, 1973.

The English wool trade in the Middle Ages, Cambridge, 1977.

Lomas, R. A., 'The priory of Durham and its demesnes in the fourteenth and fifteenth centuries', *EcHR* 31 (1978), 339–53.

Loomis, R. S., 'Ecological dimensions of medieval agrarian systems: an ecologist responds', *Agricultural History* 52 (1978), 478–83.

Lyons, M., 'The manor of Ballysax 1280–1288', *Retrospect*, new series, 1 (1981), 40–50.

Lythe, S. G. E., 'The organization of drainage and embankment in medieval Holderness', *Yorkshire Archaeological Journal* 34 (1939), 282–95.

McCloskey, D. N., 'The open fields of England: rent, risk, and the rate of interest, 1300–1815', in D. W. Galenson (ed.), *Markets in history: economic studies of the past*, Cambridge, 1989, pp. 5–51.

McCloskey, D. N. and Nash, J., 'Corn at interest: the extent and cost of grain storage in medieval England', *American Economic Review* 74 (1984), 174–87.

MacCulloch, D., 'Kett's Rebellion in context', *PP* 84 (1979), 36–59.

McDonnell, J., 'Medieval assarting hamlets in Bilsdale, north-east Yorkshire', *Northern History* 22 (1986), 269–79.

McIntosh, M. K., *Autonomy and community: the royal manor of Havering, 1200–1500*, Cambridge, 1986.

McNamee, C., *The wars of the Bruces: Scotland, England and Ireland, 1306–1328*, East Linton, 1997.

Maddicott, J. R., *The English peasantry and the demands of the crown 1294–1341*, Past and Present Supplement 1, 1975; reprinted in T. H. Aston (ed.), *Landlords, peasants and politics in medieval England*, Cambridge, 1987, pp. 285–359.

Maitland, F. W., *Domesday Book and beyond: three essays in the early history of England*, revised edition, Cambridge, 1987.

Masschaele, J., *Peasants, merchants, and markets: inland trade in medieval England, 1150–1350*, New York, 1997.

Mate, M., 'Profit and productivity on the estates of Isabella de Forz (1260–92)', *EcHR* 33 (1980), 326–34.

'The farming out of manors: a new look at the evidence from Canterbury Cathedral Priory', *Journal of Medieval History* 9 (1983), 331–44.

'Medieval agrarian practices: the determining factors?', *AHR* 33 (1985), 22–31.

'Pastoral farming in south-east England in the fifteenth century', *EcHR* 40 (1987), 523–36.

'Farming practice and techniques: Kent and Sussex', in Miller (ed.), *AHEW*, vol. III, pp. 268–85.

'The occupation of the land: Kent and Sussex', in Miller (ed.), *AHEW*, vol. III, pp. 119–36.

'Agricultural technology in south-east England, 1348–1530', in Astill and Langdon (eds.), *Medieval farming*, pp. 251–74.

May, A. N., 'An index of thirteenth-century peasant impoverishment? Manor court fines', *EcHR* 26 (1973), 389–402.

Mayhew, N. J., 'Numismatic evidence and falling prices in the fourteenth century', *EcHR* 27 (1974), 1–15.

'Modelling medieval monetisation', in Britnell and Campbell (eds.), *Commercialising economy*, pp. 55–77.

'Appendix 2: The calculation of GDP from Domesday Book', in Britnell and Campbell (eds.), *Commercialising economy*, pp. 195–6.

Merrington, J., 'Town and country in the transition to capitalism', *New Left Review* 93 (1975), 452–506; reprinted in R. H. Hilton (ed.), *The transition from feudalism to capitalism*, London, 1976, pp. 170–95.

Miller, E., *The abbey and bishopric of Ely: the social history of an ecclesiastical estate from the tenth to the early fourteenth century*, Cambridge, 1951.

War in the north: the Anglo-Scottish wars of the Middle Ages, Hull, 1960.

'Farming in northern England', *Northern History* 11 (1975), 1–16.

'War, taxation and the English economy in the late thirteenth and early fourteenth centuries', in J. M. Winter (ed.), *War and economic development: essays in memory of David Joslin*, Cambridge, 1975, pp. 11–31.

'Farming techniques: northern England', in Hallam (ed.), *AHEW*, vol. II, pp. 399–411.

'Farming practice and techniques: Yorkshire and Lancashire', in Miller (ed.), *AHEW*, vol. III, pp. 182–94.

'The occupation of the land: Yorkshire and Lancashire', in Miller (ed.), *AHEW*, vol. III, pp. 42–52.

Miller, E. (ed.), *The agrarian history of England and Wales*, vol. III, *1348–1500*, Cambridge, 1991.

Miller, E. and Hatcher, J., *Medieval England: rural society and economic change 1086–1348*, London, 1978.

Medieval England: towns, commerce and crafts 1086–1348, London, 1995.

Mokyr, J., *The lever of riches: technological creativity and economic progress*, Oxford, 1990.

Moore, J. S., *Laughton: a study in the evolution of the Wealden landscape*, Department of English Local History, University of Leicester, Occasional Paper 19, Leicester, 1965.

Morimoto, N., 'Arable farming of Durham Cathedral Priory in the fourteenth century', *Nagoya Gakuin University Review* 11 (1975), 137–331.

Munro, J. H., 'Wool-price schedules and the qualities of English wool in the later Middle Ages, circa 1270–1499', *Textile History* 9 (1978), 118–69.

'Environment, land management, and the changing qualities of English wools in the later Middle Ages', paper presented at the session on 'Pastoral resources and the medieval English economy', 20th International Congress on Medieval Studies, Kalamazoo, May 1985.

'Industrial transformations in the north-west European textile trades, *c.* 1290–*c.* 1340: economic progress or economic crisis?', in Campbell (ed.), *Before the Black Death*, pp. 110–48.

'Structural changes in late medieval textile manufacturing: the Flemish response to market adversities, 1300–1500', unpublished public lecture delivered at the Katholieke Universiteit Leuven, 5 November 1986.

Murphy, M. and Galloway, J. A., 'Marketing animals and animal products in London's hinterland *circa* 1300', *Anthropozoologica* 16 (1992), 93–100.

Myrdal, J., *Medieval arable farming in Sweden. Technical change A.D. 1000–1520*, Nordiska museets Handlingar 105, Stockholm, 1986.

National Economic Development Office, *Farm productivity: a report by the agriculture EDC on factors affecting productivity at the farm level*, London, 1973.

Newman, E. I. and Harvey, P. D. A., 'Did soil fertility decline in medieval English farms? Evidence from Cuxham, Oxfordshire, 1320–1340', *AHR* 45 (1997), 119–36.

Nichols, J., *The history and antiquities of the county of Leicester*, 4 vols., London, 1795; reprinted Wakefield, 1971.

Nightingale, P., 'The growth of London in the medieval English economy', in R. H. Britnell and J. Hatcher (eds.), *Progress and problems in medieval England: essays in honour of Edward Miller*, Cambridge, 1996, pp. 89–106.

O'Brien, P. K. and Toniolo, G., 'The poverty of Italy and the backwardness of its agriculture before 1914', in Campbell and Overton (eds.), *Land, labour and livestock*, pp. 385–409.

Ormrod, W. M., 'The English crown and the customs, 1349–63', *EcHR* 40 (1987), 27–40.

'The crown and the English economy, 1290–1348', in Campbell (ed.), *Before the Black Death*, pp. 149–83.

Oschinsky, D., 'Medieval treatises on estate accounting', *Economic History Review* 17 (1947), 52–61.

Outhwaite, R. B., 'Progress and backwardness in English agriculture, 1500–1650', *EcHR* 39 (1986), 1–18.

Overton, M., 'Agricultural change in Norfolk and Suffolk, 1580–1740', PhD thesis, University of Cambridge, 1981.

'Agricultural regions in early modern England: an example from East Anglia', University of Newcastle upon Tyne, Department of Geography, Seminar Paper 42, 1985.

'The diffusion of agricultural innovations in early modern England: turnips and clover in Norfolk and Suffolk 1580–1740', *TIBG*, new series, 10 (1985), 205–21.

'Agriculture', in J. Langton and R. J. Morris (eds.), *Atlas of industrialising Britain 1780–1914*, London, 1986, pp. 34–53.

'Depression or revolution? English agriculture, 1640–1750', *Journal of British Studies* 25 (1986), 344–52.

Agricultural revolution in England: the transformation of the agrarian economy 1500–1850, Cambridge, 1996.

Overton, M. and Campbell, B. M. S., 'Productivity change in European agricultural development', in Campbell and Overton (eds.), *Land, labour and livestock*, pp. 1–50.

'Norfolk livestock farming 1250–1740: a comparative study of manorial accounts and probate inventories', *JHG* 18 (1992), 377–96.

'Production et productivité dans l'agriculture anglaise, 1086–1871', *Histoire et Mesure*, 11, 3/4 (1996), 255–97.

Page, F. M., '"Bidentes Hoylandie": a medieval sheep farm', *Economic History* 1 (1929), 603–13.

Paul, A. A. and Southgate, D. A. T., *McCance and Widdowson's 'The composition of foods'*, London, 1978.

Pelham, R. A., 'The distribution of sheep in Sussex in the early fourteenth century', *Sussex Archaeological Collections* 125 (1934), 128–35.

'Fourteenth-century England', in Darby (ed.), *Historical geography*, pp. 230–65.

'Medieval foreign trade: eastern ports', in Darby (ed.), *Historical geography*, pp. 298–329.

Perry, P. J., 'Where was the "Great Agricultural Depression"? A geography of agricultural bankruptcy in late Victorian England and Wales', *AHR* 20 (1972), 30–45.

Persson, K. G., *Pre-industrial economic growth, social organization and technological progress in Europe*, Oxford, 1988.

'Labour productivity in medieval agriculture: Tuscany and the "Low Countries"', in Campbell and Overton (eds.), *Land, labour and livestock*, pp. 124–43.

'Was there a productivity gap between fourteenth-century Italy and England?', *EcHR* 46 (1993), 105–14.

Total factor productivity growth in English Agriculture, 1250–1450, Discussion Paper 93–11, Institute of Economics, University of Copenhagen, Copenhagen, 1994.

'The seven lean years, elasticity traps, and intervention in grain markets in pre-industrial Europe', *EcHR* 49 (1996), 698–702.

Phelps Brown, E. H. and Hopkins, S. V., 'Seven centuries of the prices of consumables compared with builders' wage-rates', *Economica* 23 (1956), 296–314; reprinted in Carus-Wilson (ed.), *Essays*, vol. II, pp. 179–96; and in E. H. Phelps Brown and S. V. Hopkins, *A perspective of wages and prices*, London, 1981, pp. 13–59.

A perspective of wages and prices, London, 1981.

Platt, C., *The monastic grange in medieval England: a reassessment*, London, 1969.

Platts, G., *Land and people in medieval Lincolnshire*, History of Lincolnshire 4, Lincoln, 1985.

Poos, L. R., 'The rural population of Essex in the later Middle Ages', *EcHR* 38 (1985), 515–30.

A rural society after the Black Death: Essex 1350–1525, Cambridge, 1991.

Postan, M. M., 'Revisions in economic history: IX. The fifteenth century', *Economic History Review* 9 (1939), 160–7.

The famulus: the estate labourer in the XIIth and XIIIth centuries, Economic History Review Supplement 2, Cambridge, 1954.

'Village livestock in the thirteenth century', *EcHR* 15 (1962), 219–49; reprinted in Postan, *Essays*, pp. 214–48.

'Medieval agrarian society in its prime: England', in Postan (ed.), *The Cambridge economic history of Europe*, vol. I, *The agrarian life of the Middle Ages*, 2nd edition, Cambridge, 1966, pp. 549–632.

The medieval economy and society: an economic history of Britain in the Middle Ages, London, 1972.

Essays on medieval agriculture and general problems of the medieval economy, Cambridge, 1973.

Postan, M. M. and Hatcher, J., 'Population and class relations in feudal society', in Aston and Philpin (eds.), *Brenner debate*, pp. 64–78.

Postan, M. M. and Titow, J. Z., with statistical Notes by J. Longden, 'Heriots and prices on Winchester manors', *EcHR* 11 (1958), 392–417; reprinted in Postan, *Essays*, pp. 150–78.

Postgate, M. R., 'Field systems of East Anglia', in Baker and Butlin (eds.), *Field systems*, pp. 281–324.

Postles, D., 'Grain issues from some properties of Oseney Abbey, 1274–1348', *Oxoniensia* 44 (1979), 100–16.

'Customary carrying services', *Journal of Transport History*, 3rd series, 5 (1984), 1–15.

'The acquisition and administration of spiritualities by Oseney Abbey', *Oxoniensia* 51 (1986), 69–77.

'The perception of profit before the leasing of demesnes', *AHR* 34 (1986), 12–28.

'Cleaning the medieval arable', *AHR* 37 (1989), 130–43.

Power, E., *The wool trade in English medieval history*, Oxford, 1941.

Power, J. P. and Campbell, B. M. S., 'Cluster analysis and the classification of medieval demesne-farming systems', *TIBG*, new series, 17 (1992), 227–45.

Poynder, N., 'Grain provision and conventual economics in medieval England', unpublished paper presented at the Annual Conference of the Economic History Society, Leeds, 1998.

Prestwich, M., *War, politics and finance under Edward I*, London, 1972.

Pretty, J. N., 'Sustainable agriculture in the Middle Ages: the English manor', *AHR* 38 (1990), 1–19.

Rackham, O., *Ancient woodland: its history, vegetation and uses in England*, London, 1980.

Raepsaet, G., 'The development of farming implements between the Seine and the Rhine from the second to the twelfth centuries', in Astill and Langdon (eds.), *Medieval farming*, pp. 41–68.

Raftis, J. A., *The estates of Ramsey Abbey: a study of economic growth and organization*, Toronto, 1957.

Tenure and mobility: studies in the social history of the medieval English village, Toronto, 1964.

'Changes in an English village after the Black Death', *Mediaeval Studies* 29 (1967), 158–77.

Assart data and land values: two studies in the east midlands 1200–1350, Toronto, 1974.

Peasant economic development within the English manorial system, Stroud, 1997.

Ravensdale, J. R., *Liable to floods: village landscape on the edge of the Fens, 450–1850*, Cambridge, 1974.

Razi, Z., *Life, marriage and death in a medieval parish: economy, society and demography in Halesowen, 1270–1400*, Cambridge, 1980.

Rees, W., *South Wales and the March 1284–1415: a social and agrarian study*, Oxford, 1924.

Ricardo, D., *The principles of political economy and taxation*, ed. O. St Clair, London, 1957.

Richard, J. M., 'Thierry d'Hireçon, agriculteur artésien', *Bibliothèque de l'école des chartes* 53 (1892), 383–416, 571–604.

Ritson, C., *Agricultural economics: principles and policy*, London, 1980.

Roberts, B. K., 'A study of medieval colonization in the Forest of Arden, Warwickshire', *AHR* 16 (1968), 101–13.

Roden, D., 'Demesne farming in the Chiltern Hills', *AHR* 17 (1969), 9–23.

Russell, E., 'The societies of the Bardi and Peruzzi and their dealings with Edward III, 1327–1345', in G. Unwin (ed.), *Finance and trade under Edward III*, Manchester, 1918, pp. 93–135.

Russell, J. C., *British medieval population*, Albuquerque, 1948.

Ryder, M. L., 'Medieval sheep and wool types', *AHR* 32 (1984), 14–28.

Sabine, E. L., 'City cleaning in medieval London', *Speculum* 12 (1937), 19–43.

Salzman, L. F., 'The inning of Pevensey Levels', *Sussex Archaeological Collections* 52 (1910), 32–60.

    'Social and economic history: medieval Cambridgeshire', in Salzman (ed.), *The Victoria history of the counties of England: Cambridgeshire and the Isle of Ely*, London, 1948, vol. II, pp. 58–72.

    Building in England down to 1540: a documentary history, Oxford, 1952.

Saul, A., 'Great Yarmouth in the fourteenth century: a study in trade, politics, and society', DPhil thesis, University of Oxford, 1975.

Saunders, H. W., *An introduction to the obedientiary and manor rolls of Norwich Cathedral Priory*, Norwich, 1930.

Schofield, P. R., 'Dearth, debt and the local land market in a late thirteenth-century village community', *AHR* 45 (1997), 1–17.

Schofield, R. S., 'The geographical distribution of wealth in England, 1334–1649', *EcHR* 18 (1965), 483–510; reprinted in R. Floud (ed.), *Essays in quantitative economic history*, Oxford, 1974, pp. 79–106.

Scott, R., 'Medieval agriculture', in E. Crittal (ed.), *The Victoria history of the counties of England: a history of Wiltshire*, London, 1959, vol. IV, pp. 7–42.

Scott, S., Duncan, S. R., and Duncan, C. J., 'The origins, interactions and causes of the cycles in grain prices in England, 1450–1812', *AHR* 46 (1998), 1–14.

Searle, E., *Lordship and community: Battle Abbey and its banlieu, 1066–1538*, Toronto, 1974.

Seebohm, F., *The English village community: examined in its relations to the manorial and tribal systems and to the common or open field system of husbandry*, London, 1883.

Shaw, R. C., *The Royal Forest of Lancaster*, Preston, 1956.

Sheppard, J. A., 'Pre-enclosure field and settlement patterns in an English township: Wheldrake, near York', *Geografiska Annaler* 48B (1966), 59–77.

Shiel, R. S., 'Improving soil fertility in the pre-fertiliser era', in Campbell and Overton (eds.), *Land, labour and livestock*, pp. 51–77.

Shrewsbury, J. F. D., *A history of bubonic plague in the British Isles*, Cambridge, 1970.

Simkhovitch, V. G., 'Hay and history', *Political Science Quarterly* 28 (1913), 385–404.

Simmons, I. G., *The ecology of natural resources*, London, 1974; 2nd edition, 1981.

Simpson, A., *The wealth of the gentry, 1540–1660: East Anglian studies*, Cambridge, 1963.

Skeel, C., 'The cattle trade between Wales and England from the fifteenth to the nineteenth century', *TRHS*, 4th series, 9 (1926), 135–58.

Slicher Van Bath, B. H., 'The rise of intensive husbandry in the Low Countries', in J. S. Bromley and E. H. Kossmann (eds.), *Britain and the Netherlands: papers delivered to the Oxford-Netherlands Historical Conference 1959*, London, 1960, pp. 130–53.

The agrarian history of western Europe A.D. 500–1850, trans. O. Ordish, London, 1963.

'The yields of different crops, mainly cereals in relation to the seed c. 810–1820', *Acta Historiae Neerlandica* 2 (Leiden, 1967), 78–97.

Smith, A., *An inquiry into the nature and causes of the wealth of nations*, new edition, Edinburgh, 1872; modern edition ed. R. Campbell and A. Skinner, 2 vols., Oxford, 1976.

Smith, R. A. L., 'The estates of Pershore Abbey', MA thesis, University of London, 1939.

'Marsh embankment and sea defence in medieval Kent', *Economic History Review* 10 (1940), 29–37.

Canterbury Cathedral Priory: a study in monastic administration, Cambridge, 1943.

Smith, R. M., 'Human resources in rural England', in Astill and Grant (eds.), *Countryside*, pp. 188–212.

'Demographic developments in rural England, 1300–48: a survey', in Campbell (ed.), *Before the Black Death*, pp. 25–78.

Snooks, G. D., *Economics without time*, London, 1992.

'The dynamic role of the market in the Anglo-Norman economy and beyond, 1086–1300', in Britnell and Campbell (eds.), *Commercialising economy*, pp. 27–54.

Stacey, R. C., 'Agricultural investment and the management of the royal demesne manors, 1236–1240', *JEH* 46 (1986), 919–34.

Stephenson, M. J., 'The productivity of medieval sheep on the great estates, 1100–1500', PhD thesis, University of Cambridge, 1987.

'Wool yields in the medieval economy', *EcHR* 41 (1988), 368–91.

Stern, D. V., 'A Hertfordshire manor of Westminster Abbey: an examination of demesne profits, corn yields, and weather evidence', PhD thesis, University of London, 1978; to be published by Hertfordshire University Press, ed. C. Thornton.

Stevenson, E. R., 'The escheator', in W. A. Morris and J. R. Strayer (eds.), *The English government at work 1327–36*, 2 vols., Cambridge, Mass., 1947, vol. II, pp. 109–67.

Stitt, F. B., 'The medieval minister's account', *Society of Local Archivists Bulletin* 11 (1953), 2–8.

Stone, D., 'The productivity of hired and customary labour: evidence from Wisbech Barton in the fourteenth century', *EcHR* 50 (1997), 640–56.

Stone, E., 'The estates of Norwich Cathedral Priory 1100–1300', DPhil thesis, University of Oxford, 1956.

'Profit and loss accountancy at Norwich Cathedral Priory', *TRHS*, 5th series, 12 (1962), 25–48.

Sutton, A. F., 'The early linen and worsted industry of Norfolk and the evolution of the London Mercers' Company', *NA* 40 (1989), 201–25.

Sweeney, D. (ed.), *Agriculture in the Middle Ages: technology, practice and representation*, Philadelphia, 1995.

Tannahill, R., *Food in history*, 2nd edition, London, 1988.

Thirsk, J., *Fenland farming in the sixteenth century*, Department of English Local History, University of Leicester, Occasional Paper 3, Leicester, 1965.

'The farming regions of England', in Thirsk (ed.), *The agrarian history of England and Wales*, vol. IV, *1500–1640*, Cambridge, 1967, pp. 1–112.

'Field systems of the east midlands', in Baker and Butlin (eds.), *Field systems*, pp. 232–80.

Thoen, E., *Landbouwekonomie en bevolking in Vlaanderen gedurende de late Middeleeuwen en het begin van de Moderne Tijden. Testregio: de kasselrijen van Oudenaarde en Aalst*, Belgisch Centrum voor Landelijke Geschiedenis 90, 2 vols., Ghent, 1988.

'The birth of "the Flemish husbandry": agricultural technology in medieval Flanders', in Astill and Langdon (eds.), *Medieval farming*, pp. 69–88.

Thomas, C., 'Thirteenth-century farm economies in North Wales', *AHR* 16 (1968), 1–14.

Thornton, C., 'The demesne of Rimpton, 938 to 1412: a study in economic development', PhD thesis, University of Leicester, 1989.

'The determinants of land productivity on the bishop of Winchester's demesne of Rimpton, 1208 to 1403', in Campbell and Overton (eds.), *Land, labour and livestock*, pp. 183–210.

'Efficiency in thirteenth-century livestock farming: the fertility and mortality of herds and flocks at Rimpton, Somerset, 1208–1349', in P. R. Coss and S. D. Lloyd (eds.), *Thirteenth century England*, vol. IV, *Proceedings of the Newcastle upon Tyne conference 1991*, Woodbridge, 1992, pp. 25–46.

'The level of land productivity at Taunton, Somerset, 1283–1348', unpublished paper.

Thünen, J. H. von, *Der isolierte Staat*, Hamburg, 1826; trans. C. M. Wartenberg, *Von Thünen's isolated state*, ed. P. Hall, Oxford, 1966.

Tingey, J. C., 'The journals of John Dernell and John Boys, carters at the lathes in Norwich', *NA* 15 (1904), 114–63.

Titow, J. Z., 'Evidence of weather in the account rolls of the bishopric of Winchester 1209–1350', *EcHR* 12 (1959), 360–407.

'Land and population on the bishop of Winchester's estates 1209–1350', PhD thesis, University of Cambridge, 1962.

English rural society 1200–1350, London, 1969.

Winchester yields: a study in medieval agricultural productivity, Cambridge, 1972.

Trow-Smith, R., *A history of British livestock husbandry to 1700*, London, 1957.

Tupling, G. H., *The economic history of Rossendale*, Chetham Society 86, Manchester, 1927.

Turner, M. E., 'Arable in England and Wales: estimates from the 1801 crop return', *JHG* 7 (1981), 291–302.

'Livestock in the agrarian economy of Counties Down and Antrim from 1803 to the Famine', *Irish Social and Economic History* 11 (1984), 19–43.

Twigg, G., *The Black Death: a biological reappraisal*, London, 1984.

Ugawa, K., 'The economic development of some Devon manors in the thirteenth century', *Reports and Transactions of the Devonshire Association* 94 (1962), 630–83.

Lay estates in medieval England, Tokyo, 1966.

Van Cauwenberghe, E. and Van der Wee, H., 'Productivity, evolution of rents and farm size in the southern Netherlands agriculture from the fourteenth to the seventeenth century', in Van der Wee and Van Cauwenberghe (eds.), *Productivity and agricultural innovation*, pp. 125–61.

Van der Wee, H., 'Introduction – the agricultural development of the Low Countries as revealed by the tithe and rent statistics, 1250–1800', in Van der Wee and Van Cauwenberghe (eds.), *Productivity and agricultural innovation*, pp. 1–23.

Van der Wee, H. and Van Cauwenberghe, E. (eds.), *Productivity of land and agricultural innovation in the Low Countries (1250–1800)*, Leuven, 1978.

Veale, E. M., 'The rabbit in England', *AHR* 5 (1957), 85–90.

Verhulst, A., 'L'Intensification et la commercialisation de l'agriculture dans les Pays-Bas méridionaux au XIIIe siècle', in Verhulst (ed.), *La Belgique rurale. Mélanges offerts à J. J. Hoebanx*, Brussels, 1985, pp. 89–100.

The Victoria history of the counties of England, London, 1900 – in progress.

Virgoe, R., 'The estates of Norwich Cathedral Priory, 1101–1538', in I. Atherton, E. Fernie, C. Harper-Bill, and H. Smith (eds.), *Norwich Cathedral: church, city and diocese, 1096–1996*, London, 1996, pp. 339–59.

Vollans, E. C., 'The evolution of farm-lands in the central Chilterns in the twelfth and thirteenth centuries', *TIBG* 26 (1959), 197–235.

Vries, J. de, *European urbanization 1500–1800*, London, 1984.

Wade-Martins, P. (ed.), *An historical atlas of Norfolk*, Norwich, 1993.

Waites, B., *Monasteries and landscape in north east England: the medieval colonisation of the North York Moors*, Oakham, 1997.

Walton, J. R., 'Agriculture and rural society 1730–1914', in R. A. Dodgshon and R. A. Butlin (eds.), *An historical geography of England and Wales*, 1st edition, London, 1978, pp. 239–65.

Watts, D. G., 'A model for the early fourteenth century', *EcHR* 20 (1967), 543–7.

Williams, M., *The draining of the Somerset Levels*, Cambridge, 1970.

Winchester, A. J. L., *Landscape and society in medieval Cumbria*, Edinburgh, 1987.

Witney, K., 'The woodland economy of Kent, 1066–1348', *AHR* 38 (1990), 20–39.

Wordie, J. R., 'The chronology of English enclosure, 1500–1914', *EcHR* 36 (1983), 483–505.

Wrigley, E. A., 'The supply of raw materials in the industrial revolution', *EcHR* 15 (1962), 1–16.

'A simple model of London's importance in changing English society and economy, 1650–1750', *PP* 37 (1967), 44–70; reprinted in *People, cities and wealth*, pp. 133–56.

'Urban growth and agricultural change: England and the Continent in the early modern period', *Journal of Interdisciplinary History* 15 (1985), 683–728; reprinted in *People, cities and wealth*, pp. 157–96.

People, cities and wealth: the transformation of traditional society, Oxford, 1987.

'Some reflections on corn yields and prices in pre-industrial economies', in *People, cities and wealth*, pp. 92–130.

'The classical economists and the Industrial Revolution', in *People, cities and wealth*, pp. 21–45.

Continuity, chance and change: the character of the industrial revolution in England, Cambridge, 1988.

Wrigley, E. A. and Schofield, R. S., *The population history of England, 1541–1871: a reconstruction*, Cambridge, 1989.

Yates, E. M., 'Dark Age and medieval settlement on the edge of wastes and forests', *Field Studies* 2 (1965), 133–53.

Yelling, J. A., 'Probate inventories and the geography of livestock farming: a study of east Worcestershire, 1540–1750', *TIBG* 51 (1970), 111–26.

Young, C. R., *The royal forests of medieval England*, Leicester, 1979.

Ziegler, P., *The Black Death*, London, 1969.

Index

Page references to figures are in *italics*.

The following abbreviations are used: ER, East Riding; FTC, 'Feeding the city'; NR, North Riding; WACY/WAGY, Weighted aggregate crop/grain yield; WR, West Riding.

Cambridge Studies in Historical Geography

Titles marked with an asterisk * are available in paperback.